Quantum Mechanics

Quantum Mechanics

Richard Fitzpatrick

University of Texas at Austin, USA

World Scientific

NEW JERSEY · LONDON · SINGAPORE · BEIJING · SHANGHAI · HONG KONG · TAIPEI · CHENNAI

Published by

World Scientific Publishing Co. Pte. Ltd.
5 Toh Tuck Link, Singapore 596224
USA office: 27 Warren Street, Suite 401-402, Hackensack, NJ 07601
UK office: 57 Shelton Street, Covent Garden, London WC2H 9HE

British Library Cataloguing-in-Publication Data
A catalogue record for this book is available from the British Library.

QUANTUM MECHANICS

ISBN 978-981-4689-94-6
ISBN 978-981-4689-95-3 (pbk)

Printed in Singapore

Preface

Quantum mechanics was developed during the first few decades of the twentieth century via a series of inspired guesses made by various physicists, including Planck, Einstein, Bohr, Schrödinger, Heisenberg, Pauli, and Dirac. The basic aim of quantum mechanics is to formulate a self-consistent theory of atomic and molecular dynamics that is compatible with experimental observations. The purpose of this book is to present (non-relativistic) quantum mechanics in a systematic fashion, starting from the fundamental postulates, and developing the theory in as logical a manner as possible. The material in this book is presented at the introductory graduate level.

The fundamental postulates and theoretical machinery of quantum mechanics are set out in Chapter 1. Chapter 2 is devoted to a discussion of the role of position and momentum in quantum mechanics, whereas Chapter 3 explains how time evolution is taken into account. Chapters 4, 5, and 6 describe the quantum mechanical theories of orbital angular momentum, spin angular momentum, and the addition of angular momentum, respectively. Time-independent and time-dependent perturbation theory are discussed in Chapters 7 and 8, respectively. Chapter 9 is devoted to the quantum mechanical treatment of identical particles. The quantum theory of scattering is set forth in Chapter 10. Finally, Chapter 11 describes the quantum theory of relativistic electrons.

Working through exercises is a vital stage in mastering any branch of physics. Hence, every chapter in this book ends with a selection of exercises that range from simply filling in the inevitable gaps in the material presented in the chapter (such exercises are tedious, but mandatory for the serious student) to interesting further applications of this material. A complete set of solutions for these exercises is given at the back of the book.

Contents

Chapter 1

Fundamental Concepts

1.1 Breakdown of Classical Physics

The necessity for a departure from classical physics at the microscopic (i.e., atomic, molecular, and particle) level is amply demonstrated by the following well-known phenomena:

Anomalous Atomic and Molecular Stability: According to classical physics, an electron orbiting an atomic nucleus undergoes acceleration and should, therefore, lose energy via the continuous emission of electromagnetic radiation [Fitzpatrick (2008)], causing it to gradually spiral in towards the nucleus. (See Exercise 1.1.) Experimentally, this is not observed to happen.

Anomalously Low Atomic and Molecular Specific Heats: According to the *equipartition theorem* of classical statistical thermodynamics, each degree of freedom of an atom or molecule should contribute $R/2$ to the molar specific heat capacity of a macroscopic system made up of a great many such atoms or molecules, where R is the molar ideal gas constant [Reif (1965)]. In fact, only the translational, and some rotational, degrees of freedom seem to contribute. The vibrational degrees of freedom appear to make no contribution at all (except at very high temperatures) [Reif (1965)]. Incidentally, this fundamental problem with classical physics was known and appreciated by the middle of the nineteenth century. Stories that physicists at the commencement of the twentieth century thought that classical physics explained everything, and that there was nothing left to discover, are largely apocryphal [Feynman *et al.* (1963)].

Ultraviolet Catastrophe: According to classical statistical thermodynamics, the equilibrium energy density of an electromagnetic field contained within a vacuum cavity whose walls are held at a fixed temperature is infinite, due to a divergence of energy carried by short-wavelength modes. This divergence is

1

called the *ultraviolet catastrophe* [Ehrenfest (1911)]. Experimentally, there is
no such divergence, and the total energy density is finite [Reif (1965)].

Wave-Particle Duality: Classical physics treats waves and microscopic particles as completely distinct phenomena. However, various experiments (e.g.,
the photoelectric effect [Einstein (1905); Millikan (1916)], Compton scattering [Compton (1923)], and electron diffraction [Davisson (1928); Thomson
(1928)]) demonstrate that waves sometimes act as if they were streams of particles, and streams of particles sometimes act as if they were waves [de Broglie
(1925)]. This behavior is completely inexplicable within the framework of
classical physics.

1.2   Photon Polarization

We know experimentally that if plane polarized light is used to eject photo-
electrons then there is a preferred direction of emission of the electrons [Bubb
(1924)]. Clearly, the polarization properties of light, which are usually associated
with its wave-like behavior, also extend to its particle-like behavior. In particular, a polarization can be ascribed to each individual *photon* (i.e., quantum of
electromagnetic radiation) in a beam of light.

Consider the following well-known experiment. A beam of plane polarized
light is passed through a thin polarizing film whose plane is normal to the beam's
direction of propagation, and which has the property that it is only transparent to
light whose direction of polarization lies perpendicular to its optic axis (which is
assumed to lie in the plane of the film). Classical electromagnetic wave theory
tells us that if the beam is polarized perpendicular to the optic axis then all of the
light is transmitted, if the beam is polarized parallel to the optic axis then none of
the light is transmitted, and if the light is polarized at an angle α to the axis then
a fraction $\sin^2\alpha$ of the beam energy is transmitted—the latter result is known as
Malus' law [Hecht and Zajac (1974)]. Let us try to account for these observations
at the individual photon level.

A beam of light that is plane polarized in a certain direction is presumably
made up of a stream of photons that are each plane polarized in that direction.
This picture leads to no difficulty if the direction of polarization lies parallel or
perpendicular to the optic axis of the polarizing film. In the former case, none of
the photons are transmitted, and, in the latter case, all of the photons are transmitted. But, what happens in the case of an obliquely polarized incident beam?

The previous question is not very precise. Let us reformulate it as a question
relating to the result of some experiment that we could perform. Suppose that we

were to fire a single photon at a polarizing film, and then look to see whether or not it emerges on the other side. The possible results of the experiment are that either a whole photon (whose energy is equal to the energy of the incident photon) is observed, or no photon is observed. Any photon that is transmitted though the film must be polarized perpendicular to the film's optic axis. Furthermore, it is impossible to imagine (in physics) finding part of a photon on the other side of the film. If we repeat the experiment a great number of times then, on average, a fraction $\sin^2\alpha$ of the photons are transmitted through the film, and a fraction $\cos^2\alpha$ are absorbed. Thus, given that the trials are statistically independent of one another, we must conclude that an individual photon has a probability $\sin^2\alpha$ of being transmitted as a photon polarized in the plane perpendicular to the optic axis, and a probability $\cos^2\alpha$ of being absorbed. These values for the probabilities lead to the correct classical limit for a beam containing a large number of photons.

Note that we have only been able to preserve the individuality of photons, in all cases, by abandoning the determinacy of classical theory, and adopting a fundamentally probabilistic approach. We have no way of knowing whether a given obliquely polarized photon is going to be absorbed by, or transmitted through, the polarizing film. We only know the probability of each event occurring. This is a fairly sweeping statement. Recall, however, that the state of a photon is fully specified once its energy, direction of propagation, and polarization are known. If we imagine performing experiments using monochromatic light, normally incident on a polarizing film, with a particular oblique polarization, then the state of each individual photon in the beam is completely specified, and nothing remains to uniquely determine whether the photon is transmitted or absorbed by the film.

The previous discussion about the possible results of an experiment with a single obliquely polarized photon incident on a polarizing film answers all that can be legitimately asked about what happens to the photon when it reaches the film. Questions as to what determines whether the photon is transmitted or not, or how it changes its direction of polarization, are illegitimate, because they do not relate to the outcome of a possible experiment. Nevertheless, some further description is needed, in order to allow the results of this experiment to be correlated with the results of other experiments that can be performed using photons.

The further description provided by quantum mechanics is as follows. It is supposed that a photon polarized obliquely to the optic axis can be regarded as being partly in a state of polarization parallel to the axis, and partly in a state of polarization perpendicular to the axis. In other words, the oblique polarization state is some sort of superposition of two states of parallel and perpendicular polarization. Because there is nothing special about the orientation of the optic axis in our experiment, we deduce that any photon polarization state can be regarded as a superposition of two mutually perpendicular polarization states. When we cause a photon to encounter a polarizing film, we are subjecting it to an observation. In fact, we are observing whether it is polarized parallel or perpendicular to the film's optic axis. The effect of making this observation is to force the photon entirely into a state of parallel or perpendicular polarization. In other words, the photon has to jump suddenly from being partly in each of these two states to being entirely in one or the other of them. Which of the two states it will jump into cannot be predicted, but is governed by probability laws. If the photon jumps into a state of parallel polarization then it is absorbed. Otherwise, it is transmitted. Note that, in this example, the introduction of indeterminacy into the problem is clearly connected with the act of observation. In other words, the indeterminacy is related to the inevitable disturbance of the system associated with the act of observation.

1.3   Fundamental Principles of Quantum Mechanics

There is nothing particularly unique about the transmission and absorption of photons through a polarizing film. Exactly the same conclusions as those outlined in the previous section are obtained by studying other simple experiments involving photons or microscopic particles. For instance, the interference of photons [Taylor (1909); Dirac (1958)], or the Stern-Gerlach experiment [Gerlach and Stern (1922); Feynman *et al.* (1963); Sakurai and Napolitano (2011)]. An examination of these experiments leads us to formulate the following fundamental principles of quantum mechanics:

Dirac's Razor: Quantum mechanics can only answer questions regarding the outcome of possible experiments. Any other questions lie beyond the realms of physics.

Principle of Superposition of States: Any microscopic system (e.g., an atom, molecule, or particle) in a given state can be regarded as being partly in each of two or more other states. In other words, any state can be regarded as a superposition of two or more other states. Such superpositions can be performed in an infinite number of different ways.

Principle of Indeterminacy: An observation made on a microscopic system causes it to jump into one or more particular states (that are related to the type of observation). In general, it is impossible to predict into which final state the system will jump. However, the probability of a given system jumping into a given final state can be predicted.

The first of these principles was formulated by quantum physicists (such as Dirac) in the 1920's to fend off awkward questions such as "How can a microscopic system suddenly jump from one state into another?" or "How does a microscopic system decide into which state to jump?" As we shall see, the second principle is the basis for the mathematical formulation of quantum mechanics. The final principle is still rather vague. We need to extend it so that we can predict into which possible states a system can jump after a particular type of observation, as well as the probability of the system making a particular jump.

1.4 Ket Space

Consider a microscopic system composed of particles, or microscopic bodies, with specific properties (mass, moment of inertia, etc.) interacting according to specific laws of force. There will be various possible motions of the particles, or bodies, consistent with these laws of force. Let us term each such motion a "state" of the system. According to the principle of superposition of states, any given state can be regarded as a superposition of two or more other states. Thus, states must be related to mathematical quantities of a kind that can be added together to give other quantities of the same kind. The most obvious examples of such quantities are vectors.

Let us consider a particular microscopic system in a particular state, which we label A. For example, a photon with a particular energy, momentum, and polarization. We can represent this state as a particular vector, which we also label A, residing in some vector space, where the other elements of the space represent all of the other possible states of the system. Such a space is called a *ket space* (after Dirac), and the vectors that reside in this space are know as *kets* [Dirac (1958)]. The ket that corresponds to state A is conventionally written

$$|A\rangle. \tag{1.1}$$

Suppose that state A is, in fact, the superposition of two different states, B and C. This inter-relation is represented in ket space by writing

$$|A\rangle = |B\rangle + |C\rangle, \tag{1.2}$$

where $|B\rangle$ is the ket relating to the state B, et cetera. For instance, state B might represent a photon propagating in the z-direction, and plane polarized in the x-direction, and state C might represent a similar photon plane polarized in the y-direction. In this case, the sum of these two states represents a photon whose direction of polarization makes an angle of 45° with both the x- and y-directions (by analogy with classical physics) [Hecht and Zajac (1974)]. This latter state is represented by the ket $|B\rangle + |C\rangle$.

Suppose that we wish to construct a state whose direction of polarization subtends an arbitrary angle α with the x-direction. We can do this via a suitably weighted superposition of states B and C. By analogy with classical physics, we require $\cos \alpha$ of state B, and $\sin \alpha$ of state C [Hecht and Zajac (1974)]. This new state is represented by the ket

$$\cos \alpha \, |B\rangle + \sin \alpha \, |C\rangle. \tag{1.3}$$

Note that we cannot form a new state by superposing a state with itself. For instance, a photon polarized in the y-direction superposed with another photon polarized in the y-direction (with the same energy and momentum) gives the same photon. This implies that the ket

$$c_1 \, |A\rangle + c_2 \, |A\rangle = (c_1 + c_2) \, |A\rangle \tag{1.4}$$

corresponds to the same state that $|A\rangle$ does. Here, c_1 and c_2 are real numbers. Thus, kets differ from conventional vectors in that their magnitudes, or lengths, are physically irrelevant. All of the possible states of the system are in one to one correspondence with all of the possible directions of vectors in the ket space, no distinction being made between the directions of the kets $|A\rangle$ and $-|A\rangle$. There is, however, one caveat to the previous statements. If $c_1 + c_2 = 0$ then the superposition process yields nothing at all. In other words, no state. The absence of a state is represented by the *null ket*, $|0\rangle$. The null ket has the fairly obvious property that

$$|A\rangle + |0\rangle = |A\rangle \tag{1.5}$$

for any ket $|A\rangle$. The reason that kets pointing in the same direction represent the same state is ultimately related to the quantization of matter. In other words, the fact that matter comes in irreducible packets called photons, electrons, atoms, et cetera. If we observe a microscopic system then we either see a state (i.e., a photon, an atom, or a molecule, etc.) or we see nothing—we can never see a fraction, or a multiple, of a state. In classical physics, on the other hand, if we observe a wave then the amplitude of the wave can take any value between zero and infinity. Thus, if we were to represent a classical wave by a vector then the magnitude, or length, of the vector would correspond to the amplitude of the wave, and the direction would correspond to the frequency and wavelength, so that two vectors of different lengths pointing in the same direction would represent different wave states.

We have seen, in Equation (1.3), that any plane polarized state of a photon can be represented as a linear superposition of two orthogonal polarization states in which the weights are real numbers. Suppose, however, that we wish to construct a circularly polarized photon state. We know from classical physics that a circularly polarized wave is a superposition of two waves of equal amplitude, plane polarized in orthogonal directions, that oscillate in phase quadrature [Hecht and Zajac (1974)]. This suggests that a circularly polarized photon is the superposition of a photon polarized in the x-direction (state B) and a photon polarized in the y-direction (state C), with equal weights given to the two states, but with the proviso that state C oscillates 90° out of phase with state B. By analogy with classical physics, we can use complex numbers to simultaneously represent the weighting and relative phase in a linear superposition [Fitzpatrick (2013)]. Thus, a circularly polarized photon is represented by the ket

$$|B\rangle + i\,|C\rangle. \tag{1.6}$$

Likewise, a general elliptically polarized photon is represented by

$$c_1\,|B\rangle + c_2\,|C\rangle, \tag{1.7}$$

where c_1 and c_2 are complex numbers [Fitzpatrick (2013)]. We conclude that a ket space must be a complex vector space if it is to properly represent the mutual inter-relations between the possible states of a microscopic system.

Suppose that the ket $|R\rangle$ is expressible linearly in terms of the kets $|A\rangle$ and $|B\rangle$, so that

$$|R\rangle = c_1\,|A\rangle + c_2\,|B\rangle. \tag{1.8}$$

We say that $|R\rangle$ is *dependent* on $|A\rangle$ and $|B\rangle$. It follows that the state R can be regarded as a linear superposition of the states A and B. So, we can also say that state R is dependent on states A and B. In fact, any ket (or state) that is expressible linearly in terms of certain others is said to be dependent on them. Likewise, a set of ket (or states) are termed *independent* if none of them are expressible linearly in terms of the others.

The *dimensionality* of a conventional vector space is defined as the number of independent vectors contained in that space. Likewise, the dimensionality of a ket space is equal to the number of independent kets that it contains. Thus, the ket space that represents the possible polarization states of a photon propagating in the z-direction is two dimensional (the two independent kets correspond to photons plane polarized in the x- and y-directions, respectively). Some microscopic systems have a finite number of independent states (e.g., the spin states of an electron in a magnetic field). If there are N independent states then the possible states of the system are represented as an N-dimensional ket space. Some microscopic systems have a denumerably infinite number of independent states (e.g., a particle trapped in an infinitely deep, one-dimensional, potential well). The possible states of such a system are represented as a ket space whose dimensions are denumerably infinite. Such a space can be treated in more or less the same manner as a finite-dimensional space. Unfortunately, some microscopic systems have a non-denumerably infinite number of independent states (e.g., a free particle). The possible states of such a system are represented as a ket space whose dimensions are non-denumerably infinite. This type of space requires a slightly different treatment to spaces of finite, or denumerably infinite, dimensions. (See Section 1.15.)

In conclusion, the states of a general microscopic system can be represented as a complex vector space of (possibly) infinite dimensions. Such a space is termed a *Hilbert space* by mathematicians.

1.5   Bra Space

A snack machine inputs coins plus some code entered on a key pad, and (hopefully) outputs a snack. It also does so in a deterministic manner: that is, the same money plus the same code produces the same snack (or the same error message) time after time. Note that the input and output of the machine have completely different natures. We can imagine building a rather abstract snack machine that inputs kets and outputs complex numbers in a deterministic fashion. Mathematicians call such a machine a *functional*. Imagine a general functional, labeled F, acting on a general ket, labeled A, and spitting out a general complex number ϕ_A. This process is represented mathematically by writing

$$\langle F|(|A\rangle) = \phi_A. \tag{1.9}$$

Let us narrow our focus to those functionals that preserve the linear dependencies of the kets upon which they operate. Not surprisingly, such functionals are termed *linear functionals*. A general linear functional, labeled F, satisfies

$$\langle F|(|A\rangle + |B\rangle) = \langle F|(|A\rangle) + \langle F|(|B\rangle), \tag{1.10}$$

where $|A\rangle$ and $|B\rangle$ are any two kets in a given ket space.

Consider an N-dimensional ket space [i.e., a finite-dimensional, or denumerably infinite dimensional (i.e., $N \to \infty$), space]. Let the $|i\rangle$ (where i runs from 1 to N) represent N independent kets in this space. A general ket can be written[1]

$$|A\rangle = \sum_{i=1,N} \alpha_i |i\rangle, \tag{1.11}$$

---

[1] Actually, this is strictly true only for finite-dimensional spaces. Only a special subset of denumerably infinite dimensional spaces have this property (i.e., they are complete). However, because a ket space must be complete if it is to represent the states of a microscopic system, we can restrict our attention to this special subset.

where the α_i are an arbitrary set of complex numbers. The functional F can satisfy Equation (1.10) for all vectors in the ket space only if

$$\langle F|(|A\rangle) = \sum_{i=1,N} f_i \, \alpha_i, \tag{1.12}$$

where the $f_i = F(|i\rangle)$ are a set of complex numbers relating to the functional.

Let us define N basis functionals $\langle i|$ that satisfy

$$\langle i|(|j\rangle) = \delta_{ij}. \tag{1.13}$$

Here, the *Kronecker delta symbol* is defined such that $\delta_{ij} = 1$ if $i = j$, and $\delta_{ij} = 0$ otherwise [Riley *et al.* (2013)]. It follows from the previous three equations that

$$\langle F| = \sum_{i=1,N} f_i \langle i|. \tag{1.14}$$

But, this implies that the set of all possible linear functionals acting on an N-dimensional ket space is itself an N-dimensional vector space. This type of vector space is called a *bra space* (after Dirac), and its constituent vectors (which are actually functionals of the ket space) are called *bras* [Dirac (1958)]. Note that bras are quite different in nature to kets (hence, they are written in mirror image notation, $\langle \cdots |$ and $| \cdots \rangle$, so that they can never be confused). Bra space is an example of what mathematicians call a *dual vector space* (i.e., it is dual to the original ket space). There is a one to one correspondence between the elements of the ket space and those of the related bra space. So, for every element A of the ket space, there is a corresponding element, which it is also convenient to label A, in the bra space. That is,

$$|A\rangle \xleftrightarrow{\text{DC}} \langle A|, \tag{1.15}$$

where DC stands for *dual correspondence*.

There are an infinite number of ways of setting up the correspondence between vectors in a ket space and those in the related bra space. However, only one of these has any physical significance. (See Section 1.11.) For a general ket $|A\rangle$, specified by Equation (1.11), the corresponding bra is written

$$\langle A| = \sum_{i=1,N} \alpha_i^* \langle i|, \tag{1.16}$$

where the α_i^* are the complex conjugates of the α_i. $\langle A|$ is termed the *dual* to $|A\rangle$. It follows, from the previous equation, that

$$c|A\rangle \xleftrightarrow{\text{DC}} c^* \langle A|, \tag{1.17}$$

where c is a complex number. More generally,

$$c_1|A\rangle + c_2|B\rangle \xleftrightarrow{\text{DC}} c_1^* \langle A| + c_2^* \langle B|. \tag{1.18}$$

Recall that a bra is a functional that acts on a general ket, and spits out a complex number. Consider the functional that is dual to the ket

$$|B\rangle = \sum_{i=1,N} \beta_i\,|i\rangle \qquad (1.19)$$

acting on the ket $|A\rangle$. This operation is denoted $\langle B|(|A\rangle)$. However, we can omit the round brackets without causing any ambiguity, so the operation can also be written $\langle B||A\rangle$. This expression can be further simplified to give $\langle B|A\rangle$. According to Equations (1.11), (1.13), (1.16), and (1.19),

$$\langle B|A\rangle = \sum_{i=1,N} \beta_i^*\,\alpha_i. \qquad (1.20)$$

Mathematicians term $\langle B|A\rangle$ the *inner product* of a bra and a ket.[2] An inner product is (almost) analogous to a scalar product between covariant and contravariant vectors in curvilinear coordinates [Riley *et al.* (2013)]. It is easily demonstrated that

$$\langle B|A\rangle = \langle A|B\rangle^*. \qquad (1.21)$$

(See Exercise 1.3.) Consider the special case where $|B\rangle \to |A\rangle$. It follows from Equations (1.20) and (1.21) that $\langle A|A\rangle$ is a real number, and that

$$\langle A|A\rangle \geq 0. \qquad (1.22)$$

The equality sign holds only if $|A\rangle$ is the null ket [i.e., if all of the α_i are zero in Equation (1.11)]. This property of bras and kets is essential for the probabilistic interpretation of quantum mechanics, as will become apparent in Section 1.11.

Two kets $|A\rangle$ and $|B\rangle$ are said to be *orthogonal* if

$$\langle A|B\rangle = 0, \qquad (1.23)$$

which also implies that $\langle B|A\rangle = 0$.

Given a ket $|A\rangle$, which is not the null ket, we can define a *normalized ket*, $|\tilde{A}\rangle$, where

$$|\tilde{A}\rangle = \left(\frac{1}{\sqrt{\langle A|A\rangle}}\right)|A\rangle, \qquad (1.24)$$

with the property

$$\langle \tilde{A}|\tilde{A}\rangle = 1. \qquad (1.25)$$

Here, $\sqrt{\langle A|A\rangle}$ is known as the *norm* or "length" of $|A\rangle$, and is analogous to the length, or magnitude, of a conventional vector. Because $|A\rangle$ and $c\,|A\rangle$ represent

---

[2]We can now appreciate the elegance of Dirac's notation. The combination of a bra and a ket yields a "bra(c)ket" (which is just a complex number).

the same physical state, it makes sense to require that all kets corresponding to physical states have unit norms.

It is possible to define a dual bra space for a ket space of non-denumerably infinite dimensions in much the same manner as that just described. The main differences are that summations over discrete labels become integrations over continuous labels, Kronecker delta symbols become Dirac delta functions, completeness must be assumed (it cannot be proved), and the normalization convention is somewhat different. (See Section 1.15.)

1.6 Operators

We have seen that a functional is a machine that inputs a ket and spits out a complex number. Consider a somewhat different machine that inputs a ket and spits out another ket in a deterministic fashion. Mathematicians call such a machine an *operator*. We are only interested in operators that preserve the linear dependencies of the kets upon which they act. Such operators are termed *linear operators*. Consider an operator labeled X. Suppose that when this operator acts on a general ket $|A\rangle$ it spits out a new ket which is denoted $X|A\rangle$. Operator X is linear provided that

$$X(|A\rangle + |B\rangle) = X|A\rangle + X|B\rangle, \qquad (1.26)$$

for all kets $|A\rangle$ and $|B\rangle$, and

$$X(c|A\rangle) = cX|A\rangle, \qquad (1.27)$$

for all complex numbers c. Operators X and Y are said to be equal if

$$X|A\rangle = Y|A\rangle \qquad (1.28)$$

for all kets in the ket space in question. Operator X is termed the *null operator* if

$$X|A\rangle = |0\rangle \qquad (1.29)$$

for all kets in the space. This operator is usually denoted 0. It follows from Equation (1.5) that

$$X + 0 = X, \qquad (1.30)$$

where X is a general operator. Operator X is termed the *identity operator* if

$$X|A\rangle = |A\rangle \qquad (1.31)$$

for all kets in the space. This operator is usually denoted 1. Operators can be added together. Such addition is defined to obey a commutative and associate algebra: that is,

$$X + Y = Y + X, \tag{1.32}$$

$$X + (Y + Z) = (X + Y) + Z. \tag{1.33}$$

Operators can also be multiplied. Operator multiplication is associative: that is,

$$X(Y|A\rangle) = (XY)|A\rangle = XY|A\rangle, \tag{1.34}$$

$$X(YZ) = (XY)Z = XYZ. \tag{1.35}$$

However, in general, operator multiplication is non-commutative: that is,

$$XY \neq YX. \tag{1.36}$$

So far, we have only considered linear operators acting on kets. We can also give a meaning to their operation on bras. Consider the inner product of a general bra $\langle B|$ with the ket $X|A\rangle$. This product is a number that depends linearly on $|A\rangle$. Thus, it may be considered to be the inner product of $|A\rangle$ with some bra. This bra depends linearly on $\langle B|$, so we may look on it as the result of some linear operator applied to $\langle B|$. This operator is uniquely determined by the original operator X, so we might as well call it the same operator acting on $\langle B|$. A suitable notation to use for the resulting bra when X operates on $\langle B|$ is $\langle B|X$. The equation which defines this vector is

$$(\langle B|X)|A\rangle = \langle B|(X|A\rangle) \tag{1.37}$$

for any $|A\rangle$ and $\langle B|$. The triple product of $\langle B|$, X, and $|A\rangle$ can be written $\langle B|X|A\rangle$ without ambiguity, provided we adopt the convention that the bra always goes on the left, the operator in the middle, and the ket on the right.

Consider the dual bra to $X|A\rangle$. This bra depends antilinearly on $|A\rangle$ (i.e., if $|A\rangle$ is multiplied by the complex number c then the corresponding bra is multiplied by c^*) and must therefore depend linearly on $\langle A|$. Thus, it may be regarded as the result of some linear operator applied to $\langle A|$. This operator is termed the *adjoint* of X, and is denoted $X^\dagger$. Thus,

$$X|A\rangle \xrightarrow{\text{DC}} \langle A|X^\dagger. \tag{1.38}$$

It is readily demonstrated that

$$\langle B|X^\dagger|A\rangle = \langle A|X|B\rangle^*, \tag{1.39}$$

plus

$$(XY)^\dagger = Y^\dagger X^\dagger. \tag{1.40}$$

It is also easily seen that the adjoint of the adjoint of a linear operator is equivalent to the original operator. (See Exercise 1.4.) An *Hermitian operator* ξ has the special property that it is its own adjoint: that is,

$$\xi = \xi^\dagger. \tag{1.41}$$

Obviously, a complex number can be regarded as a trivial operator that modifies the length and phase of a ket upon which it acts, without changing the ket's direction. Furthermore, it follows from Equations (1.17) and (1.27) that a complex number operator, c, commutes with any other operator, and that its adjoint is c^*. Finally, it is easily appreciated that the identity operator corresponds to the number unity, while the null operator corresponds to the number zero.

1.7 Outer Product

So far, we have formed the following products: $\langle B|A\rangle$, $X|A\rangle$, $\langle A|X$, XY, $\langle B|X|A\rangle$. Are there any other products we are allowed to form? How about

$$|B\rangle\langle A| ? \tag{1.42}$$

This product clearly depends linearly on the bra $\langle A|$ and the ket $|B\rangle$. Suppose that we right-multiply the previous product by the general ket $|C\rangle$. We obtain

$$|B\rangle\langle A|C\rangle = \langle A|C\rangle|B\rangle, \tag{1.43}$$

because $\langle A|C\rangle$ is just a number. Thus, $|B\rangle\langle A|$ acting on a general ket $|C\rangle$ yields another ket. Clearly, the product $|B\rangle\langle A|$ is a linear operator. This operator also acts on bras, as is easily demonstrated by left-multiplying the expression (1.42) by a general bra $\langle C|$. It is also readily shown that

$$(|B\rangle\langle A|)^\dagger = |A\rangle\langle B|. \tag{1.44}$$

(See Exercise 1.4.) Mathematicians term the operator $|B\rangle\langle A|$ the *outer product* of $|B\rangle$ and $\langle A|$. The outer product should not be confused with the inner product, $\langle A|B\rangle$, which is just a number.

1.8 Eigenvalues and Eigenvectors

In general, the ket $X|A\rangle$ is not a constant multiple of the ket $|A\rangle$. However, there are some special kets known as the *eigenkets* of operator X. These are denoted

$$|x'\rangle, |x''\rangle, |x'''\rangle, \ldots, \tag{1.45}$$

and have the property

$$X|x'\rangle = x'|x'\rangle, \qquad X|x''\rangle = x''|x''\rangle, \qquad X|x'''\rangle = x'''|x'''\rangle, \qquad \ldots, \qquad (1.46)$$

where x', x'', x''', ... are complex numbers called *eigenvalues*. Clearly, applying X to one of its eigenkets yields the same eigenket multiplied by the associated eigenvalue.

Consider the eigenkets and eigenvalues of an Hermitian operator ξ. These are denoted

$$\xi|\xi'\rangle = \xi'|\xi'\rangle, \tag{1.47}$$

where $|\xi'\rangle$ is the eigenket associated with the eigenvalue ξ'. Three important results are readily deduced:

(1) *The eigenvalues are all real numbers, and the eigenkets corresponding to different eigenvalues are orthogonal.* Because ξ is Hermitian, the dual equation to Equation (1.47) (for the eigenvalue ξ'') reads

$$\langle\xi''|\xi = \xi''^{*}\langle\xi''|. \tag{1.48}$$

If we left-multiply Equation (1.47) by $\langle\xi''|$, right-multiply the previous equation by $|\xi'\rangle$, and take the difference, then we obtain

$$(\xi' - \xi''^{*})\langle\xi''|\xi'\rangle = 0. \tag{1.49}$$

Suppose that the eigenvalues ξ' and ξ'' are the same. It follows from the previous equation that

$$\xi' = \xi'^{*}, \tag{1.50}$$

where we have used the fact that $|\xi'\rangle$ is not the null ket. This proves that the eigenvalues are real numbers. Suppose that the eigenvalues ξ' and ξ'' are different. It follows that

$$\langle\xi''|\xi'\rangle = 0, \tag{1.51}$$

which demonstrates that eigenkets corresponding to different eigenvalues are orthogonal.

(2) *The eigenvalues associated with eigenkets are the same as the eigenvalues associated with eigenbras.* An eigenbra of ξ corresponding to an eigenvalue ξ' is defined

$$\langle\xi'|\xi = \langle\xi'|\xi'. \tag{1.52}$$

(3) *The dual of any eigenket is an eigenbra belonging to the same eigenvalue, and conversely.*

1.9 Observables

We have developed a mathematical formalism that comprises three types of objects—bras, kets, and linear operators. We have already seen that kets can be used to represent the possible states of a microscopic system. However, there is a one to one correspondence between the elements of a ket space and its dual bra space, so we must conclude that bras could just as well be used to represent the states of a microscopic system. What about the dynamical variables of the system (e.g., its position, momentum, energy, spin, etc.)? How can these be represented in our formalism? Well, the only objects that we have left over are operators. We, therefore, assume that the dynamical variables of a microscopic system are represented as linear operators acting on the bras and kets that correspond to the various possible states of the system. Note that the operators have to be linear, otherwise they would, in general, spit out bras/kets pointing in different directions when fed bras/kets pointing in the same direction but differing in length. Because the lengths of bras and kets have no physical significance, it is reasonable to suppose that non-linear operators are also without physical significance.

We have seen (in Section 1.2) that if we observe the polarization state of a photon, by placing a polarizing film in its path, then the result is to cause the photon to jump into a state of polarization parallel or perpendicular to the optic axis of the film. The former state is absorbed, and the latter state is transmitted (which is how we tell them apart). In general, we cannot predict into which state a given photon will jump (except in a statistical sense). However, we do know that if the photon is initially polarized parallel to the optic axis then it will definitely be absorbed, and if it is initially polarized perpendicular to the axis then it will definitely be transmitted. We also know that, after passing though the film, a photon must be in a state of polarization perpendicular to the optic axis (otherwise it would not have been transmitted). We can make a second observation of the polarization state of such a photon by placing an identical polarizing film (with the same orientation of the optic axis) immediately behind the first film. It is clear that the photon will definitely be transmitted through the second film.

There is nothing particularly unique about the polarization states of a photon. So, more generally, we can say that if a dynamical variable of a microscopic system is measured then the system is caused to jump into one of a number of independent states (note that the perpendicular and parallel polarization states of our photon are independent). In general, each of these final states is associated with a different result of the measurement: that is, a different value of the dynamical variable. Obviously, the result of the measurement must be a real number (there are no measurement machines that output complex numbers). Finally, if an observation is made, and the system is found to be a one particular final state, with one particular value for the dynamical variable, then a second observation, made immediately after the first one, will definitely find the system in the same state, and yield the same value for the dynamical variable.

How can we represent all of these facts in our mathematical formalism? Well, by a fairly non-obvious leap of intuition, we are going to assert that a measurement of a dynamical variable corresponding to an operator X in ket space causes the system to jump into a state corresponding to one of the eigenkets of X. Not surprisingly, such a state is termed an *eigenstate*. Furthermore, the result of the measurement is the eigenvalue associated with the eigenket into which the system jumps. The fact that the result of the measurement must be a real number implies that dynamical variables can only be represented by Hermitian operators (because only Hermitian operators are guaranteed to have real eigenvalues). The fact that the eigenkets of an Hermitian operator corresponding to different eigenvalues (i.e., different results of the measurement) are orthogonal is in accordance with our earlier requirement that the states into which the system jumps should be mutually independent. We conclude that the result of a measurement of a dynamical variable represented by an Hermitian operator ξ must be one of the eigenvalues of ξ. Conversely, every eigenvalue of ξ is a possible result of a measurement made on the corresponding dynamical variable. This gives us the physical significance of the eigenvalues. (From now on, the distinction between a state and its representative ket, and a dynamical variable and its representative operator, will be dropped, for the sake of simplicity.)

It is reasonable to suppose that if a certain dynamical variable ξ is measured with the system in a particular state then the states into which the system may jump on account of the measurement are such that the original state is dependent on them. This fairly innocuous statement has two very important corollaries. First, immediately after an observation whose result is a particular eigenvalue ξ', the system is left in the associated eigenstate. However, this eigenstate is orthogonal to (i.e., independent of) any other eigenstate corresponding to a different eigenvalue. It follows that a second measurement made immediately after the first one must leave the system in an eigenstate corresponding to the eigenvalue ξ'. In other words, the second measurement is bound to give the same result as the first. Furthermore, if the system is in an eigenstate of ξ, corresponding to an eigenvalue ξ', then a measurement of ξ is bound to give the result ξ'. This follows because the system cannot jump into an eigenstate corresponding to a different eigenvalue of ξ, because such a state is not dependent on the original state. Second, it stands to reason that a measurement of ξ must always yield some result. It follows that in no matter what state the system is initially found, it must always be possible for it to jump into one of the eigenstates of ξ. In other words, a general ket must always be dependent on the eigenkets of ξ. This can only be the case if the eigenkets form a complete set (i.e., they span ket space). Thus, in order for an Hermitian operator ξ to be observable its eigenkets must form a complete set. An Hermitian operator that satisfies this condition is termed an *observable*. Conversely, any observable quantity must be an Hermitian operator with a complete set of eigenstates.

1.10   Measurements

We have seen that a measurement of some observable ξ of a microscopic system causes the system to jump into one of the eigenstates of ξ. The result of the measurement is the associated eigenvalue (or some function of this quantity). It is impossible to determine into which eigenstate a given system will jump, but it is possible to predict the probability of such a transition. So, what is the probability that a system in some initial state $|A\rangle$ makes a transition to an eigenstate $|\xi'\rangle$ of an observable ξ, as a result of a measurement made on the system? Let us start with the simplest case. If the system is initially in an eigenstate $|\xi'\rangle$ then the transition probability to a eigenstate $|\xi''\rangle$ corresponding to a different eigenvalue is zero, and the transition probability to the same eigenstate $|\xi'\rangle$ is unity. It is convenient to normalize our eigenkets such that they all have unit norms. It follows from the orthogonality property of the eigenkets that

$$\langle \xi'|\xi''\rangle = \delta_{\xi'\xi''}. \tag{1.53}$$

For the moment, we are assuming that the eigenvalues of ξ are all different.

Note that the probability of a transition from an initial eigenstate $|\xi'\rangle$ to a final eigenstate $|\xi''\rangle$ is the same as the value of the inner product $\langle \xi'|\xi''\rangle$. Can we use this correspondence to obtain a general rule for calculating transition probabilities? Well, suppose that the system is initially in a (normalized) state $|A\rangle$ which is not an eigenstate of ξ. Can we identify the transition probability to a final eigenstate $|\xi'\rangle$ with the inner product $\langle A|\xi'\rangle$? Unfortunately, this is not possible because $\langle A|\xi'\rangle$ is, in general, a complex number, and complex probabilities do not make any sense. Let us try again. Suppose that we identify the transition probability with the modulus squared of the inner product, $|\langle A|\xi'\rangle|^2$? This quantity is definitely a positive number (so it could be a probability). This guess also gives the right answer for the transition probabilities between eigenstates. In fact, it is the correct guess.

Because the eigenstates of an observable ξ form a complete set, we can express any given state $|A\rangle$ as a linear combination of them. It is easily demonstrated that

$$|A\rangle = \sum_{\xi'} |\xi'\rangle\langle\xi'|A\rangle, \tag{1.54}$$

$$\langle A| = \sum_{\xi'} \langle A|\xi'\rangle\langle\xi'|, \tag{1.55}$$

$$\langle A|A\rangle = \sum_{\xi'} \langle A|\xi'\rangle\langle\xi'|A\rangle = \sum_{\xi'} |\langle A|\xi'\rangle|^2, \tag{1.56}$$

where the summation is over all the different eigenvalues of ξ, and use has been made of Equation (1.21), as well as the fact that the eigenstates are mutually orthogonal. Note that all of the previous results follow from the extremely useful result

$$\sum_{\xi'} |\xi'\rangle\langle\xi'| = 1. \tag{1.57}$$

(See Exercise 1.9.) The relative probability of a transition to an eigenstate $|\xi'\rangle$, which is equivalent to the relative probability of a measurement of ξ yielding the result ξ', is

$$P(\xi') \propto |\langle A|\xi'\rangle|^2. \tag{1.58}$$

The absolute probability is clearly

$$P(\xi') = \frac{|\langle A|\xi'\rangle|^2}{\sum_{\xi'} |\langle A|\xi'\rangle|^2} = \frac{|\langle A|\xi'\rangle|^2}{\langle A|A\rangle}. \tag{1.59}$$

If the ket $|A\rangle$ is normalized such that its norm is unity then this probability simply reduces to

$$P(\xi') = |\langle A|\xi'\rangle|^2. \tag{1.60}$$

1.11   Expectation Values

Consider an ensemble of microscopic systems prepared in the same (normalized) initial state $|A\rangle$. Suppose that a measurement of the observable ξ is made on each system. We know that each measurement yields the value ξ' with probability $P(\xi')$. What is the mean value of the measurement? This quantity, which is generally referred to as the *expectation value* of ξ, is given by

$$\langle\xi\rangle = \sum_{\xi'} \xi' P(\xi') = \sum_{\xi'} \xi' |\langle A|\xi'\rangle|^2 = \sum_{\xi'} \xi' \langle A|\xi'\rangle\langle\xi'|A\rangle$$

$$= \sum_{\xi'} \langle A|\xi|\xi'\rangle\langle\xi'|A\rangle, \tag{1.61}$$

which reduces to

$$\langle \xi \rangle = \langle A | \xi | A \rangle \tag{1.62}$$

with the aid of Equation (1.57).

Consider the identity operator, 1. All states are eigenstates of this operator with the eigenvalue unity. Thus, the expectation value of this operator is always unity: that is,

$$\langle A | 1 | A \rangle = \langle A | A \rangle = 1, \tag{1.63}$$

for all $|A\rangle$. Note that it is only possible to normalize a given state ket $|A\rangle$, such that Equation (1.63) is satisfied, because of the more general property (1.22) of the norm. This property depends on the previously adopted correspondence, (1.11) and (1.16), between the elements of a ket space and those of its dual bra space.

1.12 Degeneracy

Suppose that two different (normalized) eigenstates $|\xi'_a\rangle$ and $|\xi'_b\rangle$ of the observable ξ correspond to the same eigenvalue ξ'. These states are termed *degenerate* eigenstates. Degenerate eigenstates are necessarily orthogonal to any eigenstates corresponding to different eigenvalues, but, in general, they are not orthogonal to each other (i.e., the proof of orthogonality given in Section 1.8 does not work in this case). This is unfortunate, because much of the previous formalism depends crucially on the mutual orthogonality of the different eigenstates of an observable. Note, however, that any linear combination of $|\xi'_a\rangle$ and $|\xi'_b\rangle$ is also an eigenstate corresponding to the eigenvalue ξ'. It follows that we can always construct two mutually orthogonal degenerate (normalized) eigenstates. For instance,

$$|\xi'_1\rangle = |\xi'_a\rangle, \tag{1.64}$$

$$|\xi'_2\rangle = \frac{|\xi'_b\rangle - \langle \xi'_a | \xi'_b \rangle | \xi'_a \rangle}{[1 - |\langle \xi'_a | \xi'_b \rangle|^2]^{1/2}}. \tag{1.65}$$

This result is easily generalized to the case of more than two degenerate eigenstates. (See Exercises 1.10.) We conclude that it is always possible to construct a complete set of mutually orthogonal eigenstates for any given observable.

1.13 Compatible Observables

Suppose that we wish to simultaneously measure two observables, ξ and η, of a microscopic system. Let us assume that we possess an apparatus that is capable

of measuring ξ, and another that can measure η. For instance, the two observables
in question might be the projection in the x- and z-directions of the spin angular
momentum of a spin one-half particle. These could be measured using appropriate Stern-Gerlach apparatuses [Sakurai and Napolitano (2011)]. Suppose that we
make a measurement of ξ, and the system is consequently thrown into one of the
eigenstates of ξ, $|\xi'\rangle$, with eigenvalue ξ'. What happens if we now make a mea-
surement of η? Well, suppose that the eigenstate $|\xi'\rangle$ is also an eigenstate of η,
with eigenvalue η'. In this case, a measurement of η will definitely give the result
η'. A second measurement of ξ will definitely give the result ξ', and so on. In
this sense, we can say that the observables ξ and η simultaneously have the values
ξ' and η', respectively. Clearly, if all eigenstates of ξ are also eigenstates of η
then it is always possible to make a simultaneous measurement of ξ and η. Such
observables are termed *compatible*.

Suppose, however, that the eigenstates of ξ are not eigenstates of η. Is it
still possible to measure both observables simultaneously? Let us again make an
observation of ξ that throws the system into an eigenstate $|\xi'\rangle$, with eigenvalue ξ'.
We can now make a second observation to determine η. This will throw the system
into one of the (many) eigenstates of η that depend on $|\xi'\rangle$. In principle, each of
these eigenstates is associated with a different result of the measurement. Suppose
that the system is thrown into an eigenstate $|\eta'\rangle$, with the eigenvalue η'. Another
measurement of ξ will throw the system into one of the (many) eigenstates of ξ
that depend on $|\eta'\rangle$. Each eigenstate is again associated with a different possible
result of the measurement. It is clear that if the observables ξ and η do not possess
simultaneous eigenstates then if the value of ξ is known (i.e., the system is in
an eigenstate of ξ) then the value of η is uncertain (i.e., the system is not in an
eigenstate of η), and vice versa. We say that the two observables are *incompatible*.

We have seen that the condition for two observables ξ and η to be simultaneously measurable is that they should possess simultaneous eigenstates (i.e., every
eigenstate of ξ should also be an eigenstate of η). Suppose that this is the case.
Let a general eigenstate of ξ, with eigenvalue ξ', also be an eigenstate of η, with
eigenvalue η'. It is convenient to denote this simultaneous eigenstate $|\xi'\eta'\rangle$. We
have

$$\xi|\xi'\eta'\rangle = \xi'\,|\xi'\eta'\rangle, \tag{1.66}$$

$$\eta|\xi'\eta'\rangle = \eta'\,|\xi'\eta'\rangle. \tag{1.67}$$

We can left-multiply the first equation by η, and the second equation by ξ, and
then take the difference. The result is

$$(\xi\eta - \eta\xi)|\xi'\eta'\rangle = |0\rangle \tag{1.68}$$

for each simultaneous eigenstate. Recall that the eigenstates of an observable must form a complete set. It follows that the simultaneous eigenstates of two observables must also form a complete set. Thus, the previous equation implies that

$$(\xi\eta - \eta\xi)|A\rangle = |0\rangle, \tag{1.69}$$

where $|A\rangle$ is a general ket. The only way that this can be true is if

$$\xi\eta = \eta\xi. \tag{1.70}$$

We conclude that the condition for two observables ξ and η to be simultaneously measurable is that they should commute.

1.14 Uncertainty Relation

We have seen that if ξ and η are two noncommuting observables then a determination of the value of ξ leaves the value of η uncertain, and vice versa. It is possible to quantify this uncertainty. For a general observable ξ, we can define an Hermitian operator

$$\Delta\xi = \xi - \langle\xi\rangle, \tag{1.71}$$

where the expectation value is taken over the particular physical state under consideration. It is obvious that the expectation value of $\Delta\xi$ is zero. The expectation value of $(\Delta\xi)^2 \equiv \Delta\xi\,\Delta\xi$ is termed the *variance* of ξ, and is, in general, non-zero. In fact, it is easily demonstrated that

$$\langle(\Delta\xi)^2\rangle = \langle\xi^2\rangle - \langle\xi\rangle^2. \tag{1.72}$$

The variance of ξ is a measure of the uncertainty in the value of ξ for the particular state in question (i.e., it is a measure of the width of the distribution of likely values of ξ about the expectation value). If the variance is zero then there is no uncertainty, and a measurement of ξ is bound to give the expectation value, $\langle\xi\rangle$.

Consider the *Schwarz inequality*,

$$\langle A|A\rangle\langle B|B\rangle \geq |\langle A|B\rangle|^2, \tag{1.73}$$

which is analogous to

$$|\mathbf{a}|^2\,|\mathbf{b}|^2 \geq |\mathbf{a}\cdot\mathbf{b}|^2 \tag{1.74}$$

in Euclidian space. This inequality can be proved by noting that

$$((\langle A| + c^*\langle B|)\,(|A\rangle + c\,|B\rangle)) \geq 0, \tag{1.75}$$

where c is any complex number. If c takes the special value $-\langle B|A\rangle/\langle B|B\rangle$ then the previous inequality reduces to

$$\langle A|A\rangle\langle B|B\rangle - |\langle A|B\rangle|^2 \geq 0, \tag{1.76}$$

which is equivalent to the Schwarz inequality.

Let us substitute

$$|A\rangle = \Delta\xi\,|\,\rangle, \tag{1.77}$$

$$|B\rangle = \Delta\eta\,|\,\rangle, \tag{1.78}$$

into the Schwarz inequality, where the blank ket $|\,\rangle$ stands for any general ket. We find

$$\langle(\Delta\xi)^2\rangle\langle(\Delta\eta)^2\rangle \geq |\langle\Delta\xi\,\Delta\eta\rangle|^2, \tag{1.79}$$

where use has been made of the fact that $\Delta\xi$ and $\Delta\eta$ are Hermitian operators. Note that

$$\Delta\xi\,\Delta\eta = \frac{1}{2}\,[\Delta\xi,\Delta\eta] + \frac{1}{2}\,\{\Delta\xi,\Delta\eta\}, \tag{1.80}$$

where the *commutator*, $[\Delta\xi,\Delta\eta]$, and the *anti-commutator*, $\{\Delta\xi,\Delta\eta\}$, are defined

$$[\Delta\xi,\Delta\eta] \equiv \Delta\xi\,\Delta\eta - \Delta\eta\,\Delta\xi, \tag{1.81}$$

$$\{\Delta\xi,\Delta\eta\} \equiv \Delta\xi\,\Delta\eta + \Delta\eta\,\Delta\xi, \tag{1.82}$$

respectively. The commutator is clearly *anti-Hermitian*,

$$([\Delta\xi,\Delta\eta])^\dagger = (\Delta\xi\,\Delta\eta - \Delta\eta\,\Delta\xi)^\dagger = \Delta\eta\,\Delta\xi - \Delta\xi\,\Delta\eta = -[\Delta\xi,\Delta\eta], \tag{1.83}$$

whereas the anti-commutator is obviously Hermitian. Now, it is easily demonstrated that the expectation value of an Hermitian operator is a real number, whereas the expectation value of an anti-Hermitian operator is an imaginary number. (See Exercise 1.11.) It follows that the right-hand side of

$$\langle\Delta\xi\,\Delta\eta\rangle = \frac{1}{2}\,\langle[\Delta\xi,\Delta\eta]\rangle + \frac{1}{2}\,\langle\{\Delta\xi,\Delta\eta\}\rangle, \tag{1.84}$$

consists of the sum of an imaginary and a real number. Taking the modulus squared of both sides gives

$$|\langle\Delta\xi\,\Delta\eta\rangle|^2 = \frac{1}{4}\,|\langle[\xi,\eta]\rangle|^2 + \frac{1}{4}\,|\langle\{\Delta\xi,\Delta\eta\}\rangle|^2, \tag{1.85}$$

where use has been made of $\langle\Delta\xi\rangle = 0$, et cetera. The final term on the right-hand side of the previous expression is positive definite, so we can write

$$\langle(\Delta\xi)^2\rangle\,\langle(\Delta\eta)^2\rangle \geq |\langle\Delta\xi\,\Delta\eta\rangle|^2 \geq \frac{1}{4}\,|\langle[\xi,\eta]\rangle|^2, \tag{1.86}$$

where use has been made of Equation (1.79). The previous expression is termed the *uncertainty relation*. According to this relation, an exact knowledge of the value of ξ (i.e., $\langle(\Delta\xi)^2\rangle \to 0$) implies no knowledge whatsoever of the value of η (i.e., $\langle(\Delta\eta)^2\rangle \to \infty$), and vice versa. The one exception to this rule is when ξ and η commute, in which case exact knowledge of ξ does not necessarily imply no knowledge of η.

1.15 Continuous Spectra

Up to now, we have studiously avoided dealing with observables possessing eigenvalues that lie in a continuous range, rather than having discrete values. The reason for this is that continuous eigenvalues imply a ket space of non-denumerably infinite dimension. Unfortunately, continuous eigenvalues are unavoidable in quantum mechanics. In fact, the most important observables of all—namely position and momentum—generally have continuous eigenvalues. (See the following chapter.) Fortunately, many of the results that we obtained previously for a finite-dimensional ket space with discrete eigenvalues can be generalized to ket spaces of non-denumerably infinite dimensions.

Suppose that ξ is an observable with continuous eigenvalues. We can still write the eigenvalue equation as

$$\xi \, |\xi'\rangle = \xi' \, |\xi'\rangle. \tag{1.87}$$

But, ξ' now takes a continuous range of values. Let us assume, for the sake of simplicity, that ξ' can take any value. The orthonormality condition (1.53) generalizes to

$$\langle \xi' | \xi'' \rangle = \delta(\xi' - \xi''), \tag{1.88}$$

where $\delta(x)$ denotes the famous *Dirac delta function* [Dirac (1958); Riley *et al.* (2013)], and satisfies

$$\delta(x \neq 0) = 0, \tag{1.89}$$

$$\int_{-\infty}^{\infty} dx \, \delta(x) = 1, \tag{1.90}$$

$$\int_{-\infty}^{\infty} dx \, f(x) \, \delta(x - x') = f(x') \tag{1.91}$$

for any function, $f(x)$, that is well behaved at $x = x'$. The Dirac delta function is a *generalized function* [Sobolev (1936)] that can be realized in many equivalent limiting forms. For instance,

$$\delta(x) = \lim_{\eta \to 0} \frac{1}{\pi} \frac{\eta}{x^2 + \eta^2}, \tag{1.92}$$

$$\delta(x) = \lim_{\eta \to 0} \frac{1}{\pi} \frac{\sin(x/\eta)}{x}. \tag{1.93}$$

Note from Equations (1.87) and (1.88) that there are a non-denumerably infinite number of mutually orthogonal eigenstates of ξ. Hence, the dimensionality of ket space is non-denumerably infinite. Furthermore, eigenstates corresponding to a continuous range of eigenvalues cannot be normalized such that they have unit norms. In fact, it is clear from Equation (1.88), together with the well-known fact that $\delta(0) \to \infty$, that these eigenstates have infinite norms. In other words, they are infinitely long. This is the major difference between eigenstates in a finite-dimensional and an infinite-dimensional ket space. The extremely useful relation (1.57) generalizes to

$$\int d\xi' \, |\xi'\rangle\langle\xi'| = 1. \tag{1.94}$$

(See Exercise 1.18.) Note that a summation over discrete eigenvalues goes over into an integral over a continuous range of eigenvalues. The eigenstates $|\xi'\rangle$ must form a complete set if ξ is to be an observable. It follows that any general ket can be expanded in terms of the $|\xi'\rangle$. In fact, the expansions (1.54)–(1.56) generalize to give

$$|A\rangle = \int d\xi' \, |\xi'\rangle\langle\xi'|A\rangle, \tag{1.95}$$

$$\langle A| = \int d\xi' \, \langle A|\xi'\rangle\langle\xi'|, \tag{1.96}$$

$$\langle A|A\rangle = \int d\xi' \, \langle A|\xi'\rangle\langle\xi'|A\rangle = \int d\xi' \, |\langle A|\xi'\rangle|^2, \tag{1.97}$$

respectively. These results also follow simply from Equation (1.94). We have seen that it is not possible to normalize the eigenstates $|\xi'\rangle$ such that they have unit norms. Fortunately, this convenient normalization is still possible for a general state ket. In fact, according to Equation (1.97), the normalization condition can be written

$$\langle A|A\rangle = \int d\xi' \, |\langle A|\xi'\rangle|^2 = 1. \tag{1.98}$$

We have now studied observables whose eigenvalues take a discrete number of values, as well as those whose eigenvalues take a continuous range of values. There are a number of other cases that we could look at. For instance, observables whose eigenvalues can take a (finite) continuous range of values, plus a set of discrete values. Such cases can be dealt with using a fairly straightforward generalization of the previous analysis [Dirac (1958)].

1.16 Exercises

1.1 According to the *Larmor formula* of classical physics, a non-relativistic electron whose instantaneous acceleration is of magnitude a radiates electromagnetic energy at the rate

$$P = \frac{e^2 a^2}{6\pi \epsilon_0 c^3},$$

where e is the magnitude of the electron charge, ϵ_0 the permittivity of the vacuum, and c the velocity of light in vacuum [Fitzpatrick (2008)]. Consider a classical electron in a circular orbit of radius r around a proton. Demonstrate that the radiated energy would cause the orbital radius to decrease in time according to

$$\frac{d}{dt}\left(\frac{r}{a_0}\right)^3 = -\frac{1}{\tau},$$

where $a_0 = 4\pi \epsilon_0 \hbar^2/(m_e e^2)$ is the Bohr radius, m_e the electron mass, $\hbar$ the reduced Planck constant, and

$$\tau = \frac{a_0}{4\,\alpha^4 c}.$$

Here, $\alpha = e^2/(4\pi \epsilon_0 \hbar c)$ is the fine structure constant. Deduce that the classical lifetime of a ground-state electron in a hydrogen atom is $\tau \simeq 1.6\times10^{-11}$ s.

1.2 Let the $|i\rangle$, for $i = 1, N$, be a set of *orthonormal* kets that span an N-dimensional ket space. By orthonormal, we mean that the kets are mutually orthogonal, and have unit norms, so that

$$\langle i|j\rangle = \delta_{ij}$$

for $i, j = 1, N$. Show that

$$\sum_{i=1,N} |i\rangle\langle i| = 1.$$

1.3 Demonstrate that

$$\langle B|A\rangle = \langle A|B\rangle^*$$

in a finite-dimensional ket space.

1.4 Demonstrate that in a finite-dimensional ket space:

(a)

$$\langle B| X^\dagger |A\rangle = \langle A| X |B\rangle^*.$$

(b)

$$(X^\dagger)^\dagger = X.$$

(c)

$$(X Y)^\dagger = Y^\dagger X^\dagger.$$

(d)
$$(X\,Y\,Z)^\dagger = Z^\dagger\,Y^\dagger\,X^\dagger.$$

(e)
$$(|B\rangle\langle A|)^\dagger = |A\rangle\langle B|.$$

Here, X, Y, Z are general operators.

1.5 If A, B are Hermitian operators then demonstrate that $A\,B$ is only Hermitian provided A and B commute. In addition, show that $(A + B)^n$ is Hermitian, where n is a positive integer. [Gasiorowicz (1996)]

1.6 Let A be a general operator. Show that $A + A^\dagger$, $i\,(A - A^\dagger)$, and $A\,A^\dagger$ are Hermitian operators. [Gasiorowicz (1996)]

1.7 Let H be an Hermitian operator. Demonstrate that the Hermitian conjugate of the operator $\exp(i\,H) \equiv \sum_{n=0,\infty}(i\,H)^n/n!$ is $\exp(-i\,H)$. [Gasiorowicz (1996)]

1.8 Suppose that A and B are two commuting operators. Demonstrate that
$$\exp(A)\,\exp(B) = \exp(A + B).$$

1.9 Let the $|\xi'\rangle$ be the normalized eigenkets of an observable ξ, whose corresponding eigenvalues, ξ', are discrete. Demonstrate that
$$\sum_{\xi'} |\xi'\rangle\langle\xi'| = 1,$$
where the sum is over all eigenvalues.

1.10 Let the $|\xi_i'\rangle$, where $i = 1, N$, and $N > 1$, be a set of degenerate unnormalized eigenkets of some observable ξ. Suppose that the $|\xi_i'\rangle$ are not mutually orthogonal. Demonstrate that a set of mutually orthogonal (but unnormalized) degenerate eigenkets, $|\xi_i''\rangle$, for $i = 1, N$, can be constructed as follows:
$$|\xi_i''\rangle = |\xi_i'\rangle - \sum_{j=1,i-1} \frac{\langle\xi_j''|\xi_i'\rangle}{\langle\xi_j''|\xi_j''\rangle}\,|\xi_j''\rangle.$$

This process is known as *Gram-Schmidt orthogonalization*.

1.11 Demonstrate that the expectation value of an Hermitian operator is a real number. Show that the expectation value of an anti-hermitian operator is an imaginary number.

1.12 Let H be an Hermitian operator. Demonstrate that $\langle H^2 \rangle \geq 0$.

1.13 Consider an Hermitian operator, H, that has the property that $H^4 = 1$. What are the eigenvalues of H? What are the eigenvalues if H is not restricted to being Hermitian? [Gasiorowicz (1996)]

1.14 An operator U is said to be *unitary* if
$$U\,U^\dagger = U^\dagger\,U = 1.$$
Show that if $\langle A|A\rangle = 1$ and $|B\rangle = U\,|A\rangle$ then $\langle B|B\rangle = 1$. [Gasiorowicz (1996)]

1.15 Show that if H is Hermitian then $\exp(\,i\,H)$ is unitary. [Gasiorowicz (1996)]

1.16 Show that if the $|u_i\rangle$, for $i = 1, N$, form a complete orthonormal set, so that

$$\langle u_i | u_j \rangle = \delta_{ij},$$

then the $|v_i\rangle = U\,|u_i\rangle$, for $i = 1, N$, where U is unitary, are also orthonormal. [Gasiorowicz (1996)]

1.17 The eigenstates of some operator A acting in an N-dimensional ket space are written

$$A\,|i\rangle = a_i\,|i\rangle,$$

for $i = 1, N$, where all of the a_i are real. Suppose that the $|i\rangle$ are orthonormal, and span the ket space. Deduce that A is Hermitian.

1.18 Let ξ be an observable whose eigenvalues, ξ', lie in a continuous range. Let the $|\xi'\rangle$, where

$$\langle \xi' | \xi'' \rangle = \delta(\xi' - \xi''),$$

be the corresponding eigenkets. Demonstrate that

$$\int d\xi'\,|\xi'\rangle\langle \xi'| = 1,$$

where the integral is over the whole range of eigenvalues.

1.19 Show that

$$\delta(-x) = \delta(x),$$

$$\delta(a\,x) = \frac{1}{a}\,\delta(x),$$

where $\delta(x)$ is a Dirac delta function, and a a constant.

Chapter 2

Position and Momentum

2.1 Introduction

So far, we have considered general dynamical variables represented by general linear operators acting in ket space. However, in classical mechanics, the most important dynamical variables are those involving position and momentum. Let us investigate the role of such variables in quantum mechanics.

In classical mechanics, the position, q, and momentum, p, of some component of a dynamical system are represented as real numbers that, by definition, commute. In quantum mechanics, these quantities are represented as non-commuting linear Hermitian operators acting in a ket space that represents all of the possible states of the system. Our first task is to discover a quantum mechanical replacement for the classical result $q\,p - p\,q = 0$.

2.2 Poisson Brackets

Consider a dynamical system whose state at a particular time, t, is fully specified by N independent classical coordinates q_i (where i runs from 1 to N). Associated with each generalized coordinate, q_i, is a classical canonical momentum, p_i [Goldstein *et al.* (2002)]. For instance, a Cartesian coordinate has an associated linear momentum, an angular coordinate has an associated angular momentum, et cetera [Fitzpatrick (2012)]. As is well known, the behavior of a classical system can be specified in terms of either Lagrangian or Hamiltonian dynamics [Goldstein *et al.* (2002)]. For instance, in Hamiltonian dynamics,

$$\frac{dq_i}{dt} = \frac{\partial H}{\partial p_i},$$ (2.1)

$$\frac{dp_i}{dt} = -\frac{\partial H}{\partial q_i},$$ (2.2)

where the function $H(q_i, p_i, t)$ is the system energy at time t expressed in terms of the classical coordinates and canonical momenta. This function is usually referred to as the *Hamiltonian* of the system [Goldstein *et al.* (2002)].

We are interested in finding some construct in classical dynamics that consists of products of dynamical variables. If such a construct exists then we hope to generalize it somehow to obtain a rule describing how dynamical variables commute with one another in quantum mechanics. There is, indeed, one well-known construct in classical dynamics that involves products of dynamical variables. The *classical Poisson bracket* of two dynamical variables, u and v, is defined [Goldstein *et al.* (2002)]

$$[u, v]_{cl} = \sum_{i=1,N} \left(\frac{\partial u}{\partial q_i} \frac{\partial v}{\partial p_i} - \frac{\partial u}{\partial p_i} \frac{\partial v}{\partial q_i} \right),$$ (2.3)

where u and v are regarded as functions of the coordinates and momenta, q_i and p_i, respectively. It is easily demonstrated that

$$[q_i, q_j]_{cl} = 0,$$ (2.4)

$$[p_i, p_j]_{cl} = 0,$$ (2.5)

$$[q_i, p_j]_{cl} = \delta_{ij}.$$ (2.6)

(See Exercise 2.1.) The time evolution of a dynamical variable can also be written in terms of a Poisson bracket by noting that

$$\frac{du}{dt} = \sum_{i=1,N} \left(\frac{\partial u}{\partial q_i} \frac{dq_i}{dt} + \frac{\partial u}{\partial p_i} \frac{dp_i}{dt} \right) = \sum_{i=1,N} \left(\frac{\partial u}{\partial q_i} \frac{\partial H}{\partial p_i} - \frac{\partial u}{\partial p_i} \frac{\partial H}{\partial q_i} \right) = [u, H]_{cl},$$ (2.7)

where use has been made of *Hamilton's equations*, Equations (2.1)–(2.2).

Let us attempt to construct a quantum mechanical Poisson bracket in which u and v are non-commuting operators instead of functions. Now, the main properties of the classical Poisson bracket are as follows:

$$[u, v]_{cl} = -[v, u]_{cl},$$ (2.8)

$$[u, c]_{cl} = 0,$$ (2.9)

$$[u_1 + u_2, v]_{cl} = [u_1, v]_{cl} + [u_2, v]_{cl},$$ (2.10)

$$[u, v_1 + v_2]_{cl} = [u, v_1]_{cl} + [u, v_2]_{cl},$$ (2.11)

$$[u_1 u_2, v]_{cl} = [u_1, v]_{cl} u_2 + u_1 [u_2, v]_{cl},$$ (2.12)

$$[u, v_1 v_2]_{cl} = [u, v_1]_{cl} v_2 + v_1 [u, v_2]_{cl},$$ (2.13)

$$[u, [v, w]_{cl}]_{cl} + [v, [w, u]_{cl}]_{cl} + [w, [u, v]_{cl}]_{cl} = 0.$$ (2.14)

(See Exercise 2.2.) The final relation is known as the *Jacobi identity*. In the previous expressions, u, v, w, et cetera, represent dynamical variables, and c represents a complex number. We wish to find some combination of non-commuting operators, u and v, that satisfies Equation (2.8)–(2.14). We shall refer to such a combination as a *quantum mechanical Poisson bracket*.

Actually, we can evaluate the quantum mechanical Poisson bracket, $[u_1 u_2, v_1 v_2]_{qm}$, in two different ways, because we can employ either of the formulae (2.12) or (2.13) first. Thus,

$$
\begin{aligned}
[u_1 u_2, v_1 v_2]_{qm} &= [u_1, v_1 v_2]_{qm} u_2 + u_1 [u_2, v_1 v_2]_{qm} \\
&= \big([u_1, v_1]_{qm} v_2 + v_1 [u_1, v_2]_{qm}\big) u_2 + u_1 \big([u_2, v_1]_{qm} v_2 + v_1 [u_2, v_2]_{qm}\big) \\
&= [u_1, v_1]_{qm} v_2 u_2 + v_1 [u_1, v_2]_{qm} u_2 + u_1 [u_2, v_1]_{qm} v_2 \\
&\quad + u_1 v_1 [u_2, v_2]_{qm},
\end{aligned}
\tag{2.15}
$$

and

$$
\begin{aligned}
[u_1 u_2, v_1 v_2]_{qm} &= [u_1 u_2, v_1]_{qm} v_2 + v_1 [u_1 u_2, v_2]_{qm} \\
&= [u_1, v_1]_{qm} u_2 v_2 + u_1 [u_2, v_1]_{qm} v_2 + v_1 [u_1, v_2]_{qm} u_2 \\
&\quad + v_1 u_1 [u_2, v_2]_{qm}.
\end{aligned}
\tag{2.16}
$$

Note that the order of the various factors has been preserved in the previous expressions, because these factors now represent non-commuting operators. Equating the previous two results yields

$$
[u_1, v_1]_{qm} (u_2 v_2 - v_2 u_2) = (u_1 v_1 - v_1 u_1) [u_2, v_2]_{qm}.
\tag{2.17}
$$

Because this relation must hold for u_1 and v_1, quite independent of u_2 and v_2, it follows that

$$
u_1 v_1 - v_1 u_1 = i\hbar [u_1, v_1]_{qm},
\tag{2.18}
$$

$$
u_2 v_2 - v_2 u_2 = i\hbar [u_2, v_2]_{qm},
\tag{2.19}
$$

where $\hbar$ does not depend on u_1, v_1, u_2, v_2, and also commutes with $(u_1 v_1 - v_1 u_1)$. Because u_1, et cetera, are general operators, it follows that $\hbar$ is just a number. Now, we need the quantum mechanical Poisson bracket of two Hermitian operators to be itself an Hermitian operator, because the classical Poisson bracket of two real dynamical variables is real. This requirement is satisfied if $\hbar$ is a real number. Thus, the quantum mechanical Poisson bracket of two dynamical variables u and v is given by

$$
[u, v]_{qm} = \frac{uv - vu}{i\hbar},
\tag{2.20}
$$

where $\hbar$ is a new universal constant of nature. Quantum mechanics agrees with experiments provided that $\hbar$ takes the value $h/2\pi$, where

$$h = 6.6261 \times 10^{-34} \text{ J s} \tag{2.21}$$

is *Planck's constant*. (The quantity $\hbar$ is usually referred to as the *reduced Planck constant*.) The notation $[u, v]$ is conventionally reserved for the commutator, $u\,v - v\,u$, in quantum mechanics. Thus,

$$[u, v]_{qm} = \frac{[u, v]}{i\,\hbar}. \tag{2.22}$$

It is easily demonstrated that the quantum mechanical Poisson bracket, as defined in the previous equation, satisfies all of the relations (2.8)–(2.14). (See Exercise 2.2.)

The strong analogy we have found between the classical Poisson bracket, defined in Equation (2.3), and the quantum mechanical Poisson bracket, defined in Equation (2.22), leads us to assume that the quantum mechanical bracket has the same value as the corresponding classical bracket, at least for the simplest cases. In other words, we are going to assume that Equations (2.4)–(2.6) hold for quantum mechanical, as well as classical, Poisson brackets. This argument yields the fundamental commutation relations

$$[q_i, q_j] = 0, \tag{2.23}$$

$$[p_i, p_j] = 0, \tag{2.24}$$

$$[q_i, p_j] = i\,\hbar\,\delta_{ij}. \tag{2.25}$$

These results provide us with the basis for calculating commutation relations between general dynamical variables. For instance, if two dynamical variables, ξ and η, can both be written as a power series in the q_i and p_i then repeated application of Equations (2.8)–(2.14) allows $[\xi, \eta]$ to be expressed in terms of the fundamental commutation relations (2.23)–(2.25).

Equations (2.23)–(2.25) provide the foundation for the analogy between quantum mechanics and classical mechanics. Note that the classical result that dynamical variables commute is obtained in the limit $\hbar \to 0$. Thus, classical mechanics can be regarded as the limiting case of quantum mechanics as $\hbar$ goes to zero. In classical mechanics, each generalized coordinate and its conjugate momentum, q_i and p_i, respectively, correspond to a different classical degree of freedom of the system. It is clear from Equations (2.23)–(2.25) that, in quantum mechanics, the dynamical variables corresponding to different degrees of freedom commute. In fact, it is only those variables corresponding to the same degree of freedom that may fail to commute.

2.3 Wavefunctions

Consider a simple system with one classical degree of freedom, which corresponds to the Cartesian coordinate x. Suppose that x is free to take any value (e.g., x could be the position of a free particle). The classical dynamical variable x is represented in quantum mechanics as a linear Hermitian operator that is also called x. Moreover, the operator x possesses eigenvalues x' lying in the continuous range $-\infty < x' < +\infty$ (because the eigenvalues correspond to all the possible results of a measurement of x). We can span ket space using the suitably normalized eigenkets of x. An eigenket corresponding to the eigenvalue x' is denoted $|x'\rangle$. Moreover,

$$\langle x'|x''\rangle = \delta(x' - x''). \tag{2.26}$$

[See Equation (1.88).] The eigenkets satisfy the extremely useful relation

$$\int_{-\infty}^{+\infty} dx' \, |x'\rangle\langle x'| = 1. \tag{2.27}$$

[See Equation (1.94).] This formula expresses the fact that the eigenkets are complete, mutually orthogonal, and suitably normalized.

A state ket $|A\rangle$ (which represents a general state A of the system) can be expressed as a linear superposition of the eigenkets of the position operator using Equation (2.27). Thus,

$$|A\rangle = \int_{-\infty}^{+\infty} dx' \, \langle x'|A\rangle|x'\rangle. \tag{2.28}$$

The quantity $\langle x'|A\rangle$ is a complex function of the position eigenvalue x'. We can write

$$\langle x'|A\rangle = \psi_A(x'). \tag{2.29}$$

Here, $\psi_A(x')$ is the famous *wavefunction* of quantum mechanics [Schrödinger (1926b); Dirac (1958)]. Note that state A is completely specified by its wavefunction $\psi_A(x')$ [because the wavefunction can be used to reconstruct the state ket $|A\rangle$ using Equation (2.28)]. It is clear that the wavefunction of state A is simply the collection of the weights of the corresponding state ket $|A\rangle$, when it is expanded in terms of the eigenkets of the position operator. Recall, from Section 1.10, that the probability of a measurement of a dynamical variable ξ yielding the result ξ' when the system is in (a properly normalized) state A is given by $|\langle \xi'|A\rangle|^2$, assuming that the eigenvalues of ξ are discrete. This result is easily generalized to dynamical variables possessing continuous eigenvalues. In fact, the probability of a measurement of x yielding a result lying in the range x' to $x' + dx'$ when the system is in a state $|A\rangle$ is $|\langle x'|A\rangle|^2 \, dx'$. In other words, the probability of a measurement of position yielding a result in the range x' to $x' + dx'$ when the wavefunction of the system is $\psi_A(x')$ is

$$P(x', dx') = |\psi_A(x')|^2 \, dx'. \tag{2.30}$$

This formula is only valid if the state ket $|A\rangle$ is properly normalized: that is, if $\langle A|A\rangle = 1$. The corresponding normalization for the wavefunction is

$$\int_{-\infty}^{+\infty} dx' \, |\psi_A(x')|^2 = 1. \tag{2.31}$$

Consider a second state B represented by a state ket $|B\rangle$ and a wavefunction $\psi_B(x')$. The inner product $\langle B|A\rangle$ can be written

$$\langle B|A\rangle = \int_{-\infty}^{+\infty} dx' \, \langle B|x'\rangle\langle x'|A\rangle = \int_{-\infty}^{+\infty} dx' \, \psi_B^*(x') \, \psi_A'(x'), \tag{2.32}$$

where use has been made of Equations (2.27) and (2.29). Thus, the inner product of two states is related to the overlap integral of their wavefunctions.

Consider a general function $f(x)$ of the observable x [e.g., $f(x) = x^2$]. If $|B\rangle = f(x)|A\rangle$ then it follows that

$$\psi_B(x') = \langle x'|f(x) \int_{-\infty}^{+\infty} dx'' \, \psi_A(x'') \, |x''\rangle$$

$$= \int_{-\infty}^{+\infty} dx'' \, f(x'') \, \psi_A(x'') \, \langle x'|x''\rangle, \tag{2.33}$$

giving

$$\psi_B(x') = f(x') \, \psi_A(x'), \tag{2.34}$$

where use has been made of Equation (2.26). (See Exercise 2.3.) Here, $f(x')$ is the same function of the position eigenvalue x' that $f(x)$ is of the position operator x. For instance, if $f(x) = x^2$ then $f(x') = x'^2$. It follows, from the previous result, that a general state ket $|A\rangle$ can be written

$$|A\rangle = \psi_A(x)\rangle, \tag{2.35}$$

where $\psi_A(x)$ is the same function of the operator x that the wavefunction $\psi_A(x')$ is of the position eigenvalue x', and the ket $\rangle$ has the wavefunction $\psi(x') = 1$. The ket $\rangle$ is termed the *standard ket*. The dual of the standard ket is termed the *standard bra*, and is denoted $\langle$. It is easily seen that

$$\langle \psi_A^*(x) \xleftrightarrow{\text{DC}} \psi_A(x)\rangle. \tag{2.36}$$

Note, finally, that $\psi_A(x)\rangle$ is often shortened to $\psi_A\rangle$, leaving the dependence on the position operator x tacitly understood.

2.4 Schrödinger Representation

Consider the simple system described in the previous section. A general state ket can be written $\psi(x)\rangle$, where $\psi(x)$ is a general function of the position operator x, and $\psi(x')$ is the associated wavefunction. Consider the ket whose wavefunction is $d\psi(x')/dx'$. This ket is denoted $d\psi/dx\rangle$. The new ket is clearly a linear function of the original ket, so we can think of it as the result of some linear operator acting on $\psi\rangle$. Let us denote this operator d/dx. It follows that

$$\frac{d}{dx}\psi\rangle = \frac{d\psi}{dx}\rangle. \tag{2.37}$$

Any linear operator that acts on ket vectors can also act on bra vectors. Consider d/dx acting on a general bra $\langle\phi(x)$. According to Equation (1.37), the bra $\langle\phi\, d/dx$ satisfies

$$\left(\langle\phi\,\frac{d}{dx}\right)\psi\rangle = \langle\phi\left(\frac{d}{dx}\,\psi\rangle\right). \tag{2.38}$$

Making use of Equations (2.27), (2.29), and (2.37) we can write

$$\int_{-\infty}^{+\infty} dx'\,\langle\phi\,\frac{d}{dx}\,|x'\rangle\,\psi(x') = \int_{-\infty}^{+\infty} dx'\,\phi(x')\,\frac{d\psi(x')}{dx'}. \tag{2.39}$$

The right-hand side can be transformed via integration by parts to give

$$\int_{-\infty}^{+\infty} dx'\,\langle\phi\,\frac{d}{dx}\,|x'\rangle\,\psi(x') = -\int_{-\infty}^{+\infty} dx'\,\frac{d\phi(x')}{dx'}\,\psi(x'), \tag{2.40}$$

assuming that the contributions from the limits of integration vanish. It follows that

$$\langle\phi\,\frac{d}{dx}\,|x'\rangle = -\frac{d\phi(x')}{dx'}, \tag{2.41}$$

which implies that

$$\langle\phi\,\frac{d}{dx} = -\langle\frac{d\phi}{dx}. \tag{2.42}$$

The neglect of contributions from the limits of integration in Equation (2.40) is reasonable because physical wavefunctions are square-integrable. [See Equation (2.31).] Note that

$$\frac{d}{dx}\,\psi\rangle = \frac{d\psi}{dx}\rangle \xrightarrow{\text{DC}} \langle\frac{d\psi^*}{dx} = -\langle\psi^*\,\frac{d}{dx}, \tag{2.43}$$

where use has been made of Equation (2.42). It follows, by comparison with Equations (1.38) and (2.36), that

$$\left(\frac{d}{dx}\right)^{\dagger} = -\frac{d}{dx}. \tag{2.44}$$

Thus, d/dx is an anti-Hermitian operator.

Let us evaluate the commutation relation between the operators x and d/dx. We have

$$\frac{d}{dx} x\psi\rangle = \frac{d(x\psi)}{dx}\rangle = x\frac{d}{dx}\psi\rangle + \psi\rangle. \tag{2.45}$$

Because this holds for any ket $\psi\rangle$, it follows that

$$\frac{d}{dx}x - x\frac{d}{dx} = 1. \tag{2.46}$$

Let p_x be the momentum conjugate to x (for the simple system under consideration, p_x is a straightforward linear momentum). According to Equation (2.25), x and p_x satisfy the commutation relation

$$x p_x - p_x x = i\hbar. \tag{2.47}$$

It can be seen, by comparison with Equation (2.46), that the Hermitian operator $-i\hbar\, d/dx$ satisfies the same commutation relation with x that p_x does. The most general conclusion which may be drawn from a comparison of Equations (2.46) and (2.47) is that

$$p_x = -i\hbar\frac{d}{dx} + f(x), \tag{2.48}$$

because (as is easily demonstrated) a general function $f(x)$ of the position operator automatically commutes with x. (See Exercise 2.3.)

We have chosen to normalize the eigenkets and eigenbras of the position operator such that they satisfy the normalization condition (2.26). However, this choice of normalization does not uniquely determine the eigenkets and eigenbras. Suppose that we transform to a new set of eigenbras which are related to the old set via

$$\langle x'|_{\text{new}} = e^{i\gamma'}\langle x'|_{\text{old}}, \tag{2.49}$$

where $\gamma' \equiv \gamma(x')$ is a real function of x'. This transformation amounts to a rearrangement of the relative phases of the eigenbras. The new normalization condition is

$$\langle x'|x''\rangle_{\text{new}} = \langle x'|e^{i\gamma'}e^{-i\gamma''}|x''\rangle_{\text{old}} = e^{i(\gamma'-\gamma'')}\langle x'|x''\rangle_{\text{old}}$$
$$= e^{i(\gamma'-\gamma'')}\delta(x'-x'') = \delta(x'-x''). \tag{2.50}$$

Thus, the new eigenbras satisfy the same normalization condition as the old eigenbras.

By definition, the standard ket $\rangle$ satisfies $\langle x'|\rangle = 1$. It follows from Equation (2.49) that the new standard ket is related to the old standard ket via

$$\rangle_{\text{new}} = e^{-i\gamma}\rangle_{\text{old}}, \tag{2.51}$$

where $\gamma \equiv \gamma(x)$ is a real function of the position operator x. The dual of the previous equation yields the transformation rule for the standard bra,

$$\langle_{\text{new}} = \langle_{\text{old}} \, e^{i\gamma}. \tag{2.52}$$

The transformation rule for a general operator A follows from Equations (2.51) and (2.52), plus the requirement that the triple product $\langle A \rangle$ remain invariant (which must be the case, otherwise the probability of a measurement yielding a certain result would depend on the choice of eigenbras). Thus,

$$A_{\text{new}} = e^{-i\gamma} A_{\text{old}} \, e^{i\gamma}. \tag{2.53}$$

Of course, if A commutes with x then A is invariant under the transformation. In fact, d/dx is the only operator (that we know of) that does not commute with x, so Equation (2.53) yields

$$\left(\frac{d}{dx}\right)_{\text{new}} = e^{-i\gamma} \frac{d}{dx} e^{i\gamma} = \frac{d}{dx} + i \frac{d\gamma}{dx}, \tag{2.54}$$

where the subscript "old" is taken as read. It follows, from Equation (2.48), that the momentum operator p_x can be written

$$p_x = -i\hbar \left(\frac{d}{dx}\right)_{\text{new}} - \hbar \frac{d\gamma}{dx} + f(x). \tag{2.55}$$

Thus, the special choice

$$\hbar \gamma(x) = \int^x dy \, f(y) \tag{2.56}$$

yields

$$p_x = -i\hbar \left(\frac{d}{dx}\right)_{\text{new}}. \tag{2.57}$$

Equation (2.56) fixes γ to within an arbitrary additive constant. In other words, the special eigenkets and eigenbras for which Equation (2.57) is true are determined to within an arbitrary common phase-factor.

In conclusion, it is possible to find a set of basis eigenkets and eigenbras of the position operator x that satisfy the normalization condition (2.26), and for which the momentum conjugate to x can be represented as the operator

$$p_x = -i\hbar \frac{d}{dx}. \tag{2.58}$$

A general state ket is written $\psi(x)\rangle$, where the standard ket $\rangle$ satisfies $\langle x'|\rangle = 1$, and where $\psi(x') = \langle x'|\psi(x)\rangle$ is the wavefunction. This scheme is known as the *Schrödinger representation*, and is the basis of wave mechanics.

2.5   Generalized Schrödinger Representation

In the preceding section, we developed the Schrödinger representation for the case of a single operator x corresponding to a classical Cartesian coordinate. However, this scheme can easily be extended. Consider a system with N generalized coordinates, $q_1 \cdots q_N$, which can all be simultaneously measured. These are represented as N commuting operators, $q_1 \cdots q_N$, each with a continuous range of eigenvalues, $q_1' \cdots q_N'$. Ket space is conveniently spanned by the simultaneous eigenkets of $q_1 \cdots q_N$, which are denoted $|q_1' \cdots q_N'\rangle$. These eigenkets must form a complete set, otherwise the $q_1 \cdots q_N$ would not be simultaneously observable.

The orthogonality condition for the eigenkets [i.e., the generalization of Equation (2.26)] is

$$\langle q_1' \cdots q_N' | q_1'' \cdots q_N'' \rangle = \delta(q_1' - q_1'')\,\delta(q_2' - q_2'') \cdots \delta(q_N' - q_N''). \tag{2.59}$$

The completeness condition [i.e., the generalization of Equation (2.27)] is

$$\int_{-\infty}^{+\infty} \cdots \int_{-\infty}^{+\infty} dq_1' \cdots dq_N' \, |q_1' \cdots q_N'\rangle\langle q_1' \cdots q_N'| = 1. \tag{2.60}$$

The standard ket $\rangle$ is defined such that

$$\langle q_1' \cdots q_N' | \rangle = 1. \tag{2.61}$$

The standard bra $\langle$ is the dual of the standard ket. A general state ket is written

$$\psi(q_1 \cdots q_N)\rangle. \tag{2.62}$$

The associated wavefunction is

$$\psi(q_1' \cdots q_N') = \langle q_1' \cdots q_N' | \psi \rangle. \tag{2.63}$$

Likewise, a general state bra is written

$$\langle \phi(q_1 \cdots q_N), \tag{2.64}$$

where

$$\phi(q_1' \cdots q_N') = \langle \phi | q_1' \cdots q_N' \rangle. \tag{2.65}$$

The probability of an observation of the system simultaneously finding the first coordinate in the range q_1' to $q_1' + dq_1'$, the second coordinate in the range q_2' to $q_2' + dq_2'$, et cetera, is

$$P(q_1' \cdots q_N'; dq_1' \cdots dq_N') = |\psi(q_1' \cdots q_N')|^2 \, dq_1' \cdots dq_N'. \tag{2.66}$$

Finally, the normalization condition for a physical wavefunction is

$$\int_{-\infty}^{+\infty} \cdots \int_{-\infty}^{+\infty} dq_1' \cdots dq_N' \, |\psi(q_1' \cdots q_N')|^2 = 1. \tag{2.67}$$

The N linear operators $\partial/\partial q_i$ (where i runs from 1 to N) are defined

$$\frac{\partial}{\partial q_i}\,\psi\rangle = \frac{\partial\psi}{\partial q_i}\rangle. \tag{2.68}$$

These linear operators can also act on bras (provided the associated wavefunctions are square integrable) in accordance with

$$\langle\phi\,\frac{\partial}{\partial q_i} = -\langle\frac{\partial\phi}{\partial q_i}. \tag{2.69}$$

[See Equation (2.42).] Corresponding to Equation (2.46), we can derive the commutation relations

$$\frac{\partial}{\partial q_i}\,q_j - q_j\,\frac{\partial}{\partial q_i} = \delta_{ij}. \tag{2.70}$$

It is also clear that

$$\frac{\partial}{\partial q_i}\frac{\partial}{\partial q_j}\,\psi\rangle = \frac{\partial^2\psi}{\partial q_i\,\partial q_j}\rangle = \frac{\partial^2\psi}{\partial q_j\,\partial q_i}\rangle = \frac{\partial}{\partial q_j}\frac{\partial}{\partial q_i}\,\psi\rangle, \tag{2.71}$$

showing that

$$\frac{\partial}{\partial q_i}\frac{\partial}{\partial q_j} = \frac{\partial}{\partial q_j}\frac{\partial}{\partial q_i}. \tag{2.72}$$

It can be seen, by comparison with Equations (2.23)–(2.25), that the linear operators $-i\hbar\,\partial/\partial q_i$ satisfy the same commutation relations with the q's and with each other that the p's do. The most general conclusion that we can draw from this coincidence of commutation relations is

$$p_i = -i\hbar\frac{\partial}{\partial q_i} + \frac{\partial F(q_1\cdots q_N)}{\partial q_i}. \tag{2.73}$$

However, the function F can be transformed away via a suitable readjustment of the phases of the basis eigenkets [Dirac (1958)]. (See Section 2.4.) Thus, we can always construct a set of simultaneous eigenkets of $q_1\cdots q_N$ for which

$$p_i = -i\hbar\frac{\partial}{\partial q_i}. \tag{2.74}$$

This is the generalized Schrödinger representation.

It follows from Equations (2.61), (2.68), and (2.74) that

$$p_i\rangle = 0. \tag{2.75}$$

Thus, the standard ket in the Schrödinger representation is a simultaneous eigenket of all the momentum operators belonging to the eigenvalue zero. Note that

$$\langle q_1'\cdots q_N'|\,\frac{\partial}{\partial q_i}\,\psi\rangle = \langle q_1'\cdots q_N'|\,\frac{\partial\psi}{\partial q_i}\rangle = \frac{\partial\psi(q_1'\cdots q_N')}{\partial q_i'} = \frac{\partial}{\partial q_i'}\langle q_1'\cdots q_N'|\psi\rangle. \tag{2.76}$$

Hence,

$$\langle q_1' \cdots q_N' | \frac{\partial}{\partial q_i} = \frac{\partial}{\partial q_i'} \langle q_1' \cdots q_N' |, \tag{2.77}$$

so that

$$\langle q_1' \cdots q_N' | p_i = -i\hbar \frac{\partial}{\partial q_i'} \langle q_1' \cdots q_N' |. \tag{2.78}$$

The dual of the previous equation gives

$$p_i | q_1' \cdots q_N' \rangle = i\hbar \frac{\partial}{\partial q_i'} | q_1' \cdots q_N' \rangle. \tag{2.79}$$

2.6   Momentum Representation

Consider a system with one degree of freedom, describable in terms of a coordinate x and its conjugate momentum p_x, both of which have a continuous range of eigenvalues. We have seen that it is possible to represent the system in terms of the eigenkets of x. This is termed the Schrödinger representation. However, it is also possible to represent the system in terms of the eigenkets of p_x.

Consider the eigenkets of p_x that belong to the eigenvalues p_x'. These are denoted $|p_x'\rangle$. The orthogonality relation for the momentum eigenkets is

$$\langle p_x' | p_x'' \rangle = \delta(p_x' - p_x''), \tag{2.80}$$

and the corresponding completeness relation is

$$\int_{-\infty}^{+\infty} dp_x' |p_x'\rangle\langle p_x'| = 1. \tag{2.81}$$

A general state ket can be written

$$\phi(p_x)\rangle \tag{2.82}$$

where the standard ket $\rangle$ satisfies

$$\langle p_x' | \rangle = 1. \tag{2.83}$$

Note that the standard ket in this representation is quite different to that in the Schrödinger representation. The momentum space wavefunction $\phi(p_x')$ satisfies

$$\phi(p_x') = \langle p_x' | \phi \rangle. \tag{2.84}$$

The probability that a measurement of the momentum yields a result lying in the range p_x' to $p_x' + dp_x'$ is given by

$$P(p_x', dp_x') = |\phi(p_x')|^2 \, dp_x'. \tag{2.85}$$

Finally, the normalization condition for a physical momentum space wavefunction is

$$\int_{-\infty}^{+\infty} dp'_x \, |\phi(p'_x)|^2 = 1. \tag{2.86}$$

The fundamental commutation relations (2.23)–(2.25) exhibit a particular symmetry between coordinates and their conjugate momenta. In fact, if all the coordinates are transformed into their conjugate momenta, and vice versa, and i is then replaced by −i, then the commutation relations are unchanged. It follows from this symmetry that we can always choose the eigenkets of p_x in such a manner that the coordinate x can be represented as

$$x = i\,\hbar\,\frac{d}{dp_x}. \tag{2.87}$$

(See Section 2.4.) This scheme is termed the *momentum representation*.

The previous result is easily generalized to a system with more than one degree of freedom. Suppose the system is specified by N coordinates, $q_1 \cdots q_N$, and N conjugate momenta, $p_1 \cdots p_N$. Then, in the momentum representation, the coordinates can be written as

$$q_i = i\,\hbar\,\frac{\partial}{\partial p_i}. \tag{2.88}$$

We also have

$$q_i\rangle = 0, \tag{2.89}$$

and

$$\langle p'_1 \cdots p'_N|\, q_i = i\,\hbar\,\frac{\partial}{\partial p'_i}\,\langle p'_1 \cdots p'_N| \tag{2.90}$$

[cf., Equation (2.78).]

Generally speaking, the momentum representation is less useful than the Schrödinger representation. The main reason for this is that the energy operators (i.e., the Hamiltonians) of most simple microscopic systems take the form of a sum of quadratic terms in the momenta (i.e., the kinetic energy) plus a complicated function of the coordinates (i.e., the potential energy). In the Schrödinger representation, the eigenvalue problem for the energy translates into a second-order partial differential equation in the coordinates, with a complicated potential function. In the momentum representation, the problem transforms into a high-order partial differential equation in the momenta, with a quadratic potential. With the mathematical tools at our disposal, we are far better able to solve the former type of differential equation than the latter.

2.7   Heisenberg Uncertainty Principle

How is a momentum space wavefunction related to the corresponding coordinate space wavefunction? To answer this question, let us consider the representative $\langle x'|p_x'\rangle$ of the momentum eigenkets $|p_x'\rangle$ in the Schrödinger representation for a system with a single degree of freedom. This representative satisfies

$$p_x'\langle x'|p_x'\rangle = \langle x'|p_x|p_x'\rangle = -\mathrm{i}\,\hbar\,\frac{d}{dx'}\,\langle x'|p_x'\rangle, \tag{2.91}$$

where use has been made of Equation (2.78) (for the case of a system with one degree of freedom). The solution of the previous differential equation is

$$\langle x'|p_x'\rangle = c'\,\exp\left(\frac{\mathrm{i}\,p_x'\,x'}{\hbar}\right), \tag{2.92}$$

where $c' = c'(p_x')$. It is easily demonstrated that

$$\langle p_x'|p_x''\rangle = \int_{-\infty}^{+\infty} dx'\,\langle p_x'|x'\rangle\langle x'|p_x''\rangle$$

$$= c'^*\,c''\int_{-\infty}^{\infty} dx'\,\exp\left[\frac{-\mathrm{i}\,(p_x' - p_x'')\,x'}{\hbar}\right]. \tag{2.93}$$

The well-known mathematical result [Riley *et al.* (2013)]

$$\int_{-\infty}^{+\infty} dx\,\exp(\mathrm{i}\,a\,x) = 2\pi\,\delta(a), \tag{2.94}$$

yields

$$\langle p_x'|p_x''\rangle = |c'|^2\,h\,\delta(p_x' - p_x''). \tag{2.95}$$

This is consistent with Equation (2.80), provided that $c' = h^{-1/2}$. Thus,

$$\langle x'|p_x'\rangle = \frac{1}{h^{1/2}}\,\exp\left(\frac{\mathrm{i}\,p_x'\,x'}{\hbar}\right). \tag{2.96}$$

Consider a general state ket $|A\rangle$ whose coordinate wavefunction is $\psi(x')$, and whose momentum wavefunction is $\Psi(p_x')$. In other words,

$$\psi(x') = \langle x'|A\rangle, \tag{2.97}$$

$$\Psi(p_x') = \langle p_x'|A\rangle. \tag{2.98}$$

It is easily demonstrated that

$$\psi(x') = \int_{-\infty}^{+\infty} dp_x'\,\langle x'|p_x'\rangle\langle p_x'|A\rangle = \frac{1}{h^{1/2}}\int_{-\infty}^{+\infty} dp_x'\,\Psi(p_x')\,\exp\left(\frac{\mathrm{i}\,p_x'\,x'}{\hbar}\right) \tag{2.99}$$

and

$$\Psi(p_x') = \int_{-\infty}^{+\infty} dx'\,\langle p_x'|x'\rangle\langle x'|A\rangle = \frac{1}{h^{1/2}}\int_{-\infty}^{+\infty} dx'\,\psi(x')\,\exp\left(\frac{-\mathrm{i}\,p_x'\,x'}{\hbar}\right), \tag{2.100}$$

where use has been made of Equations (2.27), (2.81), (2.94), and (2.96). Clearly, the momentum space wavefunction is the *Fourier transform* of the corresponding coordinate space wavefunction [Fitzpatrick (2013)].

Consider a state whose coordinate space wavefunction is a *wavepacket*. In other words, the wavefunction only has non-negligible amplitude in some spatially localized region of extent Δx. As is well known, the Fourier transform of a wavepacket fills up a wavenumber band of approximate extent $\Delta k \sim 1/\Delta x$. [Fitzpatrick (2013)]. Note that, in Equation (2.99), the role of the wavenumber k is played by the quantity $p'_x/\hbar$. It follows that the momentum space wavefunction corresponding to a wavepacket in coordinate space extends over a range of momenta $\Delta p_x \sim \hbar/\Delta x$. Clearly, a measurement of x is almost certain to give a result lying in a range of width Δx. Likewise, measurement of p_x is almost certain to yield a result lying in a range of width Δp_x. The product of these two uncertainties is

$$\Delta x \, \Delta p_x \sim \hbar. \tag{2.101}$$

This result is called the *Heisenberg uncertainty principle* [Heisenberg (1927)].

Actually, it is possible to write the Heisenberg uncertainty principle more exactly by making use of Equation (1.86) and the commutation relation (2.47). We obtain

$$\langle (\Delta x)^2 \rangle \, \langle (\Delta p_x)^2 \rangle \geq \frac{\hbar^2}{4} \tag{2.102}$$

for a general state. It is easily demonstrated that the minimum uncertainty states, for which the equality sign holds in the previous relation, correspond to Gaussian wavepackets in both coordinate and momentum space. (See Exercise 2.4.)

2.8 Displacement Operators

Consider a system with one degree of freedom corresponding to the Cartesian coordinate x. Suppose that we displace this system some distance along the x-axis. We could imagine that the system is on wheels, and we just give it a little push. The final state of the system is completely determined by its initial state, together with the direction and magnitude of the displacement. Note that the type of displacement we are considering is one in which everything to do with the system is displaced. So, if the system is subject to an external potential then the potential must be displaced.

The situation is not so clear with state kets. The final state of the system only determines the direction of the displaced state ket. Even if we adopt the convention that all state kets have unit norms, the final ket is still not completely determined, because it can be multiplied by a constant phase-factor. However, we know that the superposition relations between states remain invariant under the displacement. This follows because the superposition relations have a physical significance that is unaffected by a displacement of the system. Thus, if

$$|R\rangle = |A\rangle + |B\rangle \qquad (2.103)$$

in the undisplaced system, and the displacement causes ket $|R\rangle$ to transform to ket $|Rd\rangle$, et cetera, then in the displaced system we have

$$|Rd\rangle = |Ad\rangle + |Bd\rangle. \qquad (2.104)$$

Incidentally, this determines the displaced kets to within a single arbitrary phase-factor to be multiplied into all of them. The displaced kets cannot be multiplied by individual phase-factors, because this would wreck the superposition relations.

Given that Equation (2.104) holds in the displaced system whenever Equation (2.103) holds in the undisplaced system, it follows that the displaced ket $|Rd\rangle$ must be the result of some linear operator acting on the undisplaced ket $|R\rangle$. In other words,

$$|Rd\rangle = D\,|R\rangle, \qquad (2.105)$$

where D is an operator that depends only on the nature of the displacement. The arbitrary phase-factor by which all displaced kets may be multiplied results in D being undetermined to an arbitrary multiplicative constant of modulus unity.

We now adopt the ansatz that any combination of bras, kets, and dynamical variables that possesses a physical significance is invariant under a displacement of the system. The normalization condition

$$\langle A|A\rangle = 1 \qquad (2.106)$$

for a state ket $|A\rangle$ certainly has a physical significance. Thus, we must have

$$\langle Ad|Ad\rangle = 1. \qquad (2.107)$$

Now, $|Ad\rangle = D\,|A\rangle$ and $\langle Ad| = \langle A|\,D^\dagger$, so

$$\langle A|\,D^\dagger D\,|A\rangle = 1. \qquad (2.108)$$

Because this must hold for any state ket $|A\rangle$, it follows that

$$D^\dagger D = 1. \qquad (2.109)$$

Hence, the operator D is unitary. Note that the previous relation implies that

$$|A\rangle = D^\dagger\,|Ad\rangle. \qquad (2.110)$$

The equation

$$v|A\rangle = |B\rangle, \tag{2.111}$$

where the operator v represents a dynamical variable, has physical significance. Thus, we require that

$$v_d|Ad\rangle = |Bd\rangle, \tag{2.112}$$

where v_d is the displaced operator. It follows that

$$v_d|Ad\rangle = D|B\rangle = Dv|A\rangle = DvD^\dagger|Ad\rangle. \tag{2.113}$$

Because this is true for any ket $|Ad\rangle$, we have

$$v_d = DvD^\dagger. \tag{2.114}$$

Note that the arbitrary multiplicative factor in D does not affect either of the results (2.109) or (2.114).

Suppose, now, that the system is displaced an infinitesimal distance δx along the x-axis. Let $D_x(\delta x)$ be the operator that accomplishes this displacement. We expect that the displaced ket $|Ad\rangle$ should approach the undisplaced ket $|A\rangle$ in the limit as $\delta x \to 0$. Thus, we expect the limit

$$\lim_{\delta x \to 0} \frac{|Ad\rangle - |A\rangle}{\delta x} = \lim_{\delta x \to 0} \frac{D_x(\delta x) - 1}{\delta x}|A\rangle \tag{2.115}$$

to exist. Let

$$d_x = \lim_{\delta x \to 0} \frac{D_x(\delta x) - 1}{\delta x}, \tag{2.116}$$

where d_x is denoted the *displacement operator* along the x-axis. The fact that D_x can be replaced by $D_x \exp(i\gamma)$, where γ is a real phase-angle, implies that d_x can be replaced by

$$\lim_{\delta x \to 0} \frac{D_x(\delta x)\exp(i\gamma) - 1}{\delta x} = \lim_{\delta x \to 0} \frac{D_x(\delta x) - 1 + i\gamma}{\delta x} = d_x + i a_x, \tag{2.117}$$

where a_x is the limit of $\gamma/\delta x$. We have assumed, as seems reasonable, that γ tends to zero as $\delta x \to 0$. It is clear that the displacement operator is undetermined to an arbitrary imaginary additive constant.

For small δx, we have

$$D_x(\delta x) = 1 + \delta x\, d_x. \tag{2.118}$$

It follows from Equation (2.109) that

$$(1 + \delta x\, d_x^\dagger)(1 + \delta x\, d_x) = 1. \tag{2.119}$$

Neglecting order $(\delta x)^2$, we obtain

$$d_x^\dagger + d_x = 0. \tag{2.120}$$

Thus, the displacement operator is anti-Hermitian. Substituting into Equation (2.114), and again neglecting order $(\delta x)^2$, we find that

$$v_d = (1 + \delta x\, d_x)\, v\, (1 - \delta x\, d_x) = v + \delta x\, (d_x v - v\, d_x), \qquad (2.121)$$

which implies that

$$\lim_{\delta x \to 0} \frac{v_d - v}{\delta x} = d_x v - v\, d_x. \qquad (2.122)$$

Let us consider a specific example. Suppose that a state has a wavefunction $\psi(x')$. If the system is displaced a distance δx along the x-axis then the new wavefunction is $\psi(x' - \delta x)$ (i.e., the same function shifted in the x-direction by a distance δx). Actually, the new wavefunction can be multiplied by an arbitrary number of modulus unity. It can be seen that the new wavefunction is obtained from the old wavefunction according to the prescription $x' \to x' - \delta x$. Thus,

$$x_d = x - \delta x. \qquad (2.123)$$

A comparison with Equation (2.122), using $x = v$, yields

$$d_x x - x\, d_x = -1. \qquad (2.124)$$

It follows that $\mathrm{i}\,\hbar\, d_x$ obeys the same commutation relation with x that p_x, the momentum conjugate to x, does. [See Equation (2.25).] The most general conclusion we can draw from this observation is that

$$p_x = \mathrm{i}\,\hbar\, d_x + f(x), \qquad (2.125)$$

where $f(x)$ is Hermitian (because p_x is Hermitian). However, the fact that d_x is undetermined to an arbitrary additive imaginary constant (which could be a function of x) enables us to transform the function $f(x)$ out of the previous equation, leaving

$$p_x = \mathrm{i}\,\hbar\, d_x. \qquad (2.126)$$

Thus, the displacement operator in the x-direction is proportional to the momentum conjugate to x. We say that p_x is the *generator of translation* along the x-axis.

A finite displacement along the x-axis can be constructed from a series of very many infinitesimal displacements. Thus, the operator $D_x(\Delta x)$, which displaces the system a finite distance Δx along the x-axis, is written [Abramowitz and Stegun (1965)]

$$D_x(\Delta x) = \lim_{N \to \infty} \left(1 - \mathrm{i}\, \frac{\Delta x}{N} \frac{p_x}{\hbar} \right)^N \equiv \sum_{n=0,\infty} \frac{(-\mathrm{i}\, p_x \Delta x / \hbar)^n}{n!}, \qquad (2.127)$$

where use has been made of Equations (2.118) and (2.126). It follows that

$$D_x(\Delta x) = \exp\left(\frac{-\mathrm{i}\, p_x \Delta x}{\hbar}\right). \tag{2.128}$$

The unitary nature of the operator is now clearly apparent. (See Exercise 1.15.)

We can also construct operators that displace the system along the y- and z-axes. For instance, the operator that displaces the system a finite distance Δy along the y-axis is

$$D_y(\Delta y) = \exp\left(\frac{-\mathrm{i}\, p_y \Delta y}{\hbar}\right). \tag{2.129}$$

Note that a displacement a distance Δx along the x-axis commutes with a displacement a distance Δy along the y-axis. In other words, if a physical system is moved Δx along the x-axis, and then Δy along the y-axis, then it ends up in the same state as if it were moved Δy along the y-axis, and then Δx along the x-axis. The fact that finite translations in independent directions commute implies that the associated displacement operators also commute. For instance, $D_x(\Delta x)\, D_y(\Delta y) = D_y(\Delta y)\, D_x(\Delta x)$. This property of displacement operators is clearly associated with the fact that the corresponding momentum operators also commute. In this case, $p_x\, p_y = p_y\, p_x$. [See Equations (2.24), (2.128), and Exercise 2.3.]

2.9 Exercises

2.1 Demonstrate that

(a)
$$[q_i, q_j]_{cl} = 0.$$

(b)
$$[p_i, p_j]_{cl} = 0.$$

(c)
$$[q_i, p_j]_{cl} = \delta_{ij}.$$

Here, $[\cdots, \cdots]_{cl}$ represents a classical Poisson bracket. Moreover, the q_i and p_i are the coordinates and corresponding canonical momenta of a classical, many degree of freedom, dynamical system.

2.2 Verify that

(a)
$$[u, v] = -[v, u].$$

(b) $$[u, c] = 0.$$

(c) $$[u_1 + u_2, v] = [u_1, v] + [u_2, v].$$

(d) $$[u, v_1 + v_2] = [u, v_1] + [u, v_2].$$

(e) $$[u_1 u_2, v] = [u_1, v] u_2 + u_1 [u_2, v].$$

(f) $$[u, v_1 v_2] = [u, v_1] v_2 + v_1 [u, v_2].$$

(g) $$[u, [v, w]] + [v, [w, u]] + [w, [u, v]] = 0.$$

Here, $[\cdots, \cdots]$ represents either a classical or a quantum mechanical Poisson bracket. Moreover, u, u, w, et cetera, represent dynamical variables (i.e., in the classical case, functions of the coordinates and canonical momenta), and c represents a number.

2.3 Let ξ be an operator whose eigenvalues ξ' can take a continuous range of values. Let the $|\xi'\rangle$ be the corresponding eigenstates. Let $f(\xi)$ be a function of ξ that can be expanded as a power series. Demonstrate that

$$[f(\xi), \xi] = 0,$$

and

$$f(\xi) |\xi'\rangle = f(\xi') |\xi'\rangle,$$

where $f(\xi')$ is the same function of the eigenvalue ξ' that $f(\xi)$ is of the operator ξ. Let $g(\eta)$ be a function of the operator η that can be expanded as a power series, and let ξ and η commute. Demonstrate that

$$[f(\xi), g(\eta)] = 0.$$

2.4 Consider a Gaussian wavepacket whose corresponding wavefunction is

$$\psi(x') = \psi_0 \exp\left[-\frac{(x' - x_0)^2}{4\sigma^2}\right],$$

where ψ_0, x_0, and σ are real numbers. Demonstrate that

(a)
$$\langle x \rangle = x_0.$$

(b)
$$\langle (\Delta x)^2 \rangle = \sigma^2.$$

(c)
$$\langle p_x \rangle = 0.$$

(d)
$$\langle (\Delta p_x)^2 \rangle = \frac{\hbar^2}{4\sigma^2}.$$

Here, x and p_x are a position operator and its conjugate momentum operator, respectively.

2.5 Let $D_x(\Delta x)$ and $D_y(\Delta y)$ be operators that displace a quantum mechanical system the finite distances Δx and Δy along the x- and y-directions, respectively. Demonstrate that

$$D_x(\Delta x_2)\, D_x(\Delta x_1) = D_x(\Delta x_1)\, D_x(\Delta x_2) = D_x(\Delta x_2 + \Delta x_1),$$

and

$$D_x(\Delta x)\, D_y(\Delta y) = D_y(\Delta y)\, D_x(\Delta x).$$

What are the physical significances of these results?

2.6 Suppose that we displace a one-dimensional quantum mechanical system a finite distance a along the x-axis. The corresponding operator is

$$D_x(a) = \exp\left(\frac{-\mathrm{i}\, p_x\, a}{\hbar}\right),$$

where p_x is the momentum conjugate to the position operator x. Demonstrate that

$$D_x(a)\, x\, D_x(a)^\dagger = x - a.$$

[Hint: Use the momentum representation, $x = \mathrm{i}\,\hbar\, d/dp_x$.] Similarly, demonstrate that

$$D_x(a)\, x^m\, D_x(a)^\dagger = (x - a)^m,$$

where m is a non-negative integer. Hence, deduce that

$$D_x(a)\, V(x)\, D_x(a)^\dagger = V(x - a),$$

where $V(x)$ is a function of x that can be expanded as a power series.

Let $k = p_x/\hbar$, and let $|k'\rangle$ denote an eigenket of the k operator belonging to the eigenvalue k'. Demonstrate that

$$|A\rangle = \sum_{n=-\infty,\infty} c_n \, |k' + n\,k_a\rangle,$$

where the c_n are arbitrary complex coefficients, and $k_a = 2\pi/a$, is an eigenket of the $D_x(a)$ operator belonging to the eigenvalue $\exp(-i\,k'\,a)$. Show that the corresponding wavefunction can be written

$$\psi_A(x') = e^{i\,k'\,x'} \, u(x'),$$

where $u(x' + a) = u(x')$ for all x'.

Chapter 3

Quantum Dynamics

3.1 Introduction

Up to now, we have only considered quantum mechanical systems at one particular instant of time. Let us now investigate the time evolution of such systems.

3.2 Schrödinger Equation of Motion

Consider a system in a state A that evolves in time. At time t, the state of the system is represented by the ket $|At\rangle$. The label A is needed to distinguish this ket from any other ket ($|Bt\rangle$, say) that is evolving in time. The label t is needed to distinguish the different states of the system at different times.

The final state of the system at time t is completely determined by its initial state at time $t_0 < t$, plus the time interval $t - t_0$ (assuming that the system is left undisturbed during this time interval). However, the final state only determines the direction of the final state ket. Even if we adopt the convention that all state kets have unit norms, the final ket is still not completely determined, because it can be multiplied by an arbitrary phase-factor. However, we expect that if a superposition relation holds for certain states at time t_0 then the same relation should hold between the corresponding time-evolved states at time t, assuming that the system is left undisturbed between times t_0 and t. In other words, if

$$|Rt_0\rangle = |At_0\rangle + |Bt_0\rangle \tag{3.1}$$

for any three kets then we should have

$$|Rt\rangle = |At\rangle + |Bt\rangle. \tag{3.2}$$

This rule determines the time-evolved kets to within a single arbitrary phase-factor to be multiplied into all of them. The evolved kets cannot be multiplied by individual phase-factors because this would invalidate the superposition relation at later times.

According to Equations (3.1) and (3.2), the final ket $|Rt\rangle$ depends linearly on the initial ket $|Rt_0\rangle$. Thus, the final ket can be regarded as the result of some linear operator acting on the initial ket: that is,

$$|Rt\rangle = T\,|Rt_0\rangle, \tag{3.3}$$

where T is a linear operator that depends only on the times t and t_0. The arbitrary phase-factor by which all time-evolved kets may be multiplied results in $T(t, t_0)$ being undetermined to an arbitrary multiplicative constant of modulus unity.

Because we have adopted a convention in which the norm of any state ket is unity, it make sense to define the time evolution operator T in such a manner that it preserves the length of any ket upon which it acts (i.e., if a ket is properly normalized at time t_0 then it will remain normalized at all subsequent times $t > t_0$). This is always possible, because the length of a ket possesses no physical significance. Thus, we require that

$$\langle At_0|At_0\rangle = \langle At|At\rangle \tag{3.4}$$

for any ket A, which immediately yields

$$T^\dagger T = 1. \tag{3.5}$$

Hence, the time evolution operator T is unitary.

Up to now, the time evolution operator T looks very much like the spatial displacement operator D introduced in Section 2.8. However, there are some important differences between time evolution and spatial displacement. In general, we do expect the expectation value of a given observable ξ to evolve with time, even if the system is left in a state of undisturbed motion (after all, time evolution has no meaning unless something observable changes with time). The triple product $\langle A|\xi|A\rangle$ can evolve either because the ket $|A\rangle$ evolves and the operator ξ stays constant, the ket $|A\rangle$ stays constant and the operator ξ evolves, or both the ket $|A\rangle$ and the operator ξ evolve. Because we are already committed to evolving state kets, according to Equation (3.3), let us assume that the time evolution operator T can be chosen in such a manner that the operators representing the dynamical variables of the system do not evolve in time (unless they contain some specific time dependence).

We expect, from physical continuity, that if $t \to t_0$ then $|At\rangle \to |At_0\rangle$ for any ket A. Thus, the limit

$$\lim_{t \to t_0} \frac{|At\rangle - |At_0\rangle}{t - t_0} = \lim_{t \to t_0} \frac{T - 1}{t - t_0}\,|At_0\rangle \tag{3.6}$$

should exist. Note that this limit is simply the derivative of $|At_0\rangle$ with respect to t_0. Let

$$\tau(t_0) = \lim_{t \to t_0} \frac{T(t, t_0) - 1}{t - t_0}. \tag{3.7}$$

It is easily demonstrated from Equation (3.5) that τ is anti-Hermitian: that is,

$$\tau^\dagger + \tau = 0. \tag{3.8}$$

The fact that T can be replaced by $T \exp(i\gamma)$ (where γ is real) implies that τ is undetermined to an arbitrary imaginary additive constant. (See Section 2.8.) Let us define the Hermitian operator $H(t_0) = i\hbar\tau$. This operator is undetermined to an arbitrary real additive constant. It follows from Equations (3.6) and (3.7) that

$$i\hbar \frac{d|At_0\rangle}{dt_0} = i\hbar \lim_{t \to t_0} \frac{|At\rangle - |At_0\rangle}{t - t_0} = i\hbar\tau(t_0)|At_0\rangle = H(t_0)|At_0\rangle. \tag{3.9}$$

When written for general t, this equation becomes

$$i\hbar \frac{d|At\rangle}{dt} = H(t)|At\rangle. \tag{3.10}$$

Equation (3.10) gives the general law for the time evolution of a state ket in a scheme in which the operators representing the dynamical variables remain fixed. This equation is denoted the *Schrödinger equation of motion*. It involves a Hermitian operator $H(t)$ that is, presumably, a characteristic of the dynamical system under investigation.

We saw, in Section 2.8, that if the operator $D_x(x, x_0)$ displaces the system along the x-axis from x_0 to x then

$$p_x = i\hbar \lim_{x \to x_0} \frac{D_x(x, x_0) - 1}{x - x_0}, \tag{3.11}$$

where p_x is the operator representing the momentum conjugate to x. Furthermore, we have just shown that if the operator $T(t, t_0)$ evolves the system in time from t_0 to t then

$$H(t_0) = i\hbar \lim_{t \to t_0} \frac{T(t, t_0) - 1}{t - t_0}. \tag{3.12}$$

Thus, the dynamical variable corresponding to the operator H stands to time t as the momentum p_x stands to the coordinate x. By analogy with classical physics, this suggests that $H(t)$ is the operator representing the total energy of the system. (Recall that, in classical physics, if the equations of motion of a system are invariant under an x-displacement then this implies that the system conserves momentum in the x-direction. Likewise, if the equations of motion are invariant under a temporal displacement then this implies that the system conserves energy [Goldstein *et al.* (2002)].) The operator $H(t)$ is usually called the *Hamiltonian* of the system. The fact that the Hamiltonian is undetermined to an arbitrary real additive constant is related to the well-known phenomenon that energy is undetermined to an arbitrary additive constant in physics (i.e., the zero of potential energy is not well defined).

Substituting $|At\rangle = T\,|At_0\rangle$ into Equation (3.10) yields

$$i\hbar\,\frac{dT}{dt}\,|At_0\rangle = H(t)\,T\,|At_0\rangle. \tag{3.13}$$

Because this must hold for any initial state $|At_0\rangle$, we conclude that

$$i\hbar\,\frac{dT}{dt} = H(t)\,T. \tag{3.14}$$

This equation can be integrated to give

$$T(t,t_0) = \exp\left[-\frac{i}{\hbar}\int_{t_0}^{t} dt'\,H(t')\right], \tag{3.15}$$

where use has been made of Equations (3.5) and (3.6). (Here, we assume that Hamiltonian operators evaluated at different times commute with one another.) The fact that H is undetermined to an arbitrary real additive constant leaves T undetermined to a phase-factor. Incidentally, in the previous analysis, time is not an operator (we cannot observe time, as such), it is just a parameter (or, more accurately, a continuous label).

3.3 Heisenberg Equation of Motion

We have seen that in the Schrödinger scheme the dynamical variables of the system remain fixed during a period of undisturbed motion, whereas the state kets evolve according to Equation (3.10). However, this is not the only way in which to represent the time evolution of the system.

Suppose that a general state ket A is subject to the transformation

$$|A_t\rangle = T^\dagger(t,t_0)\,|A\rangle. \tag{3.16}$$

This is a time-dependent transformation, because the operator $T(t,t_0)$ obviously depends on time. The subscript t is used to remind us that the transformation is time-dependent. The time evolution of the transformed state ket is given by

$$|A_t t\rangle = T^\dagger(t,t_0)\,|At\rangle = T^\dagger(t,t_0)\,T(t,t_0)\,|At_0\rangle = |A_t t_0\rangle, \tag{3.17}$$

where use has been made of Equations (3.3), (3.5), and the fact that $T(t_0,t_0) = 1$. Clearly, the transformed state ket does not evolve in time. Thus, the transformation (3.16) has the effect of bringing all kets representing states of undisturbed motion of the system to rest.

The transformation must also be applied to bras. The dual of Equation (3.16) yields

$$\langle A_t| = \langle A| \, T. \tag{3.18}$$

The transformation rule for a general observable v is obtained from the requirement that the expectation value $\langle A| \, v \, |A\rangle$ should remain invariant. It is easily seen that

$$v_t = T^\dagger v T. \tag{3.19}$$

Thus, a dynamical variable, which corresponds to a fixed linear operator in the Schrödinger scheme, corresponds to a moving linear operator in this new scheme. It is clear that the transformation (3.16) leads us to a scenario in which the state of the system is represented by a fixed ket, and the dynamical variables are represented by moving linear operators. This is termed the *Heisenberg picture*, as opposed to the *Schrödinger picture*, which was outlined in Section 3.2.

Consider a dynamical variable v corresponding to a fixed linear operator in the Schrödinger picture. According to Equation (3.19), we can write

$$T v_t = v T. \tag{3.20}$$

Differentiation with respect to time yields

$$\frac{dT}{dt} v_t + T \frac{dv_t}{dt} = v \frac{dT}{dt}. \tag{3.21}$$

With the help of Equation (3.14), this reduces to

$$H T v_t + i \hbar T \frac{dv_t}{dt} = v H T, \tag{3.22}$$

or

$$i \hbar \frac{dv_t}{dt} = T^\dagger v H T - T^\dagger H T v_t = v_t H_t - H_t v_t, \tag{3.23}$$

where

$$H_t = T^\dagger H T. \tag{3.24}$$

Equation (3.23) can be written

$$i \hbar \frac{dv_t}{dt} = [v_t, H_t]. \tag{3.25}$$

Equation (3.25) shows how the dynamical variables of the system evolve in the Heisenberg picture. It is denoted the *Heisenberg equation of motion* [Born and Jordan (1925)]. The time-varying dynamical variables in the Heisenberg picture are usually called *Heisenberg dynamical variables* to distinguish them from *Schrödinger dynamical variables* (i.e., the corresponding variables in the Schrödinger picture), which do not evolve in time.

According to Equation (2.22), the Heisenberg equation of motion can be written

$$\frac{dv_t}{dt} = [v_t, H_t]_{qm}, \tag{3.26}$$

where $[\cdots]_{qm}$ denotes the quantum mechanical Poisson bracket. Let us compare this equation with the classical time evolution equation for a general dynamical variable v, which can be written in the form

$$\frac{dv}{dt} = [v, H]_{cl}. \tag{3.27}$$

[See Equation (2.7).] Here, $[\cdots]_{cl}$ is the classical Poisson bracket, and H denotes the classical Hamiltonian. The strong resemblance between Equations (3.26) and (3.27) provides us with further justification for our identification of the linear operator H with the energy of the system in quantum mechanics.

Note that if the Hamiltonian does not explicitly depend on time (i.e., the system is not subject to some time-dependent external force) then Equation (3.15) yields

$$T(t, t_0) = \exp\left[\frac{-i\,H\,(t - t_0)}{\hbar}\right]. \tag{3.28}$$

This operator manifestly commutes with H, so

$$H_t = T^\dagger\,H\,T = H. \tag{3.29}$$

(See Exercise 2.3.) Furthermore, Equation (3.25) gives

$$i\,\hbar\,\frac{dH}{dt} = [H, H] = 0. \tag{3.30}$$

Thus, if the energy of the system has no explicit time dependence then it is represented by the same non-time-varying operator, H, in both the Schrödinger and Heisenberg pictures.

Suppose that v is an observable that commutes with the Hamiltonian (and, hence, with the time evolution operator T). It follows from Equation (3.19) that $v_t = v$. Heisenberg's equation of motion yields

$$i\,\hbar\,\frac{dv}{dt} = [v, H] = 0. \tag{3.31}$$

Thus, any observable that commutes with the Hamiltonian is a constant of the motion (hence, it is represented by the same fixed operator in both the Schrödinger and Heisenberg pictures). Only those observables that do not commute with the Hamiltonian evolve in time in the Heisenberg picture.

3.4 Ehrenfest Theorem

We have now introduced all of the basic elements of quantum mechanics. The only element that is lacking is some rule to determine the form of the quantum mechanical Hamiltonian. For a physical system that possess a classical analog, we generally assume that the Hamiltonian has the same form as in classical physics (i.e., we replace the classical coordinates and conjugate momenta by the corresponding quantum mechanical operators). This scheme guarantees that quantum mechanics yields the correct classical equations of motion in the classical limit. Whenever an ambiguity arises because of non-commuting observables, this can usually be resolved by requiring the Hamiltonian H to be an Hermitian operator. For instance, we would write the quantum mechanical analog of the classical product $x\,p_x$, appearing in the Hamiltonian, as the Hermitian product $(1/2)\,(x\,p_x + p_x\,x)$. When the system in question has no classical analog then we are reduced to guessing a form for H that reproduces the observed behavior of the system.

Consider a three-dimensional system characterized by three independent Cartesian position coordinates x_i (where i runs from 1 to 3), with three corresponding conjugate momenta p_i. These are represented by three commuting position operators x_i, and three commuting momentum operators p_i, respectively. The commutation relations satisfied by the position and momentum operators are

$$[x_i, p_j] = i\,\hbar\,\delta_{ij}. \tag{3.32}$$

[See Equation (2.25).] It is helpful to denote (x_1, x_2, x_3) as $\mathbf{x}$ and (p_1, p_2, p_3) as $\mathbf{p}$. The following useful formulae,

$$[x_i, F(\mathbf{x}, \mathbf{p})] = i\,\hbar\,\frac{\partial F}{\partial p_i}, \tag{3.33}$$

$$[p_i, G(\mathbf{x}, \mathbf{p})] = -i\,\hbar\,\frac{\partial G}{\partial x_i}, \tag{3.34}$$

where $F(\mathbf{x}, \mathbf{p})$ and $G(\mathbf{x}, \mathbf{p})$ are functions that can be expanded as power series, are easily proved using the fundamental commutation relations, Equations (3.32). (See Exercise 3.1.)

Let us now consider the three-dimensional motion of a free particle of mass m in the Heisenberg picture. The Hamiltonian is assumed to have the same form as in classical physics: that is,

$$H(\mathbf{x}, \mathbf{p}) = \frac{p^2}{2\,m} \equiv \frac{1}{2\,m} \sum_{i=1,3} p_i^2. \tag{3.35}$$

In the following, all dynamical variables are assumed to be Heisenberg dynamical variables, although we will omit the subscript t for the sake of clarity. The time evolution of the momentum operator p_i follows from the Heisenberg equation of motion, Equation (3.25). We find that

$$\frac{dp_i}{dt} = \frac{1}{i\hbar}[p_i, H] = 0, \tag{3.36}$$

because p_i automatically commutes with any function of the momentum operators. Thus, for a free particle, the momentum operators are constants of the motion, which means that $p_i(t) = p_i(0)$ at all times t (for i is 1 to 3). The time evolution of the position operator x_i is given by

$$\frac{dx_i}{dt} = \frac{1}{i\hbar}[x_i, H] = \frac{1}{i\hbar}\frac{1}{2m}i\hbar\frac{\partial}{\partial p_i}\left(\sum_{j=1,3}p_j^2\right) = \frac{p_i}{m} = \frac{p_i(0)}{m}, \tag{3.37}$$

where use has been made of Equation (3.33). It follows that

$$x_i(t) = x_i(0) + \left[\frac{p_i(0)}{m}\right]t, \tag{3.38}$$

which is analogous to the equation of motion of a classical free particle. Note that, even though

$$[x_i(0), x_j(0)] = 0, \tag{3.39}$$

where the position operators are evaluated at equal times, the x_i do not commute when evaluated at different times. For instance,

$$[x_i(t), x_i(0)] = \left[\frac{p_i(0)\,t}{m}, x_i(0)\right] = \frac{-i\hbar t}{m}. \tag{3.40}$$

Combining the previous commutation relation with the uncertainty relation (1.86) yields

$$\langle(\Delta x_i)^2\rangle_t\,\langle(\Delta x_i)^2\rangle_{t=0} \geq \frac{\hbar^2 t^2}{4\,m^2}. \tag{3.41}$$

This result implies that, even if a particle is well localized at $t = 0$, its position becomes progressively more uncertain with time. This conclusion can also be obtained by studying the propagation of wavepackets in wave mechanics.

Let us now add a potential $V(\mathbf{x})$ to our free particle Hamiltonian:

$$H(\mathbf{x}, \mathbf{p}) = \frac{p^2}{2\,m} + V(\mathbf{x}). \tag{3.42}$$

Here, V is some (real) function of the x_i operators. The Heisenberg equation of motion gives

$$\frac{dp_i}{dt} = \frac{1}{i\hbar}[p_i, V(\mathbf{x})] = -\frac{\partial V(\mathbf{x})}{\partial x_i}, \tag{3.43}$$

where use has been made of Equation (3.34). On the other hand, the result

$$\frac{dx_i}{dt} = \frac{p_i}{m} \tag{3.44}$$

still holds, because the x_i all commute with the new term, $V(\mathbf{x})$, in the Hamiltonian. We can use the Heisenberg equation of motion a second time to deduce that

$$\frac{d^2 x_i}{dt^2} = \frac{1}{i\hbar}\left[\frac{dx_i}{dt}, H\right] = \frac{1}{i\hbar}\left[\frac{p_i}{m}, H\right] = \frac{1}{m}\frac{dp_i}{dt} = -\frac{1}{m}\frac{\partial V(\mathbf{x})}{\partial x_i}. \tag{3.45}$$

In vectorial form, this equation becomes

$$m\frac{d^2\mathbf{x}}{dt^2} = \frac{d\mathbf{p}}{dt} = -\nabla V(\mathbf{x}). \tag{3.46}$$

This is the quantum mechanical equivalent of Newton's second law of motion. Taking the expectation values of both sides with respect to a Heisenberg state ket that does not evolve in time, we obtain the so-called *Ehrenfest theorem* [Ehrenfest (1927)]:

$$m\frac{d^2\langle\mathbf{x}\rangle}{dt^2} = \frac{d\langle\mathbf{p}\rangle}{dt} = -\langle\nabla V(\mathbf{x})\rangle. \tag{3.47}$$

When written in terms of expectation values, this result is independent of whether we are using the Heisenberg or Schrödinger picture. By contrast, the operator equation (3.46) only holds if $\mathbf{x}$ and $\mathbf{p}$ are understood to be Heisenberg dynamical variables. Note that Equation (3.47) has no dependence on $\hbar$. In fact, it guarantees that the centroid of a wavepacket always moves like a classical particle.

3.5 Schrödinger Wave Equation

Consider the motion of a particle in three dimensions in the Schrödinger picture. The fixed dynamical variables of the system are the position operators, $\mathbf{x} \equiv (x_1, x_2, x_3)$, and the momentum operators, $\mathbf{p} \equiv (p_1, p_2, p_3)$. The state of the system is represented as some time evolving ket $|At\rangle$.

Let $|\mathbf{x}'\rangle$ represent a simultaneous eigenket of the position operators belonging to the eigenvalues $\mathbf{x}' \equiv (x_1', x_2', x_3')$. Note that, because the position operators are fixed in the Schrödinger picture, we do not expect the $|\mathbf{x}'\rangle$ to evolve in time. The wavefunction of the system at time t is defined

$$\psi(\mathbf{x}', t) = \langle\mathbf{x}'|At\rangle. \tag{3.48}$$

The Hamiltonian of the system is taken to be

$$H(\mathbf{x}, \mathbf{p}) = \frac{p^2}{2m} + V(\mathbf{x}). \tag{3.49}$$

The Schrödinger equation of motion, (3.10), yields

$$i\hbar \frac{\partial \langle \mathbf{x}'|At \rangle}{\partial t} = \langle \mathbf{x}'| H |At \rangle, \tag{3.50}$$

where use has been made of the time independence of the $|\mathbf{x}'\rangle$. We adopt the Schrödinger representation in which the momentum conjugate to the position operator x_i is written

$$p_i = -i\hbar \frac{\partial}{\partial x_i}. \tag{3.51}$$

[See Equation (2.74).] Thus,

$$\left\langle \mathbf{x}' \left| \frac{p^2}{2m} \right| At \right\rangle = -\left(\frac{\hbar^2}{2m} \right) \nabla'^2 \langle \mathbf{x}'|At \rangle, \tag{3.52}$$

where use has been made of Equation (2.78). Here, $\nabla' \equiv (\partial/\partial x', \partial/\partial y', \partial/\partial z')$ denotes the gradient operator written in terms of the position eigenvalues. We can also write

$$\langle \mathbf{x}'| V(\mathbf{x}) = V(\mathbf{x}') \langle \mathbf{x}'|, \tag{3.53}$$

where $V(\mathbf{x}')$ is a scalar function of the position eigenvalues. Combining Equations (3.49), (3.50), (3.52), and (3.53), we obtain

$$i\hbar \frac{\partial \langle \mathbf{x}'|At \rangle}{\partial t} = -\left(\frac{\hbar^2}{2m} \right) \nabla'^2 \langle \mathbf{x}'|At \rangle + V(\mathbf{x}') \langle \mathbf{x}'|At \rangle, \tag{3.54}$$

which can also be written

$$i\hbar \frac{\partial \psi(\mathbf{x}', t)}{\partial t} = -\left(\frac{\hbar^2}{2m} \right) \nabla'^2 \psi(\mathbf{x}', t) + V(\mathbf{x}') \psi(\mathbf{x}', t). \tag{3.55}$$

This is the *Schrödinger time-dependent wave equation*, and is the basis of wave mechanics [Schrödinger (1926b)]. Note, however, that the wave equation is just one of many possible representations of quantum mechanics. It just happens to give a type of equation that we know how to solve. In deriving the wave equation, we have chosen to represent the system in terms of the eigenkets of the position operators, instead of those of the momentum operators. We have also fixed the relative phases of the $|\mathbf{x}'\rangle$ according to the Schrödinger representation, so that Equation (3.51) is valid. Finally, we have chosen to work in the Schrödinger picture, in which state kets evolve and dynamical variables are fixed, instead of the Heisenberg picture, in which the opposite is true.

Suppose that the ket $|At\rangle$ is an eigenket of the Hamiltonian belonging to the eigenvalue H': that is,

$$H |At \rangle = H' |At \rangle. \tag{3.56}$$

The Schrödinger equation of motion, (3.10), yields

$$i\hbar \frac{d|At\rangle}{dt} = H'|At\rangle.$$ (3.57)

This can be integrated to give

$$|At\rangle = \exp\left[\frac{-i\,H'(t-t_0)}{\hbar}\right]|At_0\rangle.$$ (3.58)

Note that $|At\rangle$ only differs from $|At_0\rangle$ by a phase-factor. The direction of the vector remains fixed in ket space. This suggests that if the system is initially in an eigenstate of the Hamiltonian then it remains in this state for ever, as long as the system is undisturbed. Such a state is called a *stationary state*. The wavefunction of a stationary state satisfies

$$\psi(\mathbf{x}', t) = \exp\left[\frac{-i\,H'(t-t_0)}{\hbar}\right]\psi(\mathbf{x}', t_0).$$ (3.59)

Substituting the previous relation into the Schrödinger time-dependent wave equation, (3.55), we obtain

$$-\left(\frac{\hbar^2}{2m}\right)\nabla'^2\psi_0(\mathbf{x}') + [V(\mathbf{x}') - E]\psi_0(\mathbf{x}') = 0,$$ (3.60)

where $\psi_0(\mathbf{x}') \equiv \psi(\mathbf{x}', t_0)$, and $E = H'$ is the energy of the system. This is the *Schrödinger time-independent wave equation*. A *bound state* solution of the previous equation, in which the particle is confined within a finite region of space, satisfies the boundary condition

$$\psi_0(\mathbf{x}') \to 0 \quad \text{as } |\mathbf{x}'| \to \infty.$$ (3.61)

Such a solution is only possible if

$$E < \lim_{|\mathbf{x}'|\to\infty} V(\mathbf{x}').$$ (3.62)

Because it is conventional to set the potential at infinity equal to zero, the previous relation implies that bound states are equivalent to negative energy states [Fitzpatrick (2012)]. The boundary condition (3.61) is sufficient to uniquely specify the solution of Equation (3.60).

The quantity $\rho(\mathbf{x}', t)$, defined by

$$\rho(\mathbf{x}', t) = |\psi(\mathbf{x}', t)|^2,$$ (3.63)

is termed the *probability density*. Recall, from a direct generalization of Equation (2.30), that the probability of observing the particle in some volume element $d^3\mathbf{x}'$ around position $\mathbf{x}'$ is proportional to $\rho(\mathbf{x}', t)\, d^3\mathbf{x}'$. The probability is equal to $\rho(\mathbf{x}', t)\, d^3\mathbf{x}'$ if the wavefunction is properly normalized, so that

$$\int d^3\mathbf{x}'\, \rho(\mathbf{x}', t) = 1.$$ (3.64)

The Schrödinger time-dependent wave equation, (3.55), can easily be transformed into a conservation equation for the probability density:

$$\frac{\partial \rho}{\partial t} + \nabla' \cdot \mathbf{j} = 0. \tag{3.65}$$

The *probability current*, $\mathbf{j}$, takes the form

$$\mathbf{j}(\mathbf{x}', t) = -\left(\frac{i\hbar}{2m}\right)[\psi^* \nabla' \psi - (\nabla' \psi^*)\psi] = \left(\frac{\hbar}{m}\right) \mathrm{Im}(\psi^* \nabla' \psi). \tag{3.66}$$

We can integrate Equation (3.65) over all space, using the divergence theorem [Riley *et al.* (2013)], and the boundary condition $\rho \to 0$ as $|\mathbf{x}'| \to \infty$, to obtain

$$\frac{d}{dt} \int d^3\mathbf{x}'\, \rho(\mathbf{x}', t) = 0. \tag{3.67}$$

Thus, the Schrödinger time-dependent wave equation conserves probability. In particular, if the wavefunction starts off properly normalized, according to Equation (3.64), then it remains properly normalized at all subsequent times. It is easily demonstrated that

$$\int d^3\mathbf{x}'\, \mathbf{j}(\mathbf{x}', t) = \frac{\langle \mathbf{p} \rangle_t}{m}, \tag{3.68}$$

where $\langle \mathbf{p} \rangle_t$ denotes the expectation value of the momentum evaluated at time t. Clearly, the probability current is indirectly related to the particle momentum.

In deriving Equations (3.65), we have, naturally, assumed that the potential $V(\mathbf{x}')$ is real. Suppose, however, that the potential has an imaginary component. In this case, Equation (3.65) generalizes to

$$\frac{\partial \rho}{\partial t} + \nabla' \cdot \mathbf{j} = \frac{2\,\mathrm{Im}(V)}{\hbar}\,\rho, \tag{3.69}$$

giving

$$\frac{d}{dt} \int d^3\mathbf{x}'\, \rho(\mathbf{x}', t) = \frac{2}{\hbar} \int d^3\mathbf{x}'\, \mathrm{Im}[V(\mathbf{x}')]\,\rho(\mathbf{x}', t). \tag{3.70}$$

(See Exercise 3.2.) Thus, if $\mathrm{Im}(V) < 0$ then the total probability of observing the particle anywhere in space decreases monotonically with time. Hence, an imaginary potential can be used to account for the disappearance or decay of a particle. Such a potential is often employed to model nuclear reactions in which incident particles are absorbed by nuclei.

3.6 Charged Particle Motion in Electromagnetic Fields

The classical Hamiltonian for a particle of mass m and charge q moving under the influence of electromagnetic fields is [Goldstein *et al.* (2002)]

$$H = \frac{1}{2m} (\mathbf{p} - q\,\mathbf{A}) \cdot (\mathbf{p} - q\,\mathbf{A}) + q\,\phi, \tag{3.71}$$

where $\mathbf{A} = \mathbf{A}(\mathbf{x}, t)$ and $\phi = \phi(\mathbf{x}, t)$ are the *vector* and *scalar potentials*, respectively [Fitzpatrick (2008)]. These potentials are related to the familiar *electric* and *magnetic field-strengths*, $\mathbf{E}(\mathbf{x}, t)$ and $\mathbf{B}(\mathbf{x}, t)$, respectively, via [Fitzpatrick (2008)]

$$\mathbf{E} = -\nabla\phi - \frac{\partial \mathbf{A}}{\partial t}, \tag{3.72}$$

$$\mathbf{B} = \nabla \times \mathbf{A}. \tag{3.73}$$

Let us assume that expression (3.71) is also the correct quantum mechanical Hamiltonian for a charged particle moving in electromagnetic fields. Obviously, in quantum mechanics, we must treat $\mathbf{p}$, $\mathbf{A}$, and ϕ as operators that do not necessarily commute.

The Heisenberg equations of motion for the components of $\mathbf{x}$ are

$$\frac{dx_i}{dt} = \frac{[x_i, H]}{i\hbar}. \tag{3.74}$$

However,

$$[x_i, H] = i\hbar \frac{\partial H}{\partial p_i} = \frac{i\hbar}{m} (p_i - q\,A_i), \tag{3.75}$$

where use has been made of Equations (3.33) and (3.71). It follows that

$$m \frac{d\mathbf{x}}{dt} = \mathbf{\Pi}, \tag{3.76}$$

where

$$\mathbf{\Pi} = \mathbf{p} - q\,\mathbf{A}. \tag{3.77}$$

Here, $\mathbf{\Pi}$ is referred to as the *mechanical momentum*, whereas $\mathbf{p}$ is termed the *canonical momentum*.

It is easily seen that

$$[\Pi_i, \Pi_j] = q\,[p_j, A_i] - q\,[p_i, A_j]. \tag{3.78}$$

However,

$$[p_j, A_i] = -i\hbar \frac{\partial A_i}{\partial x_j}, \tag{3.79}$$

where we have employed Equation (3.34). Thus, we obtain

$$[\Pi_i, \Pi_j] = i\,\hbar\,q\left(\frac{\partial A_j}{\partial x_i} - \frac{\partial A_i}{\partial x_j}\right) = i\,\hbar\,q\,\epsilon_{ijk}\,B_k, \tag{3.80}$$

because [from Equation (3.73)]

$$B_i = \epsilon_{ijk}\,\frac{\partial A_k}{\partial x_j}. \tag{3.81}$$

Here, ϵ_{ijk} is the *totally antisymmetric tensor* (that is, $\epsilon_{ijk} = 1$ if i, j, k is a cyclic permutation of 1, 2, 3; $\epsilon_{ijk} = -1$ if i, j, k is an anti-cyclic permutation of 1, 2, 3; and $\epsilon_{ijk} = 0$ otherwise), and we have used the standard result $\epsilon_{ijk}\,\epsilon_{iab} = \delta_{ja}\,\delta_{kb} - \delta_{jb}\,\delta_{ka}$, as well as the *Einstein summation convention* (that repeated indices are implicitly summed from 1 to 3) [Riley *et al.* (2013)].

We can write the Hamiltonian (3.71) in the form

$$H = \frac{\Pi^2}{2\,m} + q\,\phi. \tag{3.82}$$

The Heisenberg equation of motions for the components of Π are

$$\frac{d\Pi_i}{dt} = \frac{[\Pi_i, H]}{i\,\hbar} + \frac{\partial \Pi_i}{\partial t}. \tag{3.83}$$

Here, we have taken into account the fact that Π_i depends explicitly on time through its dependence on $A_i(\mathbf{x}, t)$. However,

$$[\Pi_i, \Pi^2] = \Pi_j\,[\Pi_i, \Pi_j] + [\Pi_i, \Pi_j]\,\Pi_j = i\,\hbar\,q\left(\epsilon_{ijk}\,\Pi_j\,B_k - \epsilon_{ijk}\,B_j\,\Pi_k\right), \tag{3.84}$$

where use has been made of Equation (3.80). Moreover,

$$\frac{\partial \Pi_i}{\partial t} = -q\,\frac{\partial A_i}{\partial t}, \tag{3.85}$$

$$[\Pi_i, \phi] = [p_i, \phi] = -i\,\hbar\,\frac{\partial \phi}{\partial x_i}, \tag{3.86}$$

where we have employed Equation (3.33). The previous five equations yield

$$\frac{d\mathbf{\Pi}}{dt} = q\,\mathbf{E} + \frac{q}{2\,m}\,(\mathbf{\Pi} \times \mathbf{B} - \mathbf{B} \times \mathbf{\Pi}), \tag{3.87}$$

which can be combined with Equation (3.76) to give

$$m\,\frac{d^2\mathbf{x}}{dt^2} = q\,\mathbf{E} + \frac{q}{2}\left(\frac{d\mathbf{x}}{dt} \times \mathbf{B} - \mathbf{B} \times \frac{d\mathbf{x}}{dt}\right). \tag{3.88}$$

This equation of motion is a generalization of the Ehrenfest theorem that takes electromagnetic fields into account. The fact that Equation (3.88) is analogous in form to the corresponding classical equation of motion (given that $d\mathbf{x}/dt$ and $\mathbf{B}$ commute in classical mechanics) justifies our earlier assumption that Equation (3.71) is the correct quantum mechanical Hamiltonian for a charged particle moving in electromagnetic fields.

3.7 Gauge Transformations in Electromagnetism

In the Schrödinger picture, the Hamiltonian (3.71) leads to the following time-dependent wave equation:

$$i\hbar \frac{\partial \psi}{\partial t} = \frac{1}{2m}(-i\hbar \nabla' - q\mathbf{A}) \cdot (-i\hbar \nabla' - q\mathbf{A})\psi + q\phi\psi, \qquad (3.89)$$

where $\psi = \psi(\mathbf{x}', t)$, $\mathbf{A} = \mathbf{A}(\mathbf{x}', t)$, and $\phi = \phi(\mathbf{x}', t)$. Now, the Heisenberg equation of motion (3.88) only involves the electric and magnetic fields, and is independent of the vector and scalar potentials. On the other hand, the previous wave equation involves the potentials, but not the fields. As is well known, the vector and scalar potentials are not well defined, in that there are many different potentials that generate the same electric and magnetic fields. To be more exact, a transformation of the form $\mathbf{A} \to \mathbf{A}'$ and $\phi \to \phi'$, where

$$\mathbf{A}'(\mathbf{x}', t) = \mathbf{A}(\mathbf{x}', t) - \nabla' f(\mathbf{x}', t), \qquad (3.90)$$

$$\phi'(\mathbf{x}', t) = \phi(\mathbf{x}', t) + \frac{\partial f(\mathbf{x}', t)}{\partial t}, \qquad (3.91)$$

and $f(\mathbf{x}', t)$ is an arbitrary function, leaves the $\mathbf{E}$ and $\mathbf{B}$ fields unaffected. Such a transformation is known as a *gauge transformation*. It is evident that a gauge transformation would leave the Heisenberg equation of motion (3.88) unchanged, but would modify the time-dependent wave equation (3.89). However, these two equations are supposed to give results that are consistent with one another. Let us investigate how this is possible.

The previous three equations can be combined to give

$$i\hbar \frac{\partial \psi}{\partial t} = \frac{1}{2m}(-i\hbar \nabla' - q\mathbf{A}' - q\nabla' f)^2 \psi + q\phi'\psi - q\frac{\partial f}{\partial t}\psi. \qquad (3.92)$$

Let

$$\psi'(\mathbf{x}', t) = e^{i\Lambda(\mathbf{x}', t)}\psi(\mathbf{x}', t), \qquad (3.93)$$

where

$$\Lambda(\mathbf{x}', t) = -\frac{q}{\hbar}f(\mathbf{x}', t). \qquad (3.94)$$

It follows that

$$e^{i\Lambda} i\hbar \frac{\partial \psi}{\partial t} = e^{i\Lambda} \frac{\partial}{\partial t}\left(e^{-i\Lambda}\psi'\right) = i\hbar \frac{\partial \psi'}{\partial t} - q\frac{\partial f}{\partial t}\psi', \qquad (3.95)$$

$$(-i\hbar \nabla' - q\mathbf{A}' - q\nabla' f)\psi = (-i\hbar \nabla' - q\mathbf{A}' - q\nabla' f)\left(e^{-i\Lambda}\psi'\right)$$

$$= e^{-i\Lambda}(-i\hbar \nabla' - q\mathbf{A}')\psi', \qquad (3.96)$$

$$e^{i\Lambda}(-i\hbar \nabla' - q\mathbf{A}' - q\nabla' f)^2 \psi = (-i\hbar \nabla' - q\mathbf{A}')^2 \psi'. \qquad (3.97)$$

Hence, Equation (3.92) becomes

$$i\hbar\frac{\partial\psi'}{\partial t} = \frac{1}{2m}(-i\hbar\nabla' - q\mathbf{A}')^2\psi' + q\phi'\psi', \tag{3.98}$$

which is analogous in form to Equation (3.89). Thus, we deduce that if $\mathbf{A} \to \mathbf{A}'$ and $\phi \to \phi'$ then $\psi \to \psi'$. In other words, a gauge transformation introduces a position- and time-dependent phase-shift, $\Lambda(\mathbf{x}', t)$, into the wavefunction.

Now, Equation (3.88) is equivalent to Equations (3.76) and (3.87). If we take the expectation values of the latter two equations then we obtain

$$m\frac{d\langle\mathbf{x}\rangle}{dt} = \langle\mathbf{\Pi}\rangle, \tag{3.99}$$

$$\frac{d\langle\mathbf{\Pi}\rangle}{dt} = q\langle\mathbf{E}\rangle + \frac{q}{2m}\left(\langle\mathbf{\Pi}\times\mathbf{B}\rangle - \langle\mathbf{B}\times\mathbf{\Pi}\rangle\right). \tag{3.100}$$

However, the quantities $\langle\mathbf{x}\rangle$, $\langle\mathbf{\Pi}\rangle$, $\langle\mathbf{E}\rangle$, $\langle\mathbf{\Pi}\times\mathbf{B}\rangle$, and $\langle\mathbf{B}\times\mathbf{\Pi}\rangle$ are all invariant under the gauge transformation $\mathbf{A} \to \mathbf{A}'$, $\phi \to \phi'$, and $\psi \to \psi'$. This follows because

$$\langle\mathbf{x}\rangle = \int d^3\mathbf{x}'\,\psi^*(\mathbf{x}')\,\mathbf{x}'\,\psi(\mathbf{x}') = \int d^3\mathbf{x}'\,\psi'^*(\mathbf{x}')\,\mathbf{x}'\,\psi'(\mathbf{x}'), \tag{3.101}$$

$$\langle\mathbf{\Pi}\rangle = \int d^3\mathbf{x}'\,\psi^*(\mathbf{x}')\left[-i\hbar\nabla' - q\mathbf{A}(\mathbf{x}')\right]\psi(\mathbf{x}')$$

$$= \int d^3\mathbf{x}'\,\psi'^*(\mathbf{x}')\left[-i\hbar\nabla' - q\mathbf{A}'(\mathbf{x}')\right]\psi'(\mathbf{x}'), \tag{3.102}$$

et cetera. Thus, Equations (3.88) and (3.89) do indeed give consistent results under gauge transformation.

3.8   Flux Quantization and the Aharonov-Bohm Effect

Consider a situation in which the electric and magnetic fields are non-time-varying. In this case, Equation (3.89) becomes

$$i\hbar\frac{\partial\psi}{\partial t} = \frac{1}{2m}(-i\hbar\nabla' - q\mathbf{A})^2\psi + q\phi\psi, \tag{3.103}$$

where $\psi = \psi(\mathbf{x}', t)$, $\mathbf{A} = \mathbf{A}(\mathbf{x}')$, and $\phi = \phi(\mathbf{x}')$. The previous equation has the formal solution

$$\psi(\mathbf{x}', t) = e^{i\Lambda(\mathbf{x}')}\psi_0(\mathbf{x}', t), \tag{3.104}$$

where

$$\Lambda(\mathbf{x}') = \frac{q}{\hbar}\int_{\mathbf{x}'_0}^{\mathbf{x}'} d\mathbf{x}'' \cdot \mathbf{A}(\mathbf{x}''), \tag{3.105}$$

and $\psi_0(\mathbf{x}', t)$ is a solution of

$$i\,\hbar\,\frac{\partial \psi_0}{\partial t} = \frac{1}{2\,m}\,(-i\,\hbar\,\nabla')^2\,\psi_0 + q\,\phi\,\psi_0. \tag{3.106}$$

Here, $\mathbf{x}_0'$ is an arbitrary fixed point. However, the solution (3.104) only makes sense in a region in which $\mathbf{B} = 0$. To see this, let us imagine calculating the phase factor $\Lambda(\mathbf{x}')$ by evaluating the line integral specified in Equation (3.105) along two different paths, labelled 1 and 2, that join the points $\mathbf{x}_0'$ and $\mathbf{x}'$. We find that

$$\Lambda_1(\mathbf{x}') - \Lambda_2(\mathbf{x}') = \frac{q}{\hbar}\left[\int_1 d\mathbf{x}'' \cdot \mathbf{A}(\mathbf{x}'') - \int_2 d\mathbf{x}'' \cdot \mathbf{A}(\mathbf{x}'')\right]$$

$$= \frac{q}{\hbar}\oint d\mathbf{x}'' \cdot \mathbf{A}(\mathbf{x}'') = \frac{q}{\hbar}\int_S \mathbf{B} \cdot d\mathbf{S} = \frac{q}{\hbar}\,\Phi, \tag{3.107}$$

where Φ is the magnetic flux passing through a surface spanning the two paths. Here, we have made use of the curl theorem [Riley *et al.* (2013)], as well as Equation (3.73). Thus, the phase factor $\Lambda(\mathbf{x}')$ is only independent of the choice of path in the line integral when $\Phi = 0$. Such independence is required if we insist that the wavefunction be single valued.

Suppose that a charged particle moves in a magnetic-field-free region that is not simply connected, but surrounds a hole through which the magnetic flux Φ passes. Upon completing a circuit around the hole, the particle's wavefunction is multiplied by $\exp(i\,q\,\Phi/\hbar)$. The requirement that the wavefunction be single-valued, and, hence, that the multiplication factor be unity, implies that the enclosed magnetic flux is quantized. In fact.

$$\Phi = n\,\frac{h}{q}, \tag{3.108}$$

where n is an integer. This effect is known as *flux quantization*.

A situation like that just described arises in the motion of electrons in a super-conducting ring through which a magnetic field passes. Incidentally, superconductors contain no internal magnetic fields because they expel magnetic flux—this behavior is called the *Meissner effect* [Meissner and Ochsenfeld (1933)]. Experimentally, the magnetic flux passing through a superconducting ring is found to satisfy

$$\Phi = n\,\frac{h}{2\,e}, \tag{3.109}$$

where n is an integer, and e the magnitude of the electron charge [Deaver and Fairbank (1961); Döll and Naubauer (1961)]. This result is consistent with our present understanding of the phenomenon of superconductivity, according to which the fundamental charge carriers are correlated electron pairs known as *Cooper pairs* [Bardeen *et al.* (1957)].

Consider an electron interference experiment in which a small solenoid containing magnetic flux is placed directly behind the slits of a two-slit interference apparatus. The interference pattern at the screen is due to the superposition of two parts of the wavefunction,

$$\psi = \psi_1 + \psi_2, \tag{3.110}$$

where ψ_1 and ψ_2 are the wavefunctions due to electrons that originate from the source, pass through the first and second slits, respectively, and then strike the same point on the screen. When the solenoid is energized an additional phase shift $e\,\Phi/\hbar$ is introduced between ψ_1 and ψ_2, where Φ is the net flux passing through the solenoid. In other words, the magnetic field internal to the solenoid affects the interference pattern seen on the screen, despite the fact that neither electron beam ever directly experiences a magnetic field (because the field is internal to the solenoid, and the two interfering beams are assumed to pass on either side of the solenoid.) This phenomena is known as the *Aharonov-Bohm effect* [Aharonov and Bohm (1959)], and has been observed experimentally [Tonomura *et al.* (1982)].

3.9   Exercises

3.1 Let $\mathbf{x} \equiv (x_1, x_2, x_3)$ be a set of Cartesian position operators, and let $\mathbf{p} \equiv (p_1, p_2, p_3)$ be the corresponding momentum operators. Demonstrate that

$$[x_i, F(\mathbf{x}, \mathbf{p})] = i\,\hbar\,\frac{\partial F}{\partial p_i},$$

$$[p_i, G(\mathbf{x}, \mathbf{p})] = -i\,\hbar\,\frac{\partial G}{\partial x_i},$$

where $i = 1, 2, 3$, and $F(\mathbf{x}, \mathbf{p})$, $G(\mathbf{x}, \mathbf{p})$ are functions that can be expanded as power series.

3.2 Assuming that the potential $V(\mathbf{x})$ is complex, demonstrate that the Schrödinger time-dependent wave equation, (3.55), can be transformed to give

$$\frac{\partial \rho}{\partial t} + \nabla' \cdot \mathbf{j} = 2\,\frac{\text{Im}(V)}{\hbar}\,\rho,$$

where

$$\rho(\mathbf{x}', t) = |\psi(\mathbf{x}', t)|^2,$$

and

$$\mathbf{j}(\mathbf{x}', t) = \left(\frac{\hbar}{m}\right) \text{Im}(\psi^* \, \nabla' \psi).$$

3.3 Consider one-dimensional quantum harmonic oscillator whose Hamiltonian is

$$H = \frac{p_x^2}{2m} + \frac{1}{2}m\omega^2 x^2,$$

where x and p_x are conjugate position and momentum operators, respectively, and m, ω are positive constants.

(a) Demonstrate that the expectation value of H, for a general state, is positive definite.

(b) Let

$$A = \sqrt{\frac{m\omega}{2\hbar}}\, x + i\, \frac{p_x}{\sqrt{2m\omega\hbar}}.$$

Deduce that

$$[A, A^\dagger] = 1,$$

$$H = \hbar\omega\left(\frac{1}{2} + A^\dagger A\right),$$

$$[H, A] = -\hbar\omega A,$$

$$[H, A^\dagger] = \hbar\omega A^\dagger.$$

(c) Suppose that $|E\rangle$ is an eigenket of the Hamiltonian whose corresponding energy is E: that is,

$$H|E\rangle = E|E\rangle.$$

Demonstrate that

$$HA|E\rangle = (E - \hbar\omega)A|E\rangle,$$

$$HA^\dagger|E\rangle = (E + \hbar\omega)A^\dagger|E\rangle.$$

Hence, deduce that the allowed values of E are

$$E_n = (n + 1/2)\hbar\omega,$$

where $n = 0, 1, 2, \cdots$. Here, A and $A^\dagger$ are termed ladder operators. To be more exact, A is termed a lowering operator (because it lowers the energy quantum number, n, by unity), whereas $A^\dagger$ is termed a raising operator (because it raises the energy quantum number by unity).

(d) Let $|E_n\rangle$ be a properly normalized (i.e., $\langle E_n|E_n\rangle = 1$) energy eigenket corresponding to the eigenvalue E_n. Show that the kets can be defined such that

$$A\,|E_n\rangle = \sqrt{n}\,|E_{n-1}\rangle,$$

$$A^\dagger\,|E_n\rangle = \sqrt{n+1}\,|E_{n+1}\rangle.$$

Hence, deduce that

$$|E_n\rangle = \frac{1}{\sqrt{n!}}\,(A^\dagger)^n\,|E_0\rangle.$$

(e) Let the $\psi_n(x') = \langle x'|E_n\rangle$ be the wavefunctions of the properly normalized energy eigenkets. Given that

$$A\,|E_0\rangle = |0\rangle,$$

deduce that

$$\left(\frac{x'}{x_0} + x_0\,\frac{d}{dx'}\right)\psi_0(x') = 0,$$

where $x_0 = (\hbar/m\,\omega)^{1/2}$. Hence, show that

$$\psi_n(x') = \frac{1}{\pi^{1/4}\,(2^n\,n!)^{1/2}\,x_0^{n+1/2}}\left(x' - x_0^2\,\frac{d}{dx'}\right)^n \exp\left[-\frac{1}{2}\left(\frac{x'}{x_0}\right)^2\right].$$

3.4 Consider the one-dimensional quantum harmonic oscillator discussed in Exercise 3.3. Let $|n\rangle$ be a properly normalized energy eigenket belonging to the eigenvalue E_n. Show that

(a)

$$\langle n'|\,x\,|n\rangle = \sqrt{\frac{\hbar}{2\,m\,\omega}}\left(\sqrt{n}\,\delta_{n'\,n-1} + \sqrt{n+1}\,\delta_{n'\,n+1}\right).$$

(b)

$$\langle n'|\,p_x\,|n\rangle = i\,\sqrt{\frac{m\,\hbar\,\omega}{2}}\left(-\sqrt{n}\,\delta_{n'\,n-1} + \sqrt{n+1}\,\delta_{n'\,n+1}\right).$$

(c)

$$\langle n'|\,x^2\,|n\rangle = \left(\frac{\hbar}{2\,m\,\omega}\right)\left[\sqrt{n\,(n-1)}\,\delta_{n'\,n-2} + \sqrt{(n+1)\,(n+2)}\,\delta_{n'\,n+2}\right.$$

$$\left. + (2\,n+1)\,\delta_{n'\,n}\right].$$

(d)

$$\langle n'|\,p_x^2\,|n\rangle = \left(\frac{m\,\hbar\,\omega}{2}\right)\left[-\sqrt{n\,(n-1)}\,\delta_{n'\,n-2} - \sqrt{(n+1)\,(n+2)}\,\delta_{n'\,n+2}\right.$$

$$\left. + (2\,n+1)\,\delta_{n'\,n}\right].$$

(e) Hence, deduce that

$$\langle (\Delta x)^2 \rangle \langle (\Delta p_x)^2 \rangle = (n + 1/2)^2 \, \hbar^2$$

for the nth eigenstate.

3.5 Consider the one-dimensional quantum harmonic oscillator discussed in the previous two exercises. Let $|\alpha\rangle$ be a properly normalized eigenket of the lowering operator, A, corresponding to the eigenvalue α, where α can be any complex number. The corresponding state is known as a *coherent state*.

(a) Demonstrate that

$$\langle x \rangle = \sqrt{\frac{2\hbar}{m\omega}} \, \mathrm{Re}(\alpha),$$

$$\langle p_x \rangle = \sqrt{2 m \hbar \omega} \, \mathrm{Im}(\alpha),$$

$$\langle x^2 \rangle = \frac{2\hbar}{m\omega} \left(\frac{1}{4} + [\mathrm{Re}(\alpha)]^2 \right),$$

$$\langle p_x^2 \rangle = 2 m \hbar \omega \left(\frac{1}{4} + [\mathrm{Im}(\alpha)]^2 \right),$$

where the expectation values are relative to the coherent state. Hence, deduce that

$$\langle (\Delta x)^2 \rangle \langle (\Delta p_x)^2 \rangle = \frac{\hbar^2}{4}.$$

In other words, a coherent state is characterized by the minimum possible uncertainty in position and momentum.

(b) If $|n\rangle$ is the properly normalized energy eigenket belonging to the energy eigenvalue $E_n = (n + 1/2) \hbar \omega$ then show that

$$|\alpha\rangle = \sum_{n=0,\infty} c_n |n\rangle,$$

where

$$c_n = \frac{\alpha^n}{\sqrt{n!}} \, \exp\left(-\frac{|\alpha|^2}{2} \right).$$

(c) Show that the expectation value of the energy for the coherent state is

$$\langle H \rangle = (|\alpha|^2 + 1/2) \hbar \omega.$$

(d) Putting in time dependence, so that

$$|n, t\rangle = e^{-i E_n t/\hbar} |n\rangle,$$

where $|n\rangle \equiv |n, 0\rangle$, demonstrate that $|\alpha, t\rangle$ remains an eigenket of A, but that the eigenvalue evolves in time as

$$\alpha(t) = e^{-i \omega t} \alpha.$$

Hence, deduce that

$$|\alpha, t\rangle = e^{-i \omega t/2} |\alpha\rangle.$$

(e) Writing

$$\alpha = \sqrt{\frac{m\omega}{2\hbar}}\, a,$$

where a is real and positive, show that

$$\langle x \rangle = a\, \cos(\omega t),$$

$$\langle p_x \rangle = -m\omega a\, \sin(\omega t).$$

Of course, these expressions are analogous to those of a classical harmonic oscillator of amplitude a and angular frequency ω. This suggests that a coherent state of a quantum harmonic oscillator is the state that most closely imitates the behavior of a classical oscillator.

(f) Show that the properly normalized wavefunction corresponding to the state $|\alpha, t\rangle$ takes the form

$$\psi(x', t) = \psi_0\!\left(x' - \sqrt{\frac{2\hbar}{m\omega}}\,\alpha(t) \right), \qquad (3.111)$$

where $\psi_0(x')$ is the properly normalized, stationary, ground-state wavefunction.

3.6 Consider a particle in one dimension whose Hamiltonian is

$$H = \frac{p_x^2}{2m} + V(x).$$

By calculating $[[H, x], x]$, demonstrate that

$$\sum_{n'} |\langle n| x | n'\rangle|^2 \,(E_{n'} - E_n) = \frac{\hbar^2}{2m},$$

where $|n\rangle$ is a properly normalized energy eigenket corresponding to the eigenvalue E_n, and the sum is over all eigenkets.

3.7 Consider a particle in one dimension whose Hamiltonian is

$$H = \frac{p_x^2}{2m} + V(x).$$

Suppose that the potential is periodic, such that

$$V(x - a) = V(x),$$

for all x. Deduce that

$$[D_x(a), H] = 0,$$

where $D_x(a)$ is the displacement operator defined in Exercise 2.6. Hence, show that the wavefunction of an energy eigenstate has the general form

$$\psi(x') = e^{i k' x'}\, u(x'),$$

where k' is a real parameter, and $u(x' - a) = u(x')$ for all x'. This result is known as *Bloch's theorem*.

3.8 Consider the one-dimensional quantum harmonic oscillator discussed in Exercise 3.3. Show that the Heisenberg equations of motion of the ladder operators, A and $A^\dagger$, are

$$\frac{dA}{dt} = -i\,\omega\,A,$$

$$\frac{dA^\dagger}{dt} = i\,\omega\,A^\dagger,$$

respectively. Hence, deduce that the momentum and position operators evolve in time as

$$p_x(t) = \cos(\omega\,t)\,p_x(0) - m\,\omega\,\sin(\omega\,t)\,x(0),$$

$$x(t) = \cos(\omega\,t)\,x(0) + \frac{\sin(\omega\,t)}{m\,\omega}\,p_x(0),$$

respectively, in the Heisenberg picture.

3.9 Consider a particle in one dimension whose Hamiltonian is

$$H = \frac{p_x^2}{2\,m} + V(x).$$

Suppose that the particle is in a stationary bound state. Using the time-independent Schrödinger equation, prove that

$$\left\langle \frac{p_x^2}{2\,m} \right\rangle = E - \langle V \rangle,$$

and

$$\left\langle \frac{p_x^2}{2\,m} \right\rangle = -E + \langle V \rangle + \left\langle x\,\frac{dV}{dx} \right\rangle.$$

Here, E is the energy eigenvalue. [Hint: You may assume, without loss of generality, that the stationary wavefunction is real.] Hence, prove the *Virial theorem*,

$$\left\langle \frac{p_x^2}{2\,m} \right\rangle = \frac{1}{2} \left\langle x\,\frac{dV}{dx} \right\rangle.$$

3.10 Consider a particle of mass m and charge q moving in the x-y plane in the presence of the uniform perpendicular magnetic field $\mathbf{B} = B_z\,\mathbf{e}_z$. Demonstrate that the Hamiltonian of the system can be written

$$H = \hbar\,\omega \left(\frac{1}{2} + A^\dagger A \right),$$

where $\omega = q\,B_z/m$, and

$$A = \frac{\Pi_x + i\,\Pi_y}{\sqrt{2\,\hbar\,q\,B_z}}.$$

In addition, show that

$$[A, A^\dagger] = 1.$$

Hence, deduce that the possible energy eigenstates of the particle are

$$E_n = (n + 1/2)\hbar\omega,$$

where n is a non-negative integer. These energy levels are known as *Landau levels*.

3.11 Show that the time-dependent Schrödinger equation

$$i\hbar\frac{\partial\psi}{\partial t} = \frac{1}{2m}(-i\hbar\nabla' - q\mathbf{A})^2\psi + q\phi\psi,$$

where $\psi = \psi(\mathbf{x}', t)$, $\mathbf{A} = \mathbf{A}(\mathbf{x}', t)$, and $\phi = \phi(\mathbf{x}', t)$, can be written

$$i\hbar\frac{\partial\psi}{\partial t} = \frac{1}{2m}(-\hbar^2\nabla'^2 + 2i\hbar q\mathbf{A}\cdot\nabla' + i\hbar q\nabla'\cdot\mathbf{A} + q^2 A^2)\psi + q\phi\psi.$$

Hence, deduce that if the so-called *Coloumb gauge* [Fitzpatrick (2008)],

$$\nabla'\cdot\mathbf{A} = 0,$$

is adopted then the equation simplifies to

$$i\hbar\frac{\partial\psi}{\partial t} = -\frac{\hbar^2}{2m}\nabla'^2\psi + i\frac{\hbar q}{m}\mathbf{A}\cdot\nabla'\psi + \frac{q^2}{2m}A^2\psi + q\phi\psi.$$

Demonstrate that this equation is associated with a probability conservation law of the form

$$\frac{\partial\rho}{\partial t} + \nabla'\cdot\mathbf{j} = 0,$$

where

$$\rho(\mathbf{x}', t) = |\psi(\mathbf{x}', t)|^2,$$

and

$$\mathbf{j}(\mathbf{x}', t) = -\left(\frac{i\hbar}{2m}\right)[\psi^*\nabla'\psi - (\nabla'\psi)^*\psi] - \frac{q}{m}\rho(\mathbf{x}', t)\mathbf{A}(\mathbf{x}', t).$$

Finally, show that ρ and $\mathbf{j}$ are invariant under a gauge transformation.

Chapter 4

Orbital Angular Momentum

4.1 Orbital Angular Momentum

Consider a particle described by the Cartesian coordinates $(x, y, z) \equiv \mathbf{x}$ and their conjugate momenta $(p_x, p_y, p_z) \equiv \mathbf{p}$. The classical definition of the *orbital angular momentum* of such a particle about the origin is $\mathbf{L} = \mathbf{x} \times \mathbf{p}$ [Goldstein *et al.* (2002)], giving

$$L_x = y\, p_z - z\, p_y, \tag{4.1}$$

$$L_y = z\, p_x - x\, p_z, \tag{4.2}$$

$$L_z = x\, p_y - y\, p_x. \tag{4.3}$$

Let us assume that the operators $(L_x, L_y, L_z) \equiv \mathbf{L}$ that represent the components of orbital angular momentum in quantum mechanics can be defined in an analogous manner to the corresponding components of classical angular momentum. In other words, we are going to assume that the previous equations specify the angular momentum operators in terms of the position and linear momentum operators. According to Equations (4.1)–(4.3), L_x, L_y, and L_z are Hermitian operators, so they represent quantities that can, in principle, be measured. Note, incidentally, that there is no ambiguity regarding the order in which operators appear in products on the right-hand sides of Equations (4.1)–(4.3), because all of the products consist of operators that commute.

The fundamental commutation relations satisfied by the position and linear momentum operators are

$$[x_i, x_j] = 0, \tag{4.4}$$

$$[p_i, p_j] = 0, \tag{4.5}$$

$$[x_i, p_j] = i\,\hbar\,\delta_{ij}, \tag{4.6}$$

where i and j stand for either x, y, or z. [See Equations (2.23)–(2.25).] Consider the commutator of the operators L_x and L_y:

$$[L_x, L_y] = [(y\,p_z - z\,p_y), (z\,p_x - x\,p_z)] = y\,[p_z, z]\,p_x + x\,p_y\,[z, p_z]$$

$$= i\,\hbar\,(-y\,p_x + x\,p_y) = i\,\hbar\,L_z. \tag{4.7}$$

The cyclic permutations (i.e., x, y, $z \to y$, z, x, etc.) of the previous result yield the fundamental commutation relations satisfied by the components of an orbital angular momentum:

$$[L_x, L_y] = i\,\hbar\,L_z, \tag{4.8}$$

$$[L_y, L_z] = i\,\hbar\,L_x, \tag{4.9}$$

$$[L_z, L_x] = i\,\hbar\,L_y. \tag{4.10}$$

These expressions can be summed up more succinctly by writing

$$\mathbf{L} \times \mathbf{L} = i\,\hbar\,\mathbf{L}. \tag{4.11}$$

The three commutation relations (4.8)–(4.10) are the foundation for the whole theory of angular momentum in quantum mechanics. Whenever we encounter three operators having similar commutation relations, we know the dynamical variables that they represent have identical properties to those of the components of an angular momentum (which we are about to derive). In fact, we shall assume that any three operators that satisfy the commutation relations (4.8)–(4.10) represent the components of some sort of angular momentum.

Suppose that there are N particles in the system, with angular momentum vectors $\mathbf{L}_i$ (where i runs from 1 to N). Each of these vectors satisfies Equation (4.11), so that

$$\mathbf{L}_i \times \mathbf{L}_i = i\,\hbar\,\mathbf{L}_i. \tag{4.12}$$

However, we expect the angular momentum operators belonging to different particles to commute, because they represent different degrees of freedom of the system. (See Section 2.2.) So, we can write

$$\mathbf{L}_i \times \mathbf{L}_j + \mathbf{L}_j \times \mathbf{L}_i = 0, \tag{4.13}$$

for $i \neq j$. Consider the total angular momentum of the system, $\mathbf{L} = \sum_{i=1,N} \mathbf{L}_i$. It is clear from Equations (4.12) and (4.13) that

$$\mathbf{L} \times \mathbf{L} = \sum_{i=1,N} \mathbf{L}_i \times \sum_{j=1,N} \mathbf{L}_j = \sum_{i=1,N} \mathbf{L}_i \times \mathbf{L}_i + \frac{1}{2} \sum_{i,j=1,N}^{i \neq j} (\mathbf{L}_i \times \mathbf{L}_j + \mathbf{L}_j \times \mathbf{L}_i)$$

$$= i\,\hbar \sum_{i=1,N} \mathbf{L}_i = i\,\hbar\,\mathbf{L}. \tag{4.14}$$

Thus, the sum of two or more angular momentum vectors satisfies the same commutation relation as a primitive angular momentum vector. In particular, the total angular momentum of the system satisfies the commutation relation (4.11).

The immediate conclusion that can be drawn from the commutation relations (4.8)–(4.10) is that the three components of an angular momentum vector cannot be specified (or measured) simultaneously. In fact, once we have specified one component, the values of other two components become uncertain. It is conventional to specify the z-component, L_z.

Consider the magnitude squared of the angular momentum vector, $L^2 \equiv L_x^2 + L_y^2 + L_z^2$. The commutator of L^2 and L_z is written

$$[L^2, L_z] = [L_x^2, L_z] + [L_y^2, L_z] + [L_z^2, L_z]. \tag{4.15}$$

It is easily demonstrated that

$$[L_x^2, L_z] = -i\,\hbar\,(L_x L_y + L_y L_x), \tag{4.16}$$

$$[L_y^2, L_z] = +i\,\hbar\,(L_x L_y + L_y L_x), \tag{4.17}$$

$$[L_z^2, L_z] = 0, \tag{4.18}$$

which implies that

$$[L^2, L_z] = 0. \tag{4.19}$$

(See Exercise 4.1.) Because there is nothing special about the z-direction, we conclude that L^2 also commutes with L_x and L_y. It is clear from Equations (4.8)–(4.10) and (4.19) that the best we can do in quantum mechanics is to specify the magnitude of an angular momentum vector along with one of its components (by convention, the z-component).

It is convenient to define the *ladder operators*, L^+ and L^-:

$$L^+ = L_x + i\,L_y, \tag{4.20}$$

$$L^- = L_x - i\,L_y. \tag{4.21}$$

It can easily be shown that

$$[L^+, L_z] = -\hbar\,L^+, \tag{4.22}$$

$$[L^-, L_z] = +\hbar\,L^-, \tag{4.23}$$

$$[L^+, L^-] = 2\,\hbar\,L_z, \tag{4.24}$$

and also that both ladder operators commute with L^2. (See Exercise 4.1.)

4.2 Eigenvalues of Orbital Angular Momentum

Suppose that the simultaneous eigenkets of L^2 and L_z are completely specified by two (dimensionless) quantum numbers, l and m. These kets are denoted $|l, m\rangle$. The quantum number m is defined by

$$L_z |l, m\rangle = m \hbar |l, m\rangle. \tag{4.25}$$

Thus, m is the eigenvalue of L_z divided by $\hbar$. It is possible to write such an equation because $\hbar$ has the dimensions of angular momentum. Note that m is a real number, because L_z is an Hermitian operator.

We can write

$$L^2 |l, m\rangle = f(l, m) \hbar^2 |l, m\rangle, \tag{4.26}$$

without loss of generality, where $f(l, m)$ is some real dimensionless function of l and m. Later on, we will show that $f(l, m) = l(l + 1)$. Now,

$$\langle l, m| L^2 - L_z^2 |l, m\rangle = \langle l, m| f(l, m) \hbar^2 - m^2 \hbar^2 |l, m\rangle = [f(l, m) - m^2] \hbar^2, \tag{4.27}$$

assuming that the $|l, m\rangle$ have unit norms. However,

$$\langle l, m| L^2 - L_z^2 |l, m\rangle = \langle l, m| L_x^2 + L_y^2 |l, m\rangle = \langle l, m| L_x^2 |l, m\rangle + \langle l, m| L_y^2 |l, m\rangle. \tag{4.28}$$

It is readily demonstrated that

$$\langle A| \xi^2 |A\rangle \geq 0, \tag{4.29}$$

where $|A\rangle$ is a general ket, and ξ an Hermitian operator. The proof follows from the observation that

$$\langle A| \xi^2 |A\rangle = \langle A| \xi^\dagger \xi |A\rangle = \langle B|B\rangle, \tag{4.30}$$

where $|B\rangle = \xi |A\rangle$, plus the fact that $\langle B|B\rangle \geq 0$ for a general ket $|B\rangle$. [See Equation (1.22).] It follows from Equations (4.27)–(4.29) that

$$m^2 \leq f(l, m). \tag{4.31}$$

Consider the effect of the ladder operator L^+ on the eigenket $|l, m\rangle$. It is easily demonstrated that

$$L^2 (L^+ |l, m\rangle) = \hbar^2 f(l, m) (L^+ |l, m\rangle), \tag{4.32}$$

where use has been made of Equation (4.26), plus the fact that L^2 and L^+ commute. It follows that the ket $L^+ |l, m\rangle$ is an eigenstate of L^2 corresponding to the same eigenvalue as the ket $|l, m\rangle$. Thus, the ladder operator L^+ does not affect the magnitude of the angular momentum of any state that it acts upon. However,

$$L_z (L^+ |l, m\rangle) = (L^+ L_z + [L_z, L^+]) |l, m\rangle = (L^+ L_z + \hbar L^+) |l, m\rangle$$

$$= (m + 1) \hbar (L^+ |l, m\rangle), \tag{4.33}$$

where use has been made of Equation (4.22). The previous equation implies that $L^+|l, m\rangle$ is proportional to $|l, m + 1\rangle$. We can write

$$L^+|l, m\rangle = c^+_{lm} \hbar |l, m + 1\rangle, \tag{4.34}$$

where c^+_{lm} is a (dimensionless) number. It is clear that if the operator L^+ acts on a simultaneous eigenstate of L^2 and L_z then the eigenvalue of L^2 remains unchanged, but the eigenvalue of L_z is increased by $\hbar$. For this reason, L^+ is called a *raising operator*.

Using similar arguments to those just given, it is possible to demonstrate that

$$L^- |l, m\rangle = c^-_{lm} \hbar |l, m - 1\rangle, \tag{4.35}$$

where c^-_{lm} is a (dimensionless) number. Hence, L^- is called a *lowering operator*.

The ladder operators, L^+ and L^-, respectively step the value of m up and down by unity each time they operate on one of the simultaneous eigenkets of L^2 and L_z. It would appear, at first sight, that any value of m can be obtained by applying these operators a sufficient number of times. However, according to Equation (4.31), there is a definite upper bound to the values that m^2 can take. This bound is determined by the eigenvalue of L^2. [See Equation (4.26).] It follows that there is a maximum and a minimum possible value that m can take. Suppose that we attempt to raise the value of m above its maximum value, $m_{\max}$. Because there is no state with $m > m_{\max}$, we must have

$$L^+|l, m_{\max}\rangle = |0\rangle. \tag{4.36}$$

This implies that

$$L^- L^+|l, m_{\max}\rangle = |0\rangle. \tag{4.37}$$

However,

$$L^- L^+ = L_x^2 + L_y^2 + i [L_x, L_y] = L^2 - L_z^2 - \hbar L_z, \tag{4.38}$$

so Equation (4.37) yields

$$(L^2 - L_z^2 - \hbar L_z) |l, m_{\max}\rangle = |0\rangle. \tag{4.39}$$

The previous equation can be rearranged to give

$$L^2 |l, m_{\max}\rangle = (L_z^2 + \hbar L_z) |l, m_{\max}\rangle = m_{\max} (m_{\max} + 1) \hbar^2 |l, m_{\max}\rangle. \tag{4.40}$$

Comparison of this equation with Equation (4.26) yields the result

$$f(l, m_{\max}) = m_{\max}(m_{\max} + 1). \tag{4.41}$$

But, when L^- operates successively on $|n, m_{\max}\rangle$ it generates $|n, m_{\max} - 1\rangle$, $|n, m_{\max} - 2\rangle$, et cetera. Because the lowering operator does not change the eigenvalue of L^2, all of these states must correspond to the same value of $f(l, m)$; namely, $m_{\max} (m_{\max} + 1)$. Thus,

$$L^2 |l, m\rangle = m_{\max} (m_{\max} + 1)\, \hbar^2 |l, m\rangle. \qquad (4.42)$$

At this stage, we can give the unknown quantum number l the value $m_{\max}$, without loss of generality. We can also write the previous equation in the form

$$L^2 |l, m\rangle = l(l + 1)\, \hbar^2 |l, m\rangle. \qquad (4.43)$$

It is easily seen that

$$L^- L^+ |l, m\rangle = (L^2 - L_z^2 - \hbar L_z) |l, m\rangle = \hbar^2 \left[l(l + 1) - m(m + 1) \right] |l, m\rangle. \qquad (4.44)$$

Thus,

$$\langle l, m| L^- L^+ |l, m\rangle = \hbar^2 \left[l(l + 1) - m(m + 1) \right]. \qquad (4.45)$$

However, we also know that

$$\langle l, m| L^- L^+ |l, m\rangle = \langle l, m| L^- \hbar\, c_{lm}^+ |l, m + 1\rangle = \hbar^2\, c_{lm}^+\, c_{l\, m+1}^-, \qquad (4.46)$$

where use has been made of Equations (4.34) and (4.35). It follows that

$$c_{lm}^+\, c_{l\, m+1}^- = l(l + 1) - m(m + 1). \qquad (4.47)$$

Consider the following:

$$
\begin{aligned}
\langle l, m| L^- |l, m + 1\rangle &= \langle l, m| L_x |l, m + 1\rangle - \mathrm{i}\,\langle l, m| L_y |l, m + 1\rangle \\
&= \langle l, m + 1| L_x |l, m\rangle^* - \mathrm{i}\,\langle l, m + 1| L_y |l, m\rangle^* \\
&= \left(\langle l, m + 1| L_x |l, m\rangle + \mathrm{i}\,\langle l, m + 1| L_y |l, m\rangle \right)^* \\
&= \langle l, m + 1| L^+ |l, m\rangle^*,
\end{aligned}
\qquad (4.48)
$$

where use has been made of the fact that L_x and L_y are Hermitian operators. The previous equation reduces to

$$c_{l\, m+1}^- = (c_{lm}^+)^* \qquad (4.49)$$

with the aid of Equations (4.34) and (4.35).

Equations (4.47) and (4.49) can be combined to give

$$|c_{lm}^+|^2 = l(l + 1) - m(m + 1). \qquad (4.50)$$

The solution of the previous equation is

$$c_{lm}^+ = \sqrt{l(l + 1) - m(m + 1)}. \qquad (4.51)$$

Note that c_{lm}^+ is undetermined to an arbitrary phase-factor [i.e., we can replace c_{lm}^+, given previously, by $c_{lm}^+ \exp(i\gamma)$, where γ is real, and we still satisfy Equation (4.50)]. We have made the arbitrary, but convenient, choice that c_{lm}^+ is real and positive. This is equivalent to choosing the relative phases of the eigenkets $|l, m\rangle$. According to Equation (4.49),

$$c_{lm}^- = (c_{l\,m-1}^+)^* = \sqrt{l(l+1) - m(m-1)}. \tag{4.52}$$

We have already seen that the inequality (4.31) implies that there is a maximum and a minimum possible value of m. The maximum value of m is denoted l. What is the minimum value? Suppose that we try to lower the value of m below its minimum value m_{min}. Because there is no state with $m < m_{min}$, we must have

$$L^- |l, m_{min}\rangle = 0. \tag{4.53}$$

According to Equation (4.35), this implies that

$$c_{l\,m_{min}}^- = 0. \tag{4.54}$$

It can be seen from Equation (4.52) that $m_{min} = -l$. We conclude that the quantum number m can take a "ladder" of discrete values, each rung differing from its immediate neighbors by unity. The top rung is l, and the bottom rung is $-l$. There are only two possible choices for l. Either it is an integer (e.g., $l = 2$, which allows m to take the integer values $-2, -1, 0, 1, 2$), or it is a half-integer (e.g., $l = 3/2$, which allows m to take the half-integer values $-3/2, -1/2, 1/2, 3/2$). In fact, we shall prove, in the next section, that an orbital angular momentum can only take integer values of l.

In summary, just using the fundamental commutation relations (4.8)–(4.10), plus the fact that L_x, L_y, and L_z are Hermitian operators, we have shown that the eigenvalues of $L^2 \equiv L_x^2 + L_y^2 + L_z^2$ can be written $l(l+1)\hbar^2$, where l is an integer, or a half-integer. Without loss of generality, we can assume that l is non-negative. We have also demonstrated that the eigenvalues of L_z can only take the values $m\hbar$, where m lies in the range $-l, -l+1, \cdots l-1, l$. Finally, if $|l, m\rangle$ denotes a properly normalized simultaneous eigenket of L^2 and L_z, belonging to the eigenvalues $l(l+1)\hbar^2$ and $m\hbar$, respectively, then we have shown that

$$L^+ |l, m\rangle = \sqrt{l(l+1) - m(m+1)}\,\hbar\,|l, m+1\rangle, \tag{4.55}$$

$$L^- |l, m\rangle = \sqrt{l(l+1) - m(m-1)}\,\hbar\,|l, m-1\rangle, \tag{4.56}$$

where $L^\pm = L_x \pm i L_y$ are the so-called ladder operators.

4.3   Rotation Operators

Consider a particle whose position is described by the spherical coordinates
(r, θ, φ). The classical momentum conjugate to the azimuthal angle φ is the z-
component of angular momentum, L_z [Goldstein *et al.* (2002)]. According to
Section 2.5, in quantum mechanics, we can always adopt the Schrödinger representation, for which ket space is spanned by the simultaneous eigenkets of the
position operators, r, θ, and φ, and L_z takes the form

$$L_z = -i\hbar \frac{\partial}{\partial \varphi}. \tag{4.57}$$

Incidentally, it is legitimate to write the previous equation because there is nothing in Section 2.5 that specifies that we have to use Cartesian coordinates—the
representation (2.74) works for any well-defined set of coordinates.

   Consider an operator $R_z(\Delta\varphi)$ that rotates the system through an angle $\Delta\varphi$ about
the z-axis. This operator is very similar to the operator $D_x(\Delta x)$, introduced in
Section 2.8, which translates the system a distance Δx along the x-axis. We were
able to demonstrate in Section 2.8 that

$$p_x = i\hbar \lim_{\delta x \to 0} \frac{D_x(\delta x) - 1}{\delta x}, \tag{4.58}$$

where p_x is the linear momentum conjugate to x. There is nothing in our derivation
of this result that specifies that x has to be a Cartesian coordinate. Thus, the result
should apply just as well to an angular coordinate. We conclude that

$$L_z = i\hbar \lim_{\delta\varphi \to 0} \frac{R_z(\delta\varphi) - 1}{\delta\varphi}. \tag{4.59}$$

   According to Equation (4.59), we can write

$$R_z(\delta\varphi) = 1 - i L_z \, \delta\varphi/\hbar \tag{4.60}$$

in the limit $\delta\varphi \to 0$. In other words, the angular momentum operator L_z can be
used to rotate the system about the z-axis by an infinitesimal amount. We say that
L_z is the *generator of rotation* about the z-axis. The previous equation implies that

$$R_z(\Delta\varphi) = \lim_{N \to \infty} \left(1 - i \frac{\Delta\varphi}{N} \frac{L_z}{\hbar} \right)^N \equiv \sum_{n=0,\infty} \left[\frac{(-i\,\Delta\varphi\, L_z/\hbar)^n}{n!} \right], \tag{4.61}$$

which reduces to

$$R_z(\Delta\varphi) = \exp\left(\frac{-i L_z \Delta\varphi}{\hbar} \right). \tag{4.62}$$

Note that $R_z(\Delta\varphi)$ has all of the properties we would expect a rotation operator to possess: that is,

$$R_z(0) = 1, \qquad (4.63)$$

$$R_z(\Delta\varphi) R_z(-\Delta\varphi) = 1, \qquad (4.64)$$

$$R_z(\Delta\varphi_1) R_z(\Delta\varphi_2) = R_z(\Delta\varphi_1 + \Delta\varphi_2). \qquad (4.65)$$

(See Exercises 1.7 and 1.8.)

Suppose that the system is in a simultaneous eigenstate of L^2 and L_z. As before, this state is represented by the eigenket $|l, m\rangle$, where the eigenvalue of L^2 is $l(l+1)\hbar^2$, and the eigenvalue of L_z is $m\hbar$. We expect the wavefunction to remain unaltered if we rotate the system 2π degrees about the z-axis. Thus,

$$R_z(2\pi) |l, m\rangle = \exp\left(\frac{-i L_z 2\pi}{\hbar}\right) |l, m\rangle = \exp(-i 2\pi m) |l, m\rangle = |l, m\rangle. \qquad (4.66)$$

We conclude that m must be an integer. This implies, from the previous section, that l must also be an integer. Thus, an orbital angular momentum can only take integer values of the quantum numbers l and m.

Consider the action of the rotation operator $R_z(\Delta\varphi)$ on an eigenstate possessing zero angular momentum about the z-axis (i.e., an $m = 0$ state). We have

$$R_z(\Delta\varphi) |l, 0\rangle = \exp(0) |l, 0\rangle = |l, 0\rangle. \qquad (4.67)$$

Thus, the eigenstate is invariant to rotations about the z-axis. Clearly, its wavefunction must be symmetric about the z-axis.

There is nothing special about the z-axis, so we can write

$$R_x(\Delta\varphi_x) = \exp\left(\frac{-i L_x \Delta\varphi_x}{\hbar}\right), \qquad (4.68)$$

$$R_y(\Delta\varphi_y) = \exp\left(\frac{-i L_y \Delta\varphi_y}{\hbar}\right), \qquad (4.69)$$

$$R_z(\Delta\varphi_y) = \exp\left(\frac{-i L_z \Delta\varphi_z}{\hbar}\right), \qquad (4.70)$$

by analogy with Equation (4.62). Here, $R_x(\Delta\varphi_x)$ denotes an operator that rotates the system through an angle $\Delta\varphi_x$ about the x-axis, et cetera. Suppose that the system is in an eigenstate of zero overall orbital angular momentum (i.e., an $l = 0$ state). We know that the system is also in an eigenstate of zero orbital angular momentum about any particular axis. This follows because $l = 0$ implies $m = 0$, according to the previous section, and we can choose the z-axis to point in any direction. Thus,

$$R_x(\Delta\varphi_x)\,|0,0\rangle = \exp(0)\,|0,0\rangle = |0,0\rangle, \tag{4.71}$$

$$R_y(\Delta\varphi_y)\,|0,0\rangle = \exp(0)\,|0,0\rangle = |0,0\rangle, \tag{4.72}$$

$$R_z(\Delta\varphi_z)\,|0,0\rangle = \exp(0)\,|0,0\rangle = |0,0\rangle. \tag{4.73}$$

Clearly, a zero angular momentum state is invariant to rotations about any axis. Such a state must possess a spherically symmetric wavefunction.

Note that, in general, a rotation about the x-axis does not commute with a rotation about the y-axis. In other words, if a physical system is rotated through an angle $\Delta\varphi_x$ about the x-axis, and then through an angle $\Delta\varphi_y$ about the y-axis, it ends up in a different state to that obtained by rotating through an angle $\Delta\varphi_y$ about the y-axis, and then through an angle $\Delta\varphi_x$ about the x-axis. In quantum mechanics, this implies that $R_y(\Delta\varphi_y)\,R_x(\Delta\varphi_x) \neq R_x(\Delta\varphi_x)\,R_y(\Delta\varphi_y)$, or $L_y\,L_x \neq L_x\,L_y$. [See Equations (4.68)–(4.70) and Exercise 2.3.] Thus, the noncommuting nature of the angular momentum operators is a direct consequence of the fact that rotations do not commute.

4.4   Eigenfunctions of Orbital Angular Momentum

In Cartesian coordinates, the three components of orbital angular momentum can be written

$$L_x = -i\,\hbar\left(y\,\frac{\partial}{\partial z} - z\,\frac{\partial}{\partial y}\right), \tag{4.74}$$

$$L_y = -i\,\hbar\left(z\,\frac{\partial}{\partial x} - x\,\frac{\partial}{\partial z}\right), \tag{4.75}$$

$$L_z = -i\,\hbar\left(x\,\frac{\partial}{\partial y} - y\,\frac{\partial}{\partial x}\right), \tag{4.76}$$

using the Schrödinger representation. Transforming to standard spherical coordinates,

$$x = r\,\sin\theta\,\cos\varphi, \tag{4.77}$$

$$y = r\,\sin\theta\,\sin\varphi, \tag{4.78}$$

$$z = r\,\cos\theta, \tag{4.79}$$

we obtain

$$L_x = i\hbar \left(\sin\varphi \, \frac{\partial}{\partial\theta} + \cot\theta \cos\varphi \, \frac{\partial}{\partial\varphi} \right), \tag{4.80}$$

$$L_y = -i\hbar \left(\cos\varphi \, \frac{\partial}{\partial\theta} - \cot\theta \sin\varphi \, \frac{\partial}{\partial\varphi} \right), \tag{4.81}$$

$$L_z = -i\hbar \, \frac{\partial}{\partial\varphi}. \tag{4.82}$$

(See Exercise 4.2.) Note that Equation (4.82) accords with Equation (4.57). The ladder operators $L^\pm = L_x \pm i\, L_y$ become

$$L^\pm = \pm\hbar \, \exp(\pm i\,\varphi) \left(\frac{\partial}{\partial\theta} \pm i \cot\theta \, \frac{\partial}{\partial\varphi} \right). \tag{4.83}$$

(See Exercise 4.2.) Now,

$$L^2 = L_x^2 + L_y^2 + L_z^2 = L_z^2 + (L^+ L^- + L^- L^+)/2, \tag{4.84}$$

so

$$L^2 = -\hbar^2 \left(\frac{1}{\sin\theta} \frac{\partial}{\partial\theta} \sin\theta \, \frac{\partial}{\partial\theta} + \frac{1}{\sin^2\theta} \frac{\partial^2}{\partial\varphi^2} \right). \tag{4.85}$$

(See Exercise 4.2.)

The eigenvalue problem for L^2 takes the form

$$L^2 \psi = \lambda\hbar^2 \psi, \tag{4.86}$$

where $\psi(r,\theta,\varphi)$ is the wavefunction, and λ is a dimensionless number. Let us write

$$\psi(r,\theta,\varphi) = R(r)\, Y(\theta,\varphi). \tag{4.87}$$

Equation (4.86) reduces to

$$\left(\frac{1}{\sin\theta} \frac{\partial}{\partial\theta} \sin\theta \, \frac{\partial}{\partial\theta} + \frac{1}{\sin^2\theta} \frac{\partial^2}{\partial\varphi^2} \right) Y + \lambda\, Y = 0, \tag{4.88}$$

where use has been made of Equation (4.85). As is well known, square-integrable solutions to this equation only exist when λ takes the values $l(l + 1)$, where l is an integer (which we can take to be non-negative, without loss of generality) [Morse and Feschbach (1953)]. These solutions are known as *spherical harmonics* [Jackson (1975)], and can be written

$$Y_{lm}(\theta,\varphi) = \sqrt{\frac{2l+1}{4\pi} \frac{(l-m)!}{(l+m)!}} \; P_{lm}(\cos\theta)\, e^{i m\varphi}. \tag{4.89}$$

Here,

$$P_{lm}(\xi) = \frac{(-1)^{l+m}}{2^l \, l!} \, (1 - \xi^2)^{m/2} \frac{d^{l+m}}{d\xi^{l+m}} (1 - \xi^2)^l. \tag{4.90}$$

is an *associated Legendre function* [Jackson (1975)], satisfying the equation

$$\frac{d}{d\xi}\left[(1-\xi^2)\frac{dP_{lm}}{d\xi}\right] - \frac{m^2}{1-\xi^2}P_{lm} + l(l+1)P_{lm} = 0. \tag{4.91}$$

It follows that

$$P_{l-m}(\xi) = (-1)^m \frac{(l-m)!}{(l+m)!}P_{lm}(\xi), \tag{4.92}$$

and, hence, that

$$Y_{l-m} = (-1)^m Y_{lm}^*. \tag{4.93}$$

Of course, m must be an integer, so as to ensure that the Y_{lm} are single valued in φ. Moreover, it is clear from Equations (4.90) and (4.92) that $P_{lm} = 0$ unless $-l \leq m \leq l$. The spherical harmonics are orthogonal functions, and are properly normalized with respect to integration over the entire solid angle [Jackson (1975)]:

$$\int_0^{2\pi} d\varphi \int_0^{\pi} d\theta \sin\theta\, Y_{lm}^*(\theta,\varphi)\, Y_{l'm'}(\theta,\varphi) = \delta_{ll'}\,\delta_{mm'}. \tag{4.94}$$

The spherical harmonics also form a complete set for representing general functions of θ and φ. The first few spherical harmonics are [Jackson (1975)]:

$$Y_{00}(\theta,\varphi) = \frac{1}{\sqrt{4\pi}}, \tag{4.95}$$

$$Y_{10}(\theta,\varphi) = \sqrt{\frac{3}{4\pi}}\cos\theta, \tag{4.96}$$

$$Y_{1\pm1}(\theta,\varphi) = \mp\sqrt{\frac{3}{8\pi}}\sin\theta\, e^{\pm i\varphi}, \tag{4.97}$$

$$Y_{20}(\theta,\varphi) = \sqrt{\frac{5}{16\pi}}(3\cos^2\theta - 1), \tag{4.98}$$

$$Y_{2\pm1}(\theta,\varphi) = \mp\sqrt{\frac{15}{8\pi}}\sin\theta\cos\theta\, e^{\pm i\varphi}, \tag{4.99}$$

$$Y_{2\pm2}(\theta,\varphi) = \sqrt{\frac{15}{32\pi}}\sin^2\theta\, e^{\pm 2i\varphi}, \tag{4.100}$$

$$Y_{30}(\theta,\varphi) = \sqrt{\frac{7}{16\pi}}(5\cos^3\theta - 3\cos\theta), \tag{4.101}$$

$$Y_{3\pm1}(\theta,\varphi) = \mp\sqrt{\frac{21}{64\pi}}\sin\theta(5\cos^2\theta - 1)\, e^{\pm i\varphi}, \tag{4.102}$$

$$Y_{3\pm2}(\theta,\varphi) = \sqrt{\frac{105}{32\pi}}\sin^2\theta\cos\theta\, e^{\pm i\,2\varphi}, \tag{4.103}$$

$$Y_{3\pm3}(\theta,\varphi) = \mp\sqrt{\frac{35}{64\pi}}\sin^3\theta\, e^{\pm i\,3\varphi}. \tag{4.104}$$

By definition,

$$L^2 Y_{lm} = l(l + 1) \hbar^2 Y_{lm}, \tag{4.105}$$

where l is an integer. It follows from Equations (4.82) and (4.89) that

$$L_z Y_{lm} = m \hbar Y_{lm}, \tag{4.106}$$

where m is an integer lying in the range $-l \leq m \leq l$. Thus, the wavefunction $\psi(r, \theta, \varphi) = R(r) Y_{lm}(\theta, \phi)$, where R is a general function, has all of the expected features of the wavefunction of a simultaneous eigenstate of L^2 and L_z belonging to the quantum numbers l and m. The well-known formula [Abramowitz and Stegun (1965)]

$$\frac{dP_{lm}}{d\xi} = -\frac{1}{\sqrt{1 - \xi^2}} P_{l\,m+1} - \frac{m\xi}{1 - \xi^2} P_{lm}$$

$$= \frac{(l + m)(l - m + 1)}{\sqrt{1 - \xi^2}} P_{l\,m-1} + \frac{m\xi}{1 - \xi^2} P_{lm} \tag{4.107}$$

can be combined with Equations (4.83) and (4.89) to give

$$L^+ Y_{lm} = [l(l + 1) - m(m + 1)]^{1/2} \hbar Y_{l\,m+1}, \tag{4.108}$$

$$L^- Y_{lm} = [l(l + 1) - m(m - 1)]^{1/2} \hbar Y_{l\,m-1}. \tag{4.109}$$

(See Exercise 4.4.) These equations are equivalent to Equations (4.55)–(4.56). Note that a spherical harmonic wavefunction is symmetric about the z-axis (i.e., independent of φ) whenever $m = 0$, and is spherically symmetric whenever $l = 0$ (because $Y_{00} = 1/\sqrt{4\pi}$).

In summary, by solving directly for the eigenfunctions of L^2 and L_z in the Schrödinger representation, we have been able to reproduce all of the results of Section 4.2. Nevertheless, the results of Section 4.2 are more general than those obtained in this section, because they still apply when the quantum number l takes on half-integer values.

4.5 Motion in Central Field

Consider a particle of mass M moving in a spherically symmetric potential. The Hamiltonian takes the form

$$H = \frac{p^2}{2M} + V(r). \tag{4.110}$$

Adopting the Schrödinger representation, we can write $\mathbf{p} = -(i/\hbar)\nabla$. Hence,

$$H = -\frac{\hbar^2}{2M} \nabla^2 + V(r). \tag{4.111}$$

When written in spherical coordinates, the previous equation becomes [Riley *et al.* (2013)]

$$H = -\frac{\hbar^2}{2M} \left[\frac{1}{r^2} \frac{\partial}{\partial r} r^2 \frac{\partial}{\partial r} + \frac{1}{r^2 \sin\theta} \frac{\partial}{\partial\theta} \sin\theta \frac{\partial}{\partial\theta} + \frac{1}{r^2 \sin^2\theta} \frac{\partial^2}{\partial\varphi^2} \right] + V(r). \quad (4.112)$$

Comparing this equation with Equation (4.85), we find that

$$H = \frac{\hbar^2}{2M} \left(-\frac{1}{r^2} \frac{\partial}{\partial r} r^2 \frac{\partial}{\partial r} + \frac{L^2}{\hbar^2 r^2} \right) + V(r). \quad (4.113)$$

Now, we know that the three components of angular momentum commute with L^2. (See Section 4.1.) We also know, from Equations (4.80)–(4.82), that L_x, L_y, and L_z take the form of partial derivative operators involving only angular coordinates, when written in terms of spherical coordinates using the Schrödinger representation. It follows from Equation (4.113) that all three components of the angular momentum commute with the Hamiltonian: that is,

$$[\mathbf{L}, H] = \mathbf{0}. \quad (4.114)$$

It is also easily seen that L^2 (which can be expressed as a purely angular differential operator) commutes with the Hamiltonian:

$$[L^2, H] = 0. \quad (4.115)$$

According to Section 3.3, the previous two equations ensure that the angular momentum, $\mathbf{L}$, and its magnitude squared, L^2, are both constants of the motion. This is as expected for a spherically symmetric potential.

Consider the energy eigenvalue problem

$$H\psi = E\psi, \quad (4.116)$$

where E is a number (with the dimensions of energy). Because L^2 and L_z commute with each other and the Hamiltonian, it is always possible to represent the state of the system in terms of the simultaneous eigenstates of L^2, L_z, and H. But, we already know that the most general form for the wavefunction of a simultaneous eigenstate of L^2 and L_z is

$$\psi(r, \theta, \varphi) = R(r) Y_{lm}(\theta, \varphi). \quad (4.117)$$

(See the previous section.) Substituting Equation (4.117) into Equation (4.113), and making use of Equation (4.105), we obtain

$$\left[\frac{\hbar^2}{2M} \left(-\frac{1}{r^2} \frac{d}{dr} r^2 \frac{d}{dr} + \frac{l(l+1)}{r^2} \right) + V(r) - E \right] R = 0. \quad (4.118)$$

This is a *Sturm-Liouville equation* for the function $R(r)$ [Riley *et al.* (2013)]. We know, from the general properties of this type of equation, that if $R(r)$ is required to be well behaved at $r = 0$, and as $r \to \infty$, then solutions only exist for a discrete set of values of E. These are the energy eigenvalues, and are conventionally labeled using a quantum number denoted n. (Thus, $n = 1$ corresponds to the lowest energy state, $n = 2$ to the next lowest energy state, and so on.) In general, the energy eigenvalues depend on the quantum numbers n and l, but are independent of the quantum number m.

4.6 Energy Levels of Hydrogen Atom

Consider a hydrogen atom, for which the potential takes the specific form

$$V(r) = -\frac{e^2}{4\pi\,\epsilon_0\,r}. \tag{4.119}$$

The radial eigenfunction $R(r)$ satisfies Equation (4.118), which can be written

$$\left[\frac{\hbar^2}{2\mu} \left(-\frac{1}{r^2}\frac{d}{dr}\,r^2\,\frac{d}{dr} + \frac{l(l+1)}{r^2} \right) - \frac{e^2}{4\pi\,\epsilon_0\,r} - E \right] R = 0. \tag{4.120}$$

Here, $\mu = m_e\,m_p/(m_e+m_p) \simeq m_e$ is the *reduced mass*, which takes into account the fact that the electron (of mass m_e) and the proton (of mass m_p) both orbit about a common centre of mass, which is equivalent to a particle of mass μ orbiting about a fixed point [Fitzpatrick (2012)]. However, given that $m_e \ll m_p$, we can safely replace μ by m_e. (The correction involved in using μ, rather than m_e, in the analysis is actually less than that involved in neglecting the electron's relativistic mass increase.) Let us write the product $r\,R(r)$ as the function $P(r)$. The previous equation transforms to

$$\frac{d^2P}{dr^2} - \frac{2\,m_e}{\hbar^2} \left[\frac{l(l+1)\,\hbar^2}{2\,m_e\,r^2} - \frac{e^2}{4\pi\,\epsilon_0\,r} - E \right] P = 0, \tag{4.121}$$

which is the one-dimensional Schrödinger equation for a particle of mass m_e moving in the *effective potential*

$$V_{\text{eff}}(r) = -\frac{e^2}{4\pi\,\epsilon_0\,r} + \frac{l(l+1)\,\hbar^2}{2\,m_e\,r^2}. \tag{4.122}$$

The effective potential has a simple physical interpretation. The first part is the attractive Coulomb potential, and the second part corresponds to the repulsive centrifugal force.

Let

$$a = \sqrt{\frac{-\hbar^2}{2\,m_e\,E}}, \tag{4.123}$$

and $y = r/a$, with

$$P(r) = f(y) \exp(-y). \tag{4.124}$$

Here, it is assumed that the energy eigenvalue E is negative. Equation (4.121) transforms to

$$\left(\frac{d^2}{dy^2} - 2\frac{d}{dy} - \frac{l(l+1)}{y^2} + \frac{2\,m_e\,e^2\,a}{4\pi\,\epsilon_0\,\hbar^2\,y} \right) f = 0. \tag{4.125}$$

Let us look for a power-law solution of the form

$$f(y) = \sum_k c_k\, y^k. \tag{4.126}$$

Substituting this solution into Equation (4.125), we obtain

$$\sum_k c_k \left[k\,(k-1)\,y^{k-2} - 2\,k\,y^{k-1} - l\,(l+1)\,y^{k-2} + \frac{2\,m_e\,e^2\,a}{4\pi\,\epsilon_0\,\hbar^2}\,y^{k-1} \right] = 0. \tag{4.127}$$

Equating the coefficients of y^{k-2} gives

$$c_k\,[k\,(k-1) - l\,(l+1)] = c_{k-1}\left[2\,(k-1) - \frac{2\,m_e\,e^2\,a}{4\pi\,\epsilon_0\,\hbar^2} \right]. \tag{4.128}$$

Now, the power-law series (4.126) must terminate at small k, at some positive value of k, otherwise $f(y)$ would diverge unphysically as $y \to 0$. This is only possible if $[k_{min}\,(k_{min} - 1) - l\,(l+1)] = 0$, where the first term in the series is $c_{k_{min}}\,y^{k_{min}}$. There are two possibilities: $k_{min} = -l$ or $k_{min} = l + 1$. The former predicts unphysical divergence of the wavefunction at $y = 0$. Thus, we conclude that $k_{min} = l + 1$, and that the first term in the series is $c_{l+1}\,y^{l+1}$. Note that for an $l = 0$ state there is a finite probability of finding the electron at the nucleus, whereas for an $l > 0$ state there is zero probability of finding the electron at the nucleus (i.e., $|\psi|^2 = 0$ at $r = 0$, except when $l = 0$). Furthermore, it is only possible to obtain sensible behavior of the wavefunction as $r \to 0$ if l is an integer.

For large values of y, the ratio of successive terms in the series (4.126) is

$$\frac{c_k\,y}{c_{k-1}} = \frac{2\,y}{k}, \tag{4.129}$$

according to Equation (4.128). This is the same as the ratio of successive terms in the series

$$\sum_k \frac{(2\,y)^k}{k!}, \tag{4.130}$$

which converges to $\exp(2\,y)$. We conclude that $f(y) \to \exp(2\,y)$ as $y \to \infty$. It follows from Equation (4.124) that $R(r) \to \exp(r/a)/r$ as $r \to \infty$. This does not correspond to physically acceptable behavior of the wavefunction, because

$\int d^3\mathbf{x}\,|\psi|^2$ must be finite. The only way in which we can avoid this unphysical behavior is if the series (4.126) terminates at some maximum value of k. According to the recursion relation (4.128), this is only possible if

$$\frac{m_e\,e^2\,a}{4\pi\,\epsilon_0\,\hbar^2} = n, \tag{4.131}$$

where the last term in the series is $c_n\,y^n$. It follows from Equation (4.123) that the energy eigenvalues are quantized, and can only take the values

$$E = \frac{E_0}{n^2}, \tag{4.132}$$

where

$$E_0 = -\frac{m_e\,e^4}{32\pi^2\,\epsilon_0^2\,\hbar^2} = -13.6\,\text{eV} \tag{4.133}$$

is the energy of the *ground state* (i.e., the lowest energy state). Here, n is a positive integer that must exceed the quantum number l, otherwise there would be no terms in the series (4.126). Expression (4.132) is known as the *Bohr formula*, because it was derived heuristically by Niels Bohr in 1913 [Bohr (1913)].

In summary, the properly normalized wavefunction of a hydrogen atom is written

$$\psi(r,\theta,\varphi) = R_{n\,l}(r)\,Y_{lm}(\theta,\varphi), \tag{4.134}$$

where

$$R_{n\,l}(r) = \frac{u_{n\,l}(r/a)}{r/a}, \tag{4.135}$$

and

$$a = n\,a_0. \tag{4.136}$$

Here,

$$a_0 = \frac{4\pi\,\epsilon_0\,\hbar^2}{m_e\,e^2} = 5.3 \times 10^{-11}\ \text{meters} \tag{4.137}$$

is the *Bohr radius*, and $u_{n\,l}(x)$ is a well-behaved solution of the differential equation

$$\left[\frac{d^2}{dx^2} - \frac{l(l+1)}{x^2} + \frac{2n}{x} - 1\right]u_{n\,l} = 0 \tag{4.138}$$

that satisfies the normalization constraint

$$a^3\int_0^\infty dx\,[u_{n\,l}(x)]^2 = \int_0^\infty dr\,r^2\,[R_{nl}(r)]^2 = 1. \tag{4.139}$$

Finally, the $Y_{lm}(\theta, \varphi)$ are spherical harmonics. The restrictions on the quantum numbers are $|m| \leq l < n$, where n is a positive integer, l a non-negative integer, and m an integer. Incidentally, the quantum numbers n, l, and m are conventionally referred to as the *principle quantum number*, the *azimuthal quantum number*, and the *magnetic quantum number*, respectively.

The ground state of hydrogen corresponds to $n = 1$. The only permissible values of the other quantum numbers are $l = 0$ and $m = 0$. Thus, the ground state is a spherically symmetric, zero angular momentum, state. The next energy level corresponds to $n = 2$. The other quantum numbers are allowed to take the values $l = 0$, $m = 0$ or $l = 1$, $m = -1, 0, 1$. Thus, there are $n = 2$ states with non-zero angular momentum. Note that the energy levels given in Equation (4.132) are independent of the quantum number l, despite the fact that l appears in the radial eigenfunction equation (4.138). This is a special property of a $1/r$ Coulomb potential.

In addition to the quantized negative energy states of the hydrogen atom, which we have just found, there is also a continuum of unbound positive energy states.

4.7   Exercises

4.1 Demonstrate directly from the fundamental commutation relations for angular momentum, (4.11), that (a) $[L^2, L_z] = 0$, (b) $[L^\pm, L_z] = \mp \hbar L^\pm$, and (c) $[L^+, L^-] = 2\hbar L_z$.

4.2 Demonstrate from Equations (4.74)–(4.79) that

$$L_x = i\hbar \left(\sin\varphi \, \frac{\partial}{\partial\theta} + \cot\theta \cos\varphi \, \frac{\partial}{\partial\varphi} \right),$$

$$L_y = -i\hbar \left(\cos\varphi \, \frac{\partial}{\partial\theta} - \cot\theta \sin\varphi \, \frac{\partial}{\partial\varphi} \right),$$

$$L_z = -i\hbar \, \frac{\partial}{\partial\varphi},$$

where θ, φ are conventional spherical angles. In addition, show that

$$L^\pm = \pm\hbar \, \exp(\pm i\varphi) \left(\frac{\partial}{\partial\theta} \pm i \cot\theta \, \frac{\partial}{\partial\varphi} \right),$$

$$L^2 = -\hbar^2 \left(\frac{1}{\sin\theta} \frac{\partial}{\partial\theta} \, \sin\theta \, \frac{\partial}{\partial\theta} + \frac{1}{\sin^2\theta} \frac{\partial^2}{\partial\varphi^2} \right).$$

4.3 A system is in the state $\psi(\theta, \varphi) = Y_{lm}(\theta, \varphi)$. Evaluate $\langle L_x \rangle$, $\langle L_y \rangle$, $\langle L_x^2 \rangle$, and $\langle L_y^2 \rangle$.

4.4 Derive Equations (4.108) and (4.109) from Equation (4.107).

4.5 Find the eigenvalues and eigenfunctions (in terms of the angles θ and φ) of L_x. Express the $l = 1$ eigenfunctions in terms of the spherical harmonics.

4.6 Consider a beam of particles with $l = 1$. A measurement of L_x yields the result $\hbar$. What values will be obtained by a subsequent measurement of L_z, and with what probabilities? Repeat the calculation for the cases in which the measurement of L_x yields the results 0 and $-\hbar$.

4.7 The Hamiltonian for an axially symmetric rotator is given by

$$H = \frac{L_x^2 + L_y^2}{2 I_\perp} + \frac{L_z^2}{2 I_\parallel},$$

where $I_\parallel$ and $I_\perp$ are the moments of inertia about the z-axis (which corresponds to the symmetry axis), and about an axis lying in the x-y plane, respectively. What are the eigenvalues of H? [Gasiorowicz (1996)]

4.8 The expectation value of $f(\mathbf{x}, \mathbf{p})$ in any stationary state is a constant. Calculate

$$0 = \frac{d}{dt} (\langle \mathbf{x} \cdot \mathbf{p} \rangle) = \frac{i}{\hbar} \langle [H, \mathbf{x} \cdot \mathbf{p}] \rangle$$

for a Hamiltonian of the form

$$H = \frac{p^2}{2m} + V(r).$$

Hence, show that

$$\left\langle \frac{p^2}{2m} \right\rangle = \frac{1}{2} \left\langle r \frac{dV}{dr} \right\rangle$$

in a stationary state. This is another form of the *Virial theorem*. (See Exercise 3.9.) [Gasiorowicz (1996)]

4.9 Use the Virial theorem of the previous exercise to prove that

$$\left\langle \frac{1}{r} \right\rangle = \frac{1}{n^2 a_0}$$

for an energy eigenstate of the hydrogen atom whose principal quantum number is n.

4.10 Suppose that a particle's Hamiltonian is

$$H = \frac{p^2}{2m} + V(\mathbf{x}).$$

Show that $[\mathbf{L}, p^2] = 0$ and $[\mathbf{L}, V(\mathbf{x})] = -i\hbar \mathbf{x} \times \nabla V$. [Hint: Use the Schrödinger representation.] Hence, deduce that

$$\frac{d\langle \mathbf{L} \rangle}{dt} = -\langle \mathbf{x} \times \nabla V \rangle.$$

[Hint: Use the Heisenberg picture.] Demonstrate that if $V = V(r)$, where $r = |\mathbf{x}|$, then

$$\frac{d\langle \mathbf{L} \rangle}{dt} = \mathbf{0}.$$

4.11 Let

$$I_n = \int_0^\infty y^n \, e^{-y},$$

where n is a non-negative integer. Show that

$$I_n = n!.$$

4.12 Demonstrate that the first few properly normalized radial wavefunctions of the hydrogen atom take the form:

(a)

$$R_{10}(r) = \frac{2}{a_0^{3/2}} \exp\left(-\frac{r}{a_0}\right).$$

(b)

$$R_{20}(r) = \frac{2}{(2\,a_0)^{3/2}} \left(1 - \frac{r}{2\,a_0}\right) \exp\left(-\frac{r}{2\,a_0}\right).$$

(c)

$$R_{21}(r) = \frac{1}{\sqrt{3}\,(2\,a_0)^{3/2}} \frac{r}{a_0} \exp\left(-\frac{r}{2\,a_0}\right).$$

(d)

$$R_{30}(r) = \frac{2}{(3\,a_0)^{3/2}} \left(1 - \frac{2\,r}{3\,a_0} + \frac{2\,r^2}{27\,a_0^2}\right) \exp\left(-\frac{r}{3\,a_0}\right).$$

(e)

$$R_{31}(r) = \frac{4\sqrt{2}}{9\,(3\,a_0)^{3/2}} \frac{r}{a_0} \left(1 - \frac{r}{6\,a_0}\right) \exp\left(-\frac{r}{3\,a_0}\right).$$

(f)

$$R_{32}(r) = \frac{2\sqrt{2}}{27\sqrt{5}\,(3\,a_0)^{3/2}} \left(\frac{r}{a_0}\right)^2 \exp\left(-\frac{r}{3\,a_0}\right).$$

4.13 Demonstrate that

$$\langle r^k \rangle = \frac{(2+k)!}{2^{1+k}} a_0^k$$

for the hydrogen ground state. In addition, show that

$$\langle x^2 \rangle = \langle y^2 \rangle = \langle z^2 \rangle = a_0^2.$$

4.14 Show that the most probable value of r in the hydrogen ground state is a_0.

4.15 Demonstrate that

$$\langle 2,0,0| z |2,1,0\rangle = -3\,a_0,$$

where $|n, l, m\rangle$ denotes a properly normalized energy eigenket of the hydrogen atom corresponding to the standard quantum numbers n, l, and m.

4.16 Let $\langle r^k\rangle$ denote the expectation value of r^k for an energy eigenstate of the hydrogen atom characterized by the standard quantum numbers n, l, and m.

(a) Demonstrate that

$$\langle r^k\rangle = (n\,a_0)^{k+3}\,J_k,$$

where

$$J_k = \int_0^\infty dx\, x^k\, [u(x)]^2,$$

and $u(x)$ is a well-behaved solution of the differential equation

$$u'' = \left[\frac{l(l+1)}{x^2} - \frac{2n}{x} + 1\right]u.$$

(b) Integrating by parts, show that

$$\int_0^\infty dx\, x^k\, u\, u' = -\frac{k}{2}\,J_{k-1},$$

and

$$\int_0^\infty dx\, x^k\, (u')^2 = -\frac{2}{k+1}\int_0^\infty dx\, x^{k+1}\, u'\, u'',$$

as well as

$$\int_0^\infty dx\, x^k\, u\, u'' = \frac{k(k-1)}{2}\,J_{k-2} + \frac{2}{k+1}\int_0^\infty dx\, x^{k+1}\, u'\, u''.$$

(c) Demonstrate from the governing differential equation for $u(x)$ that

$$\int_0^\infty dx\, x^k\, u\, u'' = l(l+1)\,J_{k-2} - 2n\,J_{k-1} + J_k.$$

(d) Combine the final result of part (b) with the governing differential equation to prove that

$$\int_0^\infty dx\, x^k\, u\, u'' = -\frac{l(l+1)(k-1)}{k+1}\,J_{k-2} + \frac{2nk}{k+1}\,J_{k-1} - J_k + \frac{k(k-1)}{2}\,J_{k-2}.$$

(e) Combine the results of parts (c) and (d) to show that

$$\frac{k}{4}\left[(2l+1)^2 - k^2\right]J_{k-2} - n(2k+1)\,J_{k-1} + (k+1)\,J_k = 0.$$

Hence, derive *Kramers' relation*:

$$\frac{k\,a_0^2}{4}\left[(2l+1)^2 - k^2\right]\langle r^{k-2}\rangle - a_0(2k+1)\langle r^{k-1}\rangle + \frac{(k+1)}{n^2}\langle r^k\rangle = 0.$$

(f) Use Kramers' relation to prove that

$$\left\langle \frac{1}{r} \right\rangle = \frac{1}{n^2 a_0},$$

$$\langle r \rangle = \frac{a_0}{2} \left[3 n^2 - l(l+1) \right],$$

$$\langle r^2 \rangle = \frac{a_0^2 n^2}{2} \left[5 n^2 + 1 - 3 l(l+1) \right].$$

4.17 Let $R_{nl}(r) = v_{nl}(r/a_0)/(r/a_0)$, where $R_{nl}(r)$ is a properly normalized radial hydrogen wavefunction corresponding to the conventional quantum numbers n and l, and a_0 is the Bohr radius.

(a) Demonstrate that

$$\frac{d^2 v_{nl}}{dy^2} = \left[\frac{l(l+1)}{y^2} - \frac{2}{y} + \frac{1}{n^2} \right] v_{nl}.$$

(b) Show that $v_{nl} \sim y^{1+l}$ in the limit $y \to 0$.

(c) Demonstrate that

$$\left(\frac{1}{n^2} - \frac{1}{m^2} \right) \int_0^\infty dy \, v_{nl}(y) \, v_{ml}(y) = 0.$$

(d) Hence, deduce that

$$\int_0^\infty dr \, r^2 \, R_{nl}(r) \, R_{ml}(r) = 0$$

for $n \neq m$.

Chapter 5

Spin Angular Momentum

5.1 Introduction

Up to now, we have tacitly assumed that the state of a particle in quantum mechanics can be completely specified by giving its wavefunction, ψ, as a function of the spatial coordinates, x, y, and z. Unfortunately, there is a wealth of experimental evidence that suggests that this simplistic approach is inadequate.

Consider an isolated system whose center of mass is at rest. Let the eigenvalue of the system's total angular momentum be $j(j+1)\hbar^2$. According to the theory of orbital angular momentum outlined in the previous chapter, there are two possibilities. For a system consisting of a single particle, $j = 0$. For a system consisting of two (or more) particles, j is a non-negative integer. However, these predictions do not agree with experimental observations, because we often encounter single-particle systems that have $j \neq 0$. Even worse, multi-particle systems in which j has half-integer values abound in nature. In order to explain this apparent discrepancy between theory and experiments, Goudsmit and Uhlenbeck (in 1925) introduced the concept of an internal, purely quantum mechanical, angular momentum called *spin* [Goudsmit and Uhlenbeck (1925)]. For a single-particle system with spin, the total angular momentum in the rest frame is non-vanishing.

5.2 Properties of Spin Angular Momentum

Let us denote the three components of the spin angular momentum of a particle by the Hermitian operators $(S_x, S_y, S_z) \equiv \mathbf{S}$. We assume that these operators obey the fundamental commutation relations (4.8)–(4.10) for the components of an angular momentum. Thus, we can write

$$\mathbf{S} \times \mathbf{S} = i\hbar \mathbf{S}. \tag{5.1}$$

We can also define the operator

$$S^2 = S_x^2 + S_y^2 + S_z^2. \tag{5.2}$$

According to the quite general analysis of Section 4.1,

$$[\mathbf{S}, S^2] = 0. \tag{5.3}$$

Thus, it is possible to find simultaneous eigenstates of S^2 and S_z. These are denoted $|s, s_z\rangle$, where

$$S_z |s, s_z\rangle = s_z \hbar |s, s_z\rangle, \tag{5.4}$$

$$S^2 |s, s_z\rangle = s(s+1) \hbar^2 |s, s_z\rangle. \tag{5.5}$$

According to the equally general analysis of Section 4.2, the (dimensionless) quantum number s can, in principle, take (non-negative) integer or half-integer values, and the (dimensionless) quantum number s_z can only take the values $s, s - 1 \cdots - s + 1, -s$.

Spin angular momentum clearly has many properties in common with orbital angular momentum. However, there is one vitally important difference. Spin angular momentum operators cannot be expressed in terms of position and momentum operators, like in Equations (4.1)–(4.3), because this identification depends on an analogy with classical mechanics, and the concept of spin is purely quantum mechanical. In other words, spin has no analogy in classical physics. Consequently, the restriction that the quantum number of the overall angular momentum must take integer values does not apply to spin angular momentum, because this restriction (found in Sections 4.3 and 4.4) ultimately depends on Equations (4.1)–(4.3). In other words, the spin quantum number s is allowed to take half-integer values.

Consider a spin one-half particle (e.g., an electron, a proton, or a neutron), for which

$$S_z |\pm\rangle = \pm \frac{\hbar}{2} |\pm\rangle, \tag{5.6}$$

$$S^2 |\pm\rangle = \frac{3 \hbar^2}{4} |\pm\rangle. \tag{5.7}$$

Here, the $|\pm\rangle$ denote eigenkets of the S_z operator corresponding to the eigenvalues $\pm\hbar/2$. These kets are mutually orthogonal (because S_z is an Hermitian operator), so

$$\langle + | - \rangle = 0. \tag{5.8}$$

The kets are also properly normalized and complete, so that

$$\langle + | + \rangle = \langle - | - \rangle = 1, \tag{5.9}$$

and

$$|+\rangle\langle+| + |-\rangle\langle-| = 1. \tag{5.10}$$

It is easily verified that the Hermitian operators defined by

$$S_x = \frac{\hbar}{2}(|+\rangle\langle-| + |-\rangle\langle+|), \tag{5.11}$$

$$S_y = \frac{i\hbar}{2}(-|+\rangle\langle-| + |-\rangle\langle+|), \tag{5.12}$$

$$S_z = \frac{\hbar}{2}(|+\rangle\langle+| - |-\rangle\langle-|), \tag{5.13}$$

satisfy the commutation relations (4.8)–(4.10) (with the L_j replaced by the S_j). (See Exercise 5.1.) The operator S^2 takes the form

$$S^2 = \frac{3\hbar^2}{4}. \tag{5.14}$$

It is also easily demonstrated that S^2 and S_z, defined in this manner, satisfy the eigenvalue relations (5.6)–(5.7). Equations (5.11)–(5.14) constitute a realization of the spin operators $\mathbf{S}$ and S^2 (for a spin one-half particle) in *spin space* (i.e., the Hilbert sub-space consisting of kets that correspond to the different spin states of the particle).

5.3 Wavefunction of Spin One-Half Particle

The state of a spin one-half particle is represented as a vector in ket space. Let us suppose that this space is spanned by the basis kets $|x', y', z', \pm\rangle$. Here, $|x', y', z', \pm\rangle$ denotes a simultaneous eigenstate of the position operators x, y, z, and the spin operator S_z, corresponding to the eigenvalues x', y', z', and $\pm\hbar/2$, respectively. The basis kets are assumed to satisfy the completeness relation

$$\iiint dx'\,dy'\,dz'\,(|x', y', z', +\rangle\langle x', y', z', +| + |x', y', z', -\rangle\langle x', y', z', -|) = 1. \tag{5.15}$$

It is helpful to think of the ket $|x', y', z', +\rangle$ as the product of two kets—a position-space ket $|x', y', z'\rangle$, and a spin-space ket $|+\rangle$. We assume that such a product obeys the commutative and distributive axioms of multiplication:

$$|x', y', z'\rangle|+\rangle = |+\rangle|x', y', z'\rangle, \tag{5.16}$$

$$(c'\,|x', y', z'\rangle + c''\,|x'', y'', z''\rangle)\,|+\rangle = c'\,|x', y', z'\rangle|+\rangle + c''\,|x'', y'', z''\rangle|+\rangle, \tag{5.17}$$

$$|x', y', z'\rangle(c_+\,|+\rangle + c_-\,|-\rangle) = c_+\,|x', y', z'\rangle|+\rangle + c_-\,|x', y', z'\rangle|-\rangle, \tag{5.18}$$

'where the c's are numbers. We can give meaning to any position space operator (such as L_z) acting on the product $|x', y', z'\rangle|+\rangle$ by assuming that it operates only on the $|x', y', z'\rangle$ factor, and commutes with the $|+\rangle$ factor. Similarly, we can give a meaning to any spin operator (such as S_z) acting on $|x', y', z'\rangle|+\rangle$ by assuming that it operates only on $|+\rangle$, and commutes with $|x', y', z'\rangle$. This implies that every position space operator commutes with every spin operator. In this manner, we can give a meaning to the equation

$$|x', y', z', \pm\rangle = |x', y', z'\rangle|\pm\rangle = |\pm\rangle|x', y', z'\rangle. \tag{5.19}$$

The multiplication in the previous equation is of a quite different type to any that we have encountered previously. The ket vectors $|x', y', z'\rangle$ and $|\pm\rangle$ lie in two completely separate vector spaces, and their product $|x', y', z'\rangle|\pm\rangle$ lies in a third vector space. In mathematics, the latter space is termed the *product space* of the former spaces, which are termed *factor spaces*. The dimensionality a product space is equal to the product of the dimensionalities of each of the factor spaces. Actually, a general ket in the product space is not of the form (5.19), but is instead a sum, or integral, of kets of this form.

A general state, A, of a spin one-half particle is represented as a ket, $\|A\rangle\rangle$, in the product of the spin and position spaces. This state can be completely specified by two wavefunctions:

$$\psi_+(x', y', z') = \langle x', y', z'|\langle+\|A\rangle\rangle, \tag{5.20}$$

$$\psi_-(x', y', z') = \langle x', y', z'|\langle-\|A\rangle\rangle. \tag{5.21}$$

The probability of observing the particle in the region x' to $x' + dx'$, y' to $y' + dy'$, and z' to $z' + dz'$, with $s_z = +1/2$, is $|\psi_+(x', y', z')|^2\, dx'dy'dz'$. Likewise, the probability of observing the particle in the region x' to $x' + dx'$, y' to $y' + dy'$, and z' to $z' + dz'$, with $s_z = -1/2$, is $|\psi_-(x', y', z')|^2\, dx'dy'dz'$. The normalization condition for the wavefunctions is

$$\iiint dx'dy'dz' \left(|\psi_+|^2 + |\psi_-|^2 \right) = 1. \tag{5.22}$$

5.4 Rotation Operators in Spin Space

Let us, for the moment, forget about the spatial position of our spin one-half particle, and concentrate on its spin state. A general spin state A is represented by the ket

$$|A\rangle = \langle+|A\rangle|+\rangle + \langle-|A\rangle|-\rangle \tag{5.23}$$

in spin space. In Section 4.3, we were able to construct an operator $R_z(\Delta\varphi)$ that rotates the system through an angle $\Delta\varphi$ about the z-axis in position space. Can we also construct an operator $T_z(\Delta\varphi)$ that rotates the system through an angle $\Delta\varphi$ about the z-axis in spin space? By analogy with Equation (4.62), we would expect such an operator to take the form

$$T_z(\Delta\varphi) = \exp\left(\frac{-\mathrm{i}\,S_z\,\Delta\varphi}{\hbar}\right). \tag{5.24}$$

Thus, after rotation, the ket $|A\rangle$ becomes

$$|A_R\rangle = T_z(\Delta\varphi)\,|A\rangle. \tag{5.25}$$

To demonstrate that the operator (5.24) really does rotate the spin of the system, let us consider its effect on $\langle S_x\rangle$. Under rotation, this expectation value changes as follows:

$$\langle S_x\rangle \to \langle A_R|\,S_x\,|A_R\rangle = \langle A|\,T_z^\dagger\,S_x\,T_z\,|A\rangle. \tag{5.26}$$

Thus, we need to compute

$$\exp\left(\frac{\mathrm{i}\,S_z\,\Delta\varphi}{\hbar}\right)\,S_x\,\exp\left(\frac{-\mathrm{i}\,S_z\,\Delta\varphi}{\hbar}\right). \tag{5.27}$$

This goal can be achieved in two different ways.

First, we can use the explicit formula for S_x given in Equation (5.11). We find that expression (5.27) becomes

$$\frac{\hbar}{2}\,\exp\left(\frac{\mathrm{i}\,S_z\,\Delta\varphi}{\hbar}\right)\,(|+\rangle\langle-| + |-\rangle\langle+|)\,\exp\left(\frac{-\mathrm{i}\,S_z\,\Delta\varphi}{\hbar}\right), \tag{5.28}$$

or

$$\frac{\hbar}{2}\left(\mathrm{e}^{\mathrm{i}\Delta\varphi/2}\,|+\rangle\langle-|\,\mathrm{e}^{\mathrm{i}\Delta\varphi/2} + \mathrm{e}^{-\mathrm{i}\Delta\varphi/2}\,|-\rangle\langle+|\,\mathrm{e}^{-\mathrm{i}\Delta\varphi/2}\right), \tag{5.29}$$

which yields

$$\exp\left(\frac{\mathrm{i}\,S_z\,\Delta\varphi}{\hbar}\right)\,S_x\,\exp\left(\frac{-\mathrm{i}\,S_z\,\Delta\varphi}{\hbar}\right) = S_x\,\cos\Delta\varphi - S_y\,\sin\Delta\varphi, \tag{5.30}$$

where use has been made of Equations (5.11)–(5.13).

A second approach is to use the so-called *Baker-Campbell-Hausdorff lemma* [Baker (1901); Campbell (1896); Hausdorff (1906)]. This takes the form

$$\exp(\mathrm{i}\,G\,\lambda)\,A\,\exp(-\mathrm{i}\,G\,\lambda) \equiv A + \mathrm{i}\,\lambda\,[G,A] + \left(\frac{\mathrm{i}^2\lambda^2}{2!}\right)[G,[G,A]]$$

$$+ \left(\frac{\mathrm{i}^3\lambda^3}{3!}\right)[G,[G,[G,A]]] + \cdots, \tag{5.31}$$

where G and A are operators, and λ a real parameter. The proof of this lemma is left as an exercise. (See Exercise 5.2.) Applying the Baker-Campbell-Hausdorff lemma to expression (5.27), we obtain

$$S_x + \left(\frac{i\Delta\varphi}{\hbar}\right)[S_z, S_x] + \left(\frac{1}{2!}\right)\left(\frac{i\Delta\varphi}{\hbar}\right)^2 [S_z, [S_z, S_x]]$$

$$+ \left(\frac{1}{3!}\right)\left(\frac{i\Delta\varphi}{\hbar}\right)^3 [S_z, [S_z, [S_z, S_x]]] + \cdots \tag{5.32}$$

which reduces to

$$S_x\left[1 - \frac{(\Delta\varphi)^2}{2!} + \cdots\right] - S_y\left[\Delta\varphi - \frac{(\Delta\varphi)^3}{3!} + \cdots\right], \tag{5.33}$$

where use has been made of Equation (5.1). Thus,

$$\exp\left(\frac{i S_z \Delta\varphi}{\hbar}\right) S_x \exp\left(\frac{-i S_z \Delta\varphi}{\hbar}\right) = S_x \cos\Delta\varphi - S_y \sin\Delta\varphi. \tag{5.34}$$

The second proof is more general than the first, because it only makes use of the fundamental commutation relation (5.1), and is, therefore, valid for systems with spin angular momentum greater than one-half.

For a spin one-half system, both methods imply that

$$\langle S_x \rangle \to \langle S_x \rangle \cos\Delta\varphi - \langle S_y \rangle \sin\Delta\varphi \tag{5.35}$$

under the action of the rotation operator (5.24). It is straightforward to show that

$$\langle S_y \rangle \to \langle S_x \rangle \sin\Delta\varphi + \langle S_y \rangle \cos\Delta\varphi. \tag{5.36}$$

Furthermore,

$$\langle S_z \rangle \to \langle S_z \rangle, \tag{5.37}$$

because S_z commutes with the rotation operator. Equations (5.35)–(5.37) demonstrate that the operator (5.24) rotates the expectation value of $\mathbf{S}$ through an angle $\Delta\varphi$ about the z-axis. In fact, the expectation value of the spin operator behaves like a classical vector under rotation. In other words,

$$\langle S_k \rangle \to \sum_l R_{kl} \langle S_l \rangle, \tag{5.38}$$

where the R_{kl} are the elements of the conventional rotation matrix for the rotation in question [Riley *et al.* (2013)]. (Here, and in the following, k, l, et cetera, are indices that run from 1 to 3, with 1 corresponding to the x-axis, 2 to the y-axis, and so on.) It is clear, from our second derivation of the result (5.35), that this property is not restricted to the spin operators of a spin one-half system. In fact, we have effectively demonstrated that

$$\langle J_k \rangle \to \sum_l R_{kl} \langle J_l \rangle, \tag{5.39}$$

where the J_k are the generators of rotation, satisfying the fundamental commutation relation $\mathbf{J} \times \mathbf{J} = i\hbar\,\mathbf{J}$, and the rotation operator about the kth axis is written $R_k(\Delta\varphi) = \exp(-i\,J_k\,\Delta\varphi/\hbar)$.

Consider the effect of the rotation operator (5.24) on the state ket (5.23). It is easily seen that

$$T_z(\Delta\varphi)\,|A\rangle = e^{-i\,\Delta\varphi/2}\,\langle+|A\rangle|+\rangle + e^{i\,\Delta\varphi/2}\,\langle-|A\rangle|-\rangle. \tag{5.40}$$

Consider a rotation by 2π radians. We find that

$$|A\rangle \to T_z(2\pi)\,|A\rangle = -|A\rangle. \tag{5.41}$$

Note that a ket rotated by 2π radians differs from the original ket by a minus sign. In fact, a rotation by 4π radians is needed to transform a ket into itself. The minus sign does not affect the expectation value of $\mathbf{S}$, because $\mathbf{S}$ is sandwiched between $\langle A|$ and $|A\rangle$, both of which change sign. Nevertheless, the minus sign does give rise to observable consequences, as we shall see presently.

5.5 Magnetic Moments

Consider a particle of electric charge q and speed v performing a circular orbit of radius r in the x-y plane. The charge is equivalent to a current loop of radius r, lying in the x-y plane, and carrying a current $I = q\,v/(2\pi\,r)$. The *magnetic moment* μ of the loop is of magnitude $\pi\,r^2\,I$ and is directed along the z-axis (the direction is given by a right-hand rule with the fingers of the right-hand circulating in the same direction as the current) [Fitzpatrick (2008)]. Thus, we can write

$$\mu = \frac{q}{2}\,\mathbf{x} \times \mathbf{v}, \tag{5.42}$$

where $\mathbf{x}$ and $\mathbf{v}$ are the vector position and velocity of the particle, respectively. However, we know that $\mathbf{p} = \mathbf{v}/m$, where $\mathbf{p}$ is the particle's vector momentum, and m its mass. We also know that $\mathbf{L} = \mathbf{x} \times \mathbf{p}$, where $\mathbf{L}$ is the orbital angular momentum. It follows that

$$\mu = \frac{q}{2m}\,\mathbf{L}. \tag{5.43}$$

Using the standard analogy between classical and quantum mechanics, we expect the previous relation to also hold between the quantum mechanical operators, μ

and $\mathbf{L}$, which represent magnetic moment and orbital angular momentum, respectively. This is indeed found to the the case experimentally.

Spin angular momentum also gives rise to a contribution to the magnetic moment of a charged particle. In fact, relativistic quantum mechanics predicts that a charged particle possessing spin must also possess a corresponding magnetic moment [Dirac (1928, 1958); Bjorken and Drell (1964)]. We can write

$$\boldsymbol{\mu} = \frac{q}{2\,m}\,[\mathbf{L} + \mathrm{sgn}(q)\,g\,\mathbf{S}],\qquad(5.44)$$

where the parameter g is called the *g-factor*. For an electron this factor is found to be

$$g_e = -2\left(1 + \frac{\alpha}{2\pi}\right),\qquad(5.45)$$

where

$$\alpha = \frac{e^2}{4\pi\,\epsilon_0\,\hbar\,c} \simeq \frac{1}{137}\qquad(5.46)$$

is the *fine structure constant*. The factor -2 is correctly predicted by Dirac's famous relativistic theory of the electron [Dirac (1928, 1958)]. (See Chapter 11.) The small correction $1/(2\pi\,137)$ is due to quantum field effects [Schwinger (1948)]. We shall ignore this correction in the following, so

$$\boldsymbol{\mu} \simeq -\frac{e}{2\,m_e}\,(\mathbf{L} + 2\,\mathbf{S})\qquad(5.47)$$

for an electron. (Here, $e > 0$ is the magnitude of the electron charge, and m_e the electron mass.)

5.6 Spin Precession

The Hamiltonian for an electron at rest in a z-directed magnetic field, $\mathbf{B} = B\,\mathbf{e}_z$, is [Fitzpatrick (2008)]

$$H = -\boldsymbol{\mu} \cdot \mathbf{B} = \left(\frac{e}{m_e}\right)\mathbf{S} \cdot \mathbf{B} = \omega\,S_z,\qquad(5.48)$$

where

$$\omega = \frac{e\,B}{m_e},\qquad(5.49)$$

and use has been made of Equation (5.47). According to Equation (3.28), the time evolution operator for this system is

$$T(t,0) = \exp\left(\frac{-\mathrm{i}\,H\,t}{\hbar}\right) = \exp\left(\frac{-\mathrm{i}\,S_z\,\omega\,t}{\hbar}\right).\qquad(5.50)$$

It can be seen, by comparison with Equation (5.24), that the time evolution operator is precisely the same as the rotation (about the z-axis) operator for spin, with $\Delta\varphi$ set equal to ωt. It is immediately clear that the Hamiltonian (5.48) causes the electron spin to precess about the z-axis with angular frequency ω. In fact, Equations (5.35)–(5.37) imply that

$$\langle S_x \rangle_t = \langle S_x \rangle_{t=0} \cos(\omega t) - \langle S_y \rangle_{t=0} \sin(\omega t), \qquad (5.51)$$

$$\langle S_y \rangle_t = \langle S_x \rangle_{t=0} \sin(\omega t) + \langle S_y \rangle_{t=0} \cos(\omega t), \qquad (5.52)$$

$$\langle S_z \rangle_t = \langle S_z \rangle_{t=0}. \qquad (5.53)$$

Note that the expectation value of the electron spin precesses in a right-handed fashion. In other words, if the thumb of the right hand is directed along the magnetic field then the fingers of the right-hand indicate the direction of the precession. A particle with a positive g-factor would precess in a left-handed fashion. The time evolution of the state ket is given by analogy with Equation (5.40):

$$|A, t\rangle = e^{-i\omega t/2} \langle +|A, 0\rangle|+\rangle + e^{i\omega t/2} \langle -|A, 0\rangle|-\rangle. \qquad (5.54)$$

Note that it takes a time $t = 4\pi/\omega$ for the state ket to return to its original state. By contrast, it only takes a time $t = 2\pi/\omega$ for the spin vector to point in its original direction.

We shall now describe an experiment to detect the minus sign in Equation (5.41). An almost mono-energetic beam of neutrons is split in two, sent along two different paths, I and II, and then recombined. Path I passes through a magnetic-field-free region. However, path II enters a small region where a static magnetic field is present. As a result, a neutron state ket going along path II acquires a phase-shift $\exp(\mp i\,\omega\,T/2)$ (the $\mp$ signs correspond to $s_z = \pm 1/2$ states). Here, T is the time spent in the magnetic field, and

$$\omega = -\frac{g_n\,e\,B}{2\,m_p} \qquad (5.55)$$

is the spin precession frequency. Moreover, B is the magnetic field-strength, m_p the proton mass, and g_n the neutron g-factor. This factor is found experimentally to take the value $g_n = -3.83$ [Greene *et al.* (1979)]. (The magnetic moment of a neutron is entirely a quantum field effect. Incidentally, this particular experiment must be performed with electrically neutral particles because, otherwise, the particles that pass through the magnetic field would be strongly perturbed by the Lorentz force.) When neutrons from path I and path II meet they undergo interference. We expect the observed neutron intensity in the interference region to exhibit a $\cos(\mp\omega\,T/2 + \delta)$ variation, where δ is the phase difference between paths I and II in the absence of a magnetic field. In experiments, the time of flight T

through the magnetic field region is kept constant, while the field-strength B is varied. It follows that the change in magnetic field required to produce successive maxima is

$$\Delta B = \frac{8\pi\,\hbar}{e\,|g_n|\,\lambdabar\,l},\tag{5.56}$$

where l is the path-length through the magnetic field region, and $\lambdabar = \hbar/p$ is the reduced (by a factor 2π) *de Broglie wavelength* of the neutrons. The previous formula has been verified experimentally to within a fraction of a percent [Rauch *et al.* (1975); Werner *et al.* (1975)]. Note that the formula in question depends crucially on the fact that it takes a 4π rotation to return a state ket to its original state. If it only took a 2π rotation then ΔB would take half of the value given in Equation (5.56), which does not agree with the experimental data.

5.7   Pauli Two-Component Formalism

We have seen, in Section 4.4, that the eigenstates of orbital angular momentum can be conveniently represented as spherical harmonics. In this representation, the orbital angular momentum operators take the form of differential operators involving only angular coordinates. On the other hand, it is conventional to represent the eigenstates of spin angular momentum as column (or row) matrices. In this representation, the spin angular momentum operators take the form of matrices.

The matrix representation of a spin one-half system was introduced by Pauli in 1927 [Pauli (1927)]. Recall, from Section 5.4, that a general spin ket can be expressed as a linear combination of the two eigenkets of S_z belonging to the eigenvalues $\pm\hbar/2$. These are denoted $|\pm\rangle$. Let us represent these basis eigenkets as column vectors:

$$|+\rangle \rightarrow \begin{pmatrix} 1 \\ 0 \end{pmatrix} \equiv \chi_+,\tag{5.57}$$

$$|-\rangle \rightarrow \begin{pmatrix} 0 \\ 1 \end{pmatrix} \equiv \chi_-.\tag{5.58}$$

The corresponding eigenbras are represented as row vectors:

$$\langle+| \rightarrow (1,0) \equiv \chi_+^\dagger,\tag{5.59}$$

$$\langle-| \rightarrow (0,1) \equiv \chi_-^\dagger.\tag{5.60}$$

In this scheme, a general ket takes the form

$$|A\rangle = \langle+|A\rangle|+\rangle + \langle-|A\rangle|-\rangle \rightarrow \begin{pmatrix} \langle+|A\rangle \\ \langle-|A\rangle \end{pmatrix},\tag{5.61}$$

and a general bra becomes

$$\langle A| = \langle A|+\rangle\langle+| + \langle A|-\rangle\langle-| \rightarrow (\langle A|+\rangle, \langle A|-\rangle). \tag{5.62}$$

The column vector (5.61) is called a two-component *spinor*, and can be written

$$\chi \equiv \begin{pmatrix} \langle+|A\rangle \\ \langle-|A\rangle \end{pmatrix} = \begin{pmatrix} c_+ \\ c_- \end{pmatrix} = c_+ \chi_+ + c_- \chi_-, \tag{5.63}$$

where the $c_\pm$ are complex numbers. The row vector (5.62) becomes

$$\chi^\dagger \equiv (\langle A|+\rangle, \langle A|-\rangle) = (c_+^*, c_-^*) = c_+^* \chi_+^\dagger + c_-^* \chi_-^\dagger. \tag{5.64}$$

Consider the ket obtained by the action of a spin operator on ket A:

$$|A'\rangle = S_k |A\rangle. \tag{5.65}$$

This ket is represented as

$$|A'\rangle \rightarrow \begin{pmatrix} \langle+|A'\rangle \\ \langle-|A'\rangle \end{pmatrix} \equiv \chi'. \tag{5.66}$$

However,

$$\langle+|A'\rangle = \langle+|S_k|+\rangle\langle+|A\rangle + \langle+|S_k|-\rangle\langle-|A\rangle, \tag{5.67}$$

$$\langle-|A'\rangle = \langle-|S_k|+\rangle\langle+|A\rangle + \langle-|S_k|-\rangle\langle-|A\rangle, \tag{5.68}$$

or

$$\begin{pmatrix} \langle+|A'\rangle \\ \langle-|A'\rangle \end{pmatrix} = \begin{pmatrix} \langle+|S_k|+\rangle, & \langle+|S_k|-\rangle \\ \langle-|S_k|+\rangle, & \langle-|S_k|-\rangle \end{pmatrix} \begin{pmatrix} \langle+|A\rangle \\ \langle-|A\rangle \end{pmatrix}. \tag{5.69}$$

It follows that we can represent the operator/ket relation (5.65) as the matrix relation

$$\chi' = \left(\frac{\hbar}{2}\right) \sigma_k \chi, \tag{5.70}$$

where the σ_k are the matrices of the $\langle\pm|S_k|\pm\rangle$ values divided by $\hbar/2$. These matrices, which are called the *Pauli matrices*, can easily be evaluated using the explicit forms for the spin operators given in Equations (5.11)–(5.13). We find that

$$\sigma_1 = \begin{pmatrix} 0, & 1 \\ 1, & 0 \end{pmatrix}, \tag{5.71}$$

$$\sigma_2 = \begin{pmatrix} 0, & -i \\ i, & 0 \end{pmatrix}, \tag{5.72}$$

$$\sigma_3 = \begin{pmatrix} 1, & 0 \\ 0, & -1 \end{pmatrix}. \tag{5.73}$$

Here, 1, 2, and 3 refer to x, y, and z, respectively. Note that, in this scheme, we are effectively representing the spin operators in terms of the Pauli matrices:

$$S_k \rightarrow \left(\frac{\hbar}{2}\right)\sigma_k. \tag{5.74}$$

The expectation value of S_k can be written in terms of spinors and the Pauli matrices:

$$\langle S_k \rangle = \langle A | S_k | A \rangle = \sum_{\pm} \langle A | \pm \rangle \langle \pm | S_k | \pm \rangle \langle \pm | A \rangle = \left(\frac{\hbar}{2}\right) \chi^{\dagger} \sigma_k \chi. \tag{5.75}$$

The fundamental commutation relation for angular momentum, Equation (5.1), can be combined with Equation (5.74) to give the following commutation relation for the Pauli matrices:

$$\boldsymbol{\sigma} \times \boldsymbol{\sigma} = 2\,i\,\boldsymbol{\sigma}. \tag{5.76}$$

It is easily seen that the matrices (5.71)–(5.73) actually satisfy these relations (i.e., $\sigma_1\sigma_2 - \sigma_2\sigma_1 = 2\,i\,\sigma_3$, plus all cyclic permutations). (See Exercise 5.3.) It is also easily seen that the Pauli matrices satisfy the anti-commutation relations

$$\{\sigma_i, \sigma_j\} = 2\,\delta_{ij}. \tag{5.77}$$

(See Exercise 5.3.) Here, $\{a, b\} \equiv a\,b + b\,a$.

Let us examine how the Pauli scheme can be extended to take into account the position of a spin one-half particle. Recall, from Section 5.3, that we can represent a general basis ket as the product of basis kets in position space and spin space:

$$|x', y', z', \pm\rangle = |x', y', z'\rangle|\pm\rangle = |\pm\rangle|x', y', z'\rangle. \tag{5.78}$$

The ket corresponding to state A is denoted $\|A\rangle\rangle$, and resides in the product space of the position and spin ket spaces. State A is completely specified by the two wavefunctions

$$\psi_+(x', y', z') = \langle x', y', z'|\langle +\|A\rangle\rangle, \tag{5.79}$$

$$\psi_-(x', y', z') = \langle x', y', z'|\langle -\|A\rangle\rangle. \tag{5.80}$$

Consider the operator relation

$$\|A'\rangle\rangle = S_k \|A\rangle\rangle. \tag{5.81}$$

It is easily seen that

$$\langle x', y', z'|\langle +|A'\rangle\rangle = \langle +|S_k|+\rangle\langle x', y', z'|\langle +\|A\rangle\rangle + \langle +|S_k|-\rangle\langle x', y', z'|\langle -\|A\rangle\rangle, \tag{5.82}$$

$$\langle x', y', z'|\langle -|A'\rangle\rangle = \langle -|S_k|+\rangle\langle x', y', z'|\langle +\|A\rangle\rangle + \langle -|S_k|-\rangle\langle x', y', z'|\langle -\|A\rangle\rangle, \tag{5.83}$$

where use has been made of the fact that the spin operator S_k commutes with the eigenbras $\langle x', y', z'|$. It is fairly obvious that we can represent the operator relation (5.81) as a matrix relation if we generalize our definition of a spinor by writing

$$\|A\rangle\rangle \rightarrow \begin{pmatrix} \psi_+(\mathbf{x}') \\ \psi_-(\mathbf{x}') \end{pmatrix} \equiv \chi, \tag{5.84}$$

and so on. The components of the spinor are now wavefunctions, instead of complex numbers. We shall refer to such a construct as a *spinor-wavefunction*. In this scheme, the operator equation (5.81) becomes simply

$$\chi' = \left(\frac{\hbar}{2}\right) \sigma_k \chi. \tag{5.85}$$

Consider the operator relation

$$\|A'\rangle\rangle = p_k \|A\rangle\rangle. \tag{5.86}$$

In the Schrödinger representation, we have

$$\langle x', y', z'|\langle +|A'\rangle\rangle = \langle x', y', z'| p_k \langle +\|A\rangle\rangle = -i\hbar \frac{\partial}{\partial x_k'} \langle x', y', z'|\langle +\|A\rangle\rangle, \tag{5.87}$$

$$\langle x', y', z'|\langle -|A'\rangle\rangle = \langle x', y', z'| p_k \langle -\|A\rangle\rangle = -i\hbar \frac{\partial}{\partial x_k'} \langle x', y', z'|\langle -\|A\rangle\rangle, \tag{5.88}$$

where use has been made of Equation (2.78). The previous equation reduces to

$$\begin{pmatrix} \psi_+'(\mathbf{x}') \\ \psi_-'(\mathbf{x}') \end{pmatrix} = \begin{pmatrix} -i\hbar \, \partial\psi_+(\mathbf{x}')/\partial x_k' \\ -i\hbar \, \partial\psi_-(\mathbf{x}')/\partial x_k' \end{pmatrix}. \tag{5.89}$$

Thus, the operator equation (5.86) can be written

$$\chi' = p_k \chi, \tag{5.90}$$

where

$$p_k \rightarrow -i\hbar \frac{\partial}{\partial x_k'} \mathbf{1}. \tag{5.91}$$

Here, $\mathbf{1}$ is the 2×2 unit matrix. In fact, any position operator (e.g., p_k or L_k) is represented in the Pauli scheme as some differential operator of the position eigenvalues multiplied by the 2×2 unit matrix.

What about combinations of position and spin operators? The most commonly occurring combination is a dot product: for instance, $\mathbf{S} \cdot \mathbf{L} = (\hbar/2) \, \boldsymbol{\sigma} \cdot \mathbf{L}$. Consider the hybrid operator $\boldsymbol{\sigma} \cdot \mathbf{a}$, where $\mathbf{a} \equiv (a_x, a_y, a_z)$ is some vector position operator. This quantity is represented as a 2×2 matrix:

$$\boldsymbol{\sigma} \cdot \mathbf{a} \equiv \sum_k a_k \sigma_k = \begin{pmatrix} +a_3, & a_1 - i\,a_2 \\ a_1 + i\,a_2, & -a_3 \end{pmatrix}. \tag{5.92}$$

Because, in the Schrödinger representation, a general position operator takes the form of a differential operator in x', y', or z', it is clear that the previous quantity must be regarded as a matrix differential operator that acts on spinor-wavefunctions of the general form (5.84).

Finally, the important identity

$$(\sigma \cdot \mathbf{a})(\sigma \cdot \mathbf{b}) = \mathbf{a} \cdot \mathbf{b} + i\,\sigma \cdot (\mathbf{a} \times \mathbf{b}) \qquad (5.93)$$

follows from the commutation and anti-commutation relations (5.76) and (5.77). In fact,

$$\sum_j \sigma_j a_j \sum_k \sigma_k b_k = \sum_j \sum_k \left(\frac{1}{2}\{\sigma_j, \sigma_k\} + \frac{1}{2}[\sigma_j, \sigma_k]\right) a_j b_k$$

$$= \sum_j \sum_k (\delta_{jk} + i\,\epsilon_{jkl}\,\sigma_l)\,a_j b_k$$

$$= \mathbf{a} \cdot \mathbf{b} + i\,\sigma \cdot (\mathbf{a} \times \mathbf{b}). \qquad (5.94)$$

5.8   Spinor Rotation Matrices

A general rotation operator in spin space is written

$$T(\Delta\phi) = \exp\left(\frac{-i\,\mathbf{S} \cdot \mathbf{n}\,\Delta\varphi}{\hbar}\right), \qquad (5.95)$$

by analogy with Equation (5.24), where $\mathbf{n}$ is a unit vector pointing along the axis of rotation, and $\Delta\varphi$ is the angle of rotation. Here, $\mathbf{n}$ can be regarded as a trivial position operator. The rotation operator is represented

$$\exp\left(\frac{-i\,\mathbf{S} \cdot \mathbf{n}\,\Delta\varphi}{\hbar}\right) \rightarrow \exp\left(\frac{-i\,\sigma \cdot \mathbf{n}\,\Delta\varphi}{2}\right) \qquad (5.96)$$

in the Pauli scheme. The term on the right-hand side of the previous expression is the exponential of a matrix. This can easily be evaluated using the Taylor series for an exponential, plus the rules

$$(\sigma \cdot \mathbf{n})^k = 1 \qquad\qquad \text{for } k \text{ even}, \qquad (5.97)$$

$$(\sigma \cdot \mathbf{n})^k = \sigma \cdot \mathbf{n} \qquad\qquad \text{for } k \text{ odd}. \qquad (5.98)$$

These rules follow trivially from the identity (5.93). Thus, we can write

$$\exp\left(\frac{-i\,\sigma \cdot \mathbf{n}\,\Delta\varphi}{2}\right) = \left[1 - \frac{(\sigma \cdot \mathbf{n})^2}{2!}\left(\frac{\Delta\varphi}{2}\right)^2 + \frac{(\sigma \cdot \mathbf{n})^4}{4!}\left(\frac{\Delta\varphi}{2}\right)^4 + \cdots\right]$$

$$- i\left[(\sigma \cdot \mathbf{n})\left(\frac{\Delta\varphi}{2}\right) - \frac{(\sigma \cdot \mathbf{n})^3}{3!}\left(\frac{\Delta\varphi}{2}\right)^3 + \cdots\right]$$

$$= \cos(\Delta\varphi/2) - i\,\sin(\Delta\varphi/2)\,\sigma \cdot \mathbf{n}. \qquad (5.99)$$

The explicit 2×2 form of this matrix is

$$\begin{pmatrix} \cos(\varDelta\varphi/2) - \mathrm{i}\, n_z \sin(\varDelta\varphi/2), & (-\mathrm{i}\, n_x - n_y) \sin(\varDelta\varphi/2) \\ (-\mathrm{i}\, n_x + n_y) \sin(\varDelta\varphi/2), & \cos(\varDelta\varphi/2) + \mathrm{i}\, n_z \sin(\varDelta\varphi/2) \end{pmatrix}. \tag{5.100}$$

Rotation matrices act on spinors in much the same manner as the corresponding rotation operators act on state kets. Thus,

$$\chi' = \exp\left(\frac{-\mathrm{i}\,\boldsymbol{\sigma} \cdot \mathbf{n}\,\varDelta\varphi}{2}\right) \chi, \tag{5.101}$$

where χ' denotes the spinor obtained after rotating the spinor χ an angle $\varDelta\varphi$ about the axis $\mathbf{n}$. The Pauli matrices remain unchanged under rotations. However, the quantity $\chi^\dagger \sigma_k \chi$ is proportional to the expectation value of S_k [see Equation (5.75)], so we would expect it to transform like a vector under rotation. (See Section 5.4.) In fact, we require

$$(\chi^\dagger \sigma_k \chi)' \equiv (\chi^\dagger)' \sigma_k \chi' = \sum_l R_{kl} (\chi^\dagger \sigma_l \chi), \tag{5.102}$$

where the R_{kl} are the elements of a conventional rotation matrix [Riley *et al.* (2013)]. This is easily demonstrated, because

$$\exp\left(\frac{\mathrm{i}\,\sigma_3\,\varDelta\varphi}{2}\right) \sigma_1 \exp\left(\frac{-\mathrm{i}\,\sigma_3\,\varDelta\varphi}{2}\right) = \sigma_1 \cos\varDelta\varphi - \sigma_2 \sin\varDelta\varphi \tag{5.103}$$

plus all cyclic permutations. The previous expression is the 2×2 matrix analog of

$$\exp\left(\frac{\mathrm{i}\, S_z\,\varDelta\varphi}{\hbar}\right) S_x \exp\left(\frac{-\mathrm{i}\, S_z\,\varDelta\varphi}{\hbar}\right) = S_x \cos\varDelta\varphi - S_y \sin\varDelta\varphi. \tag{5.104}$$

[See Equation (5.30).] Equation (5.103) follows from the Baker-Campbell-Hausdorff lemma, (5.31), which holds for matrices, in addition to operators.

5.9 Factorization of Spinor-Wavefunctions

In the absence of spin, the Hamiltonian can be written as some function of the position and momentum operators. Using the Schrödinger representation, in which $\mathbf{p} \to -\mathrm{i}\,\hbar\,\nabla$, the energy eigenvalue problem,

$$H|E\rangle = E|E\rangle, \tag{5.105}$$

can be transformed into a partial differential equation for the wavefunction $\psi(\mathbf{x}') \equiv \langle\mathbf{x}'|E\rangle$. This function specifies the probability density for observing the particle at a given position, $\mathbf{x}'$. In general, we find

$$H\psi = E\psi, \tag{5.106}$$

where H is now a partial differential operator. The boundary conditions (for a bound state) are obtained from the normalization constraint

$$\int d^3\mathbf{x}' \, |\psi|^2 = 1. \tag{5.107}$$

This is all very familiar. However, we now know how to generalize this scheme to deal with a spin one-half particle. Instead of representing the state of the particle by a single wavefunction, we use two wavefunctions. The first, $\psi_+(\mathbf{x}')$, specifies the probability density of observing the particle at position $\mathbf{x}'$ with spin angular momentum $+\hbar/2$ in the z-direction. The second, $\psi_-(\mathbf{x}')$, specifies the probability density of observing the particle at position $\mathbf{x}'$ with spin angular momentum $-\hbar/2$ in the z-direction. In the Pauli scheme, these wavefunctions are combined into a spinor-wavefunction, χ, which is simply the column vector of ψ_+ and ψ_-. In general, the Hamiltonian is a function of the position, momentum, and spin operators. Adopting the Schrödinger representation, and the Pauli scheme, the energy eigenvalue problem reduces to

$$H\chi = E\chi, \tag{5.108}$$

where χ is a spinor-wavefunction (i.e., a 2×1 matrix of wavefunctions) and H is a 2×2 matrix partial differential operator. [See Equation (5.92).] The previous spinor equation can always be written out explicitly as two coupled partial differential equations for ψ_+ and ψ_-.

Suppose that the Hamiltonian has no dependence on the spin operators. In this case, the Hamiltonian is represented as diagonal 2×2 matrix partial differential operator in the Schrödinger/Pauli scheme. [See Equation (5.91).] In other words, the partial differential equation for ψ_+ decouples from that for ψ_-. In fact, both equations have the same form, so there is really only one differential equation. In this situation, the most general solution to Equation (5.108) can be written in the factorized form

$$\chi = \psi(\mathbf{x}') \begin{pmatrix} c_+ \\ c_- \end{pmatrix}. \tag{5.109}$$

Here, $\psi(\mathbf{x}')$ is determined by the solution of the differential equation, and the $c_\pm$ are arbitrary complex numbers. The physical significance of the previous expression is clear. The Hamiltonian determines the relative probabilities of finding the particle at various different positions, but the direction of its spin angular momentum remains undetermined.

Suppose that the Hamiltonian depends only on the spin operators. In this case, the Hamiltonian is represented as a 2×2 matrix of complex numbers in the Schrödinger/Pauli scheme [see Equation (5.74)], and the spinor eigenvalue equation (5.108) reduces to a straightforward matrix eigenvalue problem. The most general solution can again be written in the factorized form

$$\chi = \psi(\mathbf{x}') \begin{pmatrix} c_+ \\ c_- \end{pmatrix}. \tag{5.110}$$

Here, the ratio c_+/c_- is determined by the matrix eigenvalue problem, and the wavefunction $\psi(\mathbf{x}')$ is arbitrary. Clearly, the Hamiltonian determines the direction of the particle's spin angular momentum, but leaves its position undetermined.

In general, of course, the Hamiltonian is a function of both position and spin operators. In this case, it is not possible to factorize the spinor-wavefunction as in Equations (5.109) and (5.110). In other words, a general Hamiltonian causes the direction of the particle's spin angular momentum to vary with position in some specified manner. This can only be represented as a spinor-wavefunction involving different wavefunctions, ψ_+ and ψ_-.

5.10 Spin Greater Than One-Half Systems

We have seen how to deal with a spin-half particle in quantum mechanics. But, what happens if we have a spin one or a spin three-halves particle? It turns out that we can generalize the Pauli two-component scheme in a fairly straightforward manner. Consider a spin-s particle: that is, a particle for which the eigenvalue of S^2 is $s(s+1)\hbar^2$. Here, s is either an integer, or a half-integer. The eigenvalues of S_z are written $s_z\hbar$, where s_z is allowed to take the values $s, s-1, \cdots, -s+1, -s$. In fact, there are $2s+1$ distinct allowed values of s_z. Not surprisingly, we can represent the state of the particle by $2s+1$ different wavefunctions, denoted $\psi_{s_z}(\mathbf{x}')$. Here, $\psi_{s_z}(\mathbf{x}')$ specifies the probability density for observing the particle at position $\mathbf{x}'$ with spin angular momentum $s_z\hbar$ in the z-direction. More exactly,

$$\psi_{s_z}(\mathbf{x}') = \langle \mathbf{x}' | \langle s, s_z \| A \rangle \rangle, \tag{5.111}$$

where $\|A\rangle\rangle$ denotes a state ket in the product of the position and spin spaces. The state of the particle can be represented more succinctly by a spinor-wavefunction, χ, which is simply the $2s+1$ component column vector of the $\psi_{s_z}(\mathbf{x}')$. Thus, a spin one-half particle is represented by a two-component spinor-wavefunction, a spin one particle by a three-component spinor-wavefunction, a spin three-halves particle by a four-component spinor-wavefunction, and so on.

In this extended Schrödinger/Pauli scheme, position space operators take the form of diagonal $(2s+1) \times (2s+1)$ matrix differential operators. Thus, we can represent the momentum operators as

$$p_k \rightarrow -\mathrm{i}\hbar \frac{\partial}{\partial x'_k} \mathbf{1}, \tag{5.112}$$

where **1** is the $(2s+1)\times(2s+1)$ unit matrix. [See Equation (5.91).] We represent the spin operators as

$$S_k \rightarrow s\hbar\,\sigma_k, \qquad (5.113)$$

where the $(2s+1)\times(2s+1)$ extended Pauli matrix σ_k (which is, of course, Hermitian) has elements

$$(\sigma_k)_{jl} = \frac{\langle s,j|\,S_k\,|s,l\rangle}{s\hbar}. \qquad (5.114)$$

Here, j, l are integers, or half-integers, lying in the range $-s$ to $+s$. But, how can we evaluate the brackets $\langle s,j|\,S_k\,|s,l\rangle$ and, thereby, construct the extended Pauli matrices? In fact, it is trivial to construct the σ_z matrix. By definition,

$$S_z|s,j\rangle = j\hbar|s,j\rangle. \qquad (5.115)$$

Hence,

$$(\sigma_3)_{jl} = \frac{\langle s,j|\,S_z\,|s,l\rangle}{s\hbar} = \frac{j}{s}\,\delta_{jl}, \qquad (5.116)$$

where use has been made of the orthonormality property of the $|s,j\rangle$. Thus, σ_z is the suitably normalized diagonal matrix of the eigenvalues of S_z. The elements of σ_x and σ_y are most easily obtained by considering the ladder operators,

$$S^{\pm} = S_x \pm i\,S_y. \qquad (5.117)$$

We know, from Equations (4.55)–(4.56), that

$$S^{+}|s,j\rangle = [s(s+1) - j(j+1)]^{1/2}\,\hbar|s,j+1\rangle, \qquad (5.118)$$

$$S^{-}|s,j\rangle = [s(s+1) - j(j-1)]^{1/2}\,\hbar|s,j-1\rangle. \qquad (5.119)$$

It follows from Equations (5.114), and (5.117)–(5.119), that

$$(\sigma_1)_{jl} = \frac{[s(s+1) - j(j-1)]^{1/2}}{2s}\,\delta_{j\,l+1} + \frac{[s(s+1) - j(j+1)]^{1/2}}{2s}\,\delta_{j\,l-1}, \quad (5.120)$$

$$(\sigma_2)_{jl} = \frac{[s(s+1) - j(j-1)]^{1/2}}{2\,i\,s}\,\delta_{j\,l+1} - \frac{[s(s+1) - j(j+1)]^{1/2}}{2\,i\,s}\,\delta_{j\,l-1}. \quad (5.121)$$

According to Equations (5.116) and (5.120)–(5.121), the Pauli matrices for a spin one-half ($s = 1/2$) particle (e.g., an electron, a proton, or a neutron) are

$$\sigma_1 = \begin{pmatrix} 0, & 1 \\ 1, & 0 \end{pmatrix}, \qquad (5.122)$$

$$\sigma_2 = \begin{pmatrix} 0, & -i \\ i, & 0 \end{pmatrix}, \qquad (5.123)$$

$$\sigma_3 = \begin{pmatrix} 1, & 0 \\ 0, & -1 \end{pmatrix}, \qquad (5.124)$$

as we have seen previously. For a spin one ($s = 1$) particle (e.g., a Z-boson or a W-boson), we find that

$$\sigma_1 = \frac{1}{\sqrt{2}} \begin{pmatrix} 0, 1, 0 \\ 1, 0, 1 \\ 0, 1, 0 \end{pmatrix}, \tag{5.125}$$

$$\sigma_2 = \frac{1}{\sqrt{2}} \begin{pmatrix} 0, -i, 0 \\ i, 0, -i \\ 0, i, 0 \end{pmatrix}, \tag{5.126}$$

$$\sigma_3 = \begin{pmatrix} 1, 0, 0 \\ 0, 0, 0 \\ 0, 0, -1 \end{pmatrix}. \tag{5.127}$$

In fact, we can now construct the Pauli matrices for a particle of arbitrary spin. This means that we can convert the general energy eigenvalue problem for a spin-s particle, where the Hamiltonian is some function of position and spin operators, into $2\,s+1$ coupled partial differential equations involving the $2\,s+1$ wavefunctions $\psi_{s_z}(\mathbf{x}')$. Unfortunately, such a system of equations is generally too complicated to solve exactly.

5.11 Exercises

5.1 Demonstrate that the operators defined in Equations (5.11)–(5.13) are Hermitian, and satisfy the commutation relations (5.1).

5.2 Prove the Baker-Campbell-Hausdorff lemma, (5.31).

5.3 Let the σ_i, for $i = 1, 3$, be the three spin-1/2 Pauli matrices. Demonstrate that $\sigma_j \sigma_k = \delta_{jk} + i\,\epsilon_{jkl}\,\sigma_l$, and, hence, that $\boldsymbol{\sigma} \times \boldsymbol{\sigma} = 2\,i\,\boldsymbol{\sigma}$ and $\{\sigma_i, \sigma_j\} = 2\,\delta_{ij}$.

5.4 Find the Pauli representations of the normalized eigenstates of (a) S_x and (b) S_y for a spin-1/2 particle.

5.5 Suppose that a spin-1/2 particle has a spin vector that lies in the x-z plane, making an angle θ with the z-axis. Demonstrate that a measurement of S_z yields $\hbar/2$ with probability $\cos^2(\theta/2)$, and $-\hbar/2$ with probability $\sin^2(\theta/2)$.

5.6 An electron is in the spin-state

$$\chi = A \begin{pmatrix} 1 - 2i \\ 2 \end{pmatrix}$$

in the Pauli representation. (a) Determine the constant A by normalizing χ. (b) If a measurement of S_z is made, what values will be obtained, and with what probabilities? What is the expectation value of S_z? Repeat the previous calculations for (c) S_x and (d) S_y. [Griffiths (2005)]

5.7 Consider a spin-1/2 system represented by the normalized spinor

$$\chi = \begin{pmatrix} \cos\alpha \\ \sin\alpha \, \exp(i\beta) \end{pmatrix}$$

in the Pauli representation, where α and β are real. What is the probability that a measurement of S_y yields $-\hbar/2$? [Gasiorowicz (1996)]

5.8 An electron is at rest in an oscillating magnetic field

$$\mathbf{B} = B_0 \cos(\omega t)\,\mathbf{e}_z,$$

where B_0 and ω are real positive constants.

(a) Find the Hamiltonian of the system.

(b) If the electron starts in the spin-up state with respect to the x-axis, determine the spinor $\chi(t)$ that represents the state of the system in the Pauli representation at all subsequent times.

(c) Find the probability that a measurement of S_x yields the result $-\hbar/2$ as a function of time.

(d) What is the minimum value of B_0 required to force a complete flip in S_x?

5.9 In the Schrödinger/Pauli representation, the generalization of Schrödinger's time-dependent wave equation for an electron moving in electromagnetic fields is written

$$i\hbar\frac{\partial\chi}{\partial t} = \frac{1}{2m_e}\left[(-i\hbar\nabla + e\mathbf{A})\cdot(-i\hbar\nabla + e\mathbf{A}) + e\hbar\,\boldsymbol{\sigma}\cdot\mathbf{B}\right]\chi - e\phi\chi,$$

where $\mathbf{A}$ the vector potential, $\mathbf{B} = \nabla \times \mathbf{A}$ the magnetic field-strength, ϕ the scalar potential, and χ the spinor-wavefunction. The term involving the Pauli matrices comes from the electron's intrinsic magnetic moment. (See Section 5.5.) Demonstrate that this equation can also be written

$$i\hbar\frac{\partial\chi}{\partial t} = \frac{1}{2m_e}\left[\boldsymbol{\sigma}\cdot(-i\hbar\nabla + e\mathbf{A})\,\boldsymbol{\sigma}\cdot(-i\hbar\nabla + e\mathbf{A})\right]\chi - e\phi\chi.$$

The previous expression is known as the *Pauli equation* [Pauli (1927)].

5.10 Let σ_1, σ_2, and σ_3 be the Pauli matrices for a spin-1 particle.

(a) Show that

$$(\sigma_1)^k = \sigma_1' \qquad\qquad \text{for } k \text{ even,}$$
$$(\sigma_1)^k = \sigma_1 \qquad\qquad \text{for } k \text{ odd,}$$

where

$$\sigma_1' = \begin{pmatrix} 1/2, & 0, & 1/2 \\ 0, & 1, & 0 \\ 1/2, & 0, & 1/2 \end{pmatrix}.$$

(b) Show that

$$(\sigma_2)^k = \sigma_2' \qquad\qquad \text{for } k \text{ even,}$$

$$(\sigma_2)^k = \sigma_2 \qquad\qquad \text{for } k \text{ odd,}$$

where

$$\sigma_2' = \begin{pmatrix} 1/2, & 0, & -1/2 \\ 0, & 1, & 0 \\ -1/2, & 0, & 1/2 \end{pmatrix}.$$

(c) Show that

$$(\sigma_3)^k = \sigma_3' \qquad\qquad \text{for } k \text{ even,}$$

$$(\sigma_3)^k = \sigma_3 \qquad\qquad \text{for } k \text{ odd,}$$

where

$$\sigma_3' = \begin{pmatrix} 1, 0, 0 \\ 0, 0, 0 \\ 0, 0, 1 \end{pmatrix}.$$

(d) Hence, deduce that the spinor matrices for rotations through an angle $\Delta\varphi$ about the three Cartesian axes are

$$T_1(\Delta\varphi) = \begin{pmatrix} c^2, & -i\sqrt{2}\,sc, & -s^2 \\ -i\sqrt{2}\,sc, & c^2 - s^2, & -i\sqrt{2}\,sc \\ -s^2, & -i\sqrt{2}\,sc, & c^2 \end{pmatrix},$$

$$T_2(\Delta\varphi) = \begin{pmatrix} c^2, & -\sqrt{2}\,sc, & s^2 \\ \sqrt{2}\,sc, & c^2 - s^2, & -\sqrt{2}\,sc \\ s^2 & \sqrt{2}\,sc, & c^2 \end{pmatrix},$$

$$T_3(\Delta\varphi) = \begin{pmatrix} \exp(-i\Delta\varphi), & 0, & 0 \\ 0, & 1, & 0 \\ 0, & 0, & \exp(i\Delta\varphi) \end{pmatrix},$$

where $s = \sin(\Delta\varphi/2)$ and $c = \cos(\Delta\varphi/2)$.

(e) Suppose that a spin-1 particle has a spin vector that lies in the x-z plane, making an angle θ with the z-axis. Demonstrate that a measurement of S_z yields $\hbar$, 0, and $-\hbar$ with probabilities $\cos^4(\theta/2)$, $2\sin^2(\theta/2)\cos^2(\theta/2)$, and $\sin^4(\theta/2)$, respectively.

Chapter 6

Addition of Angular Momentum

6.1 Introduction

Consider a hydrogen atom whose orbiting electron is in an $l = 1$ state. The electron, consequently, possesses orbital angular momentum of magnitude $\hbar$, and spin angular momentum of magnitude $\hbar/2$. So, what is the electron's total angular momentum? In order to answer this question, we need to learn how to add angular momentum operators.

6.2 Commutation Rules

Consider the most general case. Suppose that we have two sets of angular momentum operators, $\mathbf{J}_1$ and $\mathbf{J}_2$. By definition, these operators are Hermitian, and obey the fundamental commutation relations

$$\mathbf{J}_1 \times \mathbf{J}_1 = i\,\hbar\,\mathbf{J}_1, \tag{6.1}$$

$$\mathbf{J}_2 \times \mathbf{J}_2 = i\,\hbar\,\mathbf{J}_2. \tag{6.2}$$

Let us assume that the two groups of operators correspond to different degrees of freedom of the system, so that

$$[J_{1\,i}, J_{2\,j}] = 0, \tag{6.3}$$

where i, j stand for either x, y, or z. (See Section 2.2.) For instance, $\mathbf{J}_1$ could be an orbital angular momentum operator, and $\mathbf{J}_2$ a spin angular momentum operator. Alternatively, $\mathbf{J}_1$ and $\mathbf{J}_2$ could be the orbital angular momentum operators of two different particles in a multi-particle system. We know, from the general properties of angular momentum outlined in the previous two chapters, that the eigenvalues of J_1^2 and J_2^2 can be written $j_1\,(j_1 + 1)\,\hbar^2$ and $j_2\,(j_2 + 1)\,\hbar^2$, respectively, where j_1 and j_2 are either integers, or half-integers. We also know that the eigenvalues of J_{1z} and J_{2z} take the form $m_1 \hbar$ and $m_2 \hbar$, respectively, where m_1 and m_2 are numbers lying in the ranges $j_1, j_1 - 1, \cdots, -j_1 + 1, -j_1$ and $j_2, j_2 - 1, \cdots, -j_2 + 1, -j_2$, respectively.

Let us define the total angular momentum operator

$$\mathbf{J} = \mathbf{J}_1 + \mathbf{J}_2. \tag{6.4}$$

Now, $\mathbf{J}$ is an Hermitian operator, because it is the sum of Hermitian operators. Moreover, according to Equation (4.14), $\mathbf{J}$ satisfies the fundamental commutation relation

$$\mathbf{J} \times \mathbf{J} = i \hbar \mathbf{J}. \tag{6.5}$$

Thus, $\mathbf{J}$ possesses all of the expected properties of an angular momentum operator. It follows that the eigenvalue of J^2 can be written $j(j+1) \hbar^2$, where j is an integer, or a half-integer. Moreover, the eigenvalue of J_z takes the form $m \hbar$, where m lies in the range $j, j - 1, \cdots, -j + 1, -j$. At this stage, however, we do not know the relationship between the quantum numbers of the total angular momentum, j and m, and those of the individual angular momenta, $j_1, j_2, m_1,$ and m_2.

Now,

$$J^2 = J_1^2 + J_2^2 + 2 \mathbf{J}_1 \cdot \mathbf{J}_2. \tag{6.6}$$

Furthermore, we know that

$$[J_1^2, J_{1i}] = 0, \tag{6.7}$$

$$[J_2^2, J_{2i}] = 0, \tag{6.8}$$

and also that all of the J_{1i}, J_1^2 operators commute with the J_{2i}, J_2^2 operators. It follows from Equation (6.6) that

$$[J^2, J_1^2] = [J^2, J_2^2] = 0. \tag{6.9}$$

This implies that the quantum numbers j_1, j_2, and j can all be measured simultaneously. In other words, it is possible to determine the magnitude of the total angular momentum together with the magnitudes of the component angular momenta. However, it is apparent from Equations (6.1), (6.2), and (6.6) that

$$[J^2, J_{1z}] \neq 0, \tag{6.10}$$

$$[J^2, J_{2z}] \neq 0. \tag{6.11}$$

This suggests that it is not possible to measure the quantum numbers m_1 and m_2 simultaneously with the quantum number j. Thus, we cannot determine the projections of the individual angular momenta along the z-axis together with the magnitude of the total angular momentum.

It is clear, from the preceding discussion, that we can form two alternate groups of mutually commuting operators. The first group is J_1^2, J_2^2, J_{1z}, and J_{2z}. The second group is J_1^2, J_2^2, J^2, and J_z. These two groups of operators are incompatible with one another. We can define simultaneous eigenkets of each operator group. The simultaneous eigenkets of J_1^2, J_2^2, J_{1z}, and J_{2z} are denoted $|j_1, j_2; m_1, m_2\rangle$, where

$$J_1^2 |j_1, j_2; m_1, m_2\rangle = j_1 (j_1 + 1) \hbar^2 |j_1, j_2; m_1, m_2\rangle, \tag{6.12}$$

$$J_2^2 |j_1, j_2; m_1, m_2\rangle = j_2 (j_2 + 1) \hbar^2 |j_1, j_2; m_1, m_2\rangle, \tag{6.13}$$

$$J_{1z} |j_1, j_2; m_1, m_2\rangle = m_1 \hbar |j_1, j_2; m_1, m_2\rangle, \tag{6.14}$$

$$J_{2z} |j_1, j_2; m_1, m_2\rangle = m_2 \hbar |j_1, j_2; m_1, m_2\rangle. \tag{6.15}$$

The simultaneous eigenkets of J_1^2, J_2^2, J^2 and J_z are denoted $|j_1, j_2; j, m\rangle$, where

$$J_1^2 |j_1, j_2; j, m\rangle = j_1 (j_1 + 1) \hbar^2 |j_1, j_2; j, m\rangle, \tag{6.16}$$

$$J_2^2 |j_1, j_2; j, m\rangle = j_2 (j_2 + 1) \hbar^2 |j_1, j_2; j, m\rangle, \tag{6.17}$$

$$J^2 |j_1, j_2; j, m\rangle = j (j + 1) \hbar^2 |j_1, j_2; j, m\rangle, \tag{6.18}$$

$$J_z |j_1, j_2; j, m\rangle = m \hbar |j_1, j_2; j, m\rangle. \tag{6.19}$$

Each set of eigenkets are complete, mutually orthogonal (for eigenkets corresponding to different sets of eigenvalues), and have unit norms. Because the operators J_1^2 and J_2^2 are common to both operator groups, we can assume that the quantum numbers j_1 and j_2 are known. In other words, we can always determine the magnitudes of the individual angular momenta. In addition, we can either know the quantum numbers m_1 and m_2, or the quantum numbers j and m, but we cannot know both pairs of quantum numbers at the same time. Finally, we can write a conventional completeness relation for both sets of eigenkets:

$$\sum_{m_1} \sum_{m_2} |j_1, j_2; m_1, m_2\rangle\langle j_1, j_2; m_1, m_2| = 1, \tag{6.20}$$

$$\sum_{j} \sum_{m} |j_1, j_2; j, m\rangle\langle j_1, j_2; j, m| = 1, \tag{6.21}$$

where the right-hand sides denote the identity operator in the ket space corresponding to states of given j_1 and j_2. The summation is over all allowed values of m_1, m_2, j, and m.

6.3 Clebsch-Gordon Coefficients

As we have seen, the operator group J_1^2, J_2^2, J^2, and J_z is incompatible with the group J_1^2, J_2^2, J_{1z}, and J_{2z}. This means that if the system is in a simultaneous eigenstate of the former group then, in general, it is not in a simultaneous eigenstate of the latter. In other words, if the quantum numbers j_1, j_2, j, and m are known with certainty then a measurement of the quantum numbers m_1 and m_2 will give a range of possible values. We can use the completeness relation (6.20) to write

$$|j_1, j_2; j, m\rangle = \sum_{m_1} \sum_{m_2} \langle j_1, j_2; m_1, m_2|j_1, j_2; j, m\rangle |j_1, j_2; m_1, m_2\rangle. \qquad (6.22)$$

Thus, we can write the eigenkets of the first group of operators as a weighted sum of the eigenkets of the second set. The weights, $\langle j_1, j_2; m_1, m_2|j_1, j_2; j, m\rangle$, are called the *Clebsch-Gordon coefficients*. If the system is in a state where a measurement of J_1^2, J_2^2, J^2, and J_z is bound to give the results $j_1(j_1 + 1)\hbar^2, j_2(j_2 + 1)\hbar^2, j(j + 1)\hbar^2$, and $j_z\hbar$, respectively, then a measurement of J_{1z} and J_{2z} will give the results $m_1\hbar$ and $m_2\hbar$, respectively, with probability $|\langle j_1, j_2; m_1, m_2|j_1, j_2; j, m\rangle|^2$.

The Clebsch-Gordon coefficients possess a number of very important properties. First, the coefficients are zero unless

$$m = m_1 + m_2. \qquad (6.23)$$

To prove this, we note that

$$(J_z - J_{1z} - J_{2z})|j_1, j_2; j, m\rangle = 0. \qquad (6.24)$$

Forming the inner product with $\langle j_1, j_2; m_1, m_2|$, we obtain

$$(m - m_1 - m_2)\langle j_1, j_2; m_1, m_2|j_1, j_2; j, m\rangle = 0, \qquad (6.25)$$

which proves the assertion. Thus, the z-components of different angular momenta add algebraically. So, an electron in an $l = 1$ state, with orbital angular momentum $\hbar$, and spin angular momentum $\hbar/2$, projected along the z-axis, constitutes a state whose total angular momentum projected along the z-axis is $3\hbar/2$. What is uncertain is the magnitude of the total angular momentum.

Second, the coefficients vanish unless

$$|j_1 - j_2| \leq j \leq j_1 + j_2. \qquad (6.26)$$

We can assume, without loss of generality, that $j_1 \geq j_2$. We know, from Equation (6.23), that for given j_1 and j_2 the largest possible value of m is $j_1 + j_2$ (because j_1 is the largest possible value of m_1, etc.). This implies that the largest possible value of j is $j_1 + j_2$ (because, by definition, the largest value of m is equal to j). Now, there are $(2j_1 + 1)$ allowable values of m_1, and $(2j_2 + 1)$ allowable values of m_2. Thus, there are $(2j_1 + 1)(2j_2 + 1)$ independent eigenkets, $|j_1, j_2; m_1, m_2\rangle$, needed to span the ket space corresponding to fixed j_1 and j_2.

Because the eigenkets $|j_1, j_2; j, m\rangle$ span the same space, they must also form a set of $(2j_1 + 1)(2j_2 + 1)$ independent kets. In other words, there can only be $(2j_1 + 1)(2j_2 + 1)$ distinct allowable values of the quantum numbers j and m. For each allowed value of j, there are $2j + 1$ allowed values of m. We have already seen that the maximum allowed value of j is $j_1 + j_2$. It is easily seen that if the minimum allowed value of j is $j_1 - j_2$ then the total number of allowed values of j and m is $(2j_1 + 1)(2j_2 + 1)$. In other words [Gradshteyn *et al.* (1980)],

$$\sum_{j=j_1-j_2, j_1+j_2} (2j + 1) = \left[(1 + j)^2\right]_{j=j_1-j_2-1}^{j=j_1+j_2} = (2j_1 + 1)(2j_2 + 1). \qquad (6.27)$$

This proves our assertion.

Third, the sum of the modulus squared of all of the Clebsch-Gordon coefficients is unity: that is,

$$\sum_{m_1} \sum_{m_2} |\langle j_1, j_2; m_1, m_2 | j_1, j_2; j, m\rangle|^2 = 1. \qquad (6.28)$$

This assertion is proved as follows:

$$\langle j_1, j_2; j, m | j_1, j_2; j, m\rangle =$$

$$\sum_{m_1} \sum_{m_2} \langle j_1, j_2; j, m | j_1, j_2; m_1, m_2\rangle\langle j_1, j_2; m_1, m_2 | j_1, j_2; j, m\rangle$$

$$= \sum_{m_1} \sum_{m_2} |\langle j_1, j_2; m_1, m_2 | j_1, j_2; j, m\rangle|^2 = 1, \quad (6.29)$$

where use has been made of the completeness relation (6.20).

Finally, the Clebsch-Gordon coefficients obey two recursion relations. To obtain these relations, we start from

$$J^{\pm}|j_1, j_2; j, m\rangle = (J_1^{\pm} + J_2^{\pm}) \sum_{m_1'} \sum_{m_2'} \langle j_1, j_2; m_1', m_2' | j_1, j_2; j, m\rangle | j_1, j_2; m_1', m_2'\rangle, \qquad (6.30)$$

where $J^{\pm} = J_x \pm i J_y$, et cetera, Making use of the well-known properties of the ladder operators, $J^{\pm}$, $J_1^{\pm}$, and $J_2^{\pm}$, which are specified by analogy with Equations (4.55)–(4.56), we obtain

$$\sqrt{j(j + 1) - m(m \pm 1)}\,|j_1, j_2; j, m \pm 1\rangle =$$

$$\sum_{m_1'} \sum_{m_2'} \langle j_1, j_2; m_1', m_2' | j_1, j_2; j, m\rangle \left(\sqrt{j_1(j_1 + 1) - m_1'(m_1' \pm 1)}\,|j_1, j_2; m_1' \pm 1, m_2'\rangle \right.$$

$$\left. + \sqrt{j_2(j_2 + 1) - m_2'(m_2' \pm 1)}\,|j_1, j_2; m_1', m_2' \pm 1\rangle \right). \quad (6.31)$$

Taking the inner product with $\langle j_1, j_2; m_1, m_2 |$, and making use of the orthonormality property of the basis eigenkets, we get the desired recursion relations:

$$\sqrt{j(j+1) - m(m \pm 1)} \langle j_1, j_2; m_1, m_2 | j_1, j_2; j, m \pm 1 \rangle =$$
$$\sqrt{j_1(j_1+1) - m_1(m_1 \mp 1)} \langle j_1, j_2; m_1 \mp 1, m_2 | j_1, j_2; j, m \rangle$$
$$+ \sqrt{j_2(j_2+1) - m_2(m_2 \mp 1)} \langle j_1, j_2; m_1, m_2 \mp 1 | j_1, j_2; j, m \rangle. \quad (6.32)$$

It is clear, from the absence of complex coupling coefficients in the previous relations, that we can always choose the Clebsch-Gordon coefficients to be real numbers. This is convenient, because it ensures that the inverse Clebsch-Gordon coefficients, $\langle j_1, j_2; j, m | j_1, j_2; m_1, m_2 \rangle$, are identical to the Clebsch-Gordon coefficients. In other words,

$$\langle j_1, j_2; j, m | j_1, j_2; m_1, m_2 \rangle = \langle j_1, j_2; m_1, m_2 | j_1, j_2; j, m \rangle. \quad (6.33)$$

The inverse Clebsch-Gordon coefficients are the weights in the expansion of the $|j_1, j_2; m_1, m_2\rangle$ in terms of the $|j_1, j_2; j, m\rangle$:

$$|j_1, j_2; m_1, m_2\rangle = \sum_j \sum_m \langle j_1, j_2; j, m | j_1, j_2; m_1, m_2 \rangle |j_1, j_2; j, m\rangle. \quad (6.34)$$

It turns out that the recursion relations (6.32), together with the normalization condition (6.28), are sufficient to completely determine the Clebsch-Gordon coefficients to within an arbitrary sign (multiplied into all of the coefficients). This sign is fixed by convention. [To be more exact, each Clebsch-Gordon sub-table associated with a specific value of j (see later) is undetermined to an arbitrary sign. It is conventional to give the Clebsch-Gordon coefficient with the largest value of m a positive sign.] The easiest way of demonstrating this assertion is by considering a specific example.

6.4   Calculation of Clebsch-Gordon Coefficients

Let us add the angular momenta of two spin one-half systems. For example, two electrons at rest. So, $j_1 = j_2 = 1/2$. We know, from general principles, that $|m_1| \leq 1/2$ and $|m_2| \leq 1/2$. We also know, from Equation (6.26), that $0 \leq j \leq 1$, where the allowed values of j differ by integer amounts. It follows that either $j = 0$ or $j = 1$. Thus, two spin one-half systems can be combined to form either a spin-zero system or a spin-one system. It is helpful to arrange all of the possibly non-zero Clebsch-Gordon coefficients in a table:

m_1	m_2				
1/2	1/2	?	?	?	?
1/2	−1/2	?	?	?	?
−1/2	1/2	?	?	?	?
−1/2	−1/2	?	?	?	?
$j_1=1/2$	j	1	1	1	0
$j_2=1/2$	m	1	0	−1	0

The box in this table corresponding to $m_1 = 1/2, m_2 = 1/2, j = 1, m = 1$ gives the Clebsch-Gordon coefficient $\langle 1/2, 1/2; 1/2, 1/2 | 1/2, 1/2; 1, 1 \rangle$, or the inverse Clebsch-Gordon coefficient $\langle 1/2, 1/2; 1, 1 | 1/2, 1/2; 1/2, 1/2 \rangle$. All the boxes contain question marks because, at this stage, we do not know the values of any Clebsch-Gordon coefficients.

A Clebsch-Gordon coefficient is automatically zero unless $m_1 + m_2 = m$. In other words, the z-components of angular momentum have to add algebraically. Many of the boxes in the previous table correspond to $m_1 + m_2 \neq m$. We immediately conclude that these boxes must contain zeroes. Thus,

m_1	m_2				
1/2	1/2	?	0	0	0
1/2	−1/2	0	?	0	?
−1/2	1/2	0	?	0	?
−1/2	−1/2	0	0	?	0
$j_1=1/2$	j	1	1	1	0
$j_2=1/2$	m	1	0	−1	0

The normalization condition (6.28) implies that the sum of the squares of all the rows and columns of the previous table must be unity. There are two rows and two columns that contain only a single non-zero entry. We conclude that these entries must be ±1, but we have no way of determining the signs at present. Thus,

m_1	m_2				
1/2	1/2	±1	0	0	0
1/2	−1/2	0	?	0	?
−1/2	1/2	0	?	0	?
−1/2	−1/2	0	0	±1	0
$j_1=1/2$	j	1	1	1	0
$j_2=1/2$	m	1	0	−1	0

Let us evaluate the recursion relation (6.32) for $j_1 = j_2 = 1/2$, with $j = 1$, $m = 0$, $m_1 = m_2 = \pm 1/2$, taking the upper/lower sign. We find that

$$\langle 1/2, -1/2|1, 0\rangle + \langle -1/2, 1/2|1, 0\rangle = \sqrt{2}\,\langle 1/2, 1/2|1, 1\rangle = \pm\sqrt{2}, \qquad (6.35)$$

and

$$\langle 1/2, -1/2|1, 0\rangle + \langle -1/2, 1/2|1, 0\rangle = \sqrt{2}\,\langle -1/2, -1/2|1, -1\rangle = \pm\sqrt{2}. \qquad (6.36)$$

Here, the j_1 and j_2 labels have been suppressed for ease of notation. We also know that

$$\langle 1/2, -1/2|1, 0\rangle^2 + \langle -1/2, 1/2|1, 0\rangle^2 = 1, \qquad (6.37)$$

from the normalization condition. The only real solutions to the previous set of equations are

$$\sqrt{2}\,\langle 1/2, -1/2|1, 0\rangle = \sqrt{2}\,\langle -1/2, 1/2|1, 0\rangle$$

$$= \langle 1/2, 1/2|1, 1\rangle = \langle 1/2, 1/2|1, -1\rangle = \pm 1. \qquad (6.38)$$

The choice of sign is arbitrary—the conventional choice is a positive sign. Thus, our table now reads

m_1	m_2				
1/2	1/2	1	0	0	0
1/2	−1/2	0	$1/\sqrt{2}$	0	?
−1/2	1/2	0	$1/\sqrt{2}$	0	?
−1/2	−1/2	0	0	1	0
$j_1=1/2$	j	1	1	1	0
$j_2=1/2$	m	1	0	−1	0

We could fill in the remaining unknown entries of our table by using the recursion relation again. However, an easier method is to observe that the rows and columns of the table must all be mutually orthogonal. That is, the dot product of a row with any other row must be zero. Likewise, for the dot product of a column with any other column. This follows because the entries in the table give the expansion coefficients of one of our alternative sets of eigenkets in terms of the other set, and each set of eigenkets contains mutually orthogonal vectors with unit norms. The normalization condition tells us that the dot product of a row or column with itself must be unity. The only way that the dot product of the fourth column with the second column can be zero is if the unknown entries are equal and opposite. The requirement that the dot product of the fourth column with itself is unity tells us that the magnitudes of the unknown entries have to be $1/\sqrt{2}$. The unknown entries are undetermined to an arbitrary sign multiplied into them both. Thus, the final form of our table (with the conventional choice of arbitrary signs) is

m_1	m_2				
1/2	1/2	1	0	0	0
1/2	−1/2	0	$1/\sqrt{2}$	0	$1/\sqrt{2}$
−1/2	1/2	0	$1/\sqrt{2}$	0	$-1/\sqrt{2}$
−1/2	−1/2	0	0	1	0
$j_1=1/2$	j	1	1	1	0
$j_2=1/2$	m	1	0	−1	0

The table can be read in one of two ways. The columns give the expansions of the eigenstates of overall angular momentum in terms of the eigenstates of the individual angular momenta of the two component systems. Thus, the second column tells us that

$$|1,0\rangle = \frac{1}{\sqrt{2}}\left(|1/2,-1/2\rangle + |-1/2,1/2\rangle\right). \qquad (6.39)$$

The ket on the left-hand side is a $|j,m\rangle$ ket, whereas those on the right-hand side are $|m_1,m_2\rangle$ kets. The rows give the expansions of the eigenstates of individual angular momentum in terms of those of overall angular momentum. Thus, the second row tells us that

$$|1/2,-1/2\rangle = \frac{1}{\sqrt{2}}\left(|1,0\rangle + |0,0\rangle\right). \qquad (6.40)$$

Here, the ket on the left-hand side is a $|m_1,m_2\rangle$ ket, whereas those on the right-hand side are $|j,m\rangle$ kets.

Note that our table is really a combination of two sub-tables, one involving $j=0$ states, and one involving $j=1$ states. The Clebsch-Gordon coefficients corresponding to two different choices of j are completely independent. That is, there is no recursion relation linking Clebsch-Gordon coefficients corresponding to different values of j. Thus, for every choice of j_1, j_2, and j we can construct a table of Clebsch-Gordon coefficients corresponding to the different allowed values of m_1, m_2, and m (subject to the constraint that $m_1 + m_2 = m$). A complete knowledge of angular momentum addition is equivalent to knowing all possible tables of Clebsch-Gordon coefficients. These tables are listed (for moderate values of j_1, j_2 and j) in many standard reference books and articles [Beringer *et al.* (2012)].

6.5 Exercises

6.1 Calculate the Clebsch-Gordon coefficients for adding spin one-half to spin one.

6.2 An electron in a hydrogen atom occupies the combined spin and position state whose spinor-wavefunction is

$$\chi(r, \theta, \varphi) = R_{21}(r) \left[\sqrt{1/3}\, Y_{10}(\theta, \varphi)\chi_+ + \sqrt{2/3}\, Y_{11}(\theta, \varphi)\chi_- \right].$$

Here, $\chi_\pm$ are the eigenstates of S_z corresponding to the eigenvalues $\pm\hbar/2$, respectively, and r, θ, φ are conventional spherical coordinates.

  (a) What values would a measurement of L^2 yield, and with what probabilities?
  (b) Same for L_z.
  (c) Same for S^2.
  (d) Same for S_z.
  (e) Same for J^2.
  (f) Same for J_z.
  (g) What is the probability density for finding the electron at r, θ, φ?
  (h) What is the probability density for finding the electron in the spin-up state (with respect to the z-axis) at radius r?

  [Griffiths (2005)]

6.3 In a low energy neutron-proton system (with zero orbital angular momentum) the potential energy is given by

$$V(\mathbf{x}) = V_1(r) + V_2(r) \left[3 \frac{(\boldsymbol{\sigma}_n \cdot \mathbf{x})(\boldsymbol{\sigma}_p \cdot \mathbf{x})}{r^2} - \boldsymbol{\sigma}_n \cdot \boldsymbol{\sigma}_p \right] + V_3(r)\, \boldsymbol{\sigma}_n \cdot \boldsymbol{\sigma}_p,$$

where $\mathbf{x}$ is the vector connecting the two particles, $r = |\mathbf{x}|$, $\boldsymbol{\sigma}_n$ denotes the vector of the Pauli matrices of the neutron, and $\boldsymbol{\sigma}_p$ denotes the vector of the Pauli matrices of the proton. Calculate the potential energy for the neutron-proton system:

  (a) In the spin singlet (i.e., spin zero) state.
  (b) In the spin triplet (i.e., spin one) state.

  [Hint: Calculate the expectation value of $V(\mathbf{x})$ with respect to the overall spin state.] [Gasiorowicz (1996)]

6.4 Consider two electrons in a spin singlet (i.e., spin zero) state.

  (a) If a measurement of the spin of one of the electrons shows that it is in the state with $S_z = \hbar/2$, what is the probability that a measurement of the z-component of the spin of the other electron yields $S_z = \hbar/2$?

(b) If a measurement of the spin of one of the electrons shows that it is in the state with $S_y = \hbar/2$, what is the probability that a measurement of the x-component of the spin of the other electron yields $S_x = -\hbar/2$?

(c) Finally, if electron 1 is in a spin state described by $\cos\alpha_1 \chi_{z+} + \sin\alpha_1 e^{i\beta_1} \chi_{z-}$, and electron 2 is in a spin state described by $\cos\alpha_2 \chi_{z+} + \sin\alpha_2 e^{i\beta_2} \chi_{z-}$, what is the probability that the two-electron spin state is a triplet (i.e., spin one) state? Here, $\chi_{z\pm}$ are the eigenstates of S_z corresponding to the eigenvalues, $\pm\hbar/2$, respectively, for the electron in question. [Gasiorowicz (1996)]

Chapter 7

Time-Independent Perturbation Theory

7.1 Introduction

We have developed techniques by which the general energy eigenvalue problem can be reduced to a set of coupled partial differential equations involving various wavefunctions. Unfortunately, the number of such problems that yield exactly soluble systems of equations is comparatively small. It is, therefore, necessary to develop techniques for finding approximate solutions to otherwise intractable problems.

Consider the following problem, which is very common. The Hamiltonian of some quantum mechanical system is written

$$H = H_0 + H_1. \tag{7.1}$$

Here, H_0 is a simple Hamiltonian for which we know the exact eigenvalues and eigenstates. H_1 introduces some interesting additional physics into the problem, but it is sufficiently complicated that, when we add it to H_0, we can no longer find the exact energy eigenvalues and eigenstates. However, H_1 can, in some sense (which we shall specify more exactly later on), be regarded as small compared to H_0. Let us try to find approximate eigenvalues and eigenstates of the modified Hamiltonian, $H_0 + H_1$, by performing a perturbation expansion about the eigenvalues and eigenstates of the original Hamiltonian, H_0.

We shall start, in this chapter, by considering *time-independent perturbation theory* [Schrödinger (1926a)], in which the modification to the Hamiltonian, H_1, has no explicit dependence on time. It is usually assumed that the unperturbed Hamiltonian, H_0, is also time independent.

7.2   Two-State System

Consider the simplest non-trivial system, in which there are only two independent eigenkets of the unperturbed Hamiltonian. These are denoted

$$H_0 |1\rangle = E_1 |1\rangle, \tag{7.2}$$

$$H_0 |2\rangle = E_2 |2\rangle. \tag{7.3}$$

It is assumed that these states, and their associated eigenvalues, are known. Because H_0 is, by definition, an Hermitian operator, its two eigenkets are mutually orthogonal and form a complete set. The lengths of these eigenkets are both normalized to unity. Let us now try to solve the modified energy eigenvalue problem

$$(H_0 + H_1)|E\rangle = E|E\rangle. \tag{7.4}$$

In fact, we can solve this problem exactly. Because the eigenkets of H_0 form a complete set, we can write

$$|E\rangle = \langle 1|E\rangle|1\rangle + \langle 2|E\rangle|2\rangle. \tag{7.5}$$

Substituting the previous expansion into Equation (7.4), and then right-multiplying by either $\langle 1|$ or $\langle 2|$, we get two coupled equations that can be written in matrix form:

$$\begin{pmatrix} E_1 - E + e_{11}, & e_{12} \\ e_{12}^*, & E_2 - E + e_{22} \end{pmatrix} \begin{pmatrix} \langle 1|E\rangle \\ \langle 2|E\rangle \end{pmatrix} = \begin{pmatrix} 0 \\ 0 \end{pmatrix}. \tag{7.6}$$

Here,

$$e_{11} = \langle 1| H_1 |1\rangle, \tag{7.7}$$

$$e_{22} = \langle 2| H_1 |2\rangle, \tag{7.8}$$

$$e_{12} = \langle 1| H_1 |2\rangle \tag{7.9}$$

are the so-called *matrix elements* of the perturbing Hamiltonian (with respect to the unperturbed eigenstates). In the special (but common) case of a perturbing Hamiltonian whose diagonal matrix elements are zero, so that

$$e_{11} = e_{22} = 0, \tag{7.10}$$

the non-trivial solution of Equation (7.6) (obtained by setting the determinant of the matrix equal to zero [Riley *et al.* (2013)]) is

$$E = \frac{(E_1 + E_2) \pm \sqrt{(E_1 - E_2)^2 + 4\,|e_{12}|^2}}{2}. \tag{7.11}$$

Let us expand in the supposedly small parameter

$$\epsilon = \frac{|e_{12}|}{|E_1 - E_2|}. \tag{7.12}$$

We obtain

$$E \simeq \frac{1}{2}(E_1 + E_2) \pm \frac{1}{2}(E_1 - E_2)(1 + 2\epsilon^2 + \cdots). \qquad (7.13)$$

The previous expression yields the modifications to the energy eigenvalues caused by the perturbing Hamiltonian:

$$E_1' = E_1 + \frac{|e_{12}|^2}{E_1 - E_2} + \cdots, \qquad (7.14)$$

$$E_2' = E_2 - \frac{|e_{12}|^2}{E_1 - E_2} + \cdots. \qquad (7.15)$$

Note that H_1 causes the upper eigenvalue to increase, and the lower eigenvalue to decrease. It is easily demonstrated that the modified eigenkets take the form

$$|1\rangle' = |1\rangle + \frac{e_{12}^*}{E_1 - E_2}|2\rangle + \cdots, \qquad (7.16)$$

$$|2\rangle' = |2\rangle - \frac{e_{12}}{E_1 - E_2}|1\rangle + \cdots. \qquad (7.17)$$

Thus, the modified energy eigenstates consist of one of the unperturbed eigenstates with a slight admixture of the other. Actually, the series expansion on the right-hand side of Equation (7.13) only converges when $2|\epsilon| < 1$ [Gradshteyn *et al.* (1980)]. This suggests that the condition for the validity of the perturbation expansion is

$$|e_{12}| < \frac{|E_1 - E_2|}{2}. \qquad (7.18)$$

In other words, when we say that H_1 must be small compared to H_0, what we really mean is that the previous inequality needs to be satisfied.

7.3 Non-Degenerate Perturbation Theory

Let us now generalize our perturbation analysis to deal with systems possessing more than two energy eigenstates. The energy eigenstates of the unperturbed Hamiltonian, H_0, are denoted

$$H_0|n\rangle = E_n|n\rangle, \qquad (7.19)$$

where n runs from 1 to N. The eigenkets $|n\rangle$ are orthonormal, and form a complete set. Let us now try to solve the energy eigenvalue problem for the perturbed Hamiltonian:

$$(H_0 + H_1)|E\rangle = E|E\rangle. \qquad (7.20)$$

We can express $|E\rangle$ as a linear superposition of the unperturbed energy eigenkets,

$$|E\rangle = \sum_k \langle k|E\rangle |k\rangle, \tag{7.21}$$

where the summation is from $k = 1$ to N. Substituting the previous equation into Equation (7.20), and right-multiplying by $\langle m|$, we obtain

$$(E_m + e_{mm} - E)\langle m|E\rangle + \sum_{k \neq m} e_{mk} \langle k|E\rangle = 0, \tag{7.22}$$

where

$$e_{mk} = \langle m| H_1 |k\rangle. \tag{7.23}$$

Let us now develop our perturbation expansion. We assume that

$$\frac{|e_{mk}|}{E_m - E_k} \sim O(\epsilon), \tag{7.24}$$

for all $m \neq k$, where $\epsilon \ll 1$ is our expansion parameter. We also assume that

$$\frac{|e_{mm}|}{E_m} \sim O(\epsilon), \tag{7.25}$$

for all m. Let us search for a modified version of the nth unperturbed energy eigenstate for which

$$E = E_n + O(\epsilon), \tag{7.26}$$

and

$$\langle n|E\rangle = 1, \tag{7.27}$$

$$\langle m|E\rangle \sim O(\epsilon), \tag{7.28}$$

for $m \neq n$. Suppose that we write out Equation (7.22) for $m \neq n$, neglecting terms that are $O(\epsilon^2)$ according to our expansion scheme. We find that

$$(E_m - E_n)\langle m|E\rangle + e_{mn} \simeq 0, \tag{7.29}$$

giving

$$\langle m|E\rangle \simeq -\frac{e_{mn}}{E_m - E_n}. \tag{7.30}$$

Substituting the previous expression into Equation (7.22), evaluated for $m = n$, and neglecting $O(\epsilon^3)$ terms, we obtain

$$(E_n + e_{nn} - E) - \sum_{k \neq n} \frac{|e_{nk}|^2}{E_k - E_n} \simeq 0. \tag{7.31}$$

Thus, the modified nth energy eigenstate possesses the eigenvalue

$$E'_n = E_n + e_{nn} + \sum_{k \neq n} \frac{|e_{nk}|^2}{E_n - E_k} + O(\epsilon^3), \qquad (7.32)$$

and the eigenket

$$|n\rangle' = |n\rangle + \sum_{k \neq n} \frac{e_{kn}}{E_n - E_k} |k\rangle + O(\epsilon^2). \qquad (7.33)$$

Note that

$$\langle m|n\rangle' = \delta_{mn} + \frac{e^*_{nm}}{E_m - E_n} + \frac{e_{mn}}{E_n - E_m} + O(\epsilon^2) = \delta_{mn} + O(\epsilon^2). \qquad (7.34)$$

Thus, the modified eigenkets remain orthonormal to $O(\epsilon^2)$.

Note, finally, that if the perturbing Hamiltonian, H_1, commutes with the unperturbed Hamiltonian, H_0, then

$$e_{ml} = e_{mm}\,\delta_{ml}, \qquad (7.35)$$

and

$$E'_n = E_n + e_{nn}, \qquad (7.36)$$

$$|n\rangle' = |n\rangle. \qquad (7.37)$$

The previous two equations are exact (i.e., they hold to all orders in ϵ).

7.4 Quadratic Stark Effect

The *Stark effect* is a phenomenon by which the energy eigenstates of an atomic or molecular system are modified in the presence of a static, external, electric field. This phenomenon was first observed experimentally (in hydrogen) by J. Stark in 1913 [Stark (1914)]. Let us employ perturbation theory to investigate the Stark effect.

Suppose that a hydrogen-like atom [i.e., either a hydrogen atom, or an alkali metal atom (which possesses one valance electron orbiting outside a closed, spherically symmetric, shell)] is subjected to a uniform electric field, $\mathbf{E}$, pointing in the positive z-direction. The Hamiltonian of the system can be split into two parts. The unperturbed Hamiltonian,

$$H_0 = \frac{p^2}{2\,m_e} + V(r), \qquad (7.38)$$

and the perturbing Hamiltonian,

$$H_1 = e\,|\mathbf{E}|\,z. \qquad (7.39)$$

It is assumed that the unperturbed energy eigenvalues and eigenstates are completely known. The electron spin is irrelevant in this problem (because the spin operators all commute with H_1), so we can ignore the spin degrees of freedom of the system. This implies that the system possesses no degenerate energy eigenvalues. Actually, this is not true for the $n \neq 1$ energy levels of the hydrogen atom, because of the special properties of a pure Coulomb potential. (See Section 4.6.) It is necessary to deal with this case separately, because the perturbation theory presented in Section 7.3 breaks down for degenerate unperturbed energy levels. (See Section 7.5.)

An energy eigenket of the unperturbed Hamiltonian is characterized by three quantum numbers—the principal quantum number n, and the azimuthal and magnetic quantum numbers, l and m, respectively. (See Section 4.6.) Let us denote such a ket $|n, l, m\rangle$, and let its energy be E_{nlm}. According to Equation (7.32), the change in this energy induced by a small external electric field (i.e., small compared to the typical electric field internal to the atom) is given by

$$\Delta E_{nlm} = e\,|\mathbf{E}|\,\langle n, l, m|\,z\,|n, l, m\rangle + e^2\,|\mathbf{E}|^2 \sum_{n', l', m' \neq n, l, m} \frac{|\langle n, l, m|\,z\,|n,'\,l',\,m'\rangle|^2}{E_{nlm} - E_{n'\,l'\,m'}}. \quad (7.40)$$

Now, because

$$L_z = x\,p_y - y\,p_x, \quad (7.41)$$

it follows that

$$[L_z, z] = 0. \quad (7.42)$$

(See Chapter 4.) Thus,

$$\langle n, l, m|\,[L_z, z]\,|n', l', m'\rangle = 0, \quad (7.43)$$

giving

$$(m - m')\,\langle n, l, m|\,z\,|n', l', m'\rangle = 0, \quad (7.44)$$

because $|n, l, m\rangle$ is, by definition, an eigenstate of L_z corresponding to the eigenvalue $m\hbar$. It is clear, from the previous equation, that the matrix element $\langle n, l, m|\,z\,|n', l', m'\rangle$ is zero unless $m' = m$. This is termed the *selection rule* for the magnetic quantum number, m.

Let us now determine the selection rule for the azimuthal quantum number, l. We have

$$[L^2, z] = [L_x^2, z] + [L_y^2, z] = L_x\,[L_x, z] + [L_x, z]\,L_x + L_y\,[L_y, z] + [L_y, z]\,L_y$$

$$= i\,\hbar\left(-L_x\,y - y\,L_x + L_y\,x + x\,L_y\right) = 2\,i\,\hbar\,(L_y\,x - L_x\,y + i\,\hbar\,z)$$

$$= 2\,i\,\hbar\,(L_y\,x - y\,L_x) = 2\,i\,\hbar\,(x\,L_y - L_x\,y), \quad (7.45)$$

where use has been made of Equations (4.1)–(4.6). Similarly,

$$[L^2, y] = 2\,i\,\hbar\,(L_x\,z - x\,L_z), \tag{7.46}$$

$$[L^2, x] = 2\,i\,\hbar\,(y\,L_z - L_y\,z). \tag{7.47}$$

Thus,

$$\begin{aligned}
[L^2, [L^2, z]] &= 2\,i\,\hbar\left(L^2, L_y\,x - L_x\,y + i\,\hbar\,z\right) \\
&= 2\,i\,\hbar\left(L_y\,[L^2, x] - L_x\,[L^2, y] + i\,\hbar\,[L^2, z]\right) \\
&= -4\,\hbar^2\,L_y\,(y\,L_z - L_y\,z) + 4\,\hbar^2\,L_x\,(L_x\,z - x\,L_z) \\
&\quad - 2\,\hbar^2\,(L^2\,z - z\,L^2).
\end{aligned} \tag{7.48}$$

This reduces to

$$[L^2, [L^2, z]] = -\hbar^2\left[4\,(L_x\,x + L_y\,y + L_z\,z)\,L_z - 4\,(L_x^2 + L_y^2 + L_z^2)\,z \right.$$
$$\left. + 2\,(L^2\,z - z\,L^2)\right]. \tag{7.49}$$

However, it is clear from Equations (4.1)–(4.3) that

$$L_x\,x + L_y\,y + L_z\,z = 0. \tag{7.50}$$

Hence, we obtain

$$[L^2, [L^2, z]] = 2\,\hbar^2\,(L^2\,z + z\,L^2), \tag{7.51}$$

which can be expanded to give

$$L^4\,z - 2\,L^2\,z\,L^2 + z\,L^4 - 2\,\hbar^2\,(L^2\,z + z\,L^2) = 0. \tag{7.52}$$

Equation (7.52) implies that

$$\langle n, l, m|\,L^4\,z - 2\,L^2\,z\,L^2 + z\,L^4 - 2\,\hbar^2\,(L^2\,z + z\,L^2)\,|n', l', m'\rangle = 0. \tag{7.53}$$

Because $|n, l, m\rangle$ is, by definition, an eigenstate of L^2 corresponding to the eigenvalue $l\,(l+1)\,\hbar^2$, the previous expression yields

$$\left[l^2\,(l+1)^2 - 2\,l\,(l+1)\,l'\,(l'+1) + l'^2\,(l'+1)^2 \right.$$
$$\left. - 2\,l\,(l+1) - 2\,l'\,(l'+1)\right]\langle n, l, m|\,z\,|n', l', m'\rangle = 0, \tag{7.54}$$

which reduces to

$$(l + l' + 2)\,(l + l')\,(l - l' + 1)\,(l - l' - 1)\,\langle n, l, m|\,z\,|n', l', m'\rangle = 0. \tag{7.55}$$

According to the previous formula, the matrix element $\langle n, l, m|\,z\,|n', l', m'\rangle$ vanishes unless $l = l' = 0$ or $l' = l \pm 1$. This matrix element can be written

$$\langle n, l, m|\,z\,|n', l', m'\rangle = \iiint d^3\mathbf{x}'\,\psi_{nlm}^*(r', \theta', \varphi')\,r'\cos\theta'\,\psi_{n'm'l'}(r', \theta', \varphi'), \tag{7.56}$$

where $\psi_{nlm}(\mathbf{x}') = \langle \mathbf{x}'|n,l,m\rangle$. Recall, however, that the wavefunction of an $l = 0$ state is spherically symmetric: that is, $\psi_{n00}(\mathbf{x}') = \psi_{n00}(r')$. (See Section 4.3.) It follows from Equation (7.56) that the matrix element vanishes, by symmetry, when $l = l' = 0$. In conclusion, the matrix element $\langle n,l,m|z|n',l',m'\rangle$ is zero unless $l' = l \pm 1$. This is the selection rule for the quantum number l.

Application of the previously derived selection rules for m and l to Equation (7.40) yields

$$\Delta E_{nlm} = e^2\,|\mathbf{E}|^2 \sum_{n'} \sum_{l'=l\pm1} \frac{|\langle n,l,m|z|n',l',m\rangle|^2}{E_{nlm} - E_{n'l'm}}. \tag{7.57}$$

Note that all of the terms appearing in Equation (7.40) that vary linearly with the electric field-strength vanish, by symmetry, according to the selection rules. Only those terms that vary quadratically with the field-strength survive. Hence, the energy-shift specified in the previous formula is known as the *quadratic Stark effect*.

The *electrical polarizability*, α_p, of an atom is defined in terms of the electric-field-induced energy-shift of a given atomic state as follows [Jackson (1975)]:

$$\Delta E = -\frac{1}{2}\,\alpha_p\,|\mathbf{E}|^2. \tag{7.58}$$

Consider the ground state of a hydrogen atom. (Recall, that we cannot address the $n > 1$ excited states because they are degenerate, and our theory cannot deal with this at present). The polarizability of the ground state is given by

$$\alpha_p = 2\,e^2 \sum_{n>1} \frac{|\langle 1,0,0|z|n,1,0\rangle|^2}{E_{n00} - E_{100}}. \tag{7.59}$$

Here, we have made use of the fact that $E_{n10} = E_{n00}$ for a hydrogen atom. (See Section 4.6.)

The sum in the previous expression can be evaluated approximately by noting that

$$E_{n00} = -\frac{e^2}{8\pi\,\epsilon_0\,a_0\,n^2} \tag{7.60}$$

for a hydrogen atom, where a_0 is the Bohr radius. [See Equation (4.132).] We can write

$$E_{n00} - E_{100} \geq E_{200} - E_{100} = \frac{3}{4}\,\frac{e^2}{8\pi\,\epsilon_0\,a_0}. \tag{7.61}$$

Thus,

$$\alpha_p < \frac{16}{3}\,4\pi\,\epsilon_0\,a_0 \sum_{n>1} |\langle 1,0,0|z|n,1,0\rangle|^2. \tag{7.62}$$

However,

$$\sum_{n>1} |\langle 1,0,0|z|n,1,0\rangle|^2 = \sum_{n',l',m'} \langle 1,0,0|z|n',l',m'\rangle\langle n',m',l'|z|1,0,0\rangle$$

$$= \langle 1,0,0|z^2|1,0,0\rangle, \tag{7.63}$$

where we have made use of the fact that the wavefunctions of a hydrogen atom form a complete set. It is easily demonstrated from the actual form of the ground-state wavefunction that

$$\langle 1,0,0|z^2|1,0,0\rangle = a_0^2. \tag{7.64}$$

(See Exercise 4.13.) Thus, we conclude that

$$\alpha_p < \frac{16}{3} 4\pi \epsilon_0 a_0^3 \simeq 5.3 \, 4\pi \epsilon_0 a_0^3. \tag{7.65}$$

The exact result is

$$\alpha_p = \frac{9}{2} 4\pi \epsilon_0 a_0^3 = 4.5 \, 4\pi \epsilon_0 a_0^3. \tag{7.66}$$

It is possible to obtain this result, without recourse to perturbation theory, by solving Schrödinger's equation in parabolic coordinates [Waller (1926); Epstein (1926)].

7.5 Degenerate Perturbation Theory

Let us now consider systems in which the eigenstates of the unperturbed Hamiltonian, H_0, possess degenerate energy levels. It is always possible to represent degenerate energy eigenstates as the simultaneous eigenstates of the Hamiltonian and some other Hermitian operator (or group of operators). Let us denote this operator (or group of operators) L. We can write

$$H_0|n,l\rangle = E_n|n,l\rangle, \tag{7.67}$$

and

$$L|n,l\rangle = L_{nl}|n,l\rangle, \tag{7.68}$$

where $[H_0, L] = 0$. Here, the E_n and the L_{nl} are real numbers that depend on the quantum numbers n, and n and l, respectively. It is always possible to find a sufficient number of operators that commute with the Hamiltonian in order to ensure that the L_{nl} are all different. In other words, we can choose L such that the quantum numbers n and l uniquely specify each eigenstate. Suppose that for each value of n there are N_n different values of l. In other words, the nth energy eigenstate is N_n-fold degenerate.

In general, L does not commute with the perturbing Hamiltonian, H_1. In this situation, we expect the perturbation to couple degenerate states with the same value of n, but different values of l. Let us naively attempt to use the standard perturbation theory of Section 7.3 to evaluate the modified energy eigenstates and energy levels. A direct generalization of Equations (7.32) and (7.33) yields

$$E'_{nl} = E_n + e_{nlnl} + \sum_{n',l' \neq n,l} \frac{|e_{n'l'nl}|^2}{E_n - E_{n'}} + O(\epsilon^3), \qquad (7.69)$$

and

$$|n, l\rangle' = |n, l\rangle + \sum_{n',l' \neq n,l} \frac{e_{n'l'nl}}{E_n - E_{n'}} |n', l'\rangle + O(\epsilon^2), \qquad (7.70)$$

where

$$e_{n'l'nl} = \langle n', l' | H_1 | n, l \rangle. \qquad (7.71)$$

It is fairly obvious that the summations in Equations (7.69) and (7.70) are not well behaved if the nth energy level is degenerate. The problem terms are those that involve coupling to unperturbed eigenstates labeled by the same value of n, but different values of l: that is, those states whose unperturbed energies are E_n. These terms give rise to singular factors $1/(E_n - E_n)$ in the summations. Note, however, that this problem would not exist if the matrix elements, $e_{nl'nl}$, of the perturbing Hamiltonian between distinct, degenerate, unperturbed energy eigenstates corresponding to the eigenvalue E_n were zero. In other words, if

$$\langle n, l' | H_1 | n, l \rangle = \lambda_{nl} \delta_{ll'} \qquad (7.72)$$

then all of the singular terms in Equations (7.69) and (7.70) would vanish.

In general, Equation (7.72) is not satisfied. Fortunately, we can always redefine the unperturbed energy eigenstates belonging to the eigenvalue E_n in such a manner that Equation (7.72) is satisfied. Let us define N_n new states that are linear combinations of the N_n original degenerate eigenstates corresponding to the eigenvalue E_n:

$$|n, l^{(1)}\rangle = \sum_{k=1,N_n} \langle n, k | n, l^{(1)} \rangle | n, k \rangle. \qquad (7.73)$$

Note that these new states are also degenerate energy eigenstates of the unperturbed Hamiltonian corresponding to the eigenvalue E_n. The $|n, l^{(1)}\rangle$ are chosen in such a manner that they are eigenstates of the perturbing Hamiltonian, H_1. Thus,

$$H_1 |n, l^{(1)}\rangle = \lambda_{nl} |n, l^{(1)}\rangle. \qquad (7.74)$$

The $|n, l^{(1)}\rangle$ are also chosen so that they are orthonormal. It follows that

$$\langle n, l'^{(1)} | H_1 | n, l^{(1)} \rangle = \lambda_{nl} \delta_{ll'}. \qquad (7.75)$$

Thus, if we use the new eigenstates, instead of the old ones, then we can employ Equations (7.69) and (7.70) directly, because all of the singular terms vanish. The only remaining difficulty is to determine the new eigenstates in terms of the original ones.

Now,

$$\sum_{l=1,N_n} |n, l\rangle\langle n, l| = 1, \tag{7.76}$$

where 1 denotes the identity operator in the sub-space of all unperturbed energy eigenkets corresponding to the eigenvalue E_n. Using this completeness relation, the operator eigenvalue equation (7.74) can be transformed into a straightforward matrix eigenvalue equation:

$$\sum_{l''=1,N_n} \langle n, l'|H_1|n, l''\rangle\langle n, l''|n, l^{(1)}\rangle = \lambda_{nl} \langle n, l'|n, l^{(1)}\rangle. \tag{7.77}$$

This can be written more transparently as

$$\mathbf{U}\mathbf{x} = \lambda\mathbf{x}, \tag{7.78}$$

where the elements of the $N_n \times N_n$ Hermitian matrix $\mathbf{U}$ are

$$U_{jk} = \langle n, j| H_1 |n, k\rangle. \tag{7.79}$$

Provided that the determinant of $\mathbf{U}$ is non-zero, Equation (7.78) can always be solved to give N_n eigenvalues λ_{nl} (for $l = 1$ to N_n), with N_n corresponding eigenvectors $\mathbf{x}_{nl}$ [Riley *et al.* (2013)]. The eigenvectors specify the weights of the new eigenstates in terms of the original eigenstates: that is,

$$(\mathbf{x}_{nl})_k = \langle n, k|n, l^{(1)}\rangle, \tag{7.80}$$

for $k = 1$ to N_n. In our new scheme, Equations (7.69) and (7.70) yield

$$E'_{nl} = E_n + \lambda_{nl} + \sum_{n' \neq n, l'} \frac{|e_{n'l'nl}|^2}{E_n - E_{n'}} + O(\epsilon^3), \tag{7.81}$$

and

$$|n, l^{(1)}\rangle' = |n, l^{(1)}\rangle + \sum_{n' \neq n, l'} \frac{e_{n'l'nl}}{E_n - E_{n'}} |n', l'^{(1)}\rangle + O(\epsilon^2). \tag{7.82}$$

There are no singular terms in these expressions, because the summations are over $n' \neq n$. In other words, they specifically exclude the problematic, degenerate, unperturbed energy eigenstates corresponding to the eigenvalue E_n. Note that the first-order energy-shifts are equivalent to the eigenvalues of the matrix equation (7.78).

Incidentally, it is clear, from the previous analysis, that if the perturbing Hamiltonian, H_1, commutes with H_0 and the operator (or group of operators) L then there is no danger of singular terms appearing in the perturbation expansion to second order, because $e_{n'l'nl} = e_{n'l'nl}\delta_{ll'}$. Another way of saying this is that there are no singular terms if the simultaneous eigenstates of H_0 and L are also eigenstates of H_1.

7.6 Linear Stark Effect

Let us examine the effect of a static, external, electric field on the excited energy levels of a hydrogen atom. For instance, consider the $n = 2$ states. There is a single $l = 0$ state, usually referred to as $2s$, and three $l = 1$ states (with $m = -1, 0, 1$), usually referred to as $2p$ [Gasiorowicz (1996)]. (In this notation, the 2 refers to the value of the principal quantum number n, whereas $l = 0, 1, 2, 3, \cdots$ are represented by the letters s, p, d, f, $\cdots$.) All of these states possess the same energy, $E_{200} = -e^2/(32\pi \epsilon_0 a_0)$. (See Section 4.6.) As in Section 7.4, the perturbing Hamiltonian is

$$H_1 = e\,|\mathbf{E}|\,z. \tag{7.83}$$

In order to apply perturbation theory, we have to solve the matrix eigenvalue equation

$$\mathbf{U}\,\mathbf{x} = \lambda\,\mathbf{x}, \tag{7.84}$$

where $\mathbf{U}$ is the array of the matrix elements of H_1 between the degenerate $2s$ and $2p$ states. Thus,

$$\mathbf{U} = e\,|\mathbf{E}| \begin{pmatrix} 0, & \langle 2,0,0|\,z\,|2,1,0\rangle, & 0, & 0 \\ \langle 2,1,0|\,z\,|2,0,0\rangle, & 0, & 0, & 0 \\ 0, & 0, & 0, & 0 \\ 0, & 0, & 0, & 0 \end{pmatrix}, \tag{7.85}$$

where the rows and columns correspond to the $|2,0,0\rangle$, $|2,1,0\rangle$, $|2,1,1\rangle$, and $|2,1,-1\rangle$ states, respectively. Here, we have made use of the selection rules, which tell us that the matrix element of z between two energy eigenstates of the hydrogen atom is zero unless the states possess the same magnetic quantum number, m, and azimuthal quantum numbers, l, that differ by unity. (See Section 7.4.) It is easily demonstrated, from the exact forms of the $2s$ and $2p$ wavefunctions, that

$$\langle 2,0,0|\,z\,|2,1,0\rangle = \langle 2,1,0|\,z\,|2,0,0\rangle = -3\,a_0. \tag{7.86}$$

(See Exercise 4.15.)

It can be seen, by inspection, that the eigenvalues of $\mathbf{U}$ are $\lambda_1 = 3\,e\,a_0\,|\mathbf{E}|$, $\lambda_2 = -3\,e\,a_0\,|\mathbf{E}|$, $\lambda_3 = 0$, and $\lambda_4 = 0$. The corresponding eigenvectors are

$$\mathbf{x}_1 = \begin{pmatrix} 1/\sqrt{2} \\ -1/\sqrt{2} \\ 0 \\ 0 \end{pmatrix}, \qquad \mathbf{x}_2 = \begin{pmatrix} 1/\sqrt{2} \\ 1/\sqrt{2} \\ 0 \\ 0 \end{pmatrix}, \qquad \mathbf{x}_3 = \begin{pmatrix} 0 \\ 0 \\ 1 \\ 0 \end{pmatrix}, \qquad \mathbf{x}_4 = \begin{pmatrix} 0 \\ 0 \\ 0 \\ 1 \end{pmatrix}. \tag{7.87}$$

It follows from Section 7.5 that the simultaneous eigenstates of the unperturbed Hamiltonian and the perturbing Hamiltonian take the form

$$|1\rangle = \frac{|2, 0, 0\rangle - |2, 1, 0\rangle}{\sqrt{2}}, \tag{7.88}$$

$$|2\rangle = \frac{|2, 0, 0\rangle + |2, 1, 0\rangle}{\sqrt{2}}, \tag{7.89}$$

$$|3\rangle = |2, 1, 1\rangle, \tag{7.90}$$

$$|4\rangle = |2, 1, -1\rangle. \tag{7.91}$$

In the absence of an external electric field, all of these states possess the same energy, E_{200}. The first-order energy-shifts induced by such a field are given by

$$\Delta E_1 = +3\, e\, a_0\, |\mathbf{E}|, \tag{7.92}$$

$$\Delta E_2 = -3\, e\, a_0\, |\mathbf{E}|, \tag{7.93}$$

$$\Delta E_3 = 0, \tag{7.94}$$

$$\Delta E_4 = 0. \tag{7.95}$$

Thus, the energies of states 1 and 2 are shifted upwards and downwards, respectively, by an amount $3\, e\, a_0\, |\mathbf{E}|$, in the presence of an external electric field. States 1 and 2 are orthogonal linear combinations of the original $2s$ and $2p(m = 0)$ states. Note that the energy-shifts are linear in the electric field-strength. Consequently, this phenomenon is known as the *linear Stark effect*. Of course, for weak perturbing electric fields, the linear Stark effect is much larger effect that the quadratic effect described in Section 7.4. The energies of states 3 and 4 [which are equivalent to the original $2p(m = 1)$ and $2p(m = -1)$ states, respectively] are not affected to first order in the perturbation expansion. Of course, to second order, the energies of these states are shifted by an amount that depends on the square of the electric field-strength.

The linear Stark effect depends crucially on the degeneracy of the $2s$ and $2p$ states. This degeneracy is a special property of a pure Coulomb potential, and, therefore, only applies to a hydrogen atom. (See Section 4.6.) Thus, alkali metal atoms do not exhibit the linear Stark effect.

7.7 Fine Structure

Let us now consider the energy levels of hydrogen-like atoms (i.e., alkali metal atoms) in more detail. The outermost electron moves in a spherically symmetric potential, $V(r)$, generated by the nuclear charge and the charges of the other electrons (which occupy spherically symmetric closed shells). Thus, according to Section 4.5, we can label the energy eigenstates of the outermost electron using the conventional quantum numbers n, l, and m. However, the shielding effect of the inner electrons causes $V(r)$ to depart from the pure Coulomb form. This splits the degeneracy of states characterized by the same value of n, but different values of l. In fact, higher l states have higher energies.

Let us examine a phenomenon known as *fine structure*, which caused by the interaction between the spin and orbital angular momenta of the outermost electron. This electron experiences an electric field [Fitzpatrick (2008)]

$$\mathbf{E} = \frac{\nabla V}{e}. \tag{7.96}$$

However, a non-relativistic charge moving in an electric field also experiences an effective magnetic field [Fitzpatrick (2008)]

$$\mathbf{B} = -\frac{\mathbf{v} \times \mathbf{E}}{c^2}. \tag{7.97}$$

Now, an electron possesses a spin magnetic moment

$$\boldsymbol{\mu} = -\frac{e\,\mathbf{S}}{m_e}. \tag{7.98}$$

[See Equation (5.47).] We, therefore, expect a contribution to the Hamiltonian of the form [Fitzpatrick (2008)]

$$H_{LS} = -\boldsymbol{\mu} \cdot \mathbf{B} = -\frac{e\,\mathbf{S}}{m_e\,c^2} \cdot \mathbf{v} \times \left(\frac{1}{e}\frac{\mathbf{x}}{r}\frac{dV}{dr} \right) = \frac{1}{m_e^2\,c^2\,r}\frac{dV}{dr}\,\mathbf{L} \cdot \mathbf{S}, \tag{7.99}$$

where $\mathbf{L} = m_e\,\mathbf{x} \times \mathbf{v}$ is the orbital angular momentum. This phenomenon is known as *spin-orbit coupling*. Actually, when the previous expression is compared to the experimentally observed spin-orbit interaction, it is found to be too large by a factor of two. There is a classical explanation for this discrepancy which involves a precession of the electron spin—the so-called *Thomas precession*—caused by the relativistic time dilation between the orbiting electron and the atomic nucleus [Thomas (1926)]. The quantum mechanical explanation requires a relativistically covariant treatment of the electron dynamics [Dirac (1958); Bjorken and Drell (1964)].

Let us now apply perturbation theory to a hydrogen-like atom, using

$$H_{LS} = \frac{1}{2\,m_e^2\,c^2\,r}\frac{dV}{dr}\,\mathbf{L} \cdot \mathbf{S} \tag{7.100}$$

as the perturbation (noting that H_{LS} takes one half of the value given previously), and

$$H_0 = \frac{p^2}{2\,m_e} + V(r) \tag{7.101}$$

as the unperturbed Hamiltonian. We have two choices for the energy eigenstates of H_0. We can adopt the simultaneous eigenstates of H_0, L^2, S^2, L_z and S_z, or the simultaneous eigenstates of H_0, L^2, S^2, J^2, and J_z, where $\mathbf{J} = \mathbf{L} + \mathbf{S}$ is the total angular momentum. Although the departure of $V(r)$ from a pure $1/r$ form splits the degeneracy of same n, different l, states, those states characterized by the same values of n and l, but different values of m_l, are still degenerate. (Here, m_l, m_s, and m_j are the quantum numbers corresponding to L_z, S_z, and J_z, respectively.) Moreover, with the addition of spin degrees of freedom, each state is doubly degenerate because of the two possible orientations of the electron spin (i.e., $m_s = \pm 1/2$). Thus, we are still dealing with a highly degenerate system. However, we know, from Section 7.5, that there is no danger of singular terms appearing to second order in the perturbation expansion if the degenerate eigenstates of the unperturbed Hamiltonian (and the set of commuting operators needed to uniquely label the degenerate eigenstates) are also eigenstates of the perturbing Hamiltonian. Now, the perturbing Hamiltonian, H_{LS}, is proportional to $\mathbf{L} \cdot \mathbf{S}$, where

$$\mathbf{L} \cdot \mathbf{S} = \frac{J^2 - L^2 - S^2}{2}. \tag{7.102}$$

It is fairly obvious that the first group of operators (H_0, L^2, S^2, L_z and S_z) does not commute with H_{LS}, whereas the second group (H_0, L^2, S^2, J^2, and J_z) does. In fact, $\mathbf{L} \cdot \mathbf{S}$ is just a combination of operators appearing in the second group. Thus, it is advantageous to work in terms of the eigenstates of the second group of operators, rather than those of the first group (because the former eigenstates are also eigenstates of the perturbing Hamiltonian).

We now need to find the simultaneous eigenstates of H_0, L^2, S^2, J^2, and J_z. This is equivalent to finding the eigenstates of the total angular momentum resulting from the addition of two angular momenta: $j_1 = l$, and $j_2 = s = 1/2$. According to Equation (6.26), the allowed values of the total angular momentum are $j = l + 1/2$ and $j = l - 1/2$. We can write

$$|l + 1/2, m_j\rangle = \cos \alpha \, |m_j - 1/2, 1/2\rangle + \sin \alpha \, |m_j + 1/2, -1/2\rangle, \tag{7.103}$$

$$|l - 1/2, m_j\rangle = -\sin \alpha \, |m_j - 1/2, 1/2\rangle + \cos \alpha \, |m_j + 1/2, -1/2\rangle. \tag{7.104}$$

Here, the kets on the left-hand side are $|j, m_j\rangle$ kets, whereas those on the right-hand side are $|m_l, m_s\rangle$ kets (the j_1, j_2 labels have been dropped, for the sake of clarity). We have made use of the fact that the Clebsch-Gordon coefficients are automatically zero unless $m_j = m_l + m_s$. (See Section 6.3.) We have also made use of the fact that both the $|j, m_j\rangle$ and $|m_l, m_s\rangle$ kets are orthonormal. We now need to determine

$$\cos \alpha = \langle m_j - 1/2, 1/2 | l + 1/2, m_j \rangle, \tag{7.105}$$

where the Clebsch-Gordon coefficient is written in $\langle m_l, m_s | j, m_j \rangle$ form.

Let us now employ the recursion relation for Clebsch-Gordon coefficients, Equation (6.32), with $j_1 = l$, $j_2 = 1/2$, $j = l + 1/2$, $m_1 = m_j - 1/2$, and $m_2 = 1/2$, choosing the lower sign. We obtain

$$[(l + 1/2)(l + 3/2) - m_j(m_j + 1)]^{1/2} \langle m_j - 1/2, 1/2 | l + 1/2, m_j \rangle$$
$$= [l(l + 1) - (m_j - 1/2)(m_j + 1/2)]^{1/2} \langle m_j + 1/2, 1/2 | l + 1/2, m_j + 1 \rangle, \quad (7.106)$$

which reduces to

$$\langle m_j - 1/2, 1/2 | l + 1/2, m_j \rangle = \sqrt{\frac{l + m_j + 1/2}{l + m_j + 3/2}} \, \langle m_j + 1/2, 1/2 | l + 1/2, m_j + 1 \rangle. \quad (7.107)$$

We can use this formula to successively increase the value of m_l. For instance,

$$\langle m_j - 1/2, 1/2 | l + 1/2, m_j \rangle$$
$$= \sqrt{\frac{l + m_j + 1/2}{l + m_j + 3/2}} \sqrt{\frac{l + m_j + 3/2}{l + m_j + 5/2}} \, \langle m_j + 3/2, 1/2 | l + 1/2, m_j + 2 \rangle. \quad (7.108)$$

This procedure can be continued until m_l attains its maximum possible value, l. Thus,

$$\langle m_j - 1/2, 1/2 | l + 1/2, m \rangle = \sqrt{\frac{l + m_j + 1/2}{2l + 1}} \, \langle l, 1/2 | l + 1/2, l + 1/2 \rangle. \quad (7.109)$$

Consider the situation in which m_l and m_s both take their maximum values, l and $1/2$, respectively. The corresponding value of m_j is $l + 1/2$. This value is possible when $j = l + 1/2$, but not when $j = l - 1/2$. Thus, the $|m_l, m_s\rangle$ ket $|l, 1/2\rangle$ must be equal to the $|j, m_j\rangle$ ket $|l + 1/2, l + 1/2\rangle$, up to an arbitrary phase-factor. By convention, this factor is taken to be unity, giving

$$\langle l, 1/2 | l + 1/2, l + 1/2 \rangle = 1. \quad (7.110)$$

It follows from Equation (7.109) that

$$\cos \alpha = \langle m_j - 1/2, 1/2 | l + 1/2, m_j \rangle = \sqrt{\frac{l + m_j + 1/2}{2l + 1}}. \quad (7.111)$$

Hence,

$$\sin^2 \alpha = 1 - \frac{l + m_j + 1/2}{2l + 1} = \frac{l - m_j + 1/2}{2l + 1}. \quad (7.112)$$

We now need to determine the sign of $\sin\alpha$. A careful examination of the recursion relation, Equation (6.32), shows that the plus sign is appropriate. Thus,

$$|l+1/2, m_j\rangle = \sqrt{\frac{l+m_j+1/2}{2l+1}}\, |m_j-1/2, 1/2\rangle$$

$$+ \sqrt{\frac{l-m_j+1/2}{2l+1}}\, |m_j+1/2, -1/2\rangle, \qquad (7.113)$$

$$|l-1/2, m_j\rangle = -\sqrt{\frac{l-m_j+1/2}{2l+1}}\, |m_j-1/2, 1/2\rangle$$

$$+ \sqrt{\frac{l+m_j+1/2}{2l+1}}\, |m_j+1/2, -1/2\rangle. \qquad (7.114)$$

It is convenient to define so-called *spin-angular functions* using the Pauli two-component formalism:

$$\mathcal{Y}_{l\,m_j}^{j=l\pm1/2} \equiv \mathcal{Y}_{l\,m_j}^{\pm} = \pm\sqrt{\frac{l\pm m_j+1/2}{2l+1}}\, Y_{l\,m_j-1/2}(\theta,\varphi)\chi_+$$

$$+ \sqrt{\frac{l\mp m_j+1/2}{2l+1}}\, Y_{l\,m_j+1/2}(\theta,\varphi)\chi_-$$

$$= \frac{1}{\sqrt{2l+1}}\left(\begin{array}{c} \pm\sqrt{l\pm m_j+1/2}\; Y_{l\,m_j-1/2}(\theta,\varphi) \\ \sqrt{l\mp m_j+1/2}\; Y_{l\,m_j+1/2}(\theta,\varphi) \end{array}\right). \qquad (7.115)$$

(See Section 5.7.) These functions are eigenfunctions of the total angular momentum for spin one-half particles, just as the spherical harmonics are eigenfunctions of the orbital angular momentum. A general spinor-wavefunction for an energy eigenstate in a hydrogen-like atom is written

$$\psi_{nlm_j\pm} = R_{nl}(r)\,\mathcal{Y}_{l\,m_j}^{\pm}. \qquad (7.116)$$

The radial part of the wavefunction, $R_{nl}(r)$, depends on the principal quantum number n, and the azimuthal quantum number l. The wavefunction is also labeled by m_j, which is the quantum number associated with J_z. For a given choice of l, the quantum number j (i.e., the quantum number associated with J^2) can take the values $l\pm 1/2$. (However, $j=1/2$ for the special case $l=0$.)

The $|l\pm 1/2, m_j\rangle$ kets are eigenstates of $\mathbf{L}\cdot\mathbf{S}$, according to Equation (7.102). Thus,

$$\mathbf{L}\cdot\mathbf{S}\,|j=l\pm 1/2, m_j\rangle = \frac{\hbar^2}{2}\,[j(j+1)-l(l+1)-3/4]\,|j, m_j\rangle, \qquad (7.117)$$

giving

$$\mathbf{L}\cdot\mathbf{S}\,|l+1/2, m_j\rangle = \frac{l\hbar^2}{2}\,|l+1/2, m_j\rangle, \qquad (7.118)$$

$$\mathbf{L}\cdot\mathbf{S}\,|l-1/2, m_j\rangle = -\frac{(l+1)\hbar^2}{2}\,|l-1/2, m_j\rangle. \qquad (7.119)$$

It follows that

$$\oint d\Omega \, (\mathcal{Y}_{l m_j}^+)^\dagger \, \mathbf{L} \cdot \mathbf{S} \, \mathcal{Y}_{l m_j}^+ = \frac{l \hbar^2}{2}, \tag{7.120}$$

$$\oint d\Omega \, (\mathcal{Y}_{l m_j}^-)^\dagger \, \mathbf{L} \cdot \mathbf{S} \, \mathcal{Y}_{l m_j}^- = -\frac{(l+1) \hbar^2}{2}, \tag{7.121}$$

where the integrals are over all solid angle, $d\Omega = \sin\theta \, d\theta \, d\varphi$.

Let us now apply degenerate perturbation theory to evaluate the energy-shift of a state whose spinor-wavefunction is $\psi_{n l m_j \pm}$ caused by the spin-orbit Hamiltonian, H_{LS}. To first order, the energy-shift is given by

$$\Delta E_{n l m_j \pm} = \int d^3 \mathbf{x} \, (\psi_{n l m_j \pm})^\dagger \, H_{LS} \, \psi_{n l m_j \pm}, \tag{7.122}$$

where the integral is over all space, $d^3 \mathbf{x} = r^2 \, d\Omega$. Equations (7.100), (7.116), and (7.120)–(7.121) yield

$$\Delta E_{n l m_j +} = +\frac{1}{2 \, m_e^2 \, c^2} \left\langle \frac{1}{r} \frac{dV}{dr} \right\rangle \frac{l \hbar^2}{2}, \tag{7.123}$$

$$\Delta E_{n l m_j -} = -\frac{1}{2 \, m_e^2 \, c^2} \left\langle \frac{1}{r} \frac{dV}{dr} \right\rangle \frac{(l+1) \hbar^2}{2}, \tag{7.124}$$

where

$$\left\langle \frac{1}{r} \frac{dV}{dr} \right\rangle = \int_0^\infty dr \, r^2 \, (R_{n l})^* \, \frac{1}{r} \frac{dV}{dr} \, R_{n l}. \tag{7.125}$$

Incidentally, for the special case of an $l = 0$ state, $j = l + 1/2 = 1/2$, and there is no state with $j = l - 1/2$, so Equation (7.124) is redundant. Thus, it is clear that $\Delta E_{n l m_j} = 0$ for an $l = 0$ state, which is not surprising, given that such a state possesses zero orbital angular momentum (i.e., it is characterized by $\mathbf{L} = \mathbf{0}$.)

Let us now apply the previous result to the case of a sodium atom. In chemist's notation [Gasiorowicz (1996)], the ground state is written

$$(1s)^2 (2s)^2 (2p)^6 (3s). \tag{7.126}$$

[Here, $(1s)^2$ implies that there are two electrons in the $1s$ state, et cetera.] The inner ten electrons effectively form a spherically symmetric electron cloud. We are interested in the excitation of the eleventh electron from $3s$ to some higher energy state. The closest (in energy) unoccupied state is $3p$. This state has a higher energy than $3s$ due to the deviations of the potential from the pure Coulomb form. In the absence of spin-orbit interaction, there are six degenerate $3p$ states. The spin-orbit interaction breaks the degeneracy of these states. The modified states are labeled $(3p)_{1/2}$ and $(3p)_{3/2}$, where the subscript refers to the value of j. The four $(3p)_{3/2}$ states lie at a slightly higher energy level than the two $(3p)_{1/2}$ states, because the radial integral (7.125) is positive. (See Exercise 7.14.) The splitting of the (3p) energy levels of the sodium atom can be observed using a spectroscope (which measures the frequency of spectral lines caused by transitions between quantum states of different energy—this frequency is, of course, $v = \Delta E/h$— see Section 8.9). The well-known *sodium D-line* is associated with transitions between the 3p and 3s states. The fact that there are two slightly different 3p energy levels (note that spin-orbit coupling does not split the 3s energy levels) means that the sodium D-line actually consists of two very closely spaced (in frequency) spectroscopic lines. It is easily demonstrated that the ratio of the typical spacing of *Paschen lines* (i.e., spectral lines associated with transitions to the 3s state [Griffiths (2005)]) to the splitting brought about by spin-orbit interaction is about $1 : \alpha^2$, where $\alpha \simeq 1/137$ is the fine structure constant. (See Exercise 7.14.) Actually, Equations (7.123)–(7.124) are not entirely correct, because we have neglected an effect (namely, the relativistic mass increase of the electron) that is the same order of magnitude as spin-orbit coupling. (See Exercises 7.11–7.14.)

7.8 Zeeman Effect

The *Zeeman effect* is a phenomenon by which the energy eigenstates of an atomic or molecular system are modified in the presence of a static, external, magnetic field. This phenomenon was first observed experimentally by P. Zeeman in 1897 [Zeeman (1897)]. Let us use perturbation theory to investigate the Zeeman effect.

Consider a hydrogen-like atom placed in a uniform z-directed magnetic field, **B**. The change in energy of the outermost electron is [Fitzpatrick (2008)]

$$H_B = -\mu \cdot \mathbf{B}, \tag{7.127}$$

where

$$\mu = -\frac{e}{2\,m_e}\,(\mathbf{L} + 2\,\mathbf{S}) \tag{7.128}$$

is its magnetic moment, including both the spin and orbital contributions. (See Section 5.5.) Thus,

$$H_B = \frac{e\,B}{2\,m_e}\,(L_z + 2\,S_z). \tag{7.129}$$

Suppose that the energy-shifts induced by the magnetic field are much smaller than those induced by spin-orbit interaction. In this situation, we can treat H_B as a small perturbation acting on the eigenstates of $H_0 + H_{LS}$. Of course, these states are the simultaneous eigenstates of J^2 and J_z. Let us consider one of these states, labeled by the quantum numbers j and m_j, where $j = l \pm 1/2$. From standard perturbation theory, the first-order energy-shift in the presence of a magnetic field is

$$\Delta E_{nlm_j\pm} = \langle l \pm 1/2, m_j| H_B |l \pm 1/2, m_j \rangle. \tag{7.130}$$

Because

$$L_z + 2 S_z = J_z + S_z, \tag{7.131}$$

we find that

$$\Delta E_{nlm_j\pm} = \frac{e B}{2 m_e} \left(m_j \hbar + \langle l \pm 1/2, m_j| S_z |l \pm 1/2, m_j \rangle \right). \tag{7.132}$$

Now, from Equations (7.113)–(7.114),

$$|l \pm 1/2, m_j\rangle = \pm \sqrt{\frac{l \pm m_j + 1/2}{2 l + 1}} \, |m_j - 1/2, 1/2\rangle$$

$$+ \sqrt{\frac{l \mp m_j + 1/2}{2 l + 1}} \, |m_j + 1/2, -1/2\rangle. \tag{7.133}$$

It follows that

$$\langle l \pm 1/2, m_j| S_z |l \pm 1/2, m_j \rangle = \frac{\hbar}{2 (2 l + 1)} \left[(l \pm m_j + 1/2) - (l \mp m_j + 1/2) \right]$$

$$= \pm \frac{m_j \hbar}{2 l + 1}. \tag{7.134}$$

Thus, we obtain the so-called *Landé formula* for the energy-shift induced by a weak magnetic field [Landé (1921)]:

$$\Delta E_{nlm_j\pm} = \mu_B B m_j \left(1 \pm \frac{1}{2 l + 1} \right), \tag{7.135}$$

where

$$\mu_B = \frac{e \hbar}{2 m_e} = 5.788 \times 10^{-5} \, \text{eV} \, \text{T}^{-1} \tag{7.136}$$

is called the *Bohr magnetron* [Procopiu (1911)]. Incidentally, for the special case in which $l = 0$, the plus sign applies in Equation (7.135). Note, finally, that the eigenstates of $H_0 + H_{LS}$ are not eigenstates of H_B. However, H_B only couples eigenstates with different values of j. It follows that such eigenstates are non-degenerate (because the eigenvalues of H_{LS} are different for states with different values of j—see Section 7.7). Hence, there is no danger of singular terms arising in the perturbation expansion to second order.

Let us apply this theory to the sodium atom. We have already seen that the non-Coulomb potential splits the degeneracy of the $3s$ and $3p$ states, the latter states acquiring a higher energy. The spin-orbit interaction splits the six $3p$ (i.e.,

$l = 1$) states into two groups, with four $j = 3/2$ states lying at a slightly higher energy than two $j = 1/2$ states. According to Equation (7.135), a magnetic field splits the $(3p)_{3/2}$ quadruplet of states, each state acquiring a different energy. In fact, the energy of each state becomes dependent on the quantum number m_j, which measures the projection of the total angular momentum along the z-axis. States with higher m_j values have higher energies. A magnetic field also splits the $(3p)_{1/2}$ doublet of states. However, it is evident from Equation (7.135) that these states are split by a lesser amount than the $j = 3/2$ states. In fact, the splitting is $(4/3)\mu_B B$ for the $(3p)_{3/2}$ quadruplet, and $(2/3)\mu_B B$ for the $(3p)_{1/2}$ doublet.

Suppose that we increase the strength of the magnetic field until the energy-shift due to the magnetic field becomes comparable to the energy-shift induced by spin-orbit interaction. Clearly, in this situation, it does not make much sense to think of H_B as a small interaction term operating on the eigenstates of $H_0 + H_{LS}$. In fact, this intermediate case is very difficult to analyze. Let us, instead, consider the extreme limit in which the energy-shift due to the magnetic field greatly exceeds that induced by spin-orbit effects. This is called the *Paschen-Back limit* [Paschen and Back (1921)].

In the Paschen-Back limit, we can think of the spin-orbit Hamiltonian, H_{LS}, as a small interaction term operating on the eigenstates of $H_0 + H_B$. Note that the magnetic Hamiltonian, H_B, commutes with L^2, S^2, L_z, S_z, but does not commute with L^2, S^2, J^2, J_z. Thus, in an intense magnetic field, the energy eigenstates of a hydrogen-like atom are approximate eigenstates of the spin and orbital angular momenta, but are not eigenstates of the total angular momentum. We can label each state by the quantum numbers n (the principal quantum number), l, m_l, and m_s. (Here, m_l and m_s are the quantum numbers corresponding to L_z and S_z, respectively.) Thus, our energy eigenkets are written $|n, l, m_l, m_s\rangle$. The unperturbed Hamiltonian, H_0, causes states with different values of the quantum numbers n and l to have different energies. However, states with the same value of n and l, but different values of m_l and m_s, are degenerate. The shift in energy due to the magnetic field is simply

$$\Delta E_{n l m_l m_s} = \langle n, l, m_l, m_s| H_B |n, l, m_l, m_s\rangle = \mu_B B (m_l + 2 m_s). \qquad (7.137)$$

Thus, states with different values of $m_l + 2 m_s$ acquire different energies.

Let us apply this result to a sodium atom. In the absence of a magnetic field, the six $3p$ states form two groups of four and two states, depending on the values of their total angular momentum. In the presence of an intense magnetic field, the $3p$ states are split into five groups. There is a state with $m_l + 2 m_s = 2$, a state with $m_l + 2 m_s = 1$, two states with $m_l + 2 m_s = 0$, a state with $m_l + 2 m_s = -1$, and a state with $m_l + 2 m_s = -2$. These groups are equally spaced in energy, the energy difference between adjacent groups being $\mu_B B$.

The energy-shift induced by the spin-orbit Hamiltonian is given by

$$\Delta E_{n\,l\,m_l\,m_s} = \langle n, l, m_l, m_s | \, H_{LS} \, | n, l, m_l, m_s \rangle, \qquad (7.138)$$

where

$$H_{LS} = \frac{1}{2\,m_e^2\,c^2} \frac{1}{r} \frac{dV}{dr} \, \mathbf{L} \cdot \mathbf{S}. \qquad (7.139)$$

Now,

$$\langle \mathbf{L} \cdot \mathbf{S} \rangle = \langle L_z\,S_z + (L^+\,S^- + L^-\,S^+)/2 \rangle = \hbar^2\,m_l\,m_s, \qquad (7.140)$$

because

$$\langle L^\pm \rangle = \langle S^\pm \rangle = 0 \qquad (7.141)$$

for expectation values taken between the simultaneous eigenkets of L_z and S_z. Thus,

$$\Delta E_{n\,l\,m_l\,m_s} = \frac{\hbar^2\,m_l\,m_s}{2\,m_e^2\,c^2} \left\langle \frac{1}{r}\frac{dV}{dr} \right\rangle. \qquad (7.142)$$

Note, finally, that the eigenstates of $H_0 + H_B$ are not eigenstates of H_{LS}. However, H_{LS} only couples non-degenerate eigenstates of $H_0 + H_B$. Hence, there is no danger of singular terms arising in the perturbation expansion to second order.

Let us apply the previous result to a sodium atom. In the presence of an intense magnetic field, the $3p$ states are split into five groups with (m_l, m_s) quantum numbers $(1, 1/2)$, $(0, 1/2)$, $(1, -1/2)$ or $(-1, 1/2)$, $(0, -1/2)$, and $(-1, -1/2)$, respectively, in order of decreasing energy. The spin-orbit term increases the energy of the highest energy state (because $\langle r^{-1}\,dV/dr \rangle > 0$), does not affect the next highest energy state, decreases, but does not split, the energy of the doublet, does not affect the next lowest energy state, and increases the energy of the lowest energy state. (See Exercise 7.18.) The net result is that the five groups of states are no longer equally spaced in energy.

The typical magnetic field-strength needed to access the Paschen-Back limit is

$$B_{PB} \sim \alpha^2 \, \frac{e\,m_e}{\epsilon_0\,h\,a_0} \simeq 25 \ \text{tesla}, \qquad (7.143)$$

where α is the fine structure constant, and a_0 the Bohr radius. (See Exercise 7.19.)

7.9 Hyperfine Structure

The proton in a hydrogen atom is a spin one-half charged particle, and therefore possesses a spin magnetic moment. By analogy with Equation (5.44), we can write

$$\boldsymbol{\mu}_p = \frac{g_p\,e}{2\,m_p}\,\mathbf{S}_p, \tag{7.144}$$

where $\boldsymbol{\mu}_p$ is the proton magnetic moment, $\mathbf{S}_p$ the proton spin, m_p the proton mass, and g_p the proton g-factor. The proton g-factor is found experimentally to take that value 5.59 [Winkler *et al.* (1972)]. [In writing the previous equation, we have made use of the fact that the proton is essentially stationary (in the center of mass frame), and, therefore, possesses zero orbital angular momentum.] Note that the spin magnetic moment of a proton is much smaller (by a factor of order m_e/m_p) than that of an electron.

According to classical electromagnetism, the vector potential due to a point magnetic dipole $\mathbf{M}$ located at the origin is [Fitzpatrick (2008)]

$$\mathbf{A}(\mathbf{r}) = \frac{\mu_0}{4\pi}\,\frac{\mathbf{M}\times\mathbf{x}}{r^3}, \tag{7.145}$$

where $r = |\mathbf{x}|$. The associated magnetic field takes the form [Fitzpatrick (2008)]

$$\mathbf{B} = \nabla\times\mathbf{A} = \frac{\mu_0}{4\pi}\left[\frac{3\,(\mathbf{M}\cdot\mathbf{e}_r)\,\mathbf{e}_r - \mathbf{M}}{r^3}\right], \tag{7.146}$$

where $\mathbf{e}_r = \mathbf{x}/r$. Suppose that $\mathbf{M} = M\,\mathbf{e}_z$. The Cartesian components of $\mathbf{B}$ are thus

$$B_x(r,\theta,\varphi) = \frac{\mu_0\,M}{4\pi\,r^3}\,3\,\cos\theta\,\sin\theta\,\cos\varphi, \tag{7.147}$$

$$B_y(r,\theta,\varphi) = \frac{\mu_0\,M}{4\pi\,r^3}\,3\,\cos\theta\,\sin\theta\,\sin\varphi, \tag{7.148}$$

$$B_z(r,\theta,\varphi) = \frac{\mu_0\,M}{4\pi\,r^3}\,(3\,\cos^2\theta - 1), \tag{7.149}$$

where (r,θ,φ) are conventional spherical coordinates. It is easily demonstrated that

$$\int_V d^3\mathbf{x}\,B_x = \int_V d^3\mathbf{x}\,B_y = \int_V d^3\mathbf{x}\,B_z = 0, \tag{7.150}$$

where V is a spherical volume of radius R, centered on the origin. However, we can also write [Jackson (1975)]

$$\int_V d^3\mathbf{x}\,\mathbf{B} = \int_V d^3\mathbf{x}\,\nabla\times\mathbf{A} = \int_S d\mathbf{S}\times\mathbf{A} = R^2\oint_{r=R} d\Omega\,\mathbf{e}_r\times\mathbf{A}, \tag{7.151}$$

where S is the bounding surface of volume V, and $d\Omega$ an element of solid angle. According to Equation (7.145),

$$\int_V d^3\mathbf{x}\,\mathbf{B} = \frac{\mu_0}{4\pi} \oint d\Omega\, \mathbf{e}_r \times (\mathbf{M} \times \mathbf{e}_r) = \frac{\mu_0}{4\pi} \oint d\Omega\, [\mathbf{M} - (\mathbf{M} \cdot \mathbf{e}_r)\,\mathbf{e}_r]$$

$$= \mu_0\,\mathbf{M} - \frac{\mu_0}{4\pi} \oint d\Omega\,(\mathbf{M} \cdot \mathbf{e}_r)\,\mathbf{e}_r. \tag{7.152}$$

Let $\mathbf{M} = M\,\mathbf{e}_z$, and $\mathbf{F} = \int d\Omega\,(\mathbf{M} \cdot \mathbf{e}_r)\,\mathbf{e}_r$. It follows that

$$F_x = M \oint d\Omega\,\cos\theta\,\sin\theta\,\cos\varphi = 0, \tag{7.153}$$

$$F_y = M \oint d\Omega\,\cos\theta\,\sin\theta\,\sin\varphi = 0, \tag{7.154}$$

$$F_z = M \oint d\Omega\,\cos^2\theta = \frac{4\pi}{3}\,M, \tag{7.155}$$

which implies that $\mathbf{F} = (4\pi/3)\,\mathbf{M}$. Hence, we obtain

$$\int_V d^3\mathbf{x}\,\mathbf{B} = \frac{2}{3}\mu_0\,\mathbf{M}. \tag{7.156}$$

However, the previous expression is inconsistent with Equations (7.147)–(7.149). Note that the right-hand side of Equation (7.156) is independent of the radius, R, of the integration volume V. Consequently, we can take the limit $R \to 0$ without changing the value of $\int_V d^3\mathbf{x}\,\mathbf{B}$. We deduce that the non-zero contribution to this integral originates from the origin. Hence, we can reconcile the previously mentioned inconsistency by modifying Equation (7.146) to read

$$\mathbf{B} = \frac{\mu_0}{4\pi} \left[\frac{3\,(\mathbf{M} \cdot \mathbf{e}_r)\,\mathbf{e}_r - \mathbf{M}}{r^3} \right] + \frac{2\mu_0}{3}\,\delta^3(\mathbf{x})\,\mathbf{M}. \tag{7.157}$$

Here, $\delta^3(\mathbf{x}) \equiv \delta(x)\,\delta(y)\,\delta(z)$ is a *three-dimensional Dirac delta function*. This function has the property that

$$\int_V d^3\mathbf{x}\,F(\mathbf{x})\,\delta^3(\mathbf{x} - \mathbf{x}_0) = F(\mathbf{x}_0), \tag{7.158}$$

where $F(\mathbf{x})$ is a general function that is well-behaved in the vicinity of $\mathbf{x} = \mathbf{x}_0$ (which is assumed to lie in the volume V) [Riley *et al.* (2013)].

According to the previous formula, the proton's magnetic moment, μ_p, generates a magnetic field of the form

$$\mathbf{B} = \frac{\mu_0}{4\pi\,r^3} \left[3\,(\mu_p \cdot \mathbf{e}_r)\,\mathbf{e}_r - \mu_p \right] + \frac{2\mu_0}{3}\,\delta^3(\mathbf{x})\,\mu_p \tag{7.159}$$

where **x** measures position relative to the proton. Now, the Hamiltonian of the electron in the magnetic field generated by the proton is simply [Fitzpatrick (2008)]

$$H_1 = -\boldsymbol{\mu}_e \cdot \mathbf{B}, \tag{7.160}$$

where

$$\boldsymbol{\mu}_e = -\frac{e}{m_e}\,\mathbf{S}_e. \tag{7.161}$$

Here, $\boldsymbol{\mu}_e$ is the electron magnetic moment [see Equation (7.98)], and $\mathbf{S}_e$ the electron spin. Thus, the perturbing Hamiltonian is written

$$H_1 = \frac{\mu_0\, g_p\, e^2}{8\pi\, m_p\, m_e}\left[\frac{3\,(\mathbf{S}_p\cdot\mathbf{e}_r)\,(\mathbf{S}_e\cdot\mathbf{e}_r)-\mathbf{S}_p\cdot\mathbf{S}_e}{r^3}\right] + \frac{\mu_0\, g_p\, e^2}{3\, m_p\, m_e}\,\mathbf{S}_p\cdot\mathbf{S}_e\,\delta^3(\mathbf{x}). \tag{7.162}$$

Note that, because we have neglected coupling between the proton spin and the magnetic field generated by the electron's orbital motion, the previous expression is only valid for $l = 0$ states.

According to standard first-order perturbation theory, the energy-shift induced by spin-spin coupling between the proton and the electron is the expectation value of the perturbing Hamiltonian. Hence,

$$\Delta E = \frac{\mu_0\, g_p\, e^2}{8\pi\, m_p\, m_e}\left\langle\frac{3\,(\mathbf{S}_p\cdot\mathbf{e}_r)\,(\mathbf{S}_e\cdot\mathbf{e}_r)-\mathbf{S}_p\cdot\mathbf{S}_e}{r^3}\right\rangle$$
$$+ \frac{\mu_0\, g_p\, e^2}{3\, m_p\, m_e}\,\langle\mathbf{S}_p\cdot\mathbf{S}_e\rangle\,|\psi(0)|^2. \tag{7.163}$$

In the final term on the right-hand side, the expectation value is taken over the overall spin state. For the ground state of hydrogen, which is spherically symmetric, the first term in the previous expression vanishes by symmetry. Moreover, it is easily demonstrated that $|\psi_{000}(\mathbf{0})|^2 = 1/(\pi\, a_0^3)$. Thus, we obtain

$$\Delta E_{000} = \frac{\mu_0\, g_p\, e^2}{3\pi\, m_p\, m_e\, a_0^3}\,\langle\mathbf{S}_p\cdot\mathbf{S}_e\rangle. \tag{7.164}$$

Let

$$\mathbf{S} = \mathbf{S}_e + \mathbf{S}_p \tag{7.165}$$

be the total spin. We can show that

$$\mathbf{S}_p\cdot\mathbf{S}_e = \frac{1}{2}\,(S^2 - S_e^2 - S_p^2). \tag{7.166}$$

Thus, the simultaneous eigenstates of the perturbing Hamiltonian and the main Hamiltonian are the simultaneous eigenstates of S_e^2, S_p^2, and S^2. (The use of simultaneous eigenstates of the perturbing and main Hamiltonian avoids the possibility of singular terms arising in the perturbation expansion to second order—see

Section 7.5.) However, both the proton and the electron are spin one-half particles. According to Section 6.4, when two spin one-half particles are combined (in the absence of orbital angular momentum) the resulting state has either spin 1 or spin 0. In fact, there are three spin 1 states, known as *triplet states*, and a single spin 0 state, known as the *singlet state*. For all states, the eigenvalues of S_e^2 and S_p^2 are $(3/4)\,\hbar^2$. The eigenvalue of S^2 is 0 for the singlet state, and $2\,\hbar^2$ for the triplet states. Hence,

$$\langle \mathbf{S}_p \cdot \mathbf{S}_e \rangle = -\frac{3}{4}\,\hbar^2 \tag{7.167}$$

for the singlet state, and

$$\langle \mathbf{S}_p \cdot \mathbf{S}_e \rangle = \frac{1}{4}\,\hbar^2 \tag{7.168}$$

for the triplet states.

It follows, from the previous analysis, that proton-electron spin-spin coupling breaks the degeneracy of the two $(1s)_{1/2}$ states of the hydrogen atom, lifting the energy of the triplet configuration, and lowering that of the singlet. This splitting is known as *hyperfine structure*. The net energy difference between the singlet and the triplet states is

$$\Delta E_{000} = \frac{8}{3}\,g_p\,\frac{m_e}{m_p}\,\alpha^2\,|E_0| = 5.88 \times 10^{-6}\,\mathrm{eV}, \tag{7.169}$$

where $|E_0| = 13.6\,\mathrm{eV}$ is the (magnitude of the) ground-state energy, and $\alpha = 1/137$ the fine structure constant. Note that the hyperfine energy-shift is much smaller, by a factor m_e/m_p, than a typical fine structure energy-shift. (See Exercise 7.14.) If we convert the previous energy into a wavelength (using $\lambda = c/\nu = c\,h/\Delta E_{000}$) then we obtain

$$\lambda = 21.1\,\mathrm{cm}. \tag{7.170}$$

This is the wavelength of the radiation emitted by a hydrogen atom that is collisionally excited from the singlet to the triplet state, and then decays back to the lower energy singlet state. The 21 cm line is famous in radio astronomy because it was used to map out the spiral structure of our galaxy in the 1950's [van de Hulst *et al.* (1954)].

7.10 Exercises

7.1 Consider the two-state system investigated in Section 7.2. Show that the most general expressions for the perturbed energy eigenvalues and eigenstates are

$$E_1' = E_1 + e_{11} + \frac{|e_{12}|^2}{E_1 - E_2} + O(\epsilon^3),$$

$$E_2' = E_2 + e_{22} - \frac{|e_{12}|^2}{E_1 - E_2} + O(\epsilon^3),$$

and

$$|1\rangle' = |1\rangle + \frac{e_{12}^*}{E_1 - E_2} |2\rangle + O(\epsilon^2),$$

$$|2\rangle' = |2\rangle - \frac{e_{12}}{E_1 - E_2} |1\rangle + O(\epsilon^2),$$

respectively. Here, $\epsilon = |e_{12}|/(E_1 - E_2) \ll 1$. You may assume that $|e_{11}|/(E_1 - E_2)$, $|e_{22}|/(E_1 - E_2) \sim O(\epsilon)$.

7.2 Consider the two-state system investigated in Section 7.2. Show that if the unperturbed energy eigenstates are also eigenstates of the perturbing Hamiltonian then

$$E_1' = E_1 + e_{11},$$
$$E_2' = E_2 + e_{22},$$

and

$$|1\rangle' = |1\rangle$$
$$|2\rangle' = |2\rangle$$

to all orders in the perturbation expansion.

7.3 Consider the two-state system investigated in Section 7.2. Show that if the unperturbed energy eigenstates are degenerate, so that $E_1 = E_2 = E_{12}$ then the most general expressions for the perturbed energy eigenvalues and eigenstates are

$$E^{\pm} = E_{12} + e^{\pm},$$

and

$$|E^{\pm}\rangle = \langle 1|E^{\pm}\rangle|1\rangle + \langle 2|E^{\pm}\rangle|2\rangle,$$

respectively, where

$$e^{\pm} = \frac{1}{2}(e_{11} + e_{22}) \pm \frac{1}{2}\left[(e_{11} - e_{22})^2 + 4|e_{12}|^2\right]^{1/2},$$

and

$$\frac{\langle 1|E^{\pm}\rangle}{\langle 2|E^{\pm}\rangle} = -\left(\frac{e_{12}}{e_{11} - e^{\pm}}\right) = -\left(\frac{e_{22} - e^{\pm}}{e_{12}^{*}}\right).$$

Demonstrate that the $|E^{\pm}\rangle$ are the simultaneous eigenkets of the unperturbed Hamiltonian, H_0, and the perturbed Hamiltonian, H_1, and that the $e^{\pm}$ are the corresponding eigenvalues of H_1.

7.4 Calculate the lowest-order energy-shift in the ground state of the one-dimensional harmonic oscillator when the perturbation

$$V = \lambda x^4$$

is added to

$$H = \frac{p_x^2}{2m} + \frac{1}{2} m \omega^2 x^2.$$

[Gasiorowicz (1996)]

7.5 Let $|n, l, m\rangle$ denote a properly normalized eigenstate of the hydrogen atom corresponding to the conventional quantum numbers n, l, and m. Show that the only non-zero matrix elements of the operator z between the various $n = 3$ states take the values

$$\langle 3, 1, 0| z |3, 0, 0\rangle = \langle 3, 0, 0| z |3, 1, 0\rangle = -3\sqrt{6}\, a_0,$$

$$\langle 3, 1, 0| z |3, 2, 0\rangle = \langle 3, 2, 0| z |3, 1, 0\rangle = -3\sqrt{3}\, a_0,$$

$$\langle 3, 1, \pm 1| z |3, 2, \pm 1\rangle = \langle 3, 2, \pm 1| z |3, 1, \pm 1\rangle = -\frac{9}{2}\, a_0.$$

7.6 Calculate the energy-shifts and perturbed eigenstates associated with the linear Stark effect in the $n = 3$ state of a hydrogen atom.

7.7 Suppose that the Hamiltonian, H, for a particular quantum system, is a function of some parameter, λ. Let $E_n(\lambda)$ and $|n\rangle(\lambda)$ be the eigenvalues and eigenkets of $H(\lambda)$. Prove the *Feynman-Hellmann theorem* [Feynman (1939)]

$$\frac{\partial E_n}{\partial \lambda} = \left\langle n(\lambda) \left| \frac{\partial H}{\partial \lambda} \right| n(\lambda) \right\rangle.$$

[Griffiths (2005)]

7.8 According to Section 4.6, the Hamiltonian for the radial wavefunction of an energy eigenstate of the hydrogen atom corresponding to the conventional quantum numbers n, l, and m is written

$$H = -\frac{\hbar^2}{2m_e} \frac{1}{r^2} \frac{d}{dr} r^2 \frac{d}{dr} + \frac{\hbar^2}{2m_e} \frac{l(l+1)}{r^2} - \frac{e^2}{4\pi \epsilon_0 r}.$$

Moreover, when expressed as a function of l, the energy eigenvalues are

$$E_n = \frac{E_0}{(k+l)^2},$$

where E_0 is the ground-state energy, and $k = n - l$ the number of terms in the power series (4.126). Treating l as a continuous parameter, use the Feynman-Hellmann theorem to prove that

$$\left\langle \frac{1}{r^2} \right\rangle = \frac{1}{(l+1/2)\, n^3\, a_0^2},$$

where the expectation value is taken over the energy eigenstate corresponding to the quantum numbers n and l. [Griffiths (2005)]

7.9 Demonstrate that

$$\left\langle \frac{1}{r^3} \right\rangle = \frac{1}{l\,(l+1/2)\,(l+1)\, n^3\, a_0^3},$$

where the expectation value is taken over the energy eigenstate of the hydrogen atom characterized by the standard quantum numbers n and l. (Hint: Use Kramer's relation—see Exercise 4.16.)

7.10 The relativistic definition of the kinetic energy, K, of a particle of total energy E and rest mass m is

$$K = E - m c^2.$$

Making use of the standard relativistic result [Fitzpatrick (2008)]

$$E = \left[p^2 c^2 + m^2 c^4 \right]^{1/2},$$

where p is the particle momentum, demonstrate that, in the non-relativistic limit $p \ll mc$,

$$K \simeq \frac{p^2}{2m} - \frac{p^4}{8 m^3 c^2}.$$

7.11 The Hamiltonian of the valence electron in a hydrogen-like atom can be written

$$H = \frac{p^2}{2 m_e} + V(r) + H_R,$$

where

$$H_R = -\frac{p^4}{8 m_e^3 c^2}$$

is the first-order correction due to the electron's relativistic mass increase. (See Exercise 7.10.) Treating H_R as a small perturbation, deduce that it causes an energy-shift in the energy eigenstate, characterized by the standard quantum numbers n, l, m, of

$$\Delta E_{nlm} = -\frac{1}{2\,m_e\,c^2}\left(E_n^2 - 2\,E_n\,\langle V\rangle + \langle V^2\rangle\right),$$

where E_n is the unperturbed energy. Finally, show that for the special case of a hydrogen atom, the energy-shift becomes

$$\Delta E_{nlm} = \frac{\alpha^2\,E_n}{n^2}\left(\frac{n}{l+1/2} - \frac{3}{4}\right),$$

where α is the fine structure constant.

7.12 According to Dirac's relativistic electron theory, there is an additional relativistic correction to the Hamiltonian of a valence electron in a hydrogen-like atom that takes the form

$$H_D = \frac{e^2\,\hbar^2}{8\,\epsilon_0\,m_e^2\,c^2}\,\delta^3(\mathbf{x})$$

[Bjorken and Drell (1964)]. This correction is usually referred to as the *Darwin term* [Darwin (1928)]. Treating H_D as a small perturbation, deduce that, for the special case of a hydrogen atom, it causes an energy-shift in the energy eigenstate, characterized by the standard quantum numbers n, l, m, of

$$\Delta E_{nlm} = -\frac{\alpha^2\,E_n}{n}$$

for an $l = 0$ state, and

$$\Delta E_{nlm} = 0$$

for an $l > 0$ state. Note that

$$|\psi_{nlm}(\mathbf{0})| = \frac{1}{\sqrt{\pi}\,(n\,a_0)^{3/2}}\,\delta_{l0}\,\delta_{m0},$$

where $\psi_{nlm}(\mathbf{x})$ is the properly normalized wavefunction associated with an energy eigenstate of the hydrogen atom [Sakurai and Napolitano (2011)].

7.13 Consider an energy eigenstate of the hydrogen atom characterized by the standard quantum numbers n, l, and m. Show that the energy-shift due to spin-orbit coupling takes the form

$$\Delta E_{nlm} = -\frac{\alpha^2\,E_n}{n^2}\left[\frac{n}{2\,(l+1/2)\,(l+1)}\right]$$

for $j = l + 1/2$, and

$$\Delta E_{nlm} = \frac{\alpha^2\,E_n}{n^2}\left[\frac{n}{2\,(l+1/2)\,l}\right]$$

for $j = l - 1/2$, and

$$\Delta E_{nlm} = 0$$

for the special case of an $l = 0$ state. Here, j is the standard quantum number associated with the magnitude of the sum of the electron's orbital and spin angular momenta. (See Section 7.7.)

7.14 Demonstrate that if the energy-shifts due to the electron's relativistic mass increase, the Darwin term, and spin-orbit coupling, calculated in the previous three exercises for an energy eigenstate of the hydrogen atom, characterized by the standard quantum numbers n, l, and m, are added together then the net *fine structure energy-shift* can be written

$$\Delta E_{nlm} = \frac{\alpha^2 E_n}{n^2} \left(\frac{n}{j+1/2} - \frac{3}{4} \right).$$

Here, E_n is the unperturbed energy, α the fine structure constant, and j the standard quantum number associated with the magnitude of the sum of the electron's orbital and spin angular momenta.

Show that fine structure causes the energy of the $(2p)_{3/2}$ states of a hydrogen atom to exceed those of the $(2p)_{1/2}$ and $(2s)_{1/2}$ states by 4.5×10^{-5} eV.

7.15 The linear Stark effect exhibited by the $n = 2$ states of a hydrogen atom depends crucially on the supposed degeneracy of these states. However, this degeneracy is lifted by fine structure. Consequently, the expressions for the Stark energy shifts derived in Section 7.6 are only valid when the energy splitting predicted by the linear Stark effect greatly exceeds that caused by fine structure. Deduce that this is the case when

$$|\mathbf{E}| \gg 1 \times 10^5 \, \text{V m}^{-1}.$$

7.16 Demonstrate that

$$Y_{l m_l} \chi_\pm = \left(\frac{l \pm m_l + 1}{2l+1} \right)^{1/2} \mathcal{Y}^{l+1/2}_{l \, m_l \pm 1/2} \mp \left(\frac{l \mp m_l}{2l+1} \right)^{1/2} \mathcal{Y}^{l-1/2}_{l \, m_l \pm 1/2},$$

where the $Y_{l m_l}$ are spherical harmonics, the $\chi_\pm$ are standard Pauli two-component spinors, and the $\mathcal{Y}^j_{l m_j}$ are spin-angular functions. In particular, show that

$$Y_{00} \chi_\pm = \mathcal{Y}^{1/2}_{0 \, \pm 1/2},$$

$$Y_{10} \chi_\pm = \sqrt{\frac{2}{3}} \, \mathcal{Y}^{3/2}_{1 \, \pm 1/2} \mp \sqrt{\frac{1}{3}} \, \mathcal{Y}^{1/2}_{1 \, \pm 1/2},$$

$$Y_{1 \mp 1} \chi_\pm = \sqrt{\frac{1}{3}} \, \mathcal{Y}^{3/2}_{1 \, \mp 1/2} \mp \sqrt{\frac{2}{3}} \, \mathcal{Y}^{1/2}_{1 \, \mp 1/2},$$

$$Y_{1 \pm 1} \chi_\pm = \mathcal{Y}^{3/2}_{1 \, \pm 3/2}.$$

7.17 Taking electron spin into account, the $n = 2$ energy eigenstates of the hydrogen atom in the presence of an external electric field are

$$|1\pm\rangle = \sqrt{\frac{1}{2}}\,|2,0,0\rangle\chi_\pm - \sqrt{\frac{1}{2}}\,|2,1,0\rangle\chi_\pm,$$

$$|2\pm\rangle = \sqrt{\frac{1}{2}}\,|2,0,0\rangle\chi_\pm + \sqrt{\frac{1}{2}}\,|2,1,0\rangle\chi_\pm,$$

$$|3\pm\rangle = |2,1,1\rangle\chi_\pm,$$

$$|4\pm\rangle = |2,1,-1\rangle\chi_\pm.$$

(See Section 7.6.) Here, $|n,l,m\rangle$ is an unperturbed energy eigenket corresponding to the standard quantum numbers n, l, and m. Moreover, the energies of these states are

$$E_{1\pm} = E_2 + 3\,e\,a_0\,|\mathbf{E}|,$$

$$E_{2\pm} = E_2 - 3\,e\,a_0\,|\mathbf{E}|,$$

$$E_{3\pm} = E_2,$$

$$E_{4\pm} = E_2,$$

where $\mathbf{E}$ is the external electric field-strength, and E_2 the unperturbed energy eigenvalue corresponding to $n = 2$. The fine structure Hamiltonian can be written

$$H_{FS} = H_{LS} + H_R + H_D,$$

where H_{LS} is the spin-orbit Hamiltonian (see Section 7.7), and H_R and H_D are defined in Exercises 7.11 and 7.12, respectively. Treating H_{FS} as a small perturbation, deduce that fine structure modifies the energies of the previously defined eigenstates such that

$$E_{1\pm} = E_2\left(1 + \frac{11}{48}\alpha^2\right) + 3\,e\,a_0\,|\mathbf{E}|,$$

$$E_{2\pm} = E_2\left(1 + \frac{11}{48}\alpha^2\right) - 3\,e\,a_0\,|\mathbf{E}|,$$

$$E_{3-} = E_2\left(1 + \frac{11}{48}\alpha^2\right),$$

$$E_{4+} = E_2\left(1 + \frac{11}{48}\alpha^2\right),$$

$$E_{3+} = E_2\left(1 + \frac{11}{48}\alpha^2\right) - \frac{1}{6}\alpha^2 E_2,$$

$$E_{4-} = E_2\left(1 + \frac{11}{48}\alpha^2\right) - \frac{1}{6}\alpha^2 E_2.$$

7.18 Consider the $n = 2$ energy eigenstates of the hydrogen atom in the Paschen-Back limit. (See Section 7.8.) These states are conveniently labeled using the standard quantum numbers n, l, m_l, and m_s. Treating the fine structure Hamiltonian, H_{FS}, defined in the previous exercise, as a small perturbation, show that the perturbed energies of the various states are

$$E_{2,1,1,1/2} = E_2\left(1 + \frac{5}{16}\alpha^2\right) + 2\mu_B B - \frac{1}{3}\alpha^2 E_2,$$

$$E_{2,1,0,1/2} = E_2\left(1 + \frac{5}{16}\alpha^2\right) + \mu_B B - \frac{1}{6}\alpha^2 E_2,$$

$$E_{2,0,0,1/2} = E_2\left(1 + \frac{5}{16}\alpha^2\right) + \mu_B B,$$

$$E_{2,1,1,-1/2} = E_2\left(1 + \frac{5}{16}\alpha^2\right),$$

$$E_{2,1,-1,1/2} = E_2\left(1 + \frac{5}{16}\alpha^2\right),$$

$$E_{2,0,0,-1/2} = E_2\left(1 + \frac{5}{16}\alpha^2\right) - \mu_B B,$$

$$E_{2,1,0,-1/2} = E_2\left(1 + \frac{5}{16}\alpha^2\right) - \mu_B B - \frac{1}{6}\alpha^2 E_2,$$

$$E_{2,1,-1,-1/2} = E_2\left(1 + \frac{5}{16}\alpha^2\right) - 2\mu_B B - \frac{1}{3}\alpha^2 E_2.$$

Here, μ_B is the Bohr magnetron, and B the external magnetic field-strength.

7.19 Justify Equation (7.143).

Chapter 8

Time-Dependent Perturbation Theory

8.1 Introduction

Suppose that the Hamiltonian of a given quantum mechanical system can be written

$$H = H_0 + H_1(t), \qquad (8.1)$$

where the unperturbed Hamiltonian, H_0, does not depend on time explicitly, and $H_1(t)$ is a small time-dependent perturbation. In the following, it is assumed that we are able to calculate the eigenkets and eigenvalues of the unperturbed Hamiltonian exactly. These are denoted

$$H_0 |n\rangle = E_n |n\rangle, \qquad (8.2)$$

where n is an integer. We know that if the system is initially in one of the eigenstates of H_0 then, in the absence of the external perturbation, it remains in that state for ever. (See Section 3.5.) However, the presence of a small time-dependent perturbation can, in principle, give rise to a finite probability that a system initially in some eigenstate $|i\rangle$ of the unperturbed Hamiltonian is found in some other eigenstate at a subsequent time (because $|i\rangle$ is no longer an exact eigenstate of the total Hamiltonian). In other words, a time-dependent perturbation allows the system to make transitions between its unperturbed energy eigenstates. Let us investigate such transitions.

8.2 General Analysis

Suppose that at $t = t_0$ the state of the system is represented by

$$|A\rangle = \sum_n c_n |n\rangle, \qquad (8.3)$$

165

where the c_n are complex numbers. Thus, the initial state is some linear superposition of the unperturbed energy eigenstates. In the absence of the time-dependent perturbation, the time evolution of the system is given by

$$|A, t_0, t\rangle = \sum_n c_n \exp\left[\frac{-\mathrm{i}\, E_n\, (t - t_0)}{\hbar}\right] |n\rangle. \tag{8.4}$$

(See Section 3.5.) Here, $|A, t_0, t\rangle$ is the state ket of the system at time t, given that the state ket at the initial time t_0 is $|A\rangle$. Now, the probability of finding the system in state $|n\rangle$ at time t is

$$P_n(t) = \left|c_n \exp\left[\frac{-\mathrm{i}\, E_n(t - t_0)}{\hbar}\right]\right|^2 = |c_n|^2 = P_n(t_0). \tag{8.5}$$

Clearly, with $H_1 = 0$, the probability of finding the system in state $|n\rangle$ at time t is exactly the same as the probability of finding the system in this state at the initial time, t_0. However, with $H_1 \neq 0$, we expect $P_n(t)$ to vary with time. Thus, we can write

$$|A, t_0, t\rangle = \sum_n c_n(t) \exp\left[\frac{-\mathrm{i}\, E_n\, (t - t_0)}{\hbar}\right] |n\rangle, \tag{8.6}$$

where $P_n(t) = |c_n(t)|^2$. Here, we have carefully separated the fast phase oscillation of the eigenkets, which depends on the unperturbed Hamiltonian, from the slow variation of the amplitudes $c_n(t)$, which depends entirely on the perturbation (because the c_n are all constant if $H_1 = 0$) [Dirac (1958)]. Note that the eigenkets $|n\rangle$, appearing in Equation (8.6), are time-independent (they are actually the eigenkets of H_0 evaluated at the time t_0).

Schrödinger's time evolution equation yields

$$\mathrm{i}\hbar \frac{\partial}{\partial t} |A, t_0, t\rangle = H |A, t_0, t\rangle = (H_0 + H_1)|A, t_0, t\rangle. \tag{8.7}$$

(See Section 3.2.) It follows from Equation (8.6) that

$$(H_0 + H_1)|A, t_0, t\rangle = \sum_m c_m(t) \exp\left[\frac{-\mathrm{i}\, E_m\, (t - t_0)}{\hbar}\right] (E_m + H_1) |m\rangle. \tag{8.8}$$

We also have

$$\mathrm{i}\hbar \frac{\partial}{\partial t} |A, t_0, t\rangle = \sum_m \left[\mathrm{i}\hbar \frac{dc_m}{dt} + c_m(t)\, E_m\right] \exp\left[\frac{-\mathrm{i}\, E_m\, (t - t_0)}{\hbar}\right] |m\rangle, \tag{8.9}$$

where use has been made of the time-independence of the kets $|m\rangle$. According to Equation (8.7), we can equate the right-hand sides of the previous two equations to obtain

$$\sum_m \mathrm{i}\hbar \frac{dc_m}{dt} \exp\left[\frac{-\mathrm{i}\, E_m\, (t - t_0)}{\hbar}\right] |m\rangle =$$

$$\sum_m c_m(t) \exp\left[\frac{-\mathrm{i}\, E_m\, (t - t_0)}{\hbar}\right] H_1 |m\rangle. \tag{8.10}$$

Left-multiplication by $\langle n|$ yields [Dirac (1958)]

$$i\hbar \frac{dc_n}{dt} = \sum_m H_{nm}(t) \exp[i\omega_{nm}(t-t_0)]\, c_m(t), \qquad (8.11)$$

where

$$H_{nm}(t) = \langle n|\, H_1(t)\, |m\rangle, \qquad (8.12)$$

and

$$\omega_{nm} = \frac{E_n - E_m}{\hbar}. \qquad (8.13)$$

Here, we have made use of the standard orthonormality result, $\langle n|m\rangle = \delta_{nm}$. Suppose that there are N linearly independent eigenkets of the unperturbed Hamiltonian. According to Equation (8.11), the time variation of the coefficients $c_n(t)$, which specify the probability of finding the system in state $|n\rangle$ at time t, is determined by N coupled first-order differential equations. Note that the set of equations specified in Equation (8.11) is exact—we have made no approximations at this stage. Unfortunately, we cannot generally find exact solutions to this set, so we have to obtain approximate solutions via suitable expansions in small quantities. However, for the particularly simple case of a two-state system (i.e., $N = 2$), it is actually possible to solve Equation (8.11) without approximation. This solution is of great practical importance.

8.3 Two-State System

Consider a system in which the time-independent Hamiltonian possesses two eigenstates, denoted

$$H_0\, |1\rangle = E_1\, |1\rangle, \qquad (8.14)$$

$$H_0\, |2\rangle = E_2\, |2\rangle, \qquad (8.15)$$

where $E_2 > E_1$. Suppose, for the sake of simplicity, that the diagonal matrix elements of the interaction Hamiltonian, H_1, are zero: that is,

$$\langle 1|\, H_1\, |1\rangle = \langle 2|\, H_1\, |2\rangle = 0. \qquad (8.16)$$

The off-diagonal matrix elements are assumed to oscillate sinusoidally at some angular frequency ω:

$$\langle 1|\, H_1\, |2\rangle = \langle 2|\, H_1\, |1\rangle^* = \frac{1}{2}\gamma\hbar \exp(i\omega t), \qquad (8.17)$$

where γ and ω are real. Note that it is only the off-diagonal matrix elements that give rise to the effect in which we are primarily interested—namely, transitions between states 1 and 2. (See Exercise 8.1.)

For a two-state system, Equation (8.11) reduces to

$$i\frac{dc_1}{dt} = \frac{\gamma}{2}\exp[+i(\omega-\omega_{21})t]c_2, \qquad (8.18)$$

$$i\frac{dc_2}{dt} = \frac{\gamma}{2}\exp[-i(\omega-\omega_{21})t]c_1, \qquad (8.19)$$

where $\omega_{21} = (E_2 - E_1)/\hbar$, and it is assumed that $t_0 = 0$. Equations (8.18) and (8.19) can be combined to give a second-order differential equation for the time variation of the amplitude c_2:

$$\frac{d^2c_2}{dt^2} + i(\omega-\omega_{21})\frac{dc_2}{dt} + \frac{\gamma^2}{4}c_2 = 0. \qquad (8.20)$$

Once we have solved for c_2, we can use Equation (8.19) to obtain the amplitude c_1. Let us look for a solution in which the system is certain to be in state 1 at time $t = 0$. Thus, our initial conditions are $c_1(0) = 1$ and $c_2(0) = 0$. It is easily demonstrated that the appropriate solutions are [Fitzpatrick (2013)]

$$c_2(t) = \frac{-i\gamma}{[\gamma^2+(\omega-\omega_{21})^2]^{1/2}}\exp\left[-i(\omega-\omega_{21})\frac{t}{2}\right]$$

$$\sin\left([\gamma^2+(\omega-\omega_{21})^2]^{1/2}\frac{t}{2}\right), \quad (8.21)$$

and

$$c_1(t) = \exp\left[i(\omega-\omega_{21})\frac{t}{2}\right]\cos\left([\gamma^2+(\omega-\omega_{21})^2]^{1/2}\frac{t}{2}\right)$$

$$-\frac{i(\omega-\omega_{21})}{[\gamma^2+(\omega-\omega_{21})^2]^{1/2}}\exp\left[i(\omega-\omega_{21})\frac{t}{2}\right]$$

$$\sin\left([\gamma^2+(\omega-\omega_{21})^2]^{1/2}\frac{t}{2}\right). \quad (8.22)$$

The probability of finding the system in state 1 at time t is simply $P_1(t) = |c_1(t)|^2$. Likewise, the probability of finding the system in state 2 at time t is $P_2(t) = |c_2(t)|^2$. It follows that

$$P_2(t) = \frac{\gamma^2}{\gamma^2+(\omega-\omega_{21})^2}\sin^2\left([\gamma^2+(\omega-\omega_{21})^2]^{1/2}\frac{t}{2}\right), \qquad (8.23)$$

$$P_1(t) = 1 - P_2(t). \qquad (8.24)$$

Equation (8.23) is generally known as the *Rabi formula* [Rabi (1937)].

Equation (8.23) exhibits all the features of a classic resonance [Fitzpatrick (2013)]. At resonance, when the oscillation frequency of the perturbation, ω, matches the so-called *Rabi frequency* [Rabi (1937)], ω_{21}, we find that

$$P_1(t) = \cos^2\left(\frac{\gamma t}{2}\right), \qquad (8.25)$$

$$P_2(t) = \sin^2\left(\frac{\gamma t}{2}\right). \qquad (8.26)$$

According to the previous result, the system starts off at $t = 0$ in state 1. After a time interval $\pi/|\gamma|$, it is certain to be in state 2. After a further time interval $\pi/|\gamma|$, it is certain to be in state 1, and so on. In other words, the system periodically flip-flops between states 1 and 2 under the influence of the time-dependent perturbation. This implies that the system alternatively absorbs energy from, and emits energy to, the source of the perturbation. The absorption-emission cycle also take place away from the resonance, when $\omega \neq \omega_{21}$. However, the amplitude of oscillation of the coefficient c_2 is reduced. This means that the maximum value of $P_2(t)$ is no longer unity, nor is the minimum value of $P_1(t)$ zero. In fact, if we plot the maximum value of $P_2(t)$ as a function of the applied frequency, ω, then we obtain a resonance curve whose maximum (unity) lies at the resonance, and whose full-width half-maximum (in angular frequency) is $2|\gamma|$. Thus, if the applied frequency differs from the resonant frequency by substantially more than $|\gamma|$ then the probability of the system making a transition from state 1 to state 2 is very small (i.e., $0 < P_2 \ll 1$). In other words, the time-dependent perturbation is only effective at causing transitions between states 1 and 2 if its angular frequency of oscillation lies in the approximate range $\omega_{21} \pm |\gamma|$. Clearly, the weaker the perturbation (i.e., the smaller $|\gamma|$ becomes), the narrower the resonance.

8.4 Nuclear Magnetic Resonance

Consider an atomic nucleus of spin one-half placed in a uniform z-directed magnetic field, and then subjected to a small time-dependent magnetic field rotating in the x-y plane at the angular frequency ω. The net magnetic field is

$$\mathbf{B} = B_0\,\mathbf{e}_z + B_1\left[\cos(\omega\,t)\,\mathbf{e}_x - \sin(\omega\,t)\,\mathbf{e}_y\right], \qquad (8.27)$$

where $B_0 > 0$ and $B_1 > 0$ are constants, with $B_1 \ll B_0$. Assuming that $\omega > 0$, the rotating magnetic field represents the magnetic component of a left-hand circularly polarized electromagnetic wave propagating along the z-axis [Fitzpatrick (2013)]. (Obviously, if $\omega < 0$ then the wave becomes right-hand circularly polarized.) Now, we are only interested in the effect of the wave on the nuclear spin state, so we can neglect the wave's electric component. The magnetic moment of the nucleus can be written

$$\mu = \frac{g\,\mu_N}{\hbar}\,\mathbf{S}, \qquad (8.28)$$

where g is the (dimensionless) nuclear g-factor, $\mu_N = e\,\hbar/(2\,m_p)$ the *nuclear magnetron*, and $\mathbf{S}$ the nuclear spin. (See Section 7.9.) Hence, the effective Hamiltonian of the system becomes [Fitzpatrick (2008)]

$$H = -\mu \cdot \mathbf{B} = H_0 + H_1, \qquad (8.29)$$

where

$$H_0 = -\frac{g\,\mu_N\,B_0}{\hbar}\,S_z, \qquad (8.30)$$

and

$$H_1 = -\frac{g\,\mu_N\,B_1}{\hbar}\left[\cos(\omega\,t)\,S_x - \sin(\omega\,t)\,S_y\right]. \qquad (8.31)$$

The eigenstates of the unperturbed Hamiltonian are the "spin up" and "spin down" states, denoted $|+\rangle$ and $|-\rangle$, respectively. (In other words, the eigenstates of S_z corresponding to the eigenvalues $+\hbar/2$ and $-\hbar/2$, respectively.) Thus,

$$H_0\,|\pm\rangle = \mp\frac{g\,\mu_N\,B_0}{2}\,|\pm\rangle. \qquad (8.32)$$

The time-dependent perturbation to the Hamiltonian can be written

$$H_1 = -\frac{g\,\mu_N\,B_1}{2\,\hbar}\left[\exp(\mathrm{i}\,\omega\,t)\,S^+ + \exp(-\mathrm{i}\,\omega\,t)\,S^-\right], \qquad (8.33)$$

where $S^\pm = S_x \pm \mathrm{i}\,S_y$ are the conventional raising and lowering operators for spin angular momentum. (See Section 5.10.) It follows that

$$\langle +|\,H_1\,|+\rangle = \langle -|\,H_1\,|-\rangle = 0, \qquad (8.34)$$

and

$$\langle +|\,H_1\,|-\rangle = \langle -|\,H_1\,|+\rangle^* = -\frac{g\,\mu_N\,B_1}{2}\exp(\mathrm{i}\,\omega\,t). \qquad (8.35)$$

Assuming that $g > 0$, it can be seen that this system is identical to the two-state system discussed in the previous section, provided that we make the identifications

$$|1\rangle \to |+\rangle, \qquad (8.36)$$

$$|2\rangle \to |-\rangle, \qquad (8.37)$$

$$\omega_{21} \to \frac{g\,\mu_N\,B_0}{\hbar}, \qquad (8.38)$$

$$\gamma \to -\frac{g\,\mu_N\,B_1}{\hbar}. \qquad (8.39)$$

The resonant frequency, ω_{21}, is simply the precession frequency of the nuclear spin in a uniform magnetic field of strength B_0. (See Section 5.6.) (If $g > 0$ then this precession is in the same sense as the direction of rotation of the magnetic component of a left-hand circularly polarized wave propagating along the magnetic field, and vice versa.) In the absence of the perturbation, the expectation values of S_x and S_y oscillate because of the spin precession, but the expectation value of S_z remains invariant. However, if we apply a magnetic perturbation rotating at the resonant frequency (i.e., $\omega = \omega_{21}$) then, according to the analysis of the previous section, the system undergoes a succession of spin-flops, $|+\rangle \rightleftharpoons |-\rangle$, in addition to the spin precession. We also know that if the rotation frequency of the applied field is significantly different from the resonant frequency then there is virtually zero probability of the field triggering a spin-flop. The width of the resonance (in frequency) is determined by the strength of the rotating magnetic perturbation. In fact, the relative width of the resonance is $|\gamma|/\omega_{21} = B_1/B_0$. Experimentalist are able to measure the g-factors of atomic nuclei to a high degree of accuracy by placing them in a strong magnetic field, and then subjecting them to a weak rotating magnetic field whose frequency is gradually scanned. By determining the resonant frequency (i.e., the frequency at which the nuclei absorb energy from the rotating field), it is possible to calculate the nuclear g-factor, and, hence, the nuclear magnetic moment [Rabi *et al.* (1939)]. In fact, $|g| = \hbar\,\omega_{21}/(\mu_N\,B_0)$. For a positive g-factor, the resonance occurs when the weak magnetic field rotates in a left-handed sense (with respect to the direction of the strong magnetic field), and vice versa.

8.5 Dyson Series

Let us now try to find approximate solutions to Equation (8.11) for a general system. It is convenient to work in terms of the time evolution operator, $T(t_0, t)$, which is defined

$$|A, t_0, t\rangle = T(t_0, t)|A\rangle. \tag{8.40}$$

(See Section 3.2.) As before, $|A, t_0, t\rangle$ is the state ket of the system at time t, given that the state ket at the initial time t_0 is $|A\rangle$. It is easily seen that the time evolution operator satisfies the differential equation

$$i\,\hbar\,\frac{\partial T(t_0, t)}{\partial t} = (H_0 + H_1)\,T(t_0, t), \tag{8.41}$$

subject to the initial condition

$$T(t_0, t_0) = 1. \tag{8.42}$$

(See Section 3.2.)

In the absence of the external perturbation, the time evolution operator reduces to

$$T(t_0, t) = \exp\left[\frac{-i\,H_0\,(t - t_0)}{\hbar}\right]. \tag{8.43}$$

Let us switch on the perturbation, and look for a solution of the form

$$T(t_0, t) = \exp\left[\frac{-i\,H_0\,(t - t_0)}{\hbar}\right]T_I(t_0, t). \tag{8.44}$$

It is readily demonstrated that $T_I(t_0, t)$ satisfies the differential equation

$$i\hbar \frac{\partial T_I(t_0, t)}{\partial t} = H_I(t_0, t) T_I(t_0, t), \tag{8.45}$$

where

$$H_I(t_0, t) = \exp\left[\frac{i H_0 (t - t_0)}{\hbar}\right] H_1 \exp\left[\frac{-i H_0 (t - t_0)}{\hbar}\right], \tag{8.46}$$

subject to the initial condition

$$T_I(t_0, t_0) = 1. \tag{8.47}$$

(See Exercise 8.3.) Note that T_I specifies that component of the time evolution operator that is generated by the time-dependent perturbation. Thus, we would expect T_I to contain all of the information regarding transitions between different eigenstates of H_0 caused by the perturbation.

Suppose that the system starts off at time t_0 in the eigenstate $|i\rangle$ of the unperturbed Hamiltonian. The subsequent evolution of the state ket is given by Equation (8.6),

$$|i, t_0, t\rangle = \sum_f c_f(t) \exp\left[\frac{-i E_f (t - t_0)}{\hbar}\right] |f\rangle. \tag{8.48}$$

However, we also have

$$|i, t_0, t\rangle = \exp\left[\frac{-i H_0 (t - t_0)}{\hbar}\right] T_I(t_0, t) |i\rangle. \tag{8.49}$$

It follows that

$$c_f(t) = \langle f| T_I(t_0, t) |i\rangle, \tag{8.50}$$

where use has been made of $\langle n|m \rangle = \delta_{nm}$. Thus, the probability that the system is found in some final state $|f\rangle$ at time t, given that it was definitely in the initial state $|i\rangle$ at time t_0, is simply

$$P_{i \to f}(t_0, t) = |\langle f| T_I(t_0, t) |i\rangle|^2. \tag{8.51}$$

This quantity is usually termed the *transition probability* between states $|i\rangle$ and $|f\rangle$.

Note that the differential equation (8.45), plus the initial condition (8.47), are equivalent to the following integral equation:

$$T_I(t_0, t) = 1 + \left(\frac{-i}{\hbar}\right) \int_{t_0}^{t} dt' \, H_I(t_0, t') T_I(t_0, t'). \tag{8.52}$$

We can obtain an approximate solution to this equation via iteration:

$$T_I(t_0, t) \simeq 1 + \left(\frac{-i}{\hbar}\right) \int_{t_0}^{t} dt' \, H_I(t_0, t') \left[1 + \left(\frac{-i}{\hbar}\right) \int_{t_0}^{t'} dt'' \, H_I(t_0, t'') \, T_I(t_0, t'')\right]$$

$$\simeq 1 + \left(\frac{-i}{\hbar}\right) \int_{t_0}^{t} dt' \, H_I(t_0, t')$$

$$+ \left(\frac{-i}{\hbar}\right)^2 \int_{t_0}^{t} dt' \int_{t_0}^{t'} dt'' \, H_I(t_0, t') \, H_I(t_0, t'') + \cdots . \tag{8.53}$$

This expansion is known as the *Dyson series* [Dyson (1949)].

Let

$$c_f = c_f^{(0)} + c_f^{(1)} + c_f^{(2)} + \cdots , \tag{8.54}$$

where the superscript $^{(1)}$ refers to a first-order term in the expansion, et cetera. It follows from Equations (8.50) and (8.53) that

$$c_f^{(0)}(t) = \delta_{if}, \tag{8.55}$$

$$c_f^{(1)}(t) = \left(-\frac{i}{\hbar}\right) \int_{t_0}^{t} dt' \, \langle f| \, H_I(t_0, t') \, |i\rangle, \tag{8.56}$$

$$c_f^{(2)}(t) = \left(\frac{-i}{\hbar}\right)^2 \int_{t_0}^{t} dt' \int_{t_0}^{t'} dt'' \, \langle f| \, H_I(t_0, t') \, H_I(t_0, t'') \, |i\rangle. \tag{8.57}$$

These expressions simplify to give

$$c_f^{(0)}(t) = \delta_{if}, \tag{8.58}$$

$$c_f^{(1)}(t) = \left(\frac{-i}{\hbar}\right) \int_{t_0}^{t} dt' \, \exp[\,i\,\omega_{fi}\,(t' - t_0)]\, H_{fi}(t'), \tag{8.59}$$

$$c_f^{(2)}(t) = \left(\frac{-i}{\hbar}\right)^2 \sum_m \int_{t_0}^{t} dt' \int_{t_0}^{t'} dt'' \, \exp[\,i\,\omega_{fm}\,(t' - t_0)]\, H_{fm}(t')$$

$$\exp[\,i\,\omega_{mi}\,(t'' - t_0)]\, H_{mi}(t''), \tag{8.60}$$

where

$$\omega_{nm} = \frac{E_n - E_m}{\hbar}, \tag{8.61}$$

and

$$H_{nm}(t) = \langle n| \, H_1(t) \, |m\rangle. \tag{8.62}$$

Here, use has been made of the completeness relation $\sum_m |m\rangle\langle m| = 1$, where the sum is over all energy eigenstates. The transition probability between states i and f is simply

$$P_{i \to f}(t_0, t) = \left| c_f^{(0)} + c_f^{(1)} + c_f^{(2)} + \cdots \right|^2. \tag{8.63}$$

According to the previous analysis, there is no possibility of a transition between the initial state, $|i\rangle$, and the final state, $|f\rangle$, (where $i \neq f$) to zeroth order (i.e., in the absence of the perturbation). To first order, the transition probability is proportional to the time integral of the matrix element $\langle f| H_1 |i\rangle$, weighted by an oscillatory phase-factor. Thus, if the matrix element is zero then there is no chance of a first-order transition between states $|i\rangle$ and $|f\rangle$. However, to second order, a transition between states $|i\rangle$ and $|f\rangle$ is possible even when the matrix element $\langle f| H_1 |i\rangle$ is zero.

8.6 Sudden Perturbations

Consider, for example, a constant perturbation that is suddenly switched on at time $t = 0$: that is,

$$H_1(t) = \begin{cases} 0 & \text{for } t < 0 \\ H_1 & \text{for } t \geq 0 \end{cases}, \tag{8.64}$$

where H_1 is independent of time, but is generally a function of the position, momentum, and spin operators. Suppose that the system is definitely in state $|i\rangle$ at time $t = 0$. According to Equations (8.58) and (8.59) (with $t_0 = 0$),

$$c_f^{(0)}(t) = \delta_{if}, \tag{8.65}$$

$$c_f^{(1)}(t) = -\frac{i}{\hbar} H_{fi} \int_0^t dt' \, \exp[\,i\,\omega_{fi}\,(t' - t)] = \frac{H_{fi}}{E_f - E_i} [1 - \exp(i\,\omega_{fi}\,t)], \tag{8.66}$$

giving

$$P_{i \to f}(t) \simeq \left| c_f^{(1)}(t) \right|^2 = \frac{4\,|H_{fi}|^2}{|E_f - E_i|^2} \sin^2\left[\frac{(E_f - E_i)\,t}{2\,\hbar} \right], \tag{8.67}$$

for $i \neq f$. The transition probability between states $|i\rangle$ and $|f\rangle$ can be written

$$P_{i \to f}(t) = \frac{|H_{fi}|^2\,t^2}{\hbar^2} \,\text{sinc}^2\left[\frac{(E_f - E_i)\,t}{2\,\hbar} \right], \tag{8.68}$$

where

$$\text{sinc}(x) \equiv \frac{\sin x}{x}. \tag{8.69}$$

The sinc function is highly oscillatory, and decays like $1/|x|$ at large $|x|$. In fact, it is a good approximation to say that $\text{sinc}(x)$ is small compared to unity except when $|x| \lesssim \pi$. It follows that the transition probability, $P_{i \to f}$, is negligibly small unless

$$|E_f - E_i| \lesssim \frac{2\pi\,\hbar}{t}. \tag{8.70}$$

Note that, in the limit $t \to \infty$, only those transitions that conserve energy (i.e., $E_f = E_i$) have an appreciable probability of occurrence. At finite t, is is possible to have transitions that do not exactly conserve energy, provided that

$$\Delta E \, \Delta t \lesssim h, \qquad (8.71)$$

where $\Delta E = |E_f - E_i|$ is the change in energy of the system associated with the transition, and $\Delta t = t$ is the time elapsed since the perturbation was switched on. This result is a manifestation of the well-known uncertainty relation for energy and time [Gasiorowicz (1996)]. Incidentally, the energy-time uncertainty relation is fundamentally different from the position-momentum uncertainty relation because (in non-relativistic quantum mechanics) position and momentum are operators, whereas time is merely a parameter.

The probability of a transition that conserves energy (i.e., $E_f = E_i$) is

$$P_{i \to f}(t) = \frac{|H_{fi}|^2 \, t^2}{\hbar^2}, \qquad (8.72)$$

where use has been made of $\operatorname{sinc}(0) = 1$. Note that this probability grows quadratically in time. This result is somewhat surprising, because it implies that the probability of a transition occurring in a fixed time interval, t to $t + dt$, grows linearly with t, despite the fact that H_1 is constant for $t > 0$. In practice, there is usually a group of final states, all possessing nearly the same energy as the energy of the initial state, $|i\rangle$. It is helpful to define the density of states, $\rho(E)$, where the number of final states lying in the energy range E to $E + dE$ is given by $\rho(E) \, dE$. Thus, the probability of a transition from the initial state $|i\rangle$ to one of the continuum of possible final states is

$$P_{i \to}(t) = \int_{-\infty}^{\infty} dE_f \, P_{i \to f}(t) \rho(E_f), \qquad (8.73)$$

giving

$$P_{i \to}(t) = \frac{2 \, t}{\hbar} \int_{-\infty}^{\infty} dx \, |H_{fi}|^2 \rho(E_f) \operatorname{sinc}^2(x), \qquad (8.74)$$

where

$$x = \frac{(E_f - E_i) \, t}{2 \, \hbar}, \qquad (8.75)$$

and use has been made of Equation (8.68). We know that, in the limit $t \to \infty$, the function $\operatorname{sinc}(x)$ is only non-zero in an infinitesimally narrow range of final energies centered on $E_f = E_i$. It follows that, in this limit, we can take $\rho(E_f)$ and $|H_{fi}|^2$ out of the integral in the previous formula to obtain

$$P_{i \to [f]}(t) = \frac{2\pi}{\hbar} \left. \overline{|H_{fi}|^2} \rho(E_f) \, t \right|_{E_f \simeq E_i}, \qquad (8.76)$$

where $P_{i\to[f]}$ denotes the transition probability between the initial state, $|i\rangle$, and all final states, $|f\rangle$, that have approximately the same energy as the initial state. Here, $\overline{|H_{fi}|^2}$ is the average of $|H_{fi}|^2$ over all final states with approximately the same energy as the initial state. In deriving the previous formula, we have made use of the result [Gradshteyn *et al.* (1980)]

$$\int_{-\infty}^{\infty} dx \, \mathrm{sinc}^2(x) = \pi. \tag{8.77}$$

Note that the transition probability, $P_{i\to[f]}$, is now proportional to t, instead of t^2.

It is convenient to define the *transition rate*, which is simply the transition probability per unit time. Thus,

$$w_{i\to[f]} = \frac{dP_{i\to[f]}}{dt}, \tag{8.78}$$

which implies that

$$w_{i\to[f]} = \frac{2\pi}{\hbar} \overline{|H_{fi}|^2} \rho(E_f) \Big|_{E_f \simeq E_i}. \tag{8.79}$$

This appealingly simple result is known as *Fermi's golden rule* [Dirac (1927); Fermi (1950)]. Note that the transition rate is constant in time (for $t > 0$): that is, the probability of a transition occurring in the time interval t to $t+dt$ is independent of t for fixed dt. Fermi's golden rule is more usually written

$$w_{i\to f} = \frac{2\pi}{\hbar} |H_{fi}|^2 \, \delta(E_f - E_i), \tag{8.80}$$

where it is understood that this formula must be integrated with $\int dE_f \, (\cdots) \rho(E_f)$ in order to obtain the actual transition rate.

Let us now calculate the second-order term in the Dyson series, using the constant perturbation (8.64). From Equation (8.60), we find that (with $t_0 = 0$)

$$c_f^{(2)}(t) = \left(\frac{-i}{\hbar}\right)^2 \sum_m H_{fm} H_{mi} \int_0^t dt' \, \exp(i\,\omega_{fm}\,t') \int_0^{t'} dt'' \, \exp(i\,\omega_{mi}\,t'')$$

$$= \frac{i}{\hbar} \sum_m \frac{H_{fm} H_{mi}}{E_m - E_i} \int_0^t dt' \left[\exp(i\,\omega_{fi}\,t') - \exp(i\,\omega_{fm}\,t')\right] \tag{8.81}$$

$$= \frac{i\,t}{\hbar} \sum_m \frac{H_{fm} H_{mi}}{E_m - E_i} \left[\exp\left(\frac{i\,\omega_{fi}\,t}{2}\right) \mathrm{sinc}\left(\frac{\omega_{fi}\,t}{2}\right) - \exp\left(\frac{i\,\omega_{fm}\,t}{2}\right) \mathrm{sinc}\left(\frac{\omega_{fm}\,t}{2}\right)\right].$$

Thus,

$$c_f(t) = c_f^{(1)} + c_f^{(2)} = \frac{i\,t}{\hbar} \exp\left(\frac{i\,\omega_{fi}\,t}{2}\right) \left[\left(H_{fi} + \sum_m \frac{H_{fm} H_{mi}}{E_m - E_i}\right) \mathrm{sinc}\left(\frac{\omega_{fi}\,t}{2}\right)\right.$$

$$\left. - \sum_m \frac{H_{fm} H_{mi}}{E_m - E_i} \exp\left(\frac{i\,\omega_{im}\,t}{2}\right) \mathrm{sinc}\left(\frac{\omega_{fm}\,t}{2}\right)\right], \tag{8.82}$$

where use has been made of Equation (8.66). It follows, by analogy with the previous analysis, that

$$w_{i \to [f]} = \frac{2\pi}{\hbar} \overline{\left| H_{fi} + \sum_m \frac{H_{fm} H_{mi}}{E_m - E_i} \right|^2 \rho(E_f)} \Bigg|_{E_f \simeq E_i}, \qquad (8.83)$$

where the transition rate is calculated for all final states, $|f\rangle$, with approximately the same energy as the initial state, $|i\rangle$, and for intermediate states, $|m\rangle$ whose energies differ from that of the initial state. The fact that $E_m \neq E_i$ causes the final term on the right-hand side of Equation (8.82) to average to zero during the evaluation of the transition probability. (See Exercise 8.4.)

According to Equation (8.83), a second-order transition takes place in two steps. First, the system makes a non-energy-conserving transition to some intermediate state $|m\rangle$. Subsequently, the system makes another non-energy-conserving transition to the final state $|f\rangle$. The net transition from state $|i\rangle$ to state $|f\rangle$ conserves energy. The non-energy-conserving transitions are generally termed *virtual transitions*, whereas the energy conserving first-order transition is termed a *real transition*. The previous formula clearly breaks down if $H_{fm} H_{mi} \neq 0$ when $E_m = E_i$. This problem can be avoided by gradually turning on the perturbation: that is, $H_1 \to \exp(\eta t) H_1$ (where η is very small). The net result is to change the energy denominator in Equation (8.83) from $E_i - E_m$ to $E_i - E_m + i\hbar\eta$.

8.7 Energy-Shifts and Decay-Widths

We have examined how a state $|f\rangle$, other than the initial state $|i\rangle$, becomes populated as a result of some time-dependent perturbation applied to the system. Let us now consider how the initial state becomes depopulated.

In this case, it is convenient to gradually turn on the perturbation from zero at $t = -\infty$. Thus,

$$H_1(t) = \exp(\eta t) H_1, \qquad (8.84)$$

where η is small and positive, and H_1 is a constant.

In the remote past, $t \to -\infty$, the system is assumed to be in the initial state $|i\rangle$. Thus, $c_i(t \to -\infty) = 1$, and $c_{f \neq i}(t \to -\infty) = 0$. We wish to calculate the time evolution of the coefficient $c_i(t)$. First, however, let us check that our previous Fermi golden rule result still applies when the perturbing potential is turned on slowly, instead of very suddenly. For $c_{f \neq i}(t)$, it follows from Equations (8.58) and

(8.59) that (with $t_0 = -\infty$)

$$c_f^{(0)}(t) = 0,$$ (8.85)

$$c_f^{(1)}(t) = -\frac{i}{\hbar} H_{fi} \int_{-\infty}^{t} dt' \, \exp[(\eta + i\,\omega_{fi})\,t'] = -\frac{i}{\hbar} H_{fi} \frac{\exp[(\eta + i\,\omega_{fi})\,t]}{\eta + i\,\omega_{fi}},$$ (8.86)

where $H_{fi} = \langle f | H_1 | i \rangle$. Thus, to first order, the transition probability from state $|i\rangle$ to state $|f\rangle$ is

$$P_{i \to f}(t) = \left| c_f^{(1)}(t) \right|^2 = \frac{|H_{fi}|^2}{\hbar^2} \frac{\exp(2\,\eta\,t)}{\eta^2 + \omega_{fi}^2}.$$ (8.87)

The transition rate is given by

$$w_{i \to f}(t) = \frac{dP_{i \to f}}{dt} = \frac{2 |H_{fi}|^2}{\hbar^2} \frac{\eta \exp(2\,\eta\,t)}{\eta^2 + \omega_{fi}^2}.$$ (8.88)

Consider the limit $\eta \to 0$. In this limit, $\exp(\eta\,t) \to 1$, but [Reif (1965)]

$$\lim_{\eta \to 0} \frac{\eta}{\eta^2 + \omega_{fi}^2} = \pi\,\delta(\omega_{fi}) = \pi\,\hbar\,\delta(E_f - E_i).$$ (8.89)

(See Exercise 1.19.) Thus, Equation (8.88) yields the standard Fermi golden rule result

$$w_{i \to f} = \frac{2\pi}{\hbar} |H_{fi}|^2 \, \delta(E_f - E_i).$$ (8.90)

It is clear that the delta-function in the previous formula actually represents a function that is highly peaked when its argument is close to zero. The width of the peak is determined by how fast the perturbation is switched on.

Let us now calculate $c_i(t)$ using Equations (8.58)–(8.60). We have

$$c_i^{(0)}(t) = 1,$$ (8.91)

$$c_i^{(1)}(t) = \left(\frac{-i}{\hbar} \right) H_{ii} \int_{-\infty}^{t} \exp(\eta\,t')\,dt' = -\frac{i}{\hbar} H_{ii} \frac{\exp(\eta\,t)}{\eta},$$ (8.92)

$$c_i^{(2)}(t) = \left(\frac{-i}{\hbar} \right)^2 \sum_f |H_{fi}|^2 \int_{-\infty}^{t} dt' \int_{-\infty}^{t'} dt'' \, \exp[(\eta + i\,\omega_{if})\,t'] \exp[(\eta + i\,\omega_{fi})\,t'']$$

$$= \left(\frac{-i}{\hbar} \right)^2 \sum_f |H_{fi}|^2 \frac{\exp(2\,\eta\,t)}{2\,\eta\,(\eta + i\,\omega_{fi})}.$$ (8.93)

Thus, to second order, we get

$$c_i(t) \simeq 1 + \left(\frac{-i}{\hbar} \right) H_{ii} \frac{\exp(\eta\,t)}{\eta}$$

$$+ \left(\frac{-i}{\hbar} \right)^2 |H_{ii}|^2 \frac{\exp(2\,\eta\,t)}{2\,\eta^2} + \left(\frac{-i}{\hbar} \right) \sum_{f \neq i} \frac{|H_{fi}|^2 \exp(2\,\eta\,t)}{2\,\eta\,(E_i - E_f + i\,\hbar\,\eta)}.$$ (8.94)

Let us now consider the ratio $\dot{c}_i/c_i$, where $\dot{c}_i \equiv dc_i/dt$. Using Equation (8.94), we can evaluate this ratio in the limit $\eta \to 0$. We obtain

$$\frac{\dot{c}_i}{c_i} \simeq \left[\left(\frac{-i}{\hbar}\right)H_{ii} + \left(\frac{-i}{\hbar}\right)^2 \frac{|H_{ii}|^2}{\eta} + \left(\frac{-i}{\hbar}\right)\sum_{f \neq i}\frac{|H_{fi}|^2}{E_i - E_f + i\hbar\eta}\right]\bigg/\left[1 + \left(\frac{-i}{\hbar}\right)\frac{H_{ii}}{\eta}\right]$$

$$\simeq \left(\frac{-i}{\hbar}\right)H_{ii} + \lim_{\eta \to 0}\left(\frac{-i}{\hbar}\right)\sum_{f \neq i}\frac{|H_{fi}|^2}{E_i - E_f + i\hbar\eta}. \tag{8.95}$$

This result is formally correct to second order in perturbed quantities. Note that the right-hand side of Equation (8.95) is independent of time. We can write

$$\frac{\dot{c}_i}{c_i} = \left(\frac{-i}{\hbar}\right)\Delta_i, \tag{8.96}$$

where

$$\Delta_i = H_{ii} + \lim_{\eta \to 0}\sum_{f \neq i}\frac{|H_{fi}|^2}{E_i - E_f + i\hbar\eta} \tag{8.97}$$

is a constant. According to a well-known result in pure mathematics known as the *Plemelj formula* [Plemelj (1908)],

$$\lim_{\epsilon \to 0}\frac{1}{x + i\epsilon} = \mathcal{P}\left(\frac{1}{x}\right) - i\pi\delta(x), \tag{8.98}$$

where $\epsilon > 0$, and $\mathcal{P}$ denotes the *Cauchy principal part* [Flanigan (2010)]. It follows that

$$\Delta_i = H_{ii} + \mathcal{P}\sum_{f \neq i}\frac{|H_{fi}|^2}{E_i - E_f} - i\pi\sum_{n \neq i}|H_{fi}|^2\,\delta(E_i - E_f). \tag{8.99}$$

It is convenient to normalize the solution of Equation (8.96) such that $c_i(0) = 1$. Thus, we obtain

$$c_i(t) = \exp\left(\frac{-i\Delta_i t}{\hbar}\right). \tag{8.100}$$

According to Equation (8.6), the time evolution of the initial state ket $|i\rangle$ is given by

$$|i, t\rangle = \exp\left[\frac{-i(\Delta_i + E_i)t}{\hbar}\right]|i\rangle. \tag{8.101}$$

We can rewrite this result as

$$|i, t\rangle = \exp\left(\frac{-i[E_i + \mathrm{Re}(\Delta_i)]t}{\hbar}\right)\exp\left[\frac{\mathrm{Im}(\Delta_i)t}{\hbar}\right]|i\rangle. \tag{8.102}$$

It is clear that the real part of Δ_i gives rise to a simple shift in energy of state $|i\rangle$, whereas the imaginary part of Δ_i governs the growth or decay of this state. Thus,

$$|i, t\rangle = \exp\left[\frac{-\mathrm{i}\,(E_i + \Delta E_i)\,t}{\hbar}\right] \exp\left(\frac{-\Gamma_i t}{2\hbar}\right) |i\rangle, \qquad (8.103)$$

where

$$\Delta E_i = \mathrm{Re}(\Delta_i) = H_{ii} + \mathcal{P} \sum_{f \neq i} \frac{|H_{fi}|^2}{E_i - E_f}, \qquad (8.104)$$

and

$$\frac{\Gamma_i}{\hbar} = -\frac{2\,\mathrm{Im}(\Delta_i)}{\hbar} = \frac{2\pi}{\hbar} \sum_{f \neq i} |H_{fi}|^2 \,\delta(E_i - E_f). \qquad (8.105)$$

Note that the energy-shift, ΔE_i, is the same as that predicted by standard time-independent perturbation theory. (See Section 7.3.)

The probability of observing the system in state $|i\rangle$ at time $t > 0$, given that it was definitely in state $|i\rangle$ at time $t = 0$, is given by

$$P_{i \to i}(t) = |c_i(t)|^2 = \exp\left(\frac{-\Gamma_i t}{\hbar}\right), \qquad (8.106)$$

where

$$\frac{\Gamma_i}{\hbar} = \sum_{f \neq i} w_{i \to f}. \qquad (8.107)$$

Here, use has been made of Equation (8.80). Clearly, the decay rate of the initial state, $\Gamma_i/\hbar$, is equal to the sum of the transition rates to all of the other states. Note that the system conserves probability up to second order in perturbed quantities, because

$$|c_i|^2 + \sum_{f \neq i} |c_f|^2 \simeq \left(1 - \frac{\Gamma_i t}{\hbar}\right) + \sum_{f \neq i} w_{i \to f}\, t = 1. \qquad (8.108)$$

We can write

$$\tau_i = \frac{\hbar}{\Gamma_i}, \qquad (8.109)$$

where

$$P_{i \to i}(t) = \exp\left(-\frac{t}{\tau_i}\right). \qquad (8.110)$$

The quantity Γ_i, which is called the *decay-width* of state $|i\rangle$, is thus closely related to the *mean lifetime*, τ_i, of this state. (See Exercise 8.5.) According to Equation (8.102), the amplitude of state $|i\rangle$ both oscillates and decays as time progresses. Clearly, state $|i\rangle$ is not a stationary state in the presence of the time-dependent perturbation. However, we can still represent this state as a superposition of stationary states (whose amplitudes simply oscillate in time). Thus,

$$\exp\left[\frac{-\mathrm{i}\,(E_i + \Delta E_i)\,t}{\hbar}\right] \exp\left(\frac{-\Gamma_i\,t}{2\,\hbar}\right) = \int_{-\infty}^{\infty} dE\, f(E) \exp\left(\frac{-\mathrm{i}\,E\,t}{\hbar}\right), \qquad (8.111)$$

where $f(E)$ is the weight of the stationary state with energy E in the superposition. The Fourier inversion theorem yields [Fitzpatrick (2013)]

$$|f(E)|^2 \propto \frac{1}{(E - [E_i + \mathrm{Re}(\Delta_i)])^2 + \Gamma_i^2/4}. \qquad (8.112)$$

In the absence of the perturbation, $|f(E)|^2$ is basically a delta-function centered on the unperturbed energy, E_i, of state $|i\rangle$. In other words, state $|i\rangle$ is a stationary state whose energy is completely determined. In the presence of the perturbation, the energy of state $|i\rangle$ is shifted by $\mathrm{Re}(\Delta_i)$. The fact that the state is no longer stationary (i.e., it decays in time) implies that its energy cannot be exactly determined. Indeed, the effective energy of the state is smeared over some range of values of width (in energy) Γ_i, centered on the shifted energy, $E_i + \mathrm{Re}(\Delta_i)$. The faster the decay of the state (i.e., the larger Γ_i), the more its energy is spread out. This phenomenon is clearly a manifestation of the energy-time uncertainty relation $\Delta E\,\Delta t \sim \hbar$. One consequence of this effect is the existence of a *natural width* of spectral lines associated with the decay of a given excited state of an atom to the ground state. The uncertainty in energy of the excited state, due to its propensity to decay, gives rise to a slight smearing (in wavelength) of the spectral line associated with the transition. [This follows because $\lambda = \nu/c = (E_i - E_f)/(h\,c)$, where E_i and E_f are the energies of the excited and ground states, respectively—see Section 8.9. Hence, $\delta\lambda/\lambda = \delta E_i/E_i \simeq \Gamma_i/E_i$.] Strong lines, which correspond to fast (i.e., large Γ_i) transitions, are smeared out more that weak lines. For this reason, spectroscopists generally favor so-called forbidden lines (which correspond to relatively slow transitions) for Doppler-shift measurements. (See Section 8.11.) Such lines are not as bright as those associated with strong transitions, but they are much sharper.

8.8 Harmonic Perturbations

Consider a perturbation to the Hamiltonian that oscillates sinusoidally in time. This is usually termed a *harmonic perturbation*. Thus,

$$H_1(t) = V \exp(\mathrm{i}\,\omega\,t) + V^\dagger \exp(-\mathrm{i}\,\omega\,t), \qquad (8.113)$$

where V is, in general, a function of the position, momentum, and spin operators.

Let us initiate the system in the eigenstate $|i\rangle$ of the unperturbed Hamiltonian, H_0, and then switch on the harmonic perturbation at $t = 0$. It follows from Equation (8.59) that

$$c_f^{(1)} = \left(\frac{-i}{\hbar}\right) \int_0^t dt' \left[V_{fi} \exp(i\,\omega\,t') + V_{fi}^\dagger \exp(-i\,\omega\,t') \right] \exp(i\,\omega_{fi}\,t')$$

$$= \frac{1}{\hbar} \left(\frac{1 - \exp[i\,(\omega_{fi} + \omega)\,t]}{\omega_{fi} + \omega} V_{fi} + \frac{1 - \exp[i\,(\omega_{fi} - \omega)\,t]}{\omega_{fi} - \omega} V_{fi}^\dagger \right), \qquad (8.114)$$

where

$$V_{fi} = \langle f | V | i \rangle, \qquad (8.115)$$

$$V_{fi}^\dagger = \langle f | V^\dagger | i \rangle = \langle i | V | f \rangle^*. \qquad (8.116)$$

Equation (8.114) is analogous to Equation (8.66), provided that

$$\omega_{fi} = \frac{E_f - E_i}{\hbar} \to \omega_{fi} \pm \omega. \qquad (8.117)$$

Thus, it follows from the analysis of Section 8.6 that the transition probability $P_{i \to f}(t) = \left| c_f^{(1)}(t) \right|^2$ is only appreciable in the limit $t \to \infty$ if

$$\omega_{fi} + \omega \simeq 0 \quad \text{or} \quad E_f \simeq E_i - \hbar\omega, \qquad (8.118)$$

$$\omega_{fi} - \omega \simeq 0 \quad \text{or} \quad E_f \simeq E_i + \hbar\omega. \qquad (8.119)$$

Clearly, criterion (8.118) corresponds to the first term on the right-hand side of Equation (8.114), whereas criterion (8.119) corresponds to the second. The former term describes a process by which the system gives up energy $\hbar\omega$ to the perturbing field, while making a transition to a final state whose energy is less than that of the initial state by $\hbar\omega$. This process is known as *stimulated emission*. The latter term describes a process by which the system gains energy $\hbar\omega$ from the perturbing field, while making a transition to a final state whose energy exceeds that of the initial state by $\hbar\omega$. This process is known as *absorption*. In both cases, the total energy (i.e., that of the system plus the perturbing field) is conserved.

By analogy with Equation (8.79),

$$w_{i \to [f]}^{stm} = \frac{2\pi}{\hbar} \overline{|V_{fi}|^2} \rho(E_f) \bigg|_{E_f = E_i - \hbar\omega}, \qquad (8.120)$$

$$w_{i \to [f]}^{abs} = \frac{2\pi}{\hbar} \overline{|V_{fi}^\dagger|^2} \rho(E_f) \bigg|_{E_f = E_i + \hbar\omega}. \qquad (8.121)$$

Equation (8.120) specifies the transition rate for stimulated emission, whereas Equation (8.121) gives the transition rate for absorption. These transition rates are more usually written [see Equation (8.80)]

$$w_{i \to f}^{\text{stm}} = \frac{2\pi}{\hbar} |V_{fi}|^2 \, \delta(E_f - E_i + \hbar\omega), \tag{8.122}$$

$$w_{i \to f}^{\text{abs}} = \frac{2\pi}{\hbar} |V_{fi}^\dagger|^2 \, \delta(E_f - E_i - \hbar\omega), \tag{8.123}$$

where it is understood that the previous expressions must be integrated with $\int dE_f \, (\cdots) \rho(E_f)$ to obtain the actual transition rates.

It is clear from Equations (8.115) and (8.116) that $|V_{if}^\dagger|^2 = |V_{fi}|^2$. It follows from Equations (8.120) and (8.121) that

$$\frac{w_{i \to [f]}^{\text{stm}}}{\rho(E_f)} = \frac{w_{f \to [i]}^{\text{abs}}}{\rho(E_i)}. \tag{8.124}$$

In other words, the rate of stimulated emission, divided by the density of final states for stimulated emission, is equal to the rate of absorption, divided by the density of final states for absorption. This result, which expresses a fundamental symmetry between absorption and stimulated emission, is known as *detailed balance*, and plays an important role in quantum statistical mechanics.

8.9 Absorption and Stimulated Emission of Radiation

Let us employ time-dependent perturbation theory to investigate the interaction of an electron in a hydrogen-like atom with classical (i.e., non-quantized) electromagnetic radiation. The Hamiltonian of such an electron is

$$H = \frac{(\mathbf{p} + e\,\mathbf{A})^2}{2\,m_e} - e\,\phi + \Phi(r), \tag{8.125}$$

where $\Phi(r)$ is the atomic potential energy, and $\mathbf{A}(\mathbf{x})$ and $\phi(\mathbf{x})$ are the vector and scalar potentials associated with the incident radiation. (See Section 3.6.) The previous equation can be written

$$H = \frac{\left(p^2 + e\,\mathbf{A} \cdot \mathbf{p} + e\,\mathbf{p} \cdot \mathbf{A} + e^2 A^2\right)}{2\,m_e} - e\,\phi + \Phi(r). \tag{8.126}$$

Now,

$$\mathbf{p} \cdot \mathbf{A} = \mathbf{A} \cdot \mathbf{p}, \tag{8.127}$$

provided that we adopt the Coulomb gauge, $\nabla \cdot \mathbf{A} = 0$. (See Exercise 8.6.) Hence,

$$H = \frac{p^2}{2\,m_e} + \frac{e\,\mathbf{A} \cdot \mathbf{p}}{m_e} + \frac{e^2 A^2}{2\,m_e} - e\,\phi + \Phi(r). \tag{8.128}$$

Suppose that the perturbation corresponds to a monochromatic plane-wave of angular frequency ω, for which [Fitzpatrick (2008)]

$$\phi = 0, \tag{8.129}$$

$$\mathbf{A} = 2\,A_0\,\boldsymbol{\epsilon}\,\cos\!\left(\frac{\omega}{c}\,\mathbf{n}\cdot\mathbf{x} - \omega\,t\right), \tag{8.130}$$

where $\boldsymbol{\epsilon}$ and $\mathbf{n}$ are unit vectors that specify the wave's electric polarization direction (i.e., the direction of its electric component) and its direction of propagation, respectively. [The wavevector is $\mathbf{k} = (\omega/c)\,\mathbf{n}$.] Here, c is the velocity of light in vacuum. Note that, according to standard electromagnetic theory, $\boldsymbol{\epsilon}\cdot\mathbf{n} = 0$ [Fitzpatrick (2008)]. The Hamiltonian becomes

$$H = H_0 + H_1(t), \tag{8.131}$$

with

$$H_0 = \frac{p^2}{2\,m_e} + \Phi(r), \tag{8.132}$$

and

$$H_1 \simeq \frac{e\,\mathbf{A}\cdot\mathbf{p}}{m_e}, \tag{8.133}$$

where the term involving A^2, which is second order in A_0, has been neglected.

The perturbing Hamiltonian can be written

$$H_1 = \frac{e\,A_0}{m_e}\left(\exp\!\left[\mathrm{i}\!\left(\frac{\omega}{c}\right)\mathbf{n}\cdot\mathbf{x} - \mathrm{i}\,\omega\,t\right] + \exp\!\left[-\mathrm{i}\!\left(\frac{\omega}{c}\right)\mathbf{n}\cdot\mathbf{x} + \mathrm{i}\,\omega\,t\right]\right)\boldsymbol{\epsilon}\cdot\mathbf{p}. \tag{8.134}$$

This has the same form as Equation (8.113), provided that

$$V = \frac{e\,A_0}{m_e}\,\exp\!\left[-\mathrm{i}\!\left(\frac{\omega}{c}\right)\mathbf{n}\cdot\mathbf{x}\right]\boldsymbol{\epsilon}\cdot\mathbf{p}. \tag{8.135}$$

It is clear, by analogy with the analysis of Section 8.8, that the first term on the right-hand side of Equation (8.134) describes a process by which the atom absorbs energy $\hbar\,\omega$ from the electromagnetic field. On the other hand, the second term describes a process by which the atom emits energy $\hbar\,\omega$ to the electromagnetic field. We can interpret the former and latter processes as the absorption and stimulated emission, respectively, of a photon of energy $\hbar\,\omega$ by the atom.

It is convenient to define

$$\mathbf{d}_{if} = \frac{-\mathrm{i}}{m_e\,\omega_{if}}\left\langle i\left|\exp\!\left[\mathrm{i}\!\left(\frac{\omega}{c}\right)\mathbf{n}\cdot\mathbf{x}\right]\mathbf{p}\right|f\right\rangle, \tag{8.136}$$

which has the dimensions of length. Note that

$$\boldsymbol{\epsilon}\cdot\mathbf{d}_{if} = \frac{-\mathrm{i}}{e\,A_0\,\omega_{if}}\,\langle i|\,V^{\dagger}\,|f\rangle. \tag{8.137}$$

It follows, from Equations (8.122) and (8.123), that the rates of absorption and stimulated emission are

$$w_{i \to f}^{\text{abs}} = 2\pi \, \frac{e^2 \, \omega_{fi}^2}{\hbar^2} \, |A_0|^2 \, |\boldsymbol{\epsilon} \cdot \mathbf{d}_{fi}|^2 \, \delta(\omega - \omega_{fi}), \tag{8.138}$$

and

$$w_{i \to f}^{\text{stm}} = 2\pi \, \frac{e^2 \, \omega_{if}^2}{\hbar^2} \, |A_0|^2 \, |\boldsymbol{\epsilon} \cdot \mathbf{d}_{if}|^2 \, \delta(\omega - \omega_{if}), \tag{8.139}$$

respectively. (See Exercise 1.19.) It can be seen that absorption involves the absorption by the atom of a photon of angular frequency

$$\omega_{fi} = \frac{E_f - E_i}{\hbar}, \tag{8.140}$$

and energy $\hbar \, \omega_{fi} = E_f - E_i$, causing a transition from an atomic state with an initial energy E_i to a state with a final energy $E_f > E_i$. On the other hand, stimulated emission involves the emission by the atom of a photon of angular frequency

$$\omega_{if} = \frac{E_i - E_f}{\hbar}, \tag{8.141}$$

and energy $\hbar \, \omega_{if} = E_i - E_f$, causing a transition from an atomic state with an initial energy E_i to a state with a final energy $E_f < E_i$. In both cases, $\mathbf{n}$ and $\boldsymbol{\epsilon}$ specify the direction of propagation and electric polarization direction, respectively, of the photon. For the case of stimulated emission, $\mathbf{n}$ and $\boldsymbol{\epsilon}$ also specify the direction of propagation and polarization direction, respectively, of the radiation that stimulates the atomic transition.

Now, the energy density of an electromagnetic radiation field is [Fitzpatrick (2008)]

$$U = \frac{1}{2} \left(\frac{\epsilon_0 \, E_0^2}{2} + \frac{B_0^2}{2 \, \mu_0} \right), \tag{8.142}$$

where E_0 and $B_0 = E_0/c = 2 \, A_0 \, \omega/c$ are the peak electric and magnetic field-strengths, respectively. Hence,

$$U = 2 \, \epsilon_0 \, \omega^2 \, |A_0|^2, \tag{8.143}$$

and expressions (8.138) and (8.139) become

$$w_{i \to f}^{\text{abs}} = 4\pi^2 \alpha \, \frac{c}{\hbar} \, U \, |\boldsymbol{\epsilon} \cdot \mathbf{d}_{fi}|^2 \, \delta(\omega - \omega_{fi}), \tag{8.144}$$

and

$$w_{i \to f}^{\text{stm}} = 4\pi^2 \alpha \, \frac{c}{\hbar} \, U \, |\boldsymbol{\epsilon} \cdot \mathbf{d}_{if}|^2 \, \delta(\omega - \omega_{if}), \tag{8.145}$$

respectively, where α is the fine structure constant.

Let us suppose that the incident radiation has a range of different angular frequencies, so that

$$U = \int_{-\infty}^{\infty} d\omega \, u(\omega), \tag{8.146}$$

where $u(\omega) \, d\omega$ is the energy density of radiation whose angular frequency lies in the range ω to $\omega + d\omega$. Equations (8.144) and (8.145) imply that

$$\frac{dw_{i\to f}^{\text{abs}}}{d\omega} = 4\pi^2 \, \alpha \, \frac{c}{\hbar} \, u(\omega) \left| \boldsymbol{\epsilon} \cdot \mathbf{d}_{fi} \right|^2 \delta(\omega - \omega_{fi}), \tag{8.147}$$

and

$$\frac{dw_{i\to f}^{\text{stm}}}{d\omega} = 4\pi^2 \, \alpha \, \frac{c}{\hbar} \, u(\omega) \left| \boldsymbol{\epsilon} \cdot \mathbf{d}_{if} \right|^2 \delta(\omega - \omega_{if}). \tag{8.148}$$

Here, $(dw_{i\to f}^{\text{abs}}/d\omega) \, d\omega$ is the rate of absorption associated with radiation whose angular frequency lies in the range ω to $\omega + d\omega$, et cetera. Incidentally, we are assuming that the radiation is incoherent, so that intensities can be added [Hecht and Zajac (1974)]. Of course,

$$w_{i\to f}^{\text{abs}} = \int_{-\infty}^{\infty} d\omega \, \frac{dw_{i\to f}^{\text{abs}}}{d\omega} = 4\pi^2 \, \alpha \, \frac{c}{\hbar} \, u(\omega_{fi}) \left| \boldsymbol{\epsilon} \cdot \mathbf{d}_{fi} \right|^2, \tag{8.149}$$

$$w_{i\to f}^{\text{stm}} = \int_{-\infty}^{\infty} d\omega \, \frac{dw_{i\to f}^{\text{stm}}}{d\omega} = 4\pi^2 \, \alpha \, \frac{c}{\hbar} \, u(\omega_{if}) \left| \boldsymbol{\epsilon} \cdot \mathbf{d}_{if} \right|^2. \tag{8.150}$$

The rate at which the atom gains energy from the electromagnetic field as a consequence of absorption can be written

$$P_{i\to f}^{\text{abs}} = \int_{-\infty}^{\infty} d\omega \, \hbar \, \omega \, \frac{dw_{i\to f}^{\text{abs}}}{d\omega} = \int_{-\infty}^{\infty} d\omega \, c \, u(\omega) \, \sigma_{i\to f}^{\text{abs}}(\omega), \tag{8.151}$$

where $\sigma_{i\to f}^{\text{abs}}$ is the so-called *absorption cross-section*, and $c \, u(\omega) \, d\omega$ the electromagnetic energy flux associated with radiation whose angular frequency lies in the range ω to $\omega + d\omega$ [Fitzpatrick (2008)]. It follows that

$$\sigma_{i\to f}^{\text{abs}}(\omega) = 4\pi^2 \, \alpha \, \omega \left| \boldsymbol{\epsilon} \cdot \mathbf{d}_{fi} \right|^2 \delta(\omega - \omega_{fi}). \tag{8.152}$$

Similarly, the *stimulated emission cross-section* takes the form

$$\sigma_{i\to f}^{\text{stm}}(\omega) = 4\pi^2 \, \alpha \, \omega \left| \boldsymbol{\epsilon} \cdot \mathbf{d}_{if} \right|^2 \delta(\omega - \omega_{if}). \tag{8.153}$$

Finally, comparing Equations (8.149) and (8.150) with the previous two equations, we deduce that

$$w_{i\to f}^{\text{abs}} = \int_{-\infty}^{\infty} d\omega \, c \, n(\omega) \, \sigma_{1\to f}^{\text{abs}}(\omega), \tag{8.154}$$

$$w_{i\to f}^{\text{stm}} = \int_{-\infty}^{\infty} d\omega \, c \, n(\omega) \, \sigma_{i\to f}^{\text{stm}}(\omega). \tag{8.155}$$

Here, we have written $u(\omega) = n(\omega) \, \hbar \, \omega$, where $n(\omega) \, d\omega$ is the number density of photons whose angular frequencies lie in the range ω to $\omega + d\omega$.

8.10 Spontaneous Emission of Radiation

So far, we have calculated the transition rates between two atomic states induced by external electromagnetic radiation. This process is known as absorption when the energy of the final state exceeds that of the initial state, and stimulated emission when the energy of the final state is less than that of the initial state. Now, in the absence of any external radiation, we would not expect an atom in a given state to spontaneously jump into a state with a higher energy. On the other hand, it should be possible for such an atom to spontaneously jump into an state with a lower energy via the emission of a photon whose energy is equal to the difference between the energies of the initial and final states. This process is known as *spontaneous emission.*

We can derive the rate of spontaneous emission between two atomic states from a knowledge of the corresponding absorption and stimulated emission rates using a famous thermodynamic argument initially formulated by Einstein [Einstein (1916)]. Consider a very large ensemble of similar atoms placed inside a closed cavity whose walls (which are assumed to be perfect emitters and absorbers of radiation) are held at the constant temperature T. Let the system have attained thermal equilibrium. According to statistical thermodynamics, the cavity is filled with so-called *black-body* electromagnetic radiation whose energy spectrum is [Reif (1965)]

$$u(\omega) = \frac{\hbar}{\pi^2 c^3} \frac{\omega^3}{\exp(\hbar \omega / k_B T) - 1}, \tag{8.156}$$

where k_B is the Boltzmann constant. This well-known result was first obtained by Max Planck in 1900 [Planck (1900a,b)].

Consider two atomic states, labeled i and f, with $E_i > E_f$. One of the tenants of statistical thermodynamics is that in thermal equilibrium we have so-called detailed balance. (See Section 8.8.) This means that, irrespective of any other atomic states, the rate at which atoms in the ensemble leave state i due to transitions to state f is exactly balanced by the rate at which atoms enter state i due to transitions from state f. The former rate (i.e., number of transitions per unit time in the ensemble) is written

$$W_{i \to f} = N_i (w_{i \to f}^{\text{spn}} + w_{i \to f}^{\text{stm}}), \tag{8.157}$$

where $w_{i \to f}^{\text{spn}}$ and $w_{i \to f}^{\text{stm}}$ are the rates of spontaneous and stimulated emission, respectively, (for a single atom) between states i and f, and N_i is the number of atoms in the ensemble in state i. Likewise, the latter rate takes the form

$$W_{f \to i} = N_f \, w_{f \to i}^{\text{abs}}, \tag{8.158}$$

where $w_{f \to i}^{\mathrm{abs}}$ is the rate of absorption (for a single atom) between states f and i, and N_f is the number of atoms in the ensemble in state f. The previous expressions describe how atoms in the ensemble make transitions from state i to state f via a combination of spontaneous and stimulated emission, and make the reverse transition as a consequence of absorption. In thermal equilibrium, we have $W_{i \to f} = W_{f \to i}$, which gives

$$w_{i \to f}^{\mathrm{spn}} = \frac{N_f}{N_i} w_{f \to i}^{\mathrm{abs}} - w_{i \to f}^{\mathrm{stm}}. \tag{8.159}$$

Equations (8.149) and (8.150) imply that

$$w_{i \to f}^{\mathrm{spn}} = \frac{4\pi^2 \alpha c}{\hbar} \left(\frac{N_f}{N_i} - 1 \right) u(\omega_{if}) \left\langle |\boldsymbol{\epsilon} \cdot \mathbf{d}_{if}|^2 \right\rangle, \tag{8.160}$$

where $\omega_{if} = (E_i - E_f)/\hbar$, and the large angle brackets denote an average over all possible polarization directions of the incident radiation (because, in equilibrium, the radiation inside the cavity is isotropic). In fact, it is easily demonstrated that

$$\left\langle |\boldsymbol{\epsilon} \cdot \mathbf{d}_{if}|^2 \right\rangle = \frac{|d_{if}|^2}{3}, \tag{8.161}$$

where

$$|d_{if}|^2 = \mathbf{d}_{if}^* \cdot \mathbf{d}_{if}. \tag{8.162}$$

(See Exercise 8.7.) Now, another famous result in statistical thermodynamics is that, in thermal equilibrium, the number of atoms in an ensemble occupying a state of energy E is proportional to $\exp(-E/k_B T)$ [Reif (1965)]. This implies that

$$\frac{N_f}{N_i} = \frac{\exp(-E_f/k_B T)}{\exp(-E_i/k_B T)} = \exp\left(\frac{\hbar \omega_{if}}{k_B T} \right). \tag{8.163}$$

Thus, it follows from Equations (8.156), (8.160), (8.161), and (8.163) that the rate of spontaneous emission between states i and f takes the form

$$w_{i \to f}^{\mathrm{spn}} = \frac{4 \alpha \, \omega_{if}^3 \, |d_{if}|^2}{3 \, c^2}. \tag{8.164}$$

Note, that, although the previous result has been derived for an atom in a radiation-filled cavity, it remains correct even in the absence of radiation.

The direction of propagation of a photon spontaneously emitted by an atom is specified by its normalized wavevector, $\mathbf{n}$. Likewise, the electric polarization direction of the photon is specified by the unit vector $\boldsymbol{\epsilon}$. However, $\boldsymbol{\epsilon} \cdot \mathbf{n} = 0$, which implies that $\boldsymbol{\epsilon} = \cos\alpha \, \boldsymbol{\epsilon}_1 + \sin\alpha \, \boldsymbol{\epsilon}_2$, where $0 \leq \alpha \leq 2\pi$. Here, the unit vectors $\boldsymbol{\epsilon}_1$, $\boldsymbol{\epsilon}_2$, and $\mathbf{n}$ are mutually perpendicular, and form a right-handed set (i.e., $\boldsymbol{\epsilon}_1 \times \boldsymbol{\epsilon}_2 \cdot \mathbf{n} = 1$). The vectors $\boldsymbol{\epsilon}_1$ and $\boldsymbol{\epsilon}_2$ represent the two possible independent electric polarization directions of the emitted photon. Equation (8.164) generalizes to give [Park (1974)]

$$\frac{dw_{i \to f}^{\text{spn}}}{d\Omega} = \frac{\alpha \, \omega_{if}^3}{2\pi \, c^2} \sum_{j=1,2} |\boldsymbol{\epsilon}_j \cdot \mathbf{d}_{if}|^2, \tag{8.165}$$

where $(dw_{i \to f}^{\text{spn}}/d\Omega)\, d\Omega$ is the rate of spontaneous emission of photons whose propagation directions lie in the range of solid angle $d\Omega$. Of course,

$$w_{i \to f}^{\text{spn}} = \oint d\Omega \, \frac{dw_{i \to f}^{\text{spn}}}{d\Omega}. \tag{8.166}$$

(See Exercise 8.8.)

8.11 Electric Dipole Transitions

In general, the wavelength of the type of electromagnetic radiation that induces, or is emitted during, transitions between different atomic states is much larger than the typical size of a light atom. Thus, recalling that $\omega/c = 2\pi/\lambda$, the expession

$$\exp\left[i\left(\frac{\omega}{c}\right)\mathbf{n} \cdot \mathbf{x}\right] = 1 + i\,\frac{\omega}{c}\,\mathbf{n} \cdot \mathbf{x} + \cdots, \tag{8.167}$$

is well approximated by its first term, unity. This approximation is known as the *electric dipole approximation*. It follows from Equation (8.136) that

$$\mathbf{d}_{if} \simeq \frac{-i}{m_e \, \omega_{if}} \langle i | \, \mathbf{p} \, | f \rangle. \tag{8.168}$$

It is readily demonstrated from Equations (3.33) and (8.132) that

$$[\mathbf{x}, H_0] = \frac{i\,\hbar\,\mathbf{p}}{m_e}, \tag{8.169}$$

so

$$\langle i | \, \mathbf{p} \, | f \rangle = -i\,\frac{m_e}{\hbar} \langle i | \, [\mathbf{x}, H_0] \, | f \rangle = i\, m_e \, \omega_{if} \, \langle i | \, \mathbf{x} \, | f \rangle, \tag{8.170}$$

which implies that

$$\mathbf{d}_{if} \simeq \langle i | \, \mathbf{x} \, | f \rangle. \tag{8.171}$$

Here, $\mathbf{d}_{if}$ is termed the *electric dipole matrix element*. Recall from Equations (8.149), (8.150), and (8.164) that

$$w_{i \to f}^{\text{abs}} = 4\pi^2\, \alpha\, \frac{c}{\hbar}\, u(\omega_{fi}) \left| \boldsymbol{\epsilon} \cdot \mathbf{d}_{fi} \right|^2 \tag{8.172}$$

for absorption, and

$$w_{i \to f}^{\text{stm}} = 4\pi^2\, \alpha\, \frac{c}{\hbar}\, u(\omega_{if}) \left| \boldsymbol{\epsilon} \cdot \mathbf{d}_{if} \right|^2 \tag{8.173}$$

for stimulated emission, and with

$$w_{i \to f}^{\text{spn}} = \frac{4 \, \alpha \, \omega_{if}^3 \, |d_{if}|^2}{3 \, c^2} \tag{8.174}$$

for spontaneous emission.

We have already seen, from Section 7.4, that $\langle i | z | f \rangle = 0$ for a hydrogen-like atom, unless the initial and final states satisfy $\Delta l = \pm 1$ and $\Delta m = 0$. Here, l is the azimuthal quantum number, and m the magnetic quantum number. Moreover, Δl is the difference between the azimuthal quantum numbers of the initial and final states, et cetera. It is easily demonstrated that $\langle i | x | f \rangle$ and $\langle i | y | f \rangle$ are only non-zero if $\Delta l = \pm 1$, and $\Delta m = \pm 1$. (See Exercise 8.10.) It follows that the *electric dipole matrix elements*, $\mathbf{d}_{if} = \langle i | \mathbf{x} | f \rangle$, which control the rates of so-called *electric dipole transitions*, via Equations (8.172)–(8.174), are only non-zero if

$$\Delta l = \pm 1, \tag{8.175}$$

$$\Delta m = 0, \pm 1, \tag{8.176}$$

$$\Delta m_s = 0. \tag{8.177}$$

Here, m_s is the *spin quantum number*, which is defined as the eigenvalue of S_z divided by $\hbar$. (Of course, $\Delta m_s = 0$ because $\mathbf{x}$ does not explicitly depend on spin.) These expressions are termed the *selection rules* for electric dipole transitions. It is clear, for instance, that the electric dipole selection rules permit a transition from a $2p$ state to a $1s$ state of a hydrogen-like atom, but disallow a transition from a $2s$ to a $1s$ state. The latter transition is called a *forbidden transition*. The previous selection rules can also be written in the slightly more general form

$$\Delta j = 0, \pm 1, \tag{8.178}$$

$$\Delta m_j = 0, \pm 1, \tag{8.179}$$

where j and m_j are the standard quantum numbers associated with the total angular momentum, and the projection of the total angular momentum along the z-axis, respectively [Corney (1977)]. (These new rules are evident from a perusal of the results of Exercise 8.15.) Note, however, that an electric dipole transition between two $j = 0$ states is forbidden [Corney (1977)].

Let us estimate the typical spontaneous emission rate for an electric dipole transition in a hydrogen atom. We expect the matrix element $\mathbf{d}_{if}$, defined in Equation (8.171), to be of order a_0, where a_0 is the Bohr radius. We also expect $\omega_{if} = (E_i - E_f)/\hbar$ to be of order $|E_0|/\hbar$, where E_0 is the hydrogen ground-state energy. It thus follows from Equation (8.174) that

$$w_{i \to f}^{\text{spn}} \sim \alpha^3 \, \omega_{if} \sim \alpha^5 \, \frac{m_e \, c^2}{\hbar}, \tag{8.180}$$

where $\alpha \simeq 1/137$ is the fine structure constant. This is an important result, because our perturbation expansion is based on the assumption that the transition rate between different energy eigenstates is much less than the frequency of phase oscillation of these states. In other words, $w_{i \to f}^{\text{spn}} \ll \omega_{if}$. This is indeed the case.

According to Equation (8.152), the absorption cross-section associated with atomic transitions between an initial state of energy E_i and a final state of energy $E_f > E_i$ can be written

$$\sigma_{i \to f}^{\text{abs}}(\omega) = 4\pi^2 \, \alpha \, \omega_{fi} \, |\epsilon \cdot \mathbf{d}_{fi}|^2 \, \delta(\omega - \omega_{fi}), \tag{8.181}$$

where $\omega_{fi} = (E_f - E_i)/\hbar$. Suppose, for the sake of argument, that the electromagnetic radiation is polarized in the x-direction, so that $\epsilon = \mathbf{e}_x$. According to Equation (8.181), if such radiation is incident on a hydrogen-like atom then the net absorption cross-section, summed over all final states, and integrated over all possible angular frequencies of the radiation, can be written

$$\int_{-\infty}^{\infty} d\omega \sum_f \sigma_{i \to f}^{\text{abs}}(\omega) = 2\pi^2 \, \alpha \, \frac{\hbar}{m_e} \sum_f F_{if}, \tag{8.182}$$

where the dimensionless parameter

$$F_{if} = \frac{2 \, m_e \, \omega_{fi}}{\hbar} \, |\langle i| \, x \, |f\rangle|^2 \tag{8.183}$$

is termed the *oscillator strength* associated with radiation-induced electric dipole transitions between states i and f. Note that if state i is the ground state (i.e., the lowest energy state) then the sum in Equation (8.182) is over all atomic states. On the other hand, if state i is not the ground state then the sum is restricted to states whose energies are greater than E_i (because we must have $E_f > E_i$ for absorption). Now, a straightforward generalization of the result proved in Exercise 3.6 yields the so-called *Thomas-Reiche-Kuhn sum rule* [Thomas (1925); Kuhn (1925); Reiche and Thomas (1925)]:

$$\sum_f F_{if} = 1, \tag{8.184}$$

where the sum is over all atomic states. (See Exercise 8.11.) Thus, it follows that, provided state i is the ground state,

$$\int_{-\infty}^{\infty} d\omega \sum_f \sigma_{i \to f}^{\text{abs}}(\omega) = 2\pi^2 \, \alpha \, \frac{\hbar}{m_e} = 2\pi^2 \, r_e \, c, \tag{8.185}$$

where $r_e = e^2/(4\pi \, \epsilon_0 \, m_e \, c^2)$ is the *classical electron radius*. Note that the integrated absorption cross-section is independent of Planck's constant. In fact, the previous result can also be obtained from classical physics [Jackson (1975)].

8.12   Forbidden Transitions

We saw in Section 8.10 that a spontaneous electromagnetic transition between some initial atomic state, i, and some final state, f, is mediated by the matrix element

$$\epsilon \cdot \mathbf{d}_{if} = \frac{-\mathrm{i}}{m_e \, \omega_{if}} \left\langle i \left| \exp\left[\mathrm{i}\left(\frac{\omega}{c}\right) \mathbf{n} \cdot \mathbf{x} \right] \epsilon \cdot \mathbf{p} \right| f \right\rangle. \tag{8.186}$$

Now,

$$\exp\left[\mathrm{i}\left(\frac{\omega}{c}\right) \mathbf{n} \cdot \mathbf{x} \right] = 1 + \mathrm{i}\,\frac{\omega}{c}\, \mathbf{n} \cdot \mathbf{x} + \cdots. \tag{8.187}$$

However, as explained in the previous section, the fact that the wavelength of the radiation that is emitted during spontaneous transition is generally much larger than the typical size of the atom allows us to truncated the previous expansion. Retaining the first two terms, we obtain

$$\epsilon \cdot \mathbf{d}_{if} = \langle i| \, \epsilon \cdot \mathbf{x} \, |f\rangle + \frac{1}{m_e \, c} \, \langle i| \, (\mathbf{n} \cdot \mathbf{x})\,(\epsilon \cdot \mathbf{p}) \, |f\rangle, \tag{8.188}$$

where use has been made of Equation (8.170). Moreover, we have assumed that $\omega = \omega_{if}$ (i.e., the angular frequency of the electromagnetic radiation matches that associated with the atomic transition.) Suppose, however, that the transition from state i to state f is forbidden according to the selection rules for electric dipole transitions. This implies that

$$\langle i| \, \epsilon \cdot \mathbf{x} \, |f\rangle = 0. \tag{8.189}$$

In this case, Equation (8.188) reduces to

$$\epsilon \cdot \mathbf{d}_{if} = \frac{1}{m_e \, c} \, \langle i| \, (\mathbf{n} \cdot \mathbf{x})\,(\epsilon \cdot \mathbf{p}) \, |f\rangle. \tag{8.190}$$

We deduce that a "forbidden" transition is not, strictly speaking, forbidden [i.e., Equation (8.189) does not necessarily mean that $\epsilon \cdot \mathbf{d}_{if} = 0$], but rather takes place at a significantly lower rate than an electric dipole transition [because, according to the previous expression, $|\epsilon \cdot \mathbf{d}_{if}| \sim a_0 \hbar/(m_e \, c \, a_0) \sim \alpha \, a_0 \ll a_0$, whereas $|\epsilon \cdot \mathbf{d}_{if}| \sim a_0$ for an electric dipole transition (see Section 8.11)].

According to classical electromagnetic theory, the polarization direction of the magnetic component of an electromagnetic wave propagating in the direction $\mathbf{n}$ is given by $\mathbf{b} = \mathbf{n} \times \epsilon$, where ϵ specifies the direction of the wave's electric component [Fitzpatrick (2008)]. Of course, $\mathbf{L} = \mathbf{x} \times \mathbf{p}$ represents orbital angular momentum. However,

$$\mathbf{b} \cdot \mathbf{L} = (\mathbf{n} \times \epsilon) \cdot (\mathbf{x} \times \mathbf{p}) = (\mathbf{n} \cdot \mathbf{x})\,(\epsilon \cdot \mathbf{p}) - (\epsilon \cdot \mathbf{x})\,(\mathbf{n} \cdot \mathbf{p}). \tag{8.191}$$

Furthermore, if

$$S = \frac{i\, m_e}{\hbar}\, [H,\, (\boldsymbol{\epsilon} \cdot \mathbf{x})\, (\mathbf{n} \cdot \mathbf{x})], \tag{8.192}$$

then

$$S = \frac{i\, m_e}{\hbar}\, \epsilon_i\, n_j\, [H,\, x_i\, x_j] = \frac{i\, m_e}{\hbar}\, \epsilon_i\, n_j \Big(x_i\, [H,\, x_j] + [H,\, x_i]\, x_j \Big)$$

$$= \epsilon_i\, n_j \Big(x_i\, p_j + p_i\, x_j \Big) = \epsilon_i\, n_j \Big(x_i\, p_j + x_j\, p_i - i\, \hbar\, \delta_{ij} \Big)$$

$$= (\boldsymbol{\epsilon} \cdot \mathbf{x})\, (\mathbf{n} \cdot \mathbf{p}) + (\mathbf{n} \cdot \mathbf{x})\, (\boldsymbol{\epsilon} \cdot \mathbf{p}). \tag{8.193}$$

Here, use has been made of Equations (3.32) and (3.33), as well as the fact that $\boldsymbol{\epsilon} \cdot \mathbf{n} = 0$. It follows, from the previous three equations, that

$$(\mathbf{n} \cdot \mathbf{x})\, (\boldsymbol{\epsilon} \cdot \mathbf{p}) = \frac{1}{2}\, \mathbf{b} \cdot \mathbf{L} + \frac{i\, m_e}{2\, \hbar}\, [H,\, (\boldsymbol{\epsilon} \cdot \mathbf{x})\, (\mathbf{n} \cdot \mathbf{x})]. \tag{8.194}$$

Hence, Equation (8.190) yields

$$\boldsymbol{\epsilon} \cdot \mathbf{d}_{if} = \frac{1}{2\, m_e\, c}\, \mathbf{b} \cdot \langle i|\, \mathbf{L}\, |f\rangle + \frac{i\, \omega_{if}}{2\, c}\, \boldsymbol{\epsilon} \cdot \mathbf{Q}_{if} \cdot \mathbf{n}, \tag{8.195}$$

where

$$(\mathbf{Q}_{if})_{jk} = \langle i|\, x_j\, x_k - r^2\, \delta_{jk}/3\, |f\rangle. \tag{8.196}$$

Here, $r^2 = x_j\, x_j$. Moreover, we have made use of the fact that $\boldsymbol{\epsilon} \cdot \mathbf{n} = 0$ to write $\mathbf{Q}_{if}$ as a traceless tensor. In the following, we shall treat the two terms on the right-hand side of Equation (8.195) separately, because they give rise to completely different selection rules. The first term governs so-called *magnetic dipole transitions*, whereas the second governs so-called *electric quadrupole transitions*.

8.13 Magnetic Dipole Transitions

According to Equation (8.195), the quantity that mediates spontaneous magnetic dipole transitions between different atomic states is

$$\boldsymbol{\epsilon} \cdot \mathbf{d}_{if} = \frac{1}{2\, m_e\, c}\, \mathbf{b} \cdot \langle i|\, \mathbf{L}\, |f\rangle. \tag{8.197}$$

However, it turns out that the previous expression is incomplete because, in writing the Hamiltonian (8.128), we neglected to take into account the interaction of the magnetic component of the electromagnetic wave with the electron's magnetic moment. This interaction gives rise to an additional term [Fitzpatrick (2008)]

$$\delta H = -\boldsymbol{\mu} \cdot \mathbf{B} = \frac{e}{m_e}\, \mathbf{S}, \tag{8.198}$$

where **S** is the electron spin. (See Section 5.5.) Now,

$$\mathbf{B} = \nabla \times \mathbf{A} = -\frac{2 A_0 \omega}{c} \mathbf{b} \sin\left(\frac{\omega}{c} \mathbf{n} \cdot \mathbf{x} - \omega t\right), \tag{8.199}$$

where $\mathbf{b} = \mathbf{n} \times \boldsymbol{\epsilon}$, and use has been made of Equation (8.130). It follows that

$$\delta H = \delta V \exp(\mathrm{i}\,\omega\,t) + \delta V^\dagger \exp(-\mathrm{i}\,\omega\,t), \tag{8.200}$$

where

$$\delta V = \frac{-\mathrm{i}\,e\,\omega\,A_0}{m_e\,c} \exp\left[-\mathrm{i}\,\frac{\omega}{c} \mathbf{n} \cdot \mathbf{x}\right] \mathbf{b} \cdot \mathbf{S} \simeq \frac{-\mathrm{i}\,e\,\omega\,A_0}{m_e\,c} \mathbf{b} \cdot \mathbf{S}. \tag{8.201}$$

Hence, according to Equation (8.137), there is an addition contribution to $\boldsymbol{\epsilon} \cdot \mathbf{d}_{if}$ of the form

$$\frac{1}{m_e\,c} \mathbf{b} \cdot \langle i | \mathbf{S} | f \rangle. \tag{8.202}$$

(Here, we have again assumed that $\omega = \omega_{fi}$.) In other words, the complete quantity that mediates magnetic dipole transitions between different atomic states is

$$\boldsymbol{\epsilon} \cdot \mathbf{d}_{if} = \frac{1}{2\,m_e\,c} \mathbf{b} \cdot \langle i | \mathbf{L} + 2\,\mathbf{S} | f \rangle. \tag{8.203}$$

It follows, by analogy with Equation (8.165), that the spontaneous emission rate associated with a magnetic dipole transition is

$$\frac{dw_{i \to f}^{\mathrm{spin}}}{d\Omega} = \frac{\alpha\,\omega_{if}^3}{8\pi\,m_e^2\,c^4} \sum_{j=1,2} |\mathbf{b}_j \cdot \mathbf{M}_{if}|^2, \tag{8.204}$$

where

$$\mathbf{M}_{if} = \langle i | \mathbf{L} + 2\,\mathbf{S} | f \rangle \tag{8.205}$$

is termed the *magnetic dipole matrix element*, and $\mathbf{b}_{1,2} = \mathbf{n} \times \boldsymbol{\epsilon}_{1,2}$. Here, $d\Omega$ is the solid angle associated with the direction of the emitted photon's normalized wavevector, $\mathbf{n}$. Moreover, $\boldsymbol{\epsilon}_{1,2}$ are the photon's two independent electric polarization vectors. In fact, it is easily demonstrated that $\mathbf{b}_1 = \boldsymbol{\epsilon}_2$, and $\mathbf{b}_2 = -\boldsymbol{\epsilon}_1$. Hence, Equation (8.204) can also be written

$$\frac{dw_{i \to f}^{\mathrm{spn}}}{d\Omega} = \frac{\alpha\,\omega_{if}^3}{8\pi\,m_e^2\,c^4} \sum_{j=1,2} |\boldsymbol{\epsilon}_j \cdot \mathbf{M}_{if}|^2. \tag{8.206}$$

It follows that

$$w_{i \to f}^{\mathrm{spn}} = \oint d\Omega \, \frac{dw_{i \to f}^{\mathrm{spn}}}{d\Omega} = \frac{\alpha\,\omega_{if}^3\,|M_{if}|^2}{3\,m_e^2\,c^4}. \tag{8.207}$$

(See Exercise 8.8.)

According to the analysis of Chapters 4 and 5, $\langle i| L_z + 2 S_z |\rangle f$ is only non-zero for a hydrogen-like atom if $\Delta l = 0$, $\Delta m = 0$, and $\Delta m_s = 0$. Here, l is the azimuthal quantum number, m the magnetic quantum number, and m_s the spin quantum number. Likewise, $\langle i| L_x + 2 S_x |f\rangle$ and $\langle i| L_y + 2 S_y |f\rangle$ are only non-zero if $\Delta l = 0$, $\Delta m = \pm 1$, and $\Delta m_s = \pm 1$. Hence, the selection rules for magnetic dipole transitions are

$$\Delta l = 0, \tag{8.208}$$

$$\Delta m = 0, \pm 1, \tag{8.209}$$

$$\Delta m_s = 0, \pm 1. \tag{8.210}$$

Consider the hydrogen atom. Let us calculate the magnetic dipole matrix element, M_{if}, for the case where the initial state is characterized by the quantum numbers n, l, m, and m_s, and the final state is characterized by the quantum numbers n', l', m', and m_s'. Because $\mathbf{L} + 2\,\mathbf{S}$ has no dependance on the radial variable r, the radial component of this matrix element is simply

$$\int_0^\infty dr\, r^2\, R_{nl}(r)\, R_{n'l}(r), \tag{8.211}$$

where $R_{nl}(r)$ is a standard hydrogen radial wavefunction. Here, we have made use of the fact that the matrix element is zero unless $l = l'$. However, according to Exercise 4.17, the previous integral is zero unless $n = n'$: that is, unless the initial and final states have the same energy. Hence, we conclude that magnetic dipole transitions between different energy levels (i.e., atomic states corresponding to different values of the principle quantum number, n) of a hydrogen atom are forbidden.

A more general form of the selection rules (8.208)–(8.210) is [Corney (1977)]

$$\Delta j = 0, \pm 1, \tag{8.212}$$

$$\Delta m_j = 0, \pm 1, \tag{8.213}$$

where j and m_j are the standard quantum numbers associated with the total angular momentum of the system. Note, however, that a magnetic dipole transition between two $j = 0$ states is forbidden. These new rules suggest that it is possible to have a magnetic dipole transition between hydrogen atom states whose energies are split by spin-orbit effects: for instance, between a $(2p)_{3/2}$ and a $(2p)_{1/2}$ state. (See Section 7.7.) Let us estimate the typical spontaneous emission rate for such a transition. We expect the matrix element M_{if}, defined in Equation (8.205), to be of order $\hbar$. We also expect ω_{if} to be of order $\alpha^2 |E_0|/\hbar$, where E_0 is the hydrogen ground-state energy. (See Exercise 7.14.) It thus follows from Equation (8.207) that

$$w_{i\to f}^{\text{spn}} \sim \alpha^9\, \omega_{if} \sim \alpha^{13}\, \frac{m_e\, c^2}{\hbar}, \tag{8.214}$$

which much smaller than a typical electric dipole transition rate. [See Equation (8.180) and Exercise 8.16.]

The most celebrated example of a magnetic dipole transition in physics is that which mediates the spontaneous decay of the $1s$ triplet state of a hydrogen atom to the corresponding singlet state. (See Section 7.9.) This transition, which takes place at a very slow rate (i.e., $w_{i \to f}^{\text{spn}} \sim 3 \times 10^{-15}\,\text{s}^{-1}$—see Exercise 8.18), produces the well-known hydrogen 21 cm line.

8.14   Electric Quadrupole Transitions

According to Equation (8.195), the quantity that mediates spontaneous electric quadrupole transitions is

$$\boldsymbol{\epsilon} \cdot \mathbf{d}_{if} = \frac{i\,\omega_{if}}{2\,c}\,\boldsymbol{\epsilon} \cdot \mathbf{Q}_{if} \cdot \mathbf{n}, \tag{8.215}$$

where

$$(\mathbf{Q}_{if})_{jk} = \langle i |\, x_j\, x_k - r^2\, \delta_{jk}/3\, | f \rangle \tag{8.216}$$

is the *electric quadrupole matrix element*. It follows, by analogy with with Equation (8.165), that the spontaneous emission rate associated with an electric quarrupole transition is

$$\frac{dw_{i \to f}^{\text{spn}}}{d\Omega} = \frac{\alpha\,\omega_{if}^5}{8\pi\,c^4} \sum_{j=1,2} |\boldsymbol{\epsilon}_j \cdot \mathbf{Q}_{if} \cdot \mathbf{n}|^2. \tag{8.217}$$

Here, $d\Omega$ is the solid angle associated with the direction of the emitted photon's normalized wavevector, $\mathbf{n}$. Moreover, $\boldsymbol{\epsilon}_{1,2}$ are the photon's two independent electric polarization vectors.

The selection rules for electric quadrupole transitions in a hydrogen-like atom are [Corney (1977)]

$$\Delta l = 0, \pm 2, \tag{8.218}$$

$$\Delta m = 0, \pm 1, \pm 2, \tag{8.219}$$

$$\Delta m_s = 0, \tag{8.220}$$

where l is the azimuthal quantum number, m the magnetic quantum number, and m_s the spin quantum number. A more general form of these selection rules is

$$\Delta j = 0, \pm 1, \pm 2, \tag{8.221}$$

$$\Delta m_j = 0, \pm 1, \pm 2, \tag{8.222}$$

where j and m_j are the standard quantum numbers associated with the total angular momentum of the system. Note, however, that electric quadrupole transitions between two $j = 0$ states, or between two $j = 1/2$ states, or between a $j = 1$ state and a $j = 0$ state, are forbidden [Corney (1977)].

Let us estimate the typical spontaneous emission rate for an electric quadrupole transition in a hydrogen atom. We expect the matrix element $\mathbf{Q}_{if}$, defined in Equation (8.216), to be of order a_0^2, where a_0 is the Bohr radius. We also expect ω_{if} to be of order $|E_0|/\hbar$, where E_0 is the hydrogen ground-state energy. It thus follows from Equation (8.217) that

$$\frac{dw_{i \to f}^{\text{spn}}}{d\Omega} \sim \alpha^5 \, \omega_{if} \sim \alpha^7 \frac{m_e \, c^2}{\hbar}, \tag{8.223}$$

which is of order α^2 times smaller than a typical electric dipole transition rate. [See Equations (8.180).]

8.15 Photo-Ionization

As a final example, let us investigate the photo-ionization of atomic hydrogen. In this phenomenon, a photon of angular frequency ω is absorbed by an electron that occupies the ground state of a hydrogen atom. The final energy of the electron,

$$E_f = \hbar \omega + E_i, \tag{8.224}$$

where $E_i = E_0$ is the (negative) hydrogen ground-state energy, is assumed to be positive, which corresponds to an unbound state. In other words, the absorption of the photon causes the hydrogen atom to dissociate.

Let $\mathbf{k} = (\omega/c)\,\mathbf{n}$ and $\boldsymbol{\epsilon}$ be the wavevector and electric polarization vector of the photon, respectively. (Recall that $\mathbf{n}$ and $\boldsymbol{\epsilon}$ are both unit vectors.) It follows from Equations (8.136) and (8.152) that the absorption cross-section of the hydrogen atom can be written

$$d\sigma_{i \to f}^{\text{abs}} = \frac{4\pi^2 \, \alpha}{m_e^2 \, \omega} \left| \left\langle f \, \middle| \, e^{i \, \mathbf{k} \cdot \mathbf{x}} \, \boldsymbol{\epsilon} \cdot \mathbf{p} \, \middle| \, i \right\rangle \right|^2 \delta(\omega - \omega_{fi}), \tag{8.225}$$

where $\mathbf{p}$ is the electron momentum, and $\omega_{fi} = (E_f - E_i)/\hbar$. Here, we have written $d\sigma_{i \to f}^{\text{abs}}$, rather than $\sigma_{i \to f}^{\text{abs}}$, because we are only considering final states in which the electron momentum, $\mathbf{p}_f$, is directed into the range of solid angles $d\Omega$. As discussed in Section 8.6, we must operate on the previous expression with $\int dE_f \, (\cdots)\, \rho(E_f)$, where $\rho(E_f)$ is the density of final states, to obtain the true absorption cross-section, which takes the form

$$d\sigma_{i \to f}^{\text{abs}} = \frac{4\pi^2 \, \alpha \, \hbar}{m_e^2 \, \omega} \left| \left\langle i \, \middle| \, \boldsymbol{\epsilon} \cdot \mathbf{p} \, e^{-i \, \mathbf{k} \cdot \mathbf{x}} \, \middle| \, f \right\rangle \right|^2 \rho(E_f) \Bigg|_{E_f = E_i + \hbar \omega}. \tag{8.226}$$

Here, we have made use of the fact that $\mathbf{p}$ is an Hermitian operator, and have also employed the result $\delta(a\,x) = \delta(x)/a$, where a is a constant. (See Exercise 1.19.)

The electron is initially in the hydrogen ground state, whose wavefunction takes the form

$$\psi_i(\mathbf{x}) = \frac{1}{\sqrt{\pi}\,a_0^{3/2}}\,\exp\left(-\frac{r}{a_0}\right), \tag{8.227}$$

where a_0 is the Bohr radius. (See Chapter 4.) On the other hand, the final electron state is assumed to be a plane-wave state whose wavefunction is

$$\psi_f(\mathbf{x}) = \frac{1}{\sqrt{V}}\,\exp\left(i\,\mathbf{k}_f\cdot\mathbf{x}\right). \tag{8.228}$$

In writing this expression, we have conveniently assumed that the emitted electron is contained in a cubic box of dimension $L \gg a_0$, and volume $V = L^3$. Furthermore, the electron wavefunction is required to be periodic at the boundaries of the box. Of course, we can later take the limit $L \to \infty$ to obtain the general case. (However, it turns out that this is not necessary, because the final result does not depend on L, provided that $L \gg a_0$.) The wavefunction of the final electron state is normalized such that the probability of finding the electron in the box is unity: that is,

$$\int_V d^3\mathbf{x}\,|\psi_f(\mathbf{x})|^2 = 1. \tag{8.229}$$

Note that the final state is an eigenstate of $\mathbf{p}_f$ belonging to the eigenvalue

$$\mathbf{p}_f = \hbar\,\mathbf{k}_f. \tag{8.230}$$

(This follows from the standard Schrödinger representation $\mathbf{p} = -i\,\hbar\,\nabla$.) Furthermore, in writing the wavefunction (8.228), we have neglected any interaction between the emitted electron and the hydrogen nucleus. This neglect is only reasonable provided the final electron energy,

$$E_f = \frac{p_f^2}{2\,m_e} = \frac{\hbar^2\,k_f^2}{2\,m_e} \tag{8.231}$$

is much larger than the ionization energy (i.e., the binding energy) of the hydrogen atom,

$$I = -E_0 = \frac{\hbar^2}{2\,m_e\,a_0^2}. \tag{8.232}$$

The condition $E_f \gg I$ yields

$$k_f\,a_0 \gg 1. \tag{8.233}$$

In other words, the de Broglie wavelength of the emitted electron must be much larger than the typical dimension of the hydrogen atom.

Let us calculate $\rho(E_f)$. Suppose that $\mathbf{k}_f = (k_x, k_y, k_z)$. The periodicity constraint on $\psi_f(\mathbf{x})$ at the boundaries of the box imply that

$$k_x = \frac{2\pi n_x}{L}, \tag{8.234}$$

$$k_y = \frac{2\pi n_y}{L}, \tag{8.235}$$

$$k_z = \frac{2\pi n_z}{L}, \tag{8.236}$$

where n_x, n_y, and n_z are integers. It follows that

$$dn_x \, dn_y \, dn_z = \frac{V}{(2\pi \hbar)^3} \, dp_x \, dp_y \, dp_z, \tag{8.237}$$

where $p_x = \hbar k_x$, et cetera. Thus, the number of final electron states contained in a volume $d^3\mathbf{p}_f = dp_x \, dp_y \, dp_z$ of momentum space is $\rho(\mathbf{p}_f) \, d^3\mathbf{p}_f$, where

$$\rho(\mathbf{p}_f) = \frac{V}{(2\pi \hbar)^3}. \tag{8.238}$$

Note that $\rho(\mathbf{p}_f)$ is constant. Hence, we deduce that the number of final electron states for which $p_f = |\mathbf{p}_f|$ lies between p_f and $p_f + dp_f$, and $\mathbf{p}_f$ is directed into the range of solid angles $d\Omega$, is $\rho(p_f) \, dp_f = \rho(\mathbf{p}_f) \, d^3\mathbf{p}_f$, where $d^3\mathbf{p}_f = p_f^2 \, dp_f \, d\Omega$. In other words,

$$\rho(p_f) = \frac{V p_f^2}{(2\pi \hbar)^2} \, d\Omega. \tag{8.239}$$

Finally, the number of final electron states whose energies lie between E_f and $E_f + dE_f$ is $\rho(E_f) \, dE_f = \rho(p_f) \, (dp_f/dE_f) \, dE_f$, which yields

$$\rho(E_f) = \frac{V m_e k_f}{(2\pi)^3 \hbar^2} \, d\Omega, \tag{8.240}$$

where use has been made of Equations (8.230) and (8.231).

Equations (8.226) and (8.240) imply that the differential cross-section for the photo-ionization of atomic hydrogen takes the form

$$\frac{d\sigma_{i\to f}^{\text{abs}}}{d\Omega} = \frac{\alpha V k_f}{2\pi \hbar m_e \omega} \left| \langle i | \, \boldsymbol{\epsilon} \cdot \mathbf{p} \, e^{-i\,\mathbf{k}\cdot\mathbf{x}} | f \rangle \right|^2. \tag{8.241}$$

Furthermore, making use of Equations (8.227) and (8.228), we can write

$$\langle i | \, \boldsymbol{\epsilon} \cdot \mathbf{p} \, e^{-i\,\mathbf{k}\cdot\mathbf{x}} | f \rangle = \frac{1}{\sqrt{\pi} \sqrt{V} \, a_0^{3/2}} \int_V d^3\mathbf{x} \, e^{-r/a_0} \, \boldsymbol{\epsilon} \cdot \mathbf{p} \, e^{-i\,\mathbf{q}\cdot\mathbf{x}}, \tag{8.242}$$

where

$$\mathbf{q} = \mathbf{k} - \mathbf{k}_f. \tag{8.243}$$

Note, by momentum conservation, that $\hbar \mathbf{q}$ is the recoil momentum of the hydrogen nucleus after ionization. The usual Schrödinger representation $\mathbf{p} = -i\nabla$ reveals that

$$\langle i | \boldsymbol{\epsilon} \cdot \mathbf{p} \, e^{-i\mathbf{k} \cdot \mathbf{x}} | f \rangle = \frac{\hbar \, \boldsymbol{\epsilon} \cdot \mathbf{k}_f}{\sqrt{\pi} \sqrt{V} \, a_0^{3/2}} \int_V d^3 \mathbf{x} \, e^{-r/a_0} \, e^{-i\mathbf{q} \cdot \mathbf{x}}, \tag{8.244}$$

where we have made use of the standard electromagnetic result $\boldsymbol{\epsilon} \cdot \mathbf{k} = 0$ [Fitzpatrick (2008)].

Assuming that $L \gg a_0$, we can write

$$\int_V d^3 \mathbf{x} \, e^{-r/a_0} \, e^{-i\mathbf{q} \cdot \mathbf{x}} = \int_0^\infty dr \, r^2 \, e^{-r/a_0} \int_{-1}^1 d\mu \, e^{-iqr\mu} \oint d\varphi', \tag{8.245}$$

where $\mu = \cos\theta'$. Here, θ', φ' are spherical angles whose polar axis is parallel to $\mathbf{q}$. Thus, we obtain

$$\int_V d^3 \mathbf{x} \, e^{-r/a_0} \, e^{-i\mathbf{q} \cdot \mathbf{x}} = \frac{4\pi \, a_0^2}{q} \int_0^\infty dy \, y \, e^{-y} \, \sin(q \, a_0 \, y). \tag{8.246}$$

However (see Exercise 8.20),

$$\int_0^\infty dy \, y \, e^{-y} \, \sin(b \, y) = \frac{2 \, b}{(1 + b^2)^2}, \tag{8.247}$$

which implies that

$$\int_V d^3 \mathbf{x} \, e^{-r/a_0} \, e^{-i\mathbf{q} \cdot \mathbf{x}} = \frac{8\pi \, a_0^3}{[1 + (q \, a_0)^2]^2}. \tag{8.248}$$

Thus, Equations (8.241), (8.244), and (8.248) yield

$$\frac{d\sigma_{i \to f}^{\text{abs}}}{d\Omega} = \frac{2^5 \, \alpha \, \hbar}{m_e \, k \, c} \frac{k_f \, (\boldsymbol{\epsilon} \cdot \mathbf{k}_f)^2 \, a_0^3}{[1 + (q \, a_0)^2]^4}, \tag{8.249}$$

where use has been made of the standard electromagnetic dispersion relation $\omega = kc$ [Fitzpatrick (2008)].

It is convenient to orientate our coordinate system such that $\boldsymbol{\epsilon} = \mathbf{e}_x$ and $\mathbf{n} = \mathbf{e}_z$. Thus, we can specify the direction of the emitted electron in terms of the conventional polar angles, θ and φ. In fact,

$$\mathbf{k}_f = k_f \, (\sin\theta \cos\varphi, \, \sin\theta \sin\varphi, \, \cos\theta), \tag{8.250}$$

and $\boldsymbol{\epsilon} \cdot \mathbf{k}_f = k_f \sin\theta \cos\varphi$, with $d\Omega = \sin\theta \, d\theta \, d\varphi$. Hence, we obtain

$$\frac{d\sigma_{i \to f}^{\text{abs}}}{d\Omega} = 2^6 \, \alpha \, \hat{I} \, (k_f \, a_0)^3 \frac{\sin^2\theta \cos^2\varphi}{[1 + (q \, a_0)^2]^4} \, a_0^2, \tag{8.251}$$

where

$$\hat{I} = \frac{I}{\hbar c k} = \frac{I}{\hbar \omega} \tag{8.252}$$

is the normalized ionization energy.

Now, energy conservation demands that

$$\hbar c k = \frac{\hbar^2 k_f^2}{2 m_e} + I. \tag{8.253}$$

[See Equation (8.224).] This expression can be rearranged to give

$$(k_f a_0)^2 = \frac{1 - \hat{I}}{\hat{I}}, \tag{8.254}$$

and

$$\frac{k_f}{k} = \frac{2}{\beta}(1 - \hat{I}), \tag{8.255}$$

where

$$\beta = \frac{\hbar k_f}{m_e c} = \frac{v}{c}. \tag{8.256}$$

Here, v is the speed of the emitted electron. However, we have already seen, from Equation (8.233), that the approximations made in deriving Equation (8.251) are only accurate when $k_f a_0 \gg 1$. Hence, according to Equation (8.254), we also require that $\hat{I} \ll 1$. Furthermore, because we are working in the non-relativistic limit, we need $\beta \ll 1$. From Equation (8.255), this necessitates

$$k_f \gg k. \tag{8.257}$$

Finally, the inequality $\hat{I} \ll 1$ can be combined with the previous inequality to give

$$\frac{I}{\hbar c} \ll k \ll k_f \tag{8.258}$$

as the condition for the validity of Equation (8.251).

Now,

$$q^2 = k_f^2 - 2 k \mathbf{k}_f \cdot \mathbf{n} + k^2 = k_f^2 - 2 k k_f \cos\theta + k^2, \tag{8.259}$$

which implies that

$$1 + (q a_0)^2 \simeq 1 + \hat{I}^{-1}(1 - \hat{I} - \beta \cos\theta) = \hat{I}^{-1}(1 - \beta \cos\theta), \tag{8.260}$$

where use has been made of Equation (8.255), and we have neglected terms of order β^2. Equations (8.251), (8.254), and (8.260) can be combined to give the following expression for the differential photo-ionization cross-section of atomic hydrogen:

$$\frac{d\sigma_{i \to f}^{abs}}{d\Omega} \simeq 2^6 \, \alpha \, \hat{I}^{7/2} \left(1 - \hat{I}\right)^{3/2} a_0^2 \, \frac{\sin^2\theta \cos^2\varphi}{(1 - \beta \cos\theta)^4}. \tag{8.261}$$

Integrating over all solid angle, and neglecting terms of order β^2, the total cross-section becomes

$$\sigma_{i \to f}^{\text{abs}} \simeq \frac{2^8 \pi}{3} \alpha \, \hat{I}^{7/2} \left(1 - \hat{I}\right)^{3/2} a_0^2. \tag{8.262}$$

(See Exercise 8.21.)

Note that the previous two expressions are only accurate in the limits $\hat{I} \ll 1$ and $\beta \ll 1$. Nevertheless, we have retained the factors $\left(1 - \hat{I}\right)^{3/2}$ in these formulae because they emphasize that there is a threshold photon energy for photo-ionization: namely, $\hbar \omega = I$. For $\hbar \omega < I$ (i.e., $\hat{I} > 1$), the incident photons are not energetic enough to ionize the hydrogen atom, so there are no emitted photo-electrons. Consequently, the cross-section for photo-ionization tends to zero as $\hat{I}$ approaches unity from below. We have retained the factor involving β in Equation (8.261) because it makes clear that the photoelectrons are emitted preferentially in the directions $\varphi = 0$, π and $\theta = \cos^{-1}(2\beta)$. Thus, in the non-relativistic limit, $\beta \to 0$, the electrons are emitted preferentially along the x-axis (i.e., parallel or anti-parallel to the incident photon's electric polarization vector.) On the other hand, as relativistic effects become important, and β consequently increases, the directions of preferentially emission are beamed forward (i.e., they acquire a component parallel to the wavevector of the incident photon.) An accurate expression for the photo-ionization cross-section close to the ionization threshold (i.e., $\hat{I} \to 1$) can only be obtained using unbound positive energy hydrogen atom wavefunctions as the final electron states [Strobbe (1930)]. Likewise, an accurate expression for the cross-section in the finite-β limit requires a relativistic treatment of the problem [Sauter (1931a,b)].

8.16   Exercises

8.1 Consider the two-state system examined in Section 8.3. Suppose that

$$\langle 1| H_1 |1\rangle = e_{11},$$

$$\langle 2| H_1 |2\rangle = e_{22},$$

$$\langle 1| H_1 |2\rangle = \langle 2| H_1 |1\rangle^* = \frac{1}{2} \gamma \hbar \exp(i \omega t),$$

where e_{11}, e_{22}, γ, and ω are real. Show that

$$i \frac{d\hat{c}_1}{dt} = \frac{\gamma}{2} \exp\left[+i\left(\omega - \hat{\omega}_{21}\right)t\right] \hat{c}_2,$$

$$i \frac{d\hat{c}_2}{dt} = \frac{\gamma}{2} \exp\left[-i\left(\omega - \hat{\omega}_{21}\right)t\right] \hat{c}_1,$$

where $\hat{c}_1 = c_1 \exp(i\,e_{11}\,t/\hbar)$, $\hat{c}_2 = c_2 \exp(i\,e_{22}\,t/\hbar)$, and

$$\hat{\omega}_{21} = \frac{E_2 + e_{22} - E_1 - e_{11}}{\hbar}.$$

Hence, deduce that if the system is definitely in state 1 at time $t = 0$ then the probability of finding it in state 2 at some subsequent time, t, is

$$P_2(t) = \frac{\gamma^2}{\gamma^2 + (\omega - \hat{\omega}_{21})^2} \, \sin^2\left(\left[\gamma^2 + (\omega - \hat{\omega}_{21})^2\right]^{1/2} \frac{t}{2}\right).$$

8.2 Consider an atomic nucleus of spin-s and g-factor g placed in the magnetic field

$$\mathbf{B} = B_0\,\mathbf{e}_z + B_1\left[\cos(\omega t)\,\mathbf{e}_x - \sin(\omega t)\,\mathbf{e}_y\right],$$

where $B_1 \ll B_0$. Let $|s, m\rangle$ be a properly normalized simultaneous eigenket of S^2 and S_z, where $\mathbf{S}$ is the nuclear spin. Thus, $S^2|s, m\rangle = s(s+1)\hbar^2|s, m\rangle$ and $S_z|s, m\rangle = m\hbar|s, m\rangle$, where $-s \le m \le s$. Furthermore, the instantaneous nuclear spin state is written

$$|A\rangle = \sum_{m=-s,s} c_m(r)\,|s, m\rangle,$$

where $\sum_{m=-s,s} |c_m|^2 = 1$.

(a) Demonstrate that

$$\frac{dc_m}{dt} = \frac{i\gamma}{2}\Big([s(s+1) - m(m-1)]^{1/2}\,e^{i(\omega-\omega_0)t}\,c_{m-1}$$

$$+ [s(s+1) - m(m+1)]^{1/2}\,e^{-i(\omega-\omega_0)t}\,c_{m+1}\Big)$$

for $-s \le m \le s$, where $\omega_0 = g\mu_N B_0/\hbar$ and $\gamma = g\mu_N B_1/\hbar$.

(b) Consider the case $s = 1/2$. Demonstrate that if $\omega = \omega_0$ and $c_{1/2}(0) = 1$ then

$$c_{1/2}(t) = \cos(\gamma t/2), \qquad c_{-1/2}(t) = i\,\sin(\gamma t/2).$$

(c) Consider the case $s = 1$. Demonstrate that if $\omega = \omega_0$ and $c_1(0) = 1$ then

$$c_1(t) = \cos^2(\gamma t/2),$$

$$c_0(t) = i\,\sqrt{2}\,\cos(\gamma t/2)\,\sin(\gamma t/2),$$

$$c_{-1}(t) = -\sin^2(\gamma t/2).$$

(d) Consider the case $s = 3/2$. Demonstrate that if $\omega = \omega_0$ and $c_{3/2}(0) = 1$ then

$$c_{3/2}(t) = \cos^3(\gamma t/2),$$

$$c_{1/2}(t) = i\,\sqrt{3}\,\cos(\gamma t/2)\,\sin^2(\gamma t/2),$$

$$c_{-1/2}(t) = -\sqrt{3}\,\cos^2(\gamma t/2)\,\sin(\gamma t/2),$$

$$c_{-3/2}(t) = -i\,\sin^3(\gamma t/2).$$

8.3 Derive Equation (8.45).

8.4 Derive Equation (8.83).

8.5 If

$$P_{i \to i}(t) = \exp\left(-\frac{t}{\tau_i}\right)$$

is the probability that a system initially in state $|i\rangle$ at time $t = 0$ is still in that state at some subsequent time, deduce that the mean lifetime of the state is τ_i.

8.6 Demonstrate that $\mathbf{p} \cdot \mathbf{A} = \mathbf{A} \cdot \mathbf{p}$ when $\nabla \cdot \mathbf{A} = 0$, where $\mathbf{p}$ is the momentum operator, and $\mathbf{A}(\mathbf{x})$ is a real function of the position operator, $\mathbf{x}$. Hence, show that the Hamiltonian (8.128) is Hermitian.

8.7 Demonstrate that

$$\left\langle |\boldsymbol{\epsilon} \cdot \mathbf{d}_{if}|^2 \right\rangle = \frac{\mathbf{d}_{if}^* \cdot \mathbf{d}_{if}}{3},$$

where the average is taken over all possible directions of the unit vector $\boldsymbol{\epsilon}$.

8.8 Demonstrate that

$$\oint d\Omega \, \frac{dw_{i \to f}^{\text{spn}}}{d\Omega} = \frac{4 \, \alpha \, \omega_{if}^3 \, |d_{if}|^2}{3 \, c^2},$$

where $dw_{i \to f}^{\text{spn}}/d\Omega$ is specified by Equation (8.165). [Hint: Write

$$\mathbf{n} = (\sin\theta \, \cos\varphi, \, \sin\theta \, \sin\varphi, \, \cos\theta),$$

$$\boldsymbol{\epsilon}_1 = (\cos\theta \, \cos\varphi, \, \cos\theta \, \sin\varphi, \, -\sin\theta),$$

$$\boldsymbol{\epsilon}_2 = (-\sin\varphi, \, \cos\varphi, \, 0),$$

where $d\Omega = \sin\theta \, d\theta \, d\varphi$.]

8.9 Demonstrate that a spontaneous transition between two atomic states of zero orbital angular momentum is absolutely forbidden. (Actually, a spontaneous transition between two zero orbital angular momentum states is possible via the simultaneous emission of two photons, but takes place at a very slow rate [Göppert-Mayer (1931); Breit and Teller (1940); Shapiro and Breit (1959)].)

8.10 Find the selection rules for the matrix elements $\langle n, l, m | x | n', l', m' \rangle$ and $\langle n, l, m | y | n', l', m' \rangle$ to be non-zero. Here, $|n, l, m\rangle$ denotes an energy eigenket of a hydrogen-like atom corresponding to the conventional quantum numbers, n, l, and m.

8.11 The Hamiltonian of an electron in a hydrogen-like atom is written

$$H = \frac{p^2}{2 \, m_e} + V(r),$$

where $r = |\mathbf{x}|$. Let $|n\rangle$ denote a properly normalized energy eigenket belonging to the eigenvalue E_n. By calculating $[[H, x_i], x_i]$, derive the Thomas-Reiche-Kuhn sum rule,

$$\sum_m F_{nm} = 1,$$

where

$$F_{nm} = \frac{2 m_e \omega_{mn}}{\hbar} |\langle n| x_i |m\rangle|^2,$$

and $\omega_{mn} = (E_m - E_n)/\hbar$, and the sum is over all energy eigenstates. Here, x_i stands for either x, y, or z. (See Exercise 3.6.)

8.12 (a) Show that the only non-zero $1s \to 2p$ electric dipole matrix elements for the hydrogen atom take the values

$$\langle 1, 0, 0| x |2, 1, \pm 1\rangle = \mp \frac{2^7}{3^5} a_0,$$

$$\langle 1, 0, 0| y |2, 1, \pm 1\rangle = -i \frac{2^7}{3^5} a_0,$$

$$\langle 1, 0, 0| z |2, 1, 0\rangle = \sqrt{2} \frac{2^7}{3^5} a_0,$$

where a_0 is the Bohr radius.

(b) Likewise, show that the only non-zero $1s \to 3p$ electric dipole matrix elements take the values

$$\langle 1, 0, 0| x |3, 1, \pm 1\rangle = \mp \frac{3^3}{2^7} a_0,$$

$$\langle 1, 0, 0| y |3, 1, \pm 1\rangle = -i \frac{3^3}{2^7} a_0,$$

$$\langle 1, 0, 0| z |3, 1, 0\rangle = \sqrt{2} \frac{3^3}{2^7} a_0.$$

(c) Finally, show that the only non-zero $2s \to 2p$ electric dipole matrix elements take the values

$$\langle 2, 0, 0| x |2, 1, \pm 1\rangle = \pm \frac{3}{\sqrt{2}} a_0,$$

$$\langle 2, 0, 0| y |2, 1, \pm 1\rangle = i \frac{3}{2} a_0,$$

$$\langle 2, 0, 0| z |2, 1, 0\rangle = -3 a_0.$$

8.13 (a) Demonstrate that the spontaneous decay rate (via an electric dipole transition) from any $2p$ state to a $1s$ state of a hydrogen atom is

$$w_{2p \to 1s} = \left(\frac{2}{3}\right)^8 \alpha^5 \frac{m_e c^2}{\hbar} = 6.27 \times 10^8 \, \text{s}^{-1},$$

where α is the fine structure constant. Hence, deduce that the relative natural width of the associated spectral line is

$$\frac{\Delta \lambda}{\lambda} = 4.0 \times 10^{-8},$$

where λ denotes wavelength.

(b) Likewise, show that the spontaneous decay rate from any $3p$ state to a $1s$ state is

$$w_{3p \to 1s} = \frac{\alpha^5}{3 \, 2^5} \frac{m_e c^2}{\hbar} = 1.67 \times 10^8 \, \text{s}^{-1}.$$

Hence, deduce that the relative natural width of the associated spectral line is

$$\frac{\Delta \lambda}{\lambda} = 9.1 \times 10^{-9}.$$

8.14 (a) Demonstrate that the net oscillator strength for $1s \to 2p$ transitions in a hydrogen atom is

$$F_{1s \to 2p} = \frac{2^{13}}{3^9} = 0.4162,$$

irrespective of the polarization of the radiation.

(b) Likewise, show that the net oscillator strength for $1s \to 3p$ transitions is

$$F_{1s \to 3p} = \frac{3^4}{2^{11}} = 0.07910,$$

irrespective of the polarization of the radiation.

8.15 Taking electron spin into account, the wavefunctions of the $(1s)_{1/2}$ states of a hydrogen atom are written $R_{10} \, \mathcal{Y}_{0m}^{1/2}$ where $m = \pm 1/2$. Here, the R_{nl} are standard radial wavefunctions, and the $\mathcal{Y}_{l m_j}^{j}$ are spin-angular functions. Likewise, the wavefunctions of the $(2p)_{1/2}$ states are written $R_{21} \, \mathcal{Y}_{1 m_j}^{1/2}$, where $m_j = \pm 1/2$. Finally, the wavefunctions of the $(2p)_{3/2}$ states are written $R_{21} \, \mathcal{Y}_{1 m_j}^{3/2}$, where $m_j = -3/2, -1/2, 1/2, 3/2$. As a consequence of spin-orbit interaction, the four $(2p)_{3/2}$ states lie at a slightly higher energy than the two $(2p)_{1/2}$ states. Demonstrate that the spontaneous decay rates, due to electric dipole transitions, between the various $2p$ and $1s$ states are

$$w\left(\mathcal{Y}_{1\ +3/2}^{3/2} \to \mathcal{Y}_{0\ +1/2}^{1/2}\right) = w_0,$$

$$w\left(\mathcal{Y}_{1\ +3/2}^{3/2} \to \mathcal{Y}_{0\ -1/2}^{1/2}\right) = 0,$$

$$w\left(\mathcal{Y}_{1\ +1/2}^{3/2} \to \mathcal{Y}_{0\ +1/2}^{1/2}\right) = \frac{2}{3}\,w_0,$$

$$w\left(\mathcal{Y}_{1\ +1/2}^{3/2} \to \mathcal{Y}_{0\ -1/2}^{1/2}\right) = \frac{1}{3}\,w_0,$$

$$w\left(\mathcal{Y}_{1\ -1/2}^{3/2} \to \mathcal{Y}_{0\ +1/2}^{1/2}\right) = \frac{1}{3}\,w_0,$$

$$w\left(\mathcal{Y}_{1\ -1/2}^{3/2} \to \mathcal{Y}_{0\ -1/2}^{1/2}\right) = \frac{2}{3}\,w_0,$$

$$w\left(\mathcal{Y}_{1\ -3/2}^{3/2} \to \mathcal{Y}_{0\ +1/2}^{1/2}\right) = 0,$$

$$w\left(\mathcal{Y}_{1\ -3/2}^{3/2} \to \mathcal{Y}_{0\ -1/2}^{1/2}\right) = w_0,$$

$$w\left(\mathcal{Y}_{1\ +1/2}^{1/2} \to \mathcal{Y}_{0\ +1/2}^{1/2}\right) = \frac{1}{3}\,w_0,$$

$$w\left(\mathcal{Y}_{1\ +1/2}^{1/2} \to \mathcal{Y}_{0\ -1/2}^{1/2}\right) = \frac{2}{3}\,w_0,$$

$$w\left(\mathcal{Y}_{1\ -1/2}^{1/2} \to \mathcal{Y}_{0\ +1/2}^{1/2}\right) = \frac{2}{3}\,w_0,$$

$$w\left(\mathcal{Y}_{1\ -1/2}^{1/2} \to \mathcal{Y}_{0\ -1/2}^{1/2}\right) = \frac{1}{3}\,w_0,$$

where

$$w_0 = \left(\frac{2}{3}\right)^8 \alpha^5 \frac{m_e c^2}{\hbar} = 6.27 \times 10^8 \text{ s}^{-1}.$$

Here, the states have been labeled by their spin-angular functions. Hence, deduce that for an ensemble of hydrogen atoms in thermal equilibrium, the spectral line associated with $(2p)_{3/2} \to (1s)_{1/2}$ transitions is twice as bright as that associated with $(2p)_{1/2} \to (1s)_{1/2}$ transitions.

8.16 Taking electron spin into account, the wavefunctions of the $(2s)_{1/2}$ states of a hydrogen atom are written $R_{20}\,\mathcal{Y}_{0m}^{1/2}$ where $m = \pm 1/2$. The wavefunctions of the $(2p)_{3/2}$ states are specified in the previous exercise. Demonstrate that the spontaneous decay rates, due to electric dipole transitions, between the various $(2p)_{3/2}$ and $(2s)_{1/2}$ states are

$$w\left(\mathcal{Y}^{3/2}_{1\ +3/2} \to \mathcal{Y}^{1/2}_{0\ +1/2}\right) = w_0,$$

$$w\left(\mathcal{Y}^{3/2}_{1\ +3/2} \to \mathcal{Y}^{1/2}_{0\ -1/2}\right) = 0,$$

$$w\left(\mathcal{Y}^{3/2}_{1\ +1/2} \to \mathcal{Y}^{1/2}_{0\ +1/2}\right) = \frac{2}{3}\,w_0,$$

$$w\left(\mathcal{Y}^{3/2}_{1\ +1/2} \to \mathcal{Y}^{1/2}_{0\ -1/2}\right) = \frac{1}{3}\,w_0,$$

$$w\left(\mathcal{Y}^{3/2}_{1\ -1/2} \to \mathcal{Y}^{1/2}_{0\ +1/2}\right) = \frac{1}{3}\,w_0,$$

$$w\left(\mathcal{Y}^{3/2}_{1\ -1/2} \to \mathcal{Y}^{1/2}_{0\ -1/2}\right) = \frac{2}{3}\,w_0,$$

$$w\left(\mathcal{Y}^{3/2}_{1\ -3/2} \to \mathcal{Y}^{1/2}_{0\ +1/2}\right) = 0,$$

$$w\left(\mathcal{Y}^{3/2}_{1\ -3/2} \to \mathcal{Y}^{1/2}_{0\ -1/2}\right) = w_0,$$

where

$$w_0 = \frac{3\,\alpha^{11}}{2^{13}}\frac{m_e\,c^2}{\hbar} = 3.55 \times 10^{-6}\,\text{s}^{-1}.$$

Here, the states have been labeled by their spin-angular functions.

8.17 (a) Consider the $2,1,\pm1 \to 1,0,0$ electric dipole transition of a hydrogen atom. Show that the angular distribution of spontaneously emitted photons is

$$\frac{dw^{\text{spn}}_{2,1,\pm1\to1,0,0}}{d\Omega} = \frac{2^4\,\alpha^5}{3^7\,\pi}\frac{m_e\,c^2}{\hbar}\,(1 + \cos^2\theta),$$

where $d\Omega = \sin\theta\,d\theta\,d\varphi$, and the photon's direction of motion is parallel to $\mathbf{n} = (\sin\theta\cos\varphi,\ \sin\theta\sin\varphi,\ \cos\theta)$.

(b) Consider the $2,1,0 \to 1,0,0$ electric dipole transition of a hydrogen atom. Show that the angular distribution of spontaneously emitted photons is

$$\frac{dw^{\text{spn}}_{2,1,0\to1,0,0}}{d\Omega} = \frac{2^5\,\alpha^5}{3^7\,\pi}\frac{m_e\,c^2}{\hbar}\,\sin^2\theta.$$

(c) Hence, show that for an ensemble of hydrogen atoms in thermal equilibrium, the angular distribution of spontaneously emitted photons associated with the $2p \to 1s$ electric dipole transition is

$$\frac{dw^{\text{spn}}_{2p\to1s}}{d\Omega} = \frac{2^6\,\alpha^5}{3^8\,\pi}\frac{m_e\,c^2}{\hbar}.$$

8.18 The properly normalized $1s$ triplet and singlet state kets of a hydrogen atom can be written

$$|1, 1\rangle = |+\rangle_p |+\rangle_e,$$

$$|1, 0\rangle = \frac{1}{\sqrt{2}} \left(|+\rangle_p |-\rangle_e + |-\rangle_p |+\rangle_e \right),$$

$$|1, -1\rangle = |-\rangle_p |-\rangle_e,$$

and

$$|0, 0\rangle = \frac{1}{\sqrt{2}} \left(|+\rangle_p |-\rangle_e - |-\rangle_p |+\rangle_e \right),$$

respectively. Here, the kets on the left are $|j, m_j\rangle$ kets, where j and m_j are the conventional quantum numbers that determine the overall angular momentum, and the projection of the overall angular momentum along the z-axis, respectively. Finally, $|\pm\rangle_p$ and $|\pm\rangle_e$ are the properly normalized spin-up and spin-down states for the proton and the electron, respectively. As explained in Section 7.9, the triplet states have slightly higher energies than the singlet state, as a consequence of spin-spin coupling between the proton and the electron. Spontaneous magnetic dipole transitions between the triplet and singlet states occur via the interaction of the magnetic component of the emitted photon and the electron magnetic moment. (The magnetic moment of the proton is much smaller than that of the electron, and consequently does not play a significant role in this transition.) In the following, the initial state, i, corresponds to one of the triplet states, and the final state, f, corresponds to the singlet state.

(a) Demonstrate that

$$\omega_{if} = \frac{E_i - E_f}{\hbar} = \frac{4}{3} g_p \alpha^4 \frac{m_e}{m_p} \frac{m_e c^2}{\hbar},$$

where E_i and E_f are the energies of the initial and final states, and g_p is the proton g-factor.

(b) Let $\mathbf{M}_{if} = \langle i| 2\, \mathbf{S}_e |f\rangle$ be the magnetic dipole matrix element, where $\mathbf{S}_e$ is the electron spin. Show that

$$\mathbf{M}_{if} = \frac{\hbar}{\sqrt{2}} \left(\mathbf{e}_x - i\, \mathbf{e}_y \right)$$

if the initial state is the $|1, 1\rangle$ state,

$$\mathbf{M}_{if} = -\frac{\hbar}{\sqrt{2}} \left(\mathbf{e}_x + i\, \mathbf{e}_y \right)$$

if the initial state is the $|1, -1\rangle$ state, and

$$\mathbf{M}_{if} = -\hbar\, \mathbf{e}_z$$

if the initial state is the $|1, 0\rangle$ state.

(c) Deduce that the angular distribution of the spontaneously emitted photon is

$$\frac{dw_{i \to f}^{\mathrm{spn}}}{d\Omega} = \frac{2^2}{3^3\,\pi}\,g_p^3\,\alpha^{13}\left(\frac{m_e}{m_p}\right)^3 \frac{m_e\,c^2}{\hbar}\,(1+\cos^2\theta)$$

if the initial state is either the $|1,1\rangle$ state or the $|1,-1\rangle$ state. Show that the angular distribution is

$$\frac{dw_{i \to f}^{\mathrm{spn}}}{d\Omega} = \frac{2^3}{3^3\,\pi}\,g_p^3\,\alpha^{13}\left(\frac{m_e}{m_p}\right)^3 \frac{m_e\,c^2}{\hbar}\,\sin^2\theta$$

if the initial state is the $|1,0\rangle$ state. Here, the photon's direction of motion is parallel to $\mathbf{n} = (\sin\theta\cos\varphi,\ \sin\theta\sin\varphi,\ \cos\theta)$.

(d) Finally, show that the overall spontaneous transition rate is

$$w_{i \to f}^{\mathrm{spn}} = \frac{2^6}{3^4}\,g_p^3\,\alpha^{13}\left(\frac{m_e}{m_p}\right)^3 \frac{m_e\,c^2}{\hbar} = 2.88 \times 10^{-15}\,\mathrm{s}^{-1}.$$

8.19 Consider the $3,2,0 \to 1,0,0$ electric quadrupole transition of a hydrogen atom.

(a) Show that the angular distribution of spontaneously emitted photons can be written

$$\frac{dw_{3d \to 1s}^{\mathrm{spn}}}{d\Omega} = \frac{\alpha^7}{8\pi}\left(\frac{2}{3}\right)^{10}\left(\sum_{j=1,2}\frac{|\boldsymbol{\epsilon}_j \cdot \mathbf{Q}_{3d \to 1s} \cdot \mathbf{n}|^2}{a_0^4}\right)\frac{m_e\,c^2}{\hbar}.$$

(b) Demonstrate that the only non-zero $3d(m=0) \to 1s$ electric quadrupole matrix elements takes the values $Q_{xx} = -Q_{zz}/2$, $Q_{yy} = -Q_{zz}/2$, and

$$Q_{zz} = \sqrt{\frac{2}{3}\frac{3^4}{2^7}}\,a_0^2.$$

(c) Hence, deduce that

$$\frac{dw_{3d \to 1s}^{\mathrm{spn}}}{d\Omega} = \frac{\alpha^7}{2^8\,3\pi}\frac{m_e\,c^2}{\hbar}\cos^2\theta\sin^2\theta,$$

and

$$w_{3d \to 2s}^{\mathrm{spn}} = \frac{\alpha^7}{2^5\,3^2\,5}\frac{m_e\,c^2}{\hbar} = 594.1\,\mathrm{s}^{-1}.$$

Here, the spontaneously emitted photon's direction of motion is parallel to $\mathbf{n} = (\sin\theta\cos\varphi,\ \sin\theta\sin\varphi,\ \cos\theta)$.

8.20 Let

$$J_n = \int_0^\infty dy\, y^n\, e^{-\beta y}\, \sin(b\,y),$$

where n is a non-negative integer. Demonstrate that

$$J_n = (-1)^n \left(\frac{\partial}{\partial \beta} \right)^n J_0,$$

and

$$J_0 = \frac{b}{\beta^2 + b^2}.$$

Hence, deduce that

$$J_1 = \frac{2\,b\,\beta}{(\beta^2 + b^2)^2},$$

and

$$J_2 = \frac{8\,b\,\beta^2}{(\beta^2 + b^2)^3} - \frac{2\,b}{(\beta^2 + b^2)^2}.$$

8.21 Treating β as a small parameter, and neglecting terms of order β^2, show that

$$\oint d\Omega\, \frac{\sin^2 \theta \, \cos^2 \varphi}{(1 - \beta \cos \theta)^4} = \frac{4\pi}{3}.$$

8.22 Repeat the calculation of Section 8.15 for the case where the electric polarization vector of the incident photon is $\epsilon = \mathbf{e}_y$. Hence, show that, in this case, the differential photo-ionization cross-section is

$$\frac{d\sigma_{i \to f}^{\text{abs}}}{d\Omega} \simeq 2^6\, \alpha\, \hat{I}^{7/2} \left(1 - \hat{I} \right)^{3/2} a_0^2\, \frac{\sin^2 \theta\, \sin^2 \varphi}{(1 - \beta \cos \theta)^4}.$$

Deduce that the photo-ionization cross-section for unpolarized electromagnetic radiation propagating in the z-direction is

$$\frac{d\sigma_{i \to f}^{\text{abs}}}{d\Omega} \simeq 2^5\, \alpha\, \hat{I}^{7/2} \left(1 - \hat{I} \right)^{3/2} a_0^2\, \frac{\sin^2 \theta}{(1 - \beta \cos \theta)^4}.$$

8.23 Consider the photo-ionization of a hydrogen atom in the $2s$ state. Demonstrate that formula (8.249) is replaced by

$$\frac{d\sigma_{i \to f}^{\text{abs}}}{d\Omega} = \frac{2^{10}\, \alpha\, \hbar}{m_e\, k\, c}\, k_f\, (\epsilon \cdot \mathbf{k}_f)^2\, a_0^3\, \frac{\left[1 - (2\, q\, a_0)^2 \right]^2}{\left[1 + (2\, q\, a_0)^2 \right]^6}.$$

8.24 Consider the inverse process to the photo-ionization of a hydrogen atom investigated in Section 8.15. In this process, which is known as *radiative association*, an unbound electron of energy $E_i > 0$ is captured by a (stationary)
proton to form a hydrogen atom in its ground state, with the emission of a
photon of energy

$$\hbar \omega = E_i - E_f,$$

where $E_f = E_0 = -I$. Here, E_0 is (negative) hydrogen ground-state energy,
and I is the (positive) ground-state ionization energy. Radiative association
can be thought of as a form of stimulated emission in which the initial state
is unbound. Thus, according to the analysis of Section 8.9,

$$d\sigma_{i \to f}^{\text{stm}} = \frac{4\pi^2 \alpha}{m_e^2 \omega} \sum_{j=1,2} \left| \langle f | \epsilon_j \cdot \mathbf{p}\, e^{-i\,\mathbf{k}\cdot\mathbf{x}} | i \rangle \right|^2 \rho(\omega) \Bigg|_{E_f = E_i - \hbar \omega},$$

where $\mathbf{p}$ is the electron momentum, $\mathbf{k}$ is the wavevector of the emitted photon,
and $\rho(\omega)\, d\omega$ is the number of photon states whose angular frequencies lie
between ω and $\omega + d\omega$, and whose direction of motion lie in the range of
solid angles $d\Omega$. Here, $\epsilon_{1,2}$ are two independent unit vectors normal to $\mathbf{k}$.

(a) Show that

$$\rho(\omega) = \frac{V k^2}{(2\pi)^2 c}\, d\Omega,$$

where the initial electron and final photon are both assumed to be contained in a periodic box of volume $V \gg a_0^3$.

(b) Hence, demonstrate that

$$\frac{d\sigma_{i \to f}^{\text{stm}}}{d\Omega} = \frac{2^5 \alpha \hbar^2}{m_e^2 c^2} \frac{k k_i^2 a_0^2 \sin^2 \theta}{[1 + (q a_0)^2]^4},$$

where $\mathbf{q} = \mathbf{k} - \mathbf{k}_i$, and $\cos\theta = \mathbf{k} \cdot \mathbf{k}_i / (k k_i)$.

(c) Finally, show that

$$\frac{d\sigma_{i \to f}^{\text{stm}}}{d\Omega} = 2^4 \alpha^3 \beta \frac{E_i^{1/2} I^{5/2}}{(E_i + I)^3} a_0^2 \frac{\sin^2 \theta}{(1 - \beta \cos\theta)^4},$$

where $\beta = (2 E_i / m_e c^2)^{1/2}$.

Chapter 9

Identical Particles

9.1 Introduction

Consider a system consisting of a collection of identical particles. In classical mechanics, it is, in principle, possible to continuously monitor the position of each particle as a function of time. Hence, the constituent particles can be unambiguously labeled. In quantum mechanics, on the other hand, this is not possible because continuous position measurements would disturb the system. It follows that identical particles cannot be unambiguously labeled in quantum mechanics. This chapter investigates the various ramifications of this inability.

9.2 Permutation Symmetry

Consider a quantum system consisting of two identical particles. Suppose that one of the particles—particle 1, say—is characterized by the state ket $|k'\rangle$. Here, k' represents the eigenvalues of the complete set of commuting observables associated with the particle. Suppose that the other particle—particle 2—is characterized by the state ket $|k''\rangle$. The state ket for the whole system can be written in the product form

$$|k'\rangle |k''\rangle, \tag{9.1}$$

where it is understood that the first ket corresponds to particle 1, and the second to particle 2. We can also consider the ket

$$|k''\rangle |k'\rangle, \tag{9.2}$$

which corresponds to a state in which particle 1 has the eigenvalues k'', and particle 2 the eigenvalues k'.

Suppose that we were to measure all of the simultaneously measurable properties of our two-particle system. We might obtain the results k' for one particle,

213

and k'' for the other. However, we have no way of knowing whether the corresponding state ket is $|k'\rangle|k''\rangle$ or $|k''\rangle|k'\rangle$, or any linear combination of these two kets. In other words, all state kets of the form

$$c_1\,|k'\rangle\,|k''\rangle + c_2\,|k''\rangle\,|k'\rangle, \tag{9.3}$$

where c_1 and c_2 are arbitrary complex numbers, correspond to an identical set of results when the properties of the system are measured. This phenomenon is known as *exchange degeneracy*. Such degeneracy is problematic because the specification of a complete set of observable eigenvalues in a system of identical particles does not seem to uniquely determine the corresponding state ket. Fortunately, nature has a way of avoiding this difficulty.

Consider the *two-particle permutation operator*, P_{12}, which is defined such that

$$P_{12}\,|k'\rangle\,|k''\rangle = |k''\rangle\,|k'\rangle. \tag{9.4}$$

In other words, P_{12} swaps the identities of particles 1 and 2. It is easily appreciated that

$$P_{21} = P_{12}, \tag{9.5}$$

$$P_{12}^2 = 1. \tag{9.6}$$

Now, the Hamiltonian of a system of two identical particles must necessarily be a symmetric function of each particle's observables (because exchange of identical particles could not possibly affect the overall energy of the system). An example of such a Hamiltonian is

$$H = \frac{\mathbf{p}_1^2}{2\,m} + \frac{\mathbf{p}_2^2}{2\,m} + V_{\text{pair}}(|\mathbf{x}_1 - \mathbf{x}_2|) + V_{\text{ext}}(\mathbf{x}_1) + V_{\text{ext}}(\mathbf{x}_2). \tag{9.7}$$

Here, we have separated the mutual interaction of the two particles from their interaction with an external potential. [To be more exact, $V_{\text{pair}}(|\mathbf{x}_1 - \mathbf{x}_2|)$ is the interaction potential, and $V_{\text{ext}}(\mathbf{x})$ the external potential.] It follows that if

$$H\,|k'\rangle\,|k''\rangle = E\,|k'\rangle\,|k''\rangle \tag{9.8}$$

then

$$H\,|k''\rangle\,|k'\rangle = E\,|k''\rangle\,|k'\rangle, \tag{9.9}$$

where E is the total energy. Operating on both sides of Equation (9.8) with P_{12}, and employing Equation (9.6), we obtain

$$P_{12}\,H\,P_{12}^2\,|k'\rangle\,|k''\rangle = E\,P_{12}\,|k'\rangle\,|k''\rangle, \tag{9.10}$$

or

$$P_{12}\,H\,P_{12}\,|k''\rangle\,|k'\rangle = E\,|k''\rangle\,|k'\rangle = H\,|k''\rangle\,|k'\rangle, \tag{9.11}$$

where use has been made of Equation (9.9). Because the $|k''\rangle |k'\rangle$ must form a complete set (otherwise, the properties of the system would not be fully observable), we deduce that

$$P_{12} H P_{12} = H, \tag{9.12}$$

which implies [from Equation (9.6)] that

$$[H, P_{12}] = 0. \tag{9.13}$$

In other words, an eigenstate of the Hamiltonian is a simultaneous eigenstate of the two-particle permutation operator, P_{12}. (See Section 1.13.)

Now, according to Equation (9.6), the two-particle permutation operator possesses the eigenvalues $+1$ and -1, respectively. (See Exercise 9.1.) The corresponding properly normalized (provided that $k' \neq k''$) eigenkets are

$$|k' k''\rangle_+ = \frac{1}{\sqrt{2}} \left(|k'\rangle |k''\rangle + |k''\rangle |k'\rangle \right), \tag{9.14}$$

and

$$|k' k''\rangle_- = \frac{1}{\sqrt{2}} \left(|k'\rangle |k''\rangle - |k''\rangle |k'\rangle \right). \tag{9.15}$$

(See Exercise 9.2.) Here, it is assumed that $\langle k'|k''\rangle = \delta_{k' k''}$. Note that $|k' k''\rangle_+$ is symmetric with respect to interchange of particles—that is,

$$|k'' k'\rangle_+ = +|k' k''\rangle_+, \tag{9.16}$$

whereas $|k' k''\rangle_-$ is antisymmetric—that is,

$$|k'' k'\rangle_- = -|k' k''\rangle_-. \tag{9.17}$$

Let us, now, consider a system of three identical particles. We can represent the overall state ket as

$$|k'\rangle |k''\rangle |k'''\rangle, \tag{9.18}$$

where k', k'', and k''' are the eigenvalues of particles 1, 2, and 3, respectively. We can also define two-particle permutation operators:

$$P_{12} |k'\rangle |k''\rangle |k'''\rangle = |k''\rangle |k'\rangle |k'''\rangle, \tag{9.19}$$

$$P_{23} |k'\rangle |k''\rangle |k'''\rangle = |k'\rangle |k'''\rangle |k''\rangle, \tag{9.20}$$

$$P_{31} |k'\rangle |k''\rangle |k'''\rangle = |k'''\rangle |k''\rangle |k'\rangle. \tag{9.21}$$

It is easily demonstrated that

$$P_{21} = P_{12}, \tag{9.22}$$

$$P_{32} = P_{23}, \tag{9.23}$$

$$P_{13} = P_{31}, \tag{9.24}$$

and

$$P_{12}^2 = P_{23}^2 = P_{31}^2 = 1. \tag{9.25}$$

As before, the Hamiltonian of the system must be a symmetric function of the particle's observables: that is,

$$H |k'\rangle |k''\rangle |k'''\rangle = E |k'\rangle |k''\rangle |k'''\rangle, \tag{9.26}$$

$$H |k''\rangle |k'''\rangle |k'\rangle = E |k''\rangle |k'''\rangle |k'\rangle, \tag{9.27}$$

$$H |k'''\rangle |k'\rangle |k''\rangle = E |k'''\rangle |k'\rangle |k''\rangle, \tag{9.28}$$

$$H |k''\rangle |k'\rangle |k'''\rangle = E |k''\rangle |k'\rangle |k'''\rangle, \tag{9.29}$$

$$H |k'\rangle |k'''\rangle |k''\rangle = E |k'\rangle |k'''\rangle |k''\rangle, \tag{9.30}$$

$$H |k'''\rangle |k''\rangle |k'\rangle = E |k'''\rangle |k''\rangle |k'\rangle, \tag{9.31}$$

where E is the total energy. Using analogous arguments to those employed for the two-particle system, we deduce that

$$[H, P_{12}] = [H, P_{23}] = [H, P_{31}] = 0. \tag{9.32}$$

Hence, an eigenstate of the Hamiltonian is a simultaneous eigenstate of the three two-particle permutation operators, P_{12}, P_{23}, and P_{31}. (See Section 1.13.) However, according to Equation (9.25), the possible eigenvalues of these operators are ± 1. (See Exercise 9.1.)

Let us define the *cyclic permutation operator*, P_{123}, where

$$P_{123} |k'\rangle |k''\rangle |k'''\rangle = |k'''\rangle |k'\rangle |k''\rangle. \tag{9.33}$$

It follows that

$$P_{123} = P_{12} P_{31} = P_{23} P_{12} = P_{31} P_{23}. \tag{9.34}$$

It is also clear from Equations (9.26) and (9.28) that

$$[H, P_{123}] = 0. \tag{9.35}$$

(See Exercise 9.4.) Thus, an eigenstate of the Hamiltonian is a simultaneous eigenstate of the four permutation operators P_{12}, P_{23}, P_{31}, and P_{123}. (See Section 1.13.) Let λ_{12}, λ_{23}, λ_{31} and λ_{123} represent the eigenvalues of these operators, respectively. We know that $\lambda_{12} = \pm 1$, $\lambda_{23} = \pm 1$, and $\lambda_{31} = \pm 1$. Moreover, it follows from Equation (9.34) that

$$\lambda_{123} = \lambda_{12} \lambda_{31} = \lambda_{23} \lambda_{12} = \lambda_{31} \lambda_{23}. \tag{9.36}$$

The previous equations imply that

$$\lambda_{123} = +1, \tag{9.37}$$

and either

$$\lambda_{12} = \lambda_{23} = \lambda_{31} = +1, \tag{9.38}$$

or

$$\lambda_{12} = \lambda_{23} = \lambda_{31} = -1. \tag{9.39}$$

In other words, the multi-particle state ket must be either totally symmetric, or totally antisymmetric, with respect to swapping the identities of any given pair of identical particles. Thus, in terms of properly normalized single-particle kets, the properly normalized (provided that $k' \neq k'' \neq k'''$) totally symmetric and totally antisymmetric kets are

$$|k'\,k''\,k'''\rangle_+ = \frac{1}{\sqrt{3!}}\,(|k'\rangle\,|k''\rangle\,|k'''\rangle + |k'''\rangle\,|k'\rangle\,|k''\rangle + |k''\rangle\,|k'''\rangle\,|k'\rangle$$

$$+|k'''\rangle\,|k''\rangle\,|k'\rangle + |k'\rangle\,|k'''\rangle\,|k''\rangle + |k''\rangle\,|k'\rangle\,|k'''\rangle)\,, \tag{9.40}$$

and

$$|k'\,k''\,k'''\rangle_- = \frac{1}{\sqrt{3!}}\,(|k'\rangle\,|k''\rangle\,|k'''\rangle + |k'''\rangle\,|k'\rangle\,|k''\rangle + |k''\rangle\,|k'''\rangle\,|k'\rangle$$

$$-|k'''\rangle\,|k''\rangle\,|k'\rangle - |k'\rangle\,|k'''\rangle\,|k''\rangle - |k''\rangle\,|k'\rangle\,|k'''\rangle)\,, \tag{9.41}$$

respectively.

The previous arguments can be generalized to systems of more than three identical particles in a straightforward manner [Slater (1929)].

9.3   Spin Statistics Theorem

We have seen that the exchange degeneracy of a system of identical particles is such that a specification of a complete set of observable eigenvalues does not uniquely determine the corresponding state ket. However, we have also demonstrated that there are only two possible state kets: that is, a ket that is totally symmetric with respect to particle interchange, or a ket that is totally antisymmetric. It turns out that systems of identical particles possessing integer-spin (e.g., spin 0, or spin 1) always choose the totally symmetric ket, whereas systems of identical particles possessing half-integer-spin (e.g., spin 1/2) always choose the totally antisymmetric ket. This additional piece of information—which is generally known as the *spin statistics theorem*—ensures that the specification of a complete set of observable eigenvalues of a system of identical particles does, in fact, uniquely determine the corresponding state ket.

Systems of identical particles whose state kets are totally symmetric with respect to particle interchange are said to obey *Bose-Einstein statistics* [Bose (1924); Einstein (1924)]. Moreover, such particles are termed *bosons*. On the other hand, systems of identical particles whose state kets are totally antisymmetric with respect to particle interchange are said to obey *Fermi-Dirac statistics* [Fermi (1926); Dirac (1926)], and the constituent particles are called *fermions*. In non-relativistic quantum mechanics, the rule that all integer-spin particles are bosons, whereas all half-integer spin particles are fermions, must be accepted as an empirical fact. However, in relativistic quantum mechanics, it is possible to formulate reasonably convincing arguments that half-integer-spin particles cannot be bosons, and integer-spin particles cannot be fermions [Fierz (1939); Pauli (1940)]. Incidentally, electrons, protons, and neutrons are all fermions.

The *Pauli exclusion principle* [Pauli (1925)] is an immediate consequence of the fact that electrons obey Fermi-Dirac statistics. This principle states that no two electrons in a multi-electron system can possess identical sets of observable eigenvalues. For instance, in the case of a three-electron system, the state ket is

$$|k'\,k''\,k'''\rangle_- = \frac{1}{\sqrt{3!}}\,(|k'\rangle\,|k''\rangle\,|k'''\rangle + |k'''\rangle\,|k'\rangle\,|k''\rangle + |k''\rangle\,|k'''\rangle\,|k'\rangle$$

$$-|k'''\rangle\,|k''\rangle\,|k'\rangle - |k'\rangle\,|k'''\rangle\,|k''\rangle - |k''\rangle\,|k'\rangle\,|k'''\rangle)\,. \tag{9.42}$$

[See Equation (9.41).] Note, however, that

$$|k'\,k'\,k'''\rangle_- = |k'\,k''\,k''\rangle_- = |k'''\,k''\,k'''\rangle_- = |0\rangle\,. \tag{9.43}$$

In other words, if two of the electrons in the system possess the same set of observable eigenvalues then the state ket becomes the null ket, which corresponds to the absence of a state.

9.4 Two-Electron System

Consider a system consisting of two electrons. Let x_1 and S_1 represent the position and spin operators of the first electron, respectively, and let x_2 and S_2 represent the corresponding operators of the second electron. Furthermore, let $S = S_1 + S_2$ represent the total spin operator of the system. Suppose that the Hamiltonian commutes with S^2, as is often the case. It follows that the state of the system is specified by the position eigenvalues x_1' and x_2', as well as the total spin quantum numbers s and m. As usual, the eigenvalue of S^2 is $s(s+1)\,\hbar^2$, and the eigenvalue of S_z is $m\,\hbar$. (See Chapter 5.) Moreover, s, m can only take the values $1, 1$ or $1, 0$ or $1, -1$ or $0, 0$, respectively. (See Chapter 6.) The overall wavefunction of the system can be written

$$\psi(\mathbf{x}'_1, \mathbf{x}'_2; s, m) = \phi(\mathbf{x}'_1, \mathbf{x}'_2) \chi(s, m), \tag{9.44}$$

where

$$\chi(1, 1) = \chi_+ \chi_+, \tag{9.45}$$

$$\chi(1, 0) = \frac{1}{\sqrt{2}} (\chi_+ \chi_- + \chi_- \chi_+), \tag{9.46}$$

$$\chi(1, -1) = \chi_- \chi_-, \tag{9.47}$$

$$\chi(0, 0) = \frac{1}{\sqrt{2}} (\chi_+ \chi_- - \chi_- \chi_+). \tag{9.48}$$

(See Section 6.4.) Here, the spinor $\chi_+ \chi_-$ denotes a state in which $m_1 = 1/2$ and $m_2 = -1/2$, et cetera, where $m_1 \hbar$ and $m_2 \hbar$ are the eigenvalues of S_{1z} and S_{2z}, respectively. The three $s = 1$ spinors are usually referred to as triplet spinors, whereas the single $s = 0$ spinor is called the singlet spinor. Note that the triplet spinors are all symmetric with respect to exchange of particles, whereas the singlet spinor is antisymmetric.

Fermi-Dirac statistics requires that the overall wavefunction be antisymmetric with respect to exchange of particles. Now, according to Equation (9.44), the overall wavefunction can be written as a product of a spatial wavefunction and a spinor. Moreover, when the system is in the spin-triplet state (i.e., $s = 1$), the spinor is symmetric with respect to exchange of particles. On the other hand, when the system is in the spin-singlet state (i.e., $s = 0$), the spinor is antisymmetric. It follows that, to maintain the overall antisymmetry of the wavefunction, the triplet spatial wavefunction must be antisymmetric with respect to exchange of particles, whereas the singlet spatial wavefunction must be symmetric. In other words, in the spin-triplet state the spatial wavefunction takes the form

$$\phi(\mathbf{x}_1, \mathbf{x}_2) = \frac{1}{\sqrt{2}} \left[\omega_A(\mathbf{x}_1) \, \omega_B(\mathbf{x}_2) - \omega_B(\mathbf{x}_1) \, \omega_A(\mathbf{x}_2) \right], \tag{9.49}$$

whereas in the spin-singlet state the spatial wavefunction is written

$$\phi(\mathbf{x}_1, \mathbf{x}_2) = \frac{1}{\sqrt{2}} \left[\omega_A(\mathbf{x}_1) \, \omega_B(\mathbf{x}_2) + \omega_B(\mathbf{x}_1) \, \omega_A(\mathbf{x}_2) \right]. \tag{9.50}$$

The probability of observing one electron in the volume element $d^3\mathbf{x}_1$ around position $\mathbf{x}_1$, and the other in the volume element $d^3\mathbf{x}_2$ around position $\mathbf{x}_2$, is proportional to $|\phi(\mathbf{x}_1, \mathbf{x}_2)|^2 \, d^3\mathbf{x}_1 \, d^3\mathbf{x}_2$, or

$$\frac{1}{2} \left(|\omega_A(\mathbf{x}_1)|^2 \, |\omega_B(\mathbf{x}_2)|^2 + |\omega_A(\mathbf{x}_2)|^2 \, |\omega_B(\mathbf{x}_1)|^2 \right.$$

$$\left. \pm 2 \, \mathrm{Re} \left[\omega_A(\mathbf{x}_1) \, \omega_B(\mathbf{x}_2) \, \omega_A^*(\mathbf{x}_2) \, \omega_B^*(\mathbf{x}_1) \right] \right) d^3\mathbf{x}_1 \, d^3\mathbf{x}_2. \tag{9.51}$$

Here, the plus sign corresponds to the spin-singlet state, whereas the minus sign corresponds to the spin-triplet state. We can immediately see that, in the spin-triplet state, the probability of finding the two electrons at the same point in space is zero. In other words, the two electrons have a tendency to avoid one another in this state. On the other hand, in the spin-singlet state, there is an enhanced probability of finding the two electrons at the same point in space (because of the final term in the previous expression). In other words, the two electrons are attracted to one another in this state. Note, however, that the spatial probability distributions associated with the singlet and triplet states differ substantially only when the two single-particle spatial wavefunctions, $\omega_A(\mathbf{x})$ and $\omega_B(\mathbf{x})$, overlap: that is, when there exists a region of space in which the two wavefunctions are simultaneously non-negligible.

9.5   Helium Atom

Consider the helium atom, which is a good example of a two-electron system. The Hamiltonian is written

$$H = \frac{\mathbf{p}_1^2}{2\,m_e} + \frac{\mathbf{p}_2^2}{2\,m_e} - \frac{Z\,e^2}{4\pi\,\epsilon_0\,r_1} - \frac{Z\,e^2}{4\pi\,\epsilon_0\,r_2} + \frac{e^2}{4\pi\,\epsilon_0\,r_{12}}, \tag{9.52}$$

where $Z\,e$ is the nuclear charge, $Z = 2$, $r_1 = |\mathbf{x}_1|$, $r_2 = |\mathbf{x}_2|$, and $r_{12} = |\mathbf{x}_1 - \mathbf{x}_2|$. Here, $\mathbf{p}_1$ and $\mathbf{x}_1$ are the momentum and position of the first electron, et cetera. Suppose that the final term on the right-hand side of the previous expression were absent. In this case, in which the two electrons do not interact with one another, the overall spatial wavefunction can be formed from products of hydrogen atom wavefunctions calculated with $Z = 2$, instead of $Z = 1$. Each of these wavefunctions is characterized by the usual triplet of quantum numbers, n, l, and m. (See Chapter 4.) Now, the total spin of the system is a constant of the motion (because $\mathbf{S}$ obviously commutes with the Hamiltonian), so the overall spin state is either the singlet or the triplet state. (See the previous section.) The corresponding spatial wavefunction is symmetric in the former case, and antisymmetric in the latter. Suppose that one electron has the quantum numbers n, l, m, whereas the other has the quantum numbers n', l', m'. The corresponding spatial wavefunction is

$$\phi(\mathbf{x}_1, \mathbf{x}_2) = \frac{1}{\sqrt{2}} \left[\psi_{nlm}(\mathbf{x}_1)\,\psi_{n'l'm'}(\mathbf{x}_2) \pm \psi_{nlm}(\mathbf{x}_2)\,\psi_{n'l'm'}(\mathbf{x}_1) \right], \tag{9.53}$$

where the plus and minus signs correspond to the singlet and triplet spin states, respectively. Here, $\psi_{nlm}(\mathbf{x})$ is a standard hydrogen atom wavefunction (calculated with $Z = 2$). For the special case in which the two sets of spatial quantum numbers, n, l, m and n', l', m', are the same, the triplet spin state does not exist (because the associated spatial wavefunction is null). Hence, only singlet spin state is allowed, and the spatial wavefunction reduces to

$$\phi(\mathbf{x}_1, \mathbf{x}_2) = \psi_{nlm}(\mathbf{x}_1)\,\psi_{nlm}(\mathbf{x}_2). \tag{9.54}$$

In particular, the ground state ($n = n' = 1, l = l' = 0, m = m' = 0$) can only exist as a singlet spin state (i.e., a state of overall spin 0), and has the spatial wavefunction

$$\phi(\mathbf{x}_1, \mathbf{x}_2) = \psi_{100}(\mathbf{x}_1)\,\psi_{100}(\mathbf{x}_2) = \frac{Z^3}{\pi\,a_0^3}\,\exp\left[\frac{-Z(r_1 + r_2)}{a_0}\right], \tag{9.55}$$

where a_0 is the Bohr radius. This follows because [see Equation (4.95) and Exercise 9.8]

$$\psi_{100}(\mathbf{x}) = \frac{1}{\sqrt{\pi}}\left(\frac{Z}{a_0}\right)^{3/2}\exp\left(\frac{-Z\,r}{a_0}\right). \tag{9.56}$$

The energy of this state is

$$E = 2Z^2\,E_0 = 8\,E_0 = -108.8\,\text{eV}, \tag{9.57}$$

because $Z = 2$. (See Exercise 9.8.) Here, $E_0 = -13.6\,\text{eV}$ is the ground-state energy of a hydrogen atom. In the previous expression, the factor of 2 (before the factor Z^2) is present because there are two electrons in a helium atom.

The previous estimate for the ground-state energy of a helium atom completely ignores the final term on the right-hand side of Equation (9.52), which describes the mutual electrostatic repulsion of the two electrons. We can obtain a better estimate for the ground-state energy by treating Equation (9.55) as the unperturbed wavefunction, and $e^2/(4\pi\,\epsilon_0\,r_{12})$ as a perturbation. According to standard first-order perturbation theory (see Chapter 7), the correction to the ground-state energy is

$$\Delta E = \left\langle\frac{e^2}{4\pi\,\epsilon_0\,r_{12}}\right\rangle = 2\left\langle\frac{a_0}{r_{12}}\right\rangle|E_0|, \tag{9.58}$$

because $E_0 = -e^2/(8\pi\,\epsilon_0\,a_0)$. Here, the expectation value is calculated using the wavefunction (9.55). The resulting expression for ΔE can be written

$$\frac{\Delta E}{|E_0|} = \frac{2Z^6}{\pi^2}\int\int\frac{d^3\mathbf{x}_1}{a_0^3}\frac{d^3\mathbf{x}_2}{a_0^3}\,\exp\left[-\frac{2Z(r_1 + r_2)}{a_0}\right]\frac{a_0}{r_{12}}. \tag{9.59}$$

Now,

$$\frac{1}{r_{12}} = \frac{1}{(r_1^2 + r_2^2 - 2r_1 r_2\cos\gamma)^{1/2}} = \sum_{l=0,\infty}\frac{r_<^l}{r_>^{l+1}}\,P_l(\cos\gamma), \tag{9.60}$$

where $r_>$ ($r_<$) is the larger (smaller) of r_1 and r_2, γ the angle subtended between $\mathbf{x}_1$ and $\mathbf{x}_2$, and $P_l(x)$ a *Legendre polynomial* [Fitzpatrick (2012)]. Moreover, the so-called *addition theorem* for spherical harmonics states that [Jackson (1975)]

$$P_l(\cos\gamma) = \frac{4\pi}{2l+1} \sum_{m=-l,+l} Y_{lm}^*(\theta_1,\varphi_1)\, Y_{lm}(\theta_2,\varphi_2). \tag{9.61}$$

Here, $\mathbf{x}_1 = r_1\,(\sin\theta_1\,\cos\varphi_1,\ \sin\theta_1\,\sin\varphi_1,\ \cos\theta_1)$, et cetera. However [see Equations (4.94) and (4.95)],

$$\oint d\Omega\, Y_{lm}(\theta,\phi) = \sqrt{4\pi}\,\delta_{l0}\,\delta_{m0}, \tag{9.62}$$

so, combining the previous four equations, we obtain

$$\frac{\Delta E}{|E_0|} = 32\,Z \int_0^\infty dx_1\, x_1 \left(\int_0^{x_1} dx_2\, x_2^2\, e^{-2\,(x_1+x_2)} + \int_{x_1}^\infty dx_2\, x_1\, x_2\, e^{-2\,(x_1+x_2)} \right)$$

$$= \frac{5Z}{4}. \tag{9.63}$$

(See Exercise 9.10.) Here, $x_1 = Z\,r_1/a_0$ and $x_2 = Z\,r_2/a_0$. Thus, our improved estimate for the ground-state energy of the helium atom is [Unsöld (1927)]

$$E = 2Z^2\,E_0 + \Delta E = \left(2Z^2 - \frac{5Z}{4}\right)E_0 = \left(8 - \frac{5}{2}\right)E_0 = \frac{11}{2}E_0$$

$$= -74.8\,\text{eV}, \tag{9.64}$$

because $Z = 2$ and $E_0 = -13.6\,\text{eV}$. This is much closer to the experimental value of $-78.98\,\text{eV}$ [Moore (1949)] than our previous estimate, $-108.8\,\text{eV}$. [See Equation (9.57).]

9.6 Orthohelium and Parahelium

Consider an excited state of the helium atom in which one electron is in the ground state, while the other is in a state characterized by the quantum numbers n, l, m, where $n > 1$. Making use of the Hamiltonian (9.52), and treating the final term as a perturbation, we can write the energy of this state as

$$E = Z^2\,E_{100} + Z^2\,E_{nlm} + \Delta E, \tag{9.65}$$

where $E_{nlm} = E_0/n^2$ is the energy of a hydrogen atom electron whose quantum numbers are n, l, m, E_0 the hydrogen ground-state energy, and ΔE the expectation value of $e^2/(4\pi\,\epsilon_0\,r_{12})$. It follows from Equation (9.53) (with $n, l, m = 1, 0, 0$ and $n', l', m' = n, l, m$) that

$$\Delta E = I \pm J, \tag{9.66}$$

where

$$\frac{I}{|E_0|} = 2 \int d^3\mathbf{x}_1 \int d^3\mathbf{x}_2 \, |\psi_{100}(\mathbf{x}_1)|^2 \, |\psi_{nlm}(\mathbf{x}_2)|^2 \, \frac{a_0}{r_{12}}, \qquad (9.67)$$

$$\frac{J}{|E_0|} = 2 \int d^3\mathbf{x}_1 \int d^3\mathbf{x}_2 \, \psi_{100}(\mathbf{x}_1) \, \psi_{nlm}(\mathbf{x}_2) \, \frac{a_0}{r_{12}} \, \psi_{100}^*(\mathbf{x}_2) \, \psi_{nlm}^*(\mathbf{x}_1), \qquad (9.68)$$

and a_0 is the Bohr radius. Here, the plus sign in Equation (9.66) corresponds to the spin-singlet state, whereas the minus sign corresponds to the spin-triplet state. The integral I—which is known as the *direct integral*—is obviously positive. The integral J—which is known as the *exchange integral*—can be shown also to be positive. (See Exercises 9.13 and 9.14.) Hence, we conclude that, in excited states of helium, the spin-singlet state has a higher energy than the corresponding spin-triplet state. Incidentally, helium in the spin-singlet state is known as *parahelium*, whereas helium in the triplet state is called *orthohelium*. As we have seen, only parahelium is possible in the ground state,

The fact that parahelium energy levels lie slightly above corresponding orthohelium levels is interesting because our original Hamiltonian, (9.52), does not explicitly depend on spin. Nevertheless, there is a spin dependent effect—that is, a helium atom has a lower energy when its electrons possess parallel spins—as a consequence of Fermi-Dirac statistics. To be more exact, the energy is lower in the spin-triplet state because the corresponding spatial wavefunction is antisymmetric, causing the electrons to tend to avoid one another (thereby reducing their electrostatic repulsion).

9.7 Variational Principle

Suppose that we wish to find an approximate solution to the time-independent Schrödinger equation,

$$H \, |\psi\rangle = E \, |\psi\rangle, \qquad (9.69)$$

where H is a known (presumably complicated) time-independent Hamiltonian, and E the energy eigenvalue. Let $|\psi\rangle$ be a properly normalized trial solution to the above equation that corresponds to the trial wavefunction $\psi(\mathbf{x}') \equiv \langle \mathbf{x}'|\psi\rangle$. The so-called *variational principle* [Ritz (1909)] states, quite simply, that the ground-state energy, E_0, is always less than, or equal to, the expectation value of H calculated with the trial solution: that is,

$$E_0 \le \langle \psi| \, H \, |\psi\rangle. \qquad (9.70)$$

Thus, by varying $|\psi\rangle$ until the expectation value of H is minimized, we can obtain approximations to both the energy and the wavefunction of the ground state. (Incidentally, if $|\psi\rangle$ is not properly normalized then we must minimize $\langle\psi|H|\psi\rangle/\langle\psi|\psi\rangle$. See Exercise 9.16.)

Let us prove the variational principle. Suppose that the E_n and the $|n\rangle$ are the true eigenvalues and eigenkets of H: that is,

$$H|n\rangle = E_n|n\rangle. \tag{9.71}$$

Furthermore, let

$$E_0 < E_1 < E_2 < \cdots, \tag{9.72}$$

so that $|0\rangle$ is the ground state, $|1\rangle$ the first excited state, et cetera. The $|n\rangle$ are assumed to be orthonormal: that is,

$$\langle n|m\rangle = \delta_{nm}. \tag{9.73}$$

If our trial ket $|\psi\rangle$ is properly normalized then we can write

$$|\psi\rangle = \sum_n c_n|n\rangle, \tag{9.74}$$

where

$$\sum_n |c_n|^2 = 1, \tag{9.75}$$

and the c_n are complex numbers. Here, n is summed from 0 to ∞. Now, the expectation value of H, calculated with $|\psi\rangle$, takes the form

$$\langle\psi|H|\psi\rangle = \sum_{n,m} c_n^* c_m \langle n|H|m\rangle = \sum_n c_n^* c_m E_m \langle n|m\rangle = \sum_n E_n |c_n|^2, \tag{9.76}$$

where use has been made of Equations (9.71), (9.73), and (9.74). So, we can write

$$\langle\psi|H|\psi\rangle = |c_0|^2 E_0 + \sum_{n>0} |c_n|^2 E_n. \tag{9.77}$$

However, Equation (9.75) can be rearranged to give

$$|c_0|^2 = 1 - \sum_{n>0} |c_n|^2. \tag{9.78}$$

Combining the previous two equations, we obtain

$$\langle\psi|H|\psi\rangle = E_0 + \sum_{n>0} |c_n|^2 (E_n - E_0). \tag{9.79}$$

But, the second term on the right-hand side of the previous expression is positive definite, because $E_n - E_0 > 0$ for all $n > 0$. [See Equation (9.72).] Hence, we obtain the desired result:

$$\langle\psi|H|\psi\rangle \geq E_0. \tag{9.80}$$

If $|\psi\rangle$ is a properly normalized trial ket that is orthogonal to the true ground-state of the system (i.e., $\langle 0|\psi\rangle = 0$) then, by repeating the previous analysis, we can easily demonstrate that

$$\langle\psi| H |\psi\rangle \geq E_1. \tag{9.81}$$

Thus, by varying $|\psi\rangle$ until the expectation value of H is minimized, we can obtain approximations to both the energy and wavefunction of the first excited state. In reality, because we do not generally know the exact ground state, we have to use a trial ket that is orthogonal to the approximate ground state obtained via the variational method described in the preceding paragraph. Obviously, we can continue this process until we have approximations to all of the stationary eigenstates. Note, however, that the errors are clearly cumulative in this approach, so that any approximations to highly excited states are likely to inaccurate. For this reason, the variational method is generally used only to calculate the ground state, and the first few excited states, of complicated quantum systems.

We can employ the variational principle to obtain an improved estimate for the ground-state energy of a helium atom. The Hamiltonian is written [see Equation (9.52)]

$$H = \frac{\mathbf{p}_1^2}{2\,m_e} + \frac{\mathbf{p}_2^2}{2\,m_e} - \frac{2\,e^2}{4\pi\,\epsilon_0\,r_1} - \frac{2\,e^2}{4\pi\,\epsilon_0\,r_2} + \frac{e^2}{4\pi\,\epsilon_0\,r_{12}}, \tag{9.82}$$

where we have now explicitly taken into account the fact that the nuclear charge is $2\,e$. Let us take [see Equation (9.54)]

$$\phi(\mathbf{x}_1, \mathbf{x}_2) = \psi(\mathbf{x}_1)\,\psi(\mathbf{x}_2) \tag{9.83}$$

as our properly normalized trial wavefunction, where [see Equation (9.56)]

$$\psi(\mathbf{x}) = \frac{1}{\sqrt{\pi}}\left(\frac{Z}{a_0}\right)^{3/2} \exp\left(\frac{-Z\,r}{a_0}\right), \tag{9.84}$$

and a_0 is the Bohr radius. In the following, we shall treat Z as a variable parameter. In fact, $Z\,e$ is the nuclear charge experienced by each electron. We would expect this quantity to be somewhat less that the true nuclear charge, $2\,e$, because the electrons partially shield one another from the nucleus.

It is convenient to rewrite the Hamiltonian in the form

$$H = H_0 + H_1, \tag{9.85}$$

where

$$H_0 = \frac{\mathbf{p}_1^2}{2\,m_e} + \frac{\mathbf{p}_2^2}{2\,m_e} - \frac{Z\,e^2}{4\pi\,\epsilon_0\,r_1} - \frac{Z\,e^2}{4\pi\,\epsilon_0\,r_2}, \tag{9.86}$$

and

$$H_1 = \frac{(Z-2)\,e^2}{4\pi\,\epsilon_0\,r_1} + \frac{(Z-2)\,e^2}{4\pi\,\epsilon_0\,r_2} + \frac{e^2}{4\pi\,\epsilon_0\,r_{12}}. \tag{9.87}$$

The trial wavefunction, (9.83), is an exact eigenstate of H_0 belonging to the eigenvalue $2Z^2 E_0$, where E_0 is the ground-state energy of hydrogen. [See Equation (9.57).] If we treat H_1 as a perturbation (see Chapter 7) then we obtain the estimate

$$E = 2Z^2 E_0 + \langle H_1 \rangle, \tag{9.88}$$

for the ground-state energy of helium. Here, the expectation value is calculated using the trial wavefunction.

Given that $E_0 = -e^2/(8\pi\,\epsilon_0\,a_0)$, we get

$$\langle H_1 \rangle = \left[2\,(Z-2)\left\langle \frac{a_0}{r_1} \right\rangle + 2\,(Z-2)\left\langle \frac{a_0}{r_2} \right\rangle + 2\left\langle \frac{a_0}{r_{12}} \right\rangle \right] |E_0|. \tag{9.89}$$

However, we already proved in Section 9.5 that

$$2\left\langle \frac{a_0}{r_{12}} \right\rangle = \frac{5Z}{4}. \tag{9.90}$$

[See Equations (9.58) and (9.63).] Furthermore,

$$\left\langle \frac{a_0}{r_1} \right\rangle = \left\langle \frac{a_0}{r_2} \right\rangle = \frac{1}{\pi}\left(\frac{Z}{a_0}\right)^3 \int d^3\mathbf{x}\,\exp\left(-\frac{2Zr}{a_0}\right)\frac{a_0}{r}$$

$$= 4\left(\frac{Z}{a_0}\right)^3 \int_0^\infty dr\,r^2 \exp\left(-\frac{2Zr}{a_0}\right)\frac{a_0}{r} = Z \int_0^\infty dx\,x\,\mathrm{e}^{-x} = Z, \tag{9.91}$$

where $x = 2Zr/a_0$, and use has been made of Exercise 4.11. Thus,

$$\frac{\langle H_1 \rangle}{|E_0|} = 4Z\,(Z-2) + \frac{5Z}{4} = 4Z^2 - \frac{27Z}{4}. \tag{9.92}$$

According to the previous analysis, when expressed as a function of Z, the ground-state energy of a helium atom takes the form

$$E(Z) = \left(2Z^2 - 4Z^2 + \frac{27Z}{4}\right) E_0 = \left(\frac{27Z}{4} - 2Z^2\right) E_0. \tag{9.93}$$

We must now minimize this expression with respect to Z to obtain our new estimate for the actual ground-state energy. It is easily seen that

$$\frac{dE}{dZ} = \left(\frac{27}{4} - 4Z\right) E_0, \tag{9.94}$$

and

$$\frac{d^2E}{dZ^2} = -4E_0 > 0. \tag{9.95}$$

Thus, E clearly attains a minimum value when

$$Z = \frac{27}{16} = 1.6875. \tag{9.96}$$

The fact that $Z < 2$ confirms our earlier conjecture that the electrons partially shield the nuclear charge from one another. Our new estimate for the ground-state energy of helium is [Kellner (1927)]

$$E(Z = 27/4) = \frac{27^2}{2^7} E_0 = 5.695 E_0 = -77.5 \, \text{eV}. \tag{9.97}$$

This is clearly an improvement on our previous estimate, -74.8 eV [see Equation (9.64)], recalling that the correct result is -78.98 eV [Moore (1949)].

9.8 Hydrogen Molecule Ion

The hydrogen molecule ion consists of an electron orbiting about two protons, and is the simplest imaginable molecule. Let us investigate whether this molecule possesses a bound state: that is, whether it possesses a ground-state whose energy is less than that of a ground-state hydrogen atom plus a free proton. In fact, according to the variational principle (see the previous section), we can prove that the H_2^+ ion possesses a bound state if we can find any trial wavefunction for which the total Hamiltonian of the system has an expectation value that is less than that of a ground-state hydrogen atom plus a free proton.

Suppose that the first and second protons lie at $+\mathbf{X}/2$ and $-\mathbf{X}/2$, respectively. Let $\mathbf{x}$ be the position vector of the electron. The position vectors of the electron relative to the first and second protons are thus $\mathbf{x}_1 = \mathbf{x} - \mathbf{X}/2$ and $\mathbf{x}_2 = \mathbf{x} + \mathbf{X}/2$, respectively. The Hamiltonian of the system is written

$$H = \frac{\mathbf{p}^2}{2\,m_e} - \frac{e^2}{4\pi\,\epsilon_0} \left(\frac{1}{r_1} + \frac{1}{r_2} \right) + \frac{e^2}{4\pi\,\epsilon_0\,R}, \tag{9.98}$$

where $\mathbf{p}$ is the electron momentum, $r_1 = |\mathbf{x}_1|$, $r_2 = |\mathbf{x}_2|$, and $R = |\mathbf{X}|$. Here, we are treating the protons as essentially stationary, which is a reasonable approximation because the electron's motion is much more rapid than that of the protons. Of course, this is the case because the electron mass is very much less than the proton mass. Incidentally, the neglect of nuclear motion when calculating the electronic structure of a molecule is known as the *Born-Oppenheimer approximation* [Born and Oppenheimer (1927)].

The Hamiltonian (9.98) is manifestly invariant under the transformation $\mathbf{X} \rightarrow -\mathbf{X}$, which simply swaps the positions of the two identical protons. This transformation is equivalent to $\mathbf{x}_1 \leftrightarrow \mathbf{x}_2$. If P_{12} is the operator that swaps the proton positions then it is clear that $P_{12}^2 = 1$. (See Exercise 9.1.) Hence, the eigenvalues of P_{12} are ± 1. Furthermore, the fact that the Hamiltonian is invariant under exchange of proton positions implies that $[P_{12}, H] = 0$. (See Section 9.2.) Thus, the eigenkets of the Hamiltonian are simultaneous eigenkets of P_{12}. Now, it is easily shown that the eigenkets of P_{12} corresponding to the eigenvalues ± 1 are, respectively, even and odd under the transformation $\mathbf{x}_1 \leftrightarrow \mathbf{x}_2$. (See Exercise 9.2.) It follows that the ground-state wavefunction of the H_2^+ ion has the general form

$$\psi_\pm(\mathbf{x}) = A\left[\psi_0(\mathbf{x}_1) \pm \psi_0(\mathbf{x}_2)\right], \tag{9.99}$$

where A is a complex number.

Let us adopt

$$\psi_0(\mathbf{x}) = \frac{1}{\sqrt{\pi}\, a_0^{3/2}}\, e^{-r/a_0} \tag{9.100}$$

as our trial single-proton wavefunction, where $r = |\mathbf{x}|$. Here, $\psi_0(\mathbf{x})$ is a properly normalized hydrogen ground-state wavefunction, and a_0 is the Bohr radius. Thus, our trial molecular wavefunction, which is specified in Equations (9.99) and (9.100), is simply a linear combination of hydrogen ground-state wavefunctions centered on each proton [Burrau (1927)].

Our first task is to normalize the trial wavefunction. We require that

$$\int d^3\mathbf{x}\, |\psi_\pm|^2 = 1. \tag{9.101}$$

Hence, from Equation (9.99), $A = I^{-1/2}$, where

$$I = \int d^3\mathbf{x}\left[|\psi_0(\mathbf{x}_1)|^2 + |\psi_0(\mathbf{x}_2)|^2 \pm 2\,\psi_0(\mathbf{x}_1)\,\psi_0(\mathbf{x}_2)\right]. \tag{9.102}$$

It follows that

$$I = 2\,(1 \pm J), \tag{9.103}$$

where

$$J = \int d^3\mathbf{x}\,\psi_0(\mathbf{x}_1)\,\psi_0(\mathbf{x}_2). \tag{9.104}$$

Without loss of generality, we can perform the previous integral using a modified coordinate system in which the first proton lies at the origin, and the second at $\mathbf{X} = R\,\mathbf{e}_z$. Let $\mathbf{x} = r\,(\sin\theta\cos\varphi,\, \sin\theta\sin\varphi,\, \cos\theta)$ be the position vector of the electron. It follows that $r_1 = r$ and $r_2 = (r^2 + R^2 - 2rR\cos\theta)^{1/2}$. Hence,

$$J = 2\int_0^\infty dx\, x^2 \int_0^\pi d\theta\,\sin\theta\,\exp\left[-x - (x^2 + y^2 - 2xy\cos\theta)^{1/2}\right], \tag{9.105}$$

where $x = r/a_0$ and $y = R/a_0$. Here, we have already performed the trivial φ integral. Let $z(\theta) = (x^2 + y^2 - 2xy \cos\theta)^{1/2}$. It follows that $d(z^2) = 2z\,dz = 2xy \sin\theta\,d\theta$, giving

$$\int_0^\pi d\theta \sin\theta\, e^{-(x^2+y^2-2xy\cos\theta)^{1/2}} = \frac{1}{xy} \int_{|x-y|}^{x+y} dz\, z\, e^{-z} \tag{9.106}$$

$$= -\frac{1}{xy}\left[e^{-(x+y)}(1+x+y) - e^{-|x-y|}(1+|x-y|)\right].$$

Thus,

$$J = -\frac{2}{y} e^{-y} \int_0^y dx\, x\left[e^{-2x}(1+y+x) - (1+y-x)\right]$$

$$- \frac{2}{y} \int_y^\infty dx\, x\, e^{-2x}\left[e^{-y}(1+y+x) - e^y(1-y+x)\right], \tag{9.107}$$

which evaluates to

$$J = e^{-y}\left(1 + y + \frac{y^2}{3}\right). \tag{9.108}$$

(See Exercise 9.20.)

Now, the Hamiltonian of the electron is written

$$H_e = \frac{\mathbf{p}^2}{2m_e} - \frac{e^2}{4\pi\,\epsilon_0}\left(\frac{1}{r_1} + \frac{1}{r_2}\right). \tag{9.109}$$

Note, however, that

$$\left(\frac{\mathbf{p}^2}{2m_e} - \frac{e^2}{4\pi\,\epsilon_0\,r_{1,2}}\right)\psi_0(\mathbf{x}_{1,2}) = E_0\,\psi_0(\mathbf{x}_{1,2}), \tag{9.110}$$

where E_0 is the hydrogen ground-state energy, because the $\psi_0(\mathbf{x}_{1,2})$ are hydrogen ground-state wavefunctions. It follows that

$$H_e\,\psi_\pm = A\left[\frac{\mathbf{p}^2}{2m_e} - \frac{e^2}{4\pi\,\epsilon_0}\left(\frac{1}{r_1} + \frac{1}{r_2}\right)\right][\psi_0(\mathbf{x}_1) \pm \psi_0(\mathbf{x}_2)]$$

$$= E_0\,\psi_\pm - A\left(\frac{e^2}{4\pi\,\epsilon_0}\right)\left[\frac{\psi_0(\mathbf{x}_1)}{r_2} \pm \frac{\psi_0(\mathbf{x}_2)}{r_1}\right]. \tag{9.111}$$

Hence,

$$\langle H_e \rangle \equiv \langle \psi_\pm | H_e | \psi_\pm \rangle = E_0 + 4A^2(D \pm E)E_0, \tag{9.112}$$

where

$$D = \left\langle \psi_0(\mathbf{x}_1)\left|\frac{a_0}{r_2}\right|\psi_0(\mathbf{x}_1)\right\rangle, \tag{9.113}$$

$$E = \left\langle \psi_0(\mathbf{x}_1)\left|\frac{a_0}{r_1}\right|\psi_0(\mathbf{x}_2)\right\rangle. \tag{9.114}$$

Here, use has been made of the fact that $e^2/4\pi \epsilon_0 a_0 = -2 E_0$, as well as the fact that the Hamiltonian is invariant under the transformation $\mathbf{x}_1 \leftrightarrow \mathbf{x}_2$.

Now,

$$D = 2 \int_0^\infty dx\, x^2 \int_0^\pi d\theta\, \sin\theta\, \frac{e^{-2x}}{(x^2 + y^2 - 2xy\cos\theta)^{1/2}}, \tag{9.115}$$

where $x = r/a_0$ and $y = R/a_0$, which reduces to

$$D = \frac{4}{y} \int_0^y dx\, x^2\, e^{-2x} + 4 \int_y^\infty dx\, x\, e^{-2x}, \tag{9.116}$$

giving

$$D = \frac{1}{y}\left[1 - (1+y)\,e^{-2y}\right]. \tag{9.117}$$

(See Exercise 9.21.) Furthermore,

$$E = 2 \int_0^\infty dx\, x \int_0^\pi d\theta\, \sin\theta\, \exp\left[-x - (x^2 + y^2 - 2xy\cos\theta)^{1/2}\right], \tag{9.118}$$

which reduces to

$$E = -\frac{2}{y}\, e^{-y} \int_0^y dx \left[e^{-2x}(1 + y + x) - (1 + y - x)\right]$$
$$- \frac{2}{y} \int_y^\infty dx\, e^{-2x} \left[e^{-y}(1 + y + x) - e^y(1 - y + x)\right], \tag{9.119}$$

yielding

$$E = (1 + y)\,e^{-y}. \tag{9.120}$$

(See Exercise 9.22.)

Our final expression for the expectation value of the electron Hamiltonian is

$$\langle H_e \rangle = \left[1 + 2\,\frac{(D \pm E)}{(1 \pm J)}\right] E_0, \tag{9.121}$$

where J, D, and E are specified as functions of $y = R/a_0$ in Equations (9.108), (9.117), and (9.120), respectively. In order to obtain the total energy of the molecule, we must add the potential energy of the two protons to this expectation value. Thus,

$$E_{\text{total}} = \langle H_e \rangle + \frac{e^2}{4\pi \epsilon_0 R} = \langle H_e \rangle - \frac{2}{y}\, E_0, \tag{9.122}$$

because $E_0 = -e^2/(8\pi \epsilon_0 a_0)$. Hence, we can write

$$E_{\text{total}} = -F_{\pm}(R/a_0)\, E_0, \tag{9.123}$$

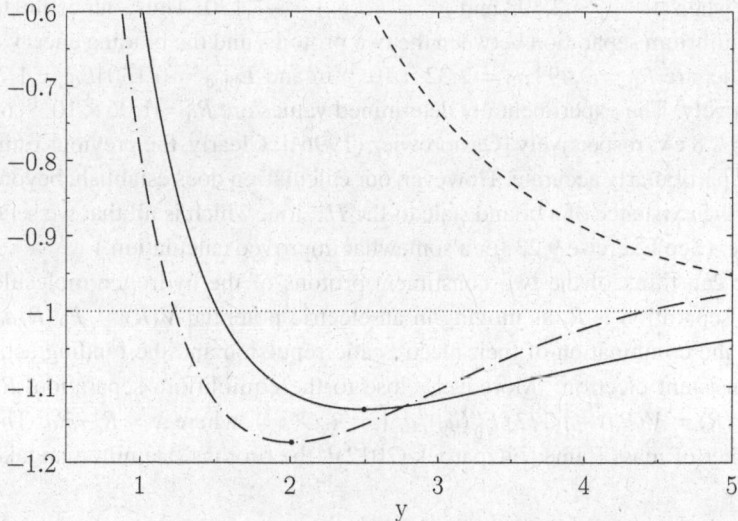

Fig. 9.1 The functions $F_+(y)$ (solid curve) and $F_-(y)$ (short-dashed curve), where $y = R/a_0$. The long-dash-dotted curve shows the improved $F_+(y)$ curve obtained from Exercise 9.23. The dots indicate the minimum values of $F_+(y)$.

where

$$F_\pm(y) = -1 + \frac{2}{y}\left[\frac{(1+y)\,e^{-2y} \pm (1-2y^2/3)\,e^{-y}}{1 \pm (1+y+y^2/3)\,e^{-y}}\right]. \qquad (9.124)$$

The functions $F_+(y)$ and $F_-(y)$ are plotted in Fig. 9.1. Recall that, in order for the H_2^+ ion possess a bound state, it must have a lower energy than a hydrogen atom and a free proton. In other words, $E_{total} < E_0$. It follows, from Equation (9.123), that a bound state corresponds to $F_\pm < -1$. It is clear, from the figure, that the even trial wavefunction, ψ_+, possesses a bound state, whereas the odd trial wavefunction, ψ_-, does not. [See Equation (9.99).] This is hardly surprising, because the even wavefunction maximizes the electron probability density between the two protons, thereby reducing their mutual electrostatic repulsion. On the other hand, the odd wavefunction does exactly the opposite. The binding energy of the H_2^+ ion is defined as minus the difference between its energy and that of a hydrogen atom plus a free proton: that is,

$$E_{bind} = E_0 - E_{total} = -(F_+ + 1)|E_0|. \qquad (9.125)$$

According to the variational principle, the binding energy is greater than, or equal to, the maximum binding energy that can be inferred from Figure 9.1. (A maximum in the binding energy corresponds to a minimum of F_+.) This maximum occurs when $y = y_0 = 2.493$ and $F_+ = F_+(y_0) = -1.130$. Thus, our estimates for the equilibrium separation between the two protons, and the binding energy of the molecule, are $R_0 = 2.493\,a_0 = 1.32 \times 10^{-10}$ m and $E_{bind} = 0.130\,|E_0| = 1.77$ eV, respectively. The experimentally determined values are $R_0 = 1.06 \times 10^{-10}$ m, and $E_{bind} = 2.8$ eV, respectively [Gasiorowicz (1996)]. Clearly, the previous estimates are not particularly accurate. However, our calculation does establish, beyond any doubt, the existence of a bound state of the H_2^+ ion, which is all that we set out to achieve. (See Exercise 9.23 for a somewhat improved calculation.)

We can think of the two constituent protons of the hydrogen molecule ion, whose separation is R, as moving in an electric potential $V(R) = F_+(R/a_0)\,|E_0|$ due to the combination of their electrostatic repulsion and the binding action of the consistent electron. Moreover, close to the equilibrium separation, R_0, we have $V(R) \simeq V(R_0) + \left[(1/2)\,F_0''(y_0)/a_0^2\right] x^2 + O(x^3)$, where $x = R - R_0$. Thus, in the center of mass frame [Fitzpatrick (2012)], the protons' Hamiltonian takes the form

$$H_p \simeq \frac{\mathbf{p}_x^2}{2\mu} + \frac{1}{2}\mu\omega^2 x^2, \qquad (9.126)$$

where $\mathbf{p}_x$ is the momentum conjugate to x, $\mu = 2\,m_p$ the reduced mass [Fitzpatrick (2012)], m_p the proton mass, $\omega^2 = F_0''(y_0)/(a_0^2\,\mu)$, and we have neglected an unimportant constant term. Here, $'$ denotes d/dy. It can be seen, by comparison with Exercise 3.3, that the previous Hamiltonian is identical to that of a harmonic oscillator. Thus, the allowed radial oscillation energies of the molecule are

$$E_n = \left(n + \frac{1}{2}\right)\hbar\,\omega, \qquad (9.127)$$

where n is a non-negative integer. In particular, there is a non-zero lowest oscillation energy—the so-called *zero-point energy* [Einstein and Stern (1913)],

$$E_{zero} = \frac{1}{2}\hbar\,\omega = \sqrt{F_+''(y_0)}\left(\frac{m_e}{m_p}\right)^{1/2}|E_0| = 0.11\,\text{eV}, \qquad (9.128)$$

that must be subtracted from the previously determined electric binding energy of 1.77 eV to give the true binding energy. In calculating, E_{zero} we have made use of the numerically determined value $F_0''(y_0) = 0.1257$ [derived from Equation (9.124.)]

Of course, the hydrogen molecule ion can rotate, as well as vibrate. According to Exercise 4.7, the rotational component of the molecular Hamiltonian can be written

$$H_{rot} = \frac{L^2}{2\,I_\perp}, \qquad (9.129)$$

where $\mathbf{L}$ is the angular momentum of rotation, and $I_\perp = m_p R_0^2/3$ is the molecular moment of inertia [about a perpendicular (to the line joining the two protons) axis passing through the center of mass]. In calculating the moment of inertia, we have neglected the electron mass, and have treated the protons as point particles. The analysis of Exercise 4.7 reveals that the rotational energy levels are

$$E_l = l(l+1)\,\frac{\hbar^2}{2I_\perp}, \tag{9.130}$$

where the eigenvalues of L^2 are $l(l+1)\hbar^2$, and l is a non-negative integer. It follows that

$$E_l = l(l+1)\,\frac{3}{y_0^2}\,\frac{m_e}{m_p}\,|E_0| = l(l+1)\,E_r, \tag{9.131}$$

where

$$E_r = \frac{3}{y_0^2}\,\frac{m_e}{m_p}\,|E_0| = 3.58 \times 10^{-3}\,\text{eV}. \tag{9.132}$$

Note that our estimate for the electric binding energy of the hydrogen molecule ion, 1.77 eV, is significantly greater than our estimate for the zero-point oscillation energy, 0.11 eV, which, in turn, is very much greater than a typical rotational energy, $\sim 7 \times 10^{-3}$ eV. This separation in energy scales lies at the heart of the previously mentioned Born-Oppenheimer approximation [Born and Oppenheimer (1927)], according to which the electric, vibrational, and rotational energy levels of molecules can all be calculated independently of one another.

9.9 Exercises

9.1 Demonstrate that the eigenvalues of the two-particle permutation operator, P_{12}, are ± 1.

9.2 Derive Equations (9.14) and (9.15).

9.3 Demonstrate that the two-particle permutation operator, P_{12}, is Hermitian.

9.4 Justify Equation (9.35).

9.5 Consider the cyclic permutation operator

$$P_{123}\,|k'\rangle\,|k''\rangle\,|k'''\rangle = |k'''\rangle\,|k'\rangle\,|k''\rangle.$$

Demonstrate that

$$P_{123}^2 = P_{123}^\dagger,$$

and

$$P_{123}\,P_{123}^\dagger = 1.$$

9.6 Consider two identical spin-1/2 particles of mass m confined in a cubic box of dimension L. Find the possible energies and wavefunctions of this system in the case of no interaction between the particles.

9.7 Consider a system of two spin-1 particles with no orbital angular momentum (i.e., both particles are in the lowest energy s-state). What are the possible eigenvalues of the total angular momentum of the system, as well as its projection along the z-direction, in the cases in which the particles are non-identical and identical?

9.8 Consider a hydrogen-like atom consisting of a single electron of charge $-e$ orbiting about a massive nucleus of charge $Z e$ (where $Z > 0$). The eigenstates of the Hamiltonian can be labelled by the conventional quantum numbers n, l, and m.

(a) Show that the energy levels are

$$E_n = \frac{Z^2 E_0}{n^2},$$

where E_0 is the ground-state energy of a conventional hydrogen atom.

(b) Demonstrate that the first few properly normalized radial wavefunctions, $R_{nl}(r)$, take the form:

$$R_{10}(r) = \frac{2 Z^{3/2}}{a_0^{3/2}} \exp\left(-\frac{Z r}{a_0}\right),$$

$$R_{20}(r) = \frac{2 Z^{3/2}}{(2 a_0)^{3/2}} \left(1 - \frac{Z r}{2 a_0}\right) \exp\left(-\frac{Z r}{2 a_0}\right),$$

$$R_{21}(r) = \frac{Z^{3/2}}{\sqrt{3} (2 a_0)^{3/2}} \frac{Z r}{a_0} \exp\left(-\frac{Z r}{2 a_0}\right),$$

where a_0 is the Bohr radius.

9.9 Demonstrate that

$$\int_0^{x_1} dx_2\, x_2^n\, e^{-\beta x_2} = \frac{n!}{\beta^{n+1}} - \frac{e^{-\beta x_1}}{\beta^{n+1}} \sum_{k=0,n} \frac{n!}{k!} (\beta x_1)^k,$$

and

$$\int_{x_1}^{\infty} dx_2\, x_2^n\, e^{-\beta x_2} = \frac{e^{-\beta x_1}}{\beta^{n+1}} \sum_{k=0,n} \frac{n!}{k!} (\beta x_1)^k$$

Here, n is a non-negative integer, and $\beta > 0$.

9.10 Justify Equation (9.63).

9.11 Given that the ground-state energy of a helium atom is $-78.98\,\text{eV}$, deduce that the ground-state ionization energy (i.e., the minimum energy that must be supplied to remove a single electron from the atom in the ground-state) is $24.56\,\text{eV}$.

9.12 Consider a helium atom in which both electrons are in the $n = 2$, $l = 1$, $m = 0$ state (i.e., the $2p$ state). Use the techniques of Section 9.5 to obtain the following estimate for the energy of this state:

$$E(2p\,2p) = \left(-2 + \frac{3 \cdot 31}{2^7}\right)|E_0| = -17.33\,\text{eV},$$

where E_0 is the hydrogen ground-state energy. Note that this energy lies well above the energy of the ground-state of singly-ionized helium, which is $E_{\text{He+}} = 4E_0 = -54.42\,\text{eV}$. This means that a helium atom excited to the $2p\,2p$ state has the option of decaying into a free electron and a singly-ionized helium ion, with the energy of the ejected electron determined by energy conservation. This process is known as *autoionization*.

9.13 Consider an excited state of the helium atom in which one electron is in the ground state, and the other is in the n, l, m state, where $n > 1$.

(a) Demonstrate that the expressions for the direct and exchange integrals given in Section 9.6 reduce to

$$\frac{I}{|E_0|} = 2\,a_0 \int_0^\infty dr_1\,r_1\,R_{10}(r_1)\,R_{10}(r_1)\left[\int_0^{r_1} dr_2\,r_2^2\,R_{nl}(r_2)\,R_{nl}(r_2)\right.$$

$$\left. + \int_{r_1}^\infty dr_2\,r_1\,r_2\,R_{nl}(r_2)\,R_{nl}(r_2)\right],$$

and

$$\frac{J}{|E_0|} = \frac{2\,a_0}{2l+1} \int_0^\infty dr_1\,r_1\,R_{10}(r_1)\,R_{nl}(r_1)\left[\int_0^{r_1} dr_2\,r_2^2\left(\frac{r_2}{r_1}\right)^l R_{10}(r_2)\,R_{nl}(r_2)\right.$$

$$\left. + \int_{r_1}^\infty dr_2\,r_1\,r_2\left(\frac{r_1}{r_2}\right)^l R_{10}(r_2)\,R_{nl}(r_2)\right],$$

respectively, where $R_{nl}(r)$ is a hydrogen radial wavefunction calculated with a nuclear charge $2\,e$.

(b) Suppose that the excited electron is in the $n = 2$, $l = 1$, $m = 0$ state (i.e., the $2p$ state). Show that

$$\frac{I}{|E_0|} = \frac{2^2 \cdot 59}{3^5} = 0.9712,$$

$$\frac{J}{|E_0|} = \frac{2^6 \cdot 7}{3^8} = 0.0683.$$

Hence, deduce that the energies of the $1s\,2p$ states of orthohelium and parahelium are

$$E(1s\,2p)_\pm = (-5 + 0.9712 \pm 0.0683)\,|E_0| = (-4.0288 \pm 0.0683)\,|E_0|$$

$$= (-54.81 \pm 0.93)\,\text{eV},$$

where the upper sign corresponds to parahelium.

9.14 The observed energies of the $1s\,2p$ states of orthohelium and parahelium are

$$E(1s\,2p)_{\pm} = (-57.89 \pm 0.13)\,\text{eV},$$

where the upper sign corresponds to parahelium [Park (1974)]. It can be seen that the calculation in the previous exercise has considerably overestimated the size of the exchange integral. The main reason for this is that we neglected to take into account the fact that the $1s$ electron largely shields the $2p$ electron from the nuclear charge (because, on average, the former electron lies much closer to the nucleus that the latter). We can arrive at a better estimate for the exchange integral by using a $R_{10}(r)$ radial function calculated with a nuclear charge $2\,e$, and a $R_{21}(r)$ radial function calculated with a nuclear charge e. Here, we are assuming that the inner ($1s$) electron experiences the full nuclear charge, whereas the outer ($2p$) electron only experiences half the nuclear charge, as a consequence of the shielding action of the inner electron. Demonstrate that, in this case, the exchange integral becomes

$$\frac{J}{|E_0|} = \frac{2^8 \cdot 7}{5^7 \cdot 3} = 0.00765,$$

which yields

$$J = 0.10\,\text{eV}.$$

Note that this estimate is much closer to the experimental value $(0.13\,\text{eV})$ than our previous estimate $(0.93\,\text{eV})$.

9.15 Let

$$J_n(\beta) = \int_{-\infty}^{\infty} dx\, x^{2n}\, e^{-\beta x^2},$$

where n is a non-negative integer, and $\beta > 0$. Demonstrate that

$$J_0(\beta) = \left(\frac{\pi}{\beta}\right)^{1/2},$$

and

$$J_{n>0}(\beta) = \frac{1 \cdot 3 \cdot 5 \cdots (2n-1)}{2^n} \left(\frac{\pi}{\beta^{2n+1}}\right)^{1/2}.$$

9.16 Consider the general energy eigenvalue problem

$$H\,|\psi\rangle = E\,|\psi\rangle.$$

Suppose that $|\psi\rangle$ is a trial solution to the previous equation that is not properly normalized. Prove that

$$E_0 \leq \frac{\langle \psi|\,H\,|\psi\rangle}{\langle \psi|\psi\rangle},$$

where E_0 is the lowest energy eigenvalue.

9.17 Consider a particle of mass m moving in the one-dimensional potential

$$V(x) = \lambda x^4,$$

where $\lambda > 0$. Let

$$\psi_0(x) = \frac{\alpha^{1/2}}{\pi^{1/4}} e^{-\alpha^2 x^2/2},$$

$$\psi_1(x) = \frac{\sqrt{2}\beta^{3/2}}{\pi^{1/4}} x e^{-\beta^2 x^2/2}.$$

Verify that these wavefunctions are properly normalized. Use the variational principle, combined with the plausible trial wavefunctions $\psi_0(x)$ and $\psi_1(x)$ (these wavefunctions are, in fact, the exact ground-state and first-excited-state wavefunctions for a particle moving in the potential λx^2) to obtain the following estimates for the energies of the ground state, and the first excited state, of the system:

$$E_0 = \frac{3^{4/3}}{4}\left(\frac{\hbar^2}{2m}\right)^{2/3}\lambda^{1/3} = 1.082\left(\frac{\hbar^2}{2m}\right)^{2/3}\lambda^{1/3},$$

$$E_1 = \frac{9 \cdot 5^{1/3}}{4}\left(\frac{\hbar^2}{2m}\right)^{2/3}\lambda^{1/3} = 3.847\left(\frac{\hbar^2}{2m}\right)^{2/3}\lambda^{1/3}.$$

The exact numerical factors that should appear in the previous two equations are 1.060 and 3.800, respectively [Park (1974)]. Hence, it is clear that our approximation to E_0 and E_1 are fairly accurate.

9.18 Use the variational technique outlined in Section 9.7 to derive the following estimate the ground-state energy of a two-electron atom with nuclear charge $Z_0 e$ in the spin-singlet state:

$$E = \frac{(16Z_0 - 5)^2}{2^7} E_0,$$

where E_0 is the hydrogen ground-state energy. For the case of a negative hydrogen ion (i.e., $Z_0 = 1$), this formula gives $E_{H^-} = 0.9453 E_0 = -12.86\,\text{eV}$. The experimental value of this energy is $E_{H^-} = -14.36\,\text{eV}$ [Haynes and Lide (2011)]. For the case of a singly-ionized lithium ion (i.e., $Z_0 = 3$), the previous formula gives $E_{Li^+} = 14.45 E_0 = -196.54\,\text{eV}$. The experimental value of this energy is $E_{Li^+} = -198.09\,\text{eV}$ [Haynes and Lide (2011)].

9.19 It can be seen from Section 9.7, as well as the previous exercise, that the variational technique described in Section 9.7 yields approximations to the ground-state energies of two-electron atoms in the spin-singlet state that are approximately 1.5 eV too high. This is not a particular problem for the helium atom, or the singly-ionized lithium ion. However, for the negative hydrogen

Z	ϵ	Z_1	Z_2	E (eV)	E_{expt} (eV)
1	+1	1.04	0.28	−13.97	−14.35
2	+1	2.18	1.19	−78.25	−78.98
2	−1	1.97	0.32	−58.79	−59.18
3	+1	3.29	2.08	−197.25	−198.10
3	−1	2.93	0.60	−138.01	−139.06

ion, our estimate for the ground-state energy, −12.86 eV, is slightly higher than the ground-state energy of a neutral hydrogen atom, −13.61 eV, giving the erroneous impression that it is not energetically favorable for a neutral hydrogen atom to absorb an additional electron to form a negative hydrogen ion (i.e., that the negative hydrogen ion has a negative binding energy). Obviously, we need to perform a more accurate calculation for the case of a negative hydrogen ion. Following Chandrasekhar [Chandrasekhar (1944)], let us adopt the following trial wavefunction:

$$\phi(\mathbf{x}_1, \mathbf{x}_2) = \frac{1}{\sqrt{2}} \left[\psi_1(\mathbf{x}_1)\psi_2(\mathbf{x}_2) + \epsilon\,\psi_2(\mathbf{x}_1)\psi_1(\mathbf{x}_2) \right],$$

where

$$\psi_1(\mathbf{x}) = \frac{1}{\sqrt{\pi}} \left(\frac{Z_1}{a_0} \right)^{3/2} \exp\left(\frac{-Z_1\,r}{a_0} \right),$$

$$\psi_2(\mathbf{x}) = \frac{1}{\sqrt{\pi}} \left(\frac{Z_2}{a_0} \right)^{3/2} \exp\left(\frac{-Z_2\,r}{a_0} \right).$$

Here, $r = |\mathbf{x}|$, a_0 is the Bohr radius, and Z_1, Z_2 are adjustable parameters. Moreover, ϵ takes the values +1 and −1 for the spin-singlet and spin-triplet states, respectively. Given that the Hamiltonian of a two-electron atom of nuclear charge $Z\,e$ is

$$H = \frac{\mathbf{p}_1^2}{2\,m_e} + \frac{\mathbf{p}_2^2}{2\,m_e} - \frac{Z\,e^2}{4\pi\,\epsilon_0\,r_1} - \frac{Z\,e^2}{4\pi\,\epsilon_0\,r_2} + \frac{e^2}{4\pi\,\epsilon_0\,r_{12}},$$

show that the expectation value of H (i.e., $\langle H \rangle = \langle \phi| H |\phi\rangle / \langle \phi|\phi\rangle$) is

$$\frac{\langle H \rangle}{|E_0|} = \left[x^8 - 2Z x^7 - \frac{1}{2} x^6 y^2 + \frac{1}{2} x^5 y^2 + \frac{1}{8} x^3 y^4 \right.$$

$$\left. - \epsilon\left(2Z - \frac{5}{8} \right) x y^6 + \frac{1}{2}\epsilon y^8 \right] \bigg/ \left(x^6 + \epsilon y^6 \right),$$

where E_0 is the hydrogen ground-state energy, $x = Z_1 + Z_2$, and $y = 2\sqrt{Z_1 Z_2}$. We now need to minimize $\langle H \rangle$ with respect to variations in Z_1 and Z_2 to obtain an estimate for the ground-state energy. Unfortunately, this can only be achieved numerically.

The previous table shows the numerically determined values of Z_1 and Z_2 that minimize $\langle H \rangle$ for various choices of Z and ϵ. The table also shows the estimate for the ground-state energy (E), as well as the corresponding experimentally measured ground-state energy (E_{expt}) [Haynes and Lide (2011); Høgaasen *et al.* (2010)]. It can be seen that our new estimate for the ground-state energy of the negative hydrogen ion is now less than the ground-state energy of a neutral hydrogen atom, which demonstrates that the negative hydrogen ion has a positive (albeit, small) binding energy. Incidentally, the case $Z = 2$, $\epsilon = -1$ yields a good estimate for the energy of the lowest-energy spin-triplet state of a helium atom (i.e., the $1s\,2s$ spin-triplet state).

9.20 Justify Equation (9.108).

9.21 Justify Equations (9.116) and (9.117).

9.22 Justify Equations (9.119) and (9.120).

9.23 Repeat the calculation of Section 9.8 using the the trial single-proton wavefunction

$$\psi_0(\mathbf{x}) = \frac{1}{\sqrt{\pi}\,a^{3/2}}\,e^{-r/a},$$

where $a = a_0/Z$, $r = |\mathbf{x}|$, a_0 is the Bohr radius, and Z an adjustable parameter [Finkelstein and Horowitz (1928)]. Show that the energy of the hydrogen molecule ion, assuming a molecular wavefunction that is even under exchange of proton positions, can be written

$$E_{\text{total}} = -F_+(R/a)\,E_0,$$

where R is the proton separation, E_0 the hydrogen ground-state energy, and

$$F_+(y) = -Z^2 + \frac{2Z}{y}\left[\frac{(1+y)\,e^{-2y} + (1 - 2y^2/3)\,e^{-y} + (Z-1)\,y\,(1 + [1+y]\,e^{-y})}{1 + (1 + y + y^2/3)\,e^{-y}}\right].$$

It can be shown, numerically, that the previous function attains its minimum value, $F_+ = -1.173$, when $Z = 1.238$ and $y = 2.480$. This leads to predictions for the equilibrium separation between the two protons, and the binding energy of the molecule, of $R_0 = (2.480/1.238)\,a_0 = 2.003\,a_0 = 1.06 \times 10^{-10}$ m and $E_{\text{bind}} = 0.173\,|E_0| = 2.35$ eV, respectively. (See Figure 9.1.) These values are far closer to the experimentally determined values, $R_0 = 1.06 \times 10^{-10}$ m and $E_{\text{bind}} = 2.8$ eV [Gasiorowicz (1996)], than those derived in Section 9.8.

Chapter 10

Scattering Theory

10.1 Introduction

Historically, data regarding quantum phenomena was obtained from two main sources. Firstly, the study of spectroscopic lines, and, secondly, the analysis of data from scattering experiments. We have already developed theories that account for some aspects of the spectra of hydrogen-like atoms. Let us now examine the quantum theory of scattering.

In the following, we shall treat scattering as an essentially two-particle effect. As is well known, when viewed in the *center of mass frame*, two particles of masses m_1 and m_2, and position vector $\mathbf{x}_1$ and $\mathbf{x}_2$, respectively, interacting via the potential $V(\mathbf{x}_1 - \mathbf{x}_2)$, can be treated as a single body of *reduced mass* $\mu_{12} = m_1 m_2/(m_1 + m_2)$, and position vector $\mathbf{x} = \mathbf{x}_1 - \mathbf{x}_2$, moving in the fixed potential $V(\mathbf{x})$ [Fitzpatrick (2012)]. For this reason, we can, without loss of generality, focus our study on the quantum theory of particles scattered by fixed potentials.

10.2 Fundamental Equations

Consider time-independent scattering theory, for which the Hamiltonian of the system is written

$$H = H_0 + H_1, \tag{10.1}$$

where

$$H_0 = \frac{p^2}{2m} \tag{10.2}$$

is the Hamiltonian of a free particle of mass m, and H_1 represents the non-time-varying source of the scattering. Let $|\phi\rangle$ be an energy eigenket of H_0,

$$H_0 |\phi\rangle = E |\phi\rangle, \tag{10.3}$$

whose wavefunction is $\phi(\mathbf{x}') \equiv \langle \mathbf{x}'|\phi\rangle$. This wavefunction is assumed to be a plane wave. Schrödinger's equation for the scattering problem is

$$(H_0 + H_1)|\psi\rangle = E|\psi\rangle, \tag{10.4}$$

where $|\psi\rangle$ is an energy eigenstate of the total Hamiltonian whose wavefunction is $\psi(\mathbf{x}') \equiv \langle \mathbf{x}'|\psi\rangle$. In general, both H_0 and $H_0 + H_1$ have continuous energy spectra: that is, their energy eigenstates are unbound. We require a solution of Equation (10.4) that satisfies the boundary condition $|\psi\rangle \to |\phi\rangle$ as $H_1 \to 0$. Here, $|\phi\rangle$ is a solution of the free-particle Schrödinger equation, (10.3), that corresponds to the same energy eigenvalue as $|\psi\rangle$.

Adopting the Schrödinger representation (see Section 2.4), we can write the scattering equation, (10.4), in the form

$$(\nabla^2 + k^2)\,\psi(\mathbf{x}) = \frac{2\,m}{\hbar^2}\,\langle\mathbf{x}|\,H_1\,|\psi\rangle, \tag{10.5}$$

where

$$E = \frac{\hbar^2\,k^2}{2\,m}. \tag{10.6}$$

(See Exercise 10.1.) Equation (10.5) is known as the *Helmholtz equation*, and can be inverted using standard Green's function techniques [Fitzpatrick (2008)]. Thus,

$$\psi(\mathbf{x}) = \phi(\mathbf{x}) + \frac{2\,m}{\hbar^2}\int d^3\mathbf{x}'\,G(\mathbf{x},\mathbf{x}')\,\langle\mathbf{x}'|\,H_1\,|\psi\rangle, \tag{10.7}$$

where

$$(\nabla^2 + k^2)\,G(\mathbf{x},\mathbf{x}') = \delta^3(\mathbf{x} - \mathbf{x}'). \tag{10.8}$$

(See Exercise 10.2.) Here, $\delta^3(\mathbf{x})$ is a three-dimensional Dirac delta function. Note that the solution (10.7) satisfies the previously mentioned constraint $|\psi\rangle \to |\phi\rangle$ as $H_1 \to 0$. As is well known, the Green's function for the Helmholtz equation is given by

$$G(\mathbf{x},\mathbf{x}') = -\frac{\exp(\pm i\,k\,|\mathbf{x} - \mathbf{x}'|)}{4\pi\,|\mathbf{x} - \mathbf{x}'|}. \tag{10.9}$$

(See Exercise 10.3.) Thus, Equation (10.7) becomes

$$\psi^{\pm}(\mathbf{x}) = \phi(\mathbf{x}) - \frac{2\,m}{\hbar^2}\int d^3\mathbf{x}'\,\frac{\exp(\pm i\,k\,|\mathbf{x} - \mathbf{x}'|)}{4\pi\,|\mathbf{x} - \mathbf{x}'|}\,\langle\mathbf{x}'|\,H_1\,|\psi^{\pm}\rangle. \tag{10.10}$$

Let us suppose that the scattering Hamiltonian, H_1, is a function only of the position operators. This implies that

$$\langle\mathbf{x}'|\,H_1\,|\mathbf{x}\rangle = V(\mathbf{x})\,\delta^3(\mathbf{x} - \mathbf{x}'). \tag{10.11}$$

We can write

$$\langle \mathbf{x}' | H_1 | \psi^{\pm} \rangle = \int d^3 \mathbf{x}'' \langle \mathbf{x}' | H_1 | \mathbf{x}'' \rangle \langle \mathbf{x}'' | \psi^{\pm} \rangle$$

$$= \int d^3 \mathbf{x}'' \, V(\mathbf{x}') \, \delta^3(\mathbf{x}' - \mathbf{x}'') \, \psi(\mathbf{x}'') = V(\mathbf{x}') \, \psi^{\pm}(\mathbf{x}'), \qquad (10.12)$$

where use has been made of the standard completeness relation $\int d^3 \mathbf{x}'' |\mathbf{x}''\rangle\langle \mathbf{x}''| = 1$. (See Section 1.15.) Thus, the integral equation (10.10) simplifies to give

$$\psi^{\pm}(\mathbf{x}) = \phi(\mathbf{x}) - \frac{2\,m}{\hbar^2} \int d^3 \mathbf{x}' \, \frac{\exp(\pm i\,k\,|\mathbf{x} - \mathbf{x}'|)}{4\pi\,|\mathbf{x} - \mathbf{x}'|} \, V(\mathbf{x}') \, \psi^{\pm}(\mathbf{x}'). \qquad (10.13)$$

Suppose that the initial state, $|\phi\rangle$, possesses a plane-wave wavefunction with wavevector $\mathbf{k}$ (i.e., it corresponds to a stream of particles of definite momentum $\mathbf{p} = \hbar\,\mathbf{k}$). The ket corresponding to this state is denoted $|\mathbf{k}\rangle$. Thus,

$$\phi(\mathbf{x}) \equiv \langle \mathbf{x} | \mathbf{k} \rangle = \frac{\exp(i\,\mathbf{k} \cdot \mathbf{x})}{(2\pi)^{3/2}}. \qquad (10.14)$$

The preceding wavefunction is conveniently normalized such that

$$\langle \mathbf{k} | \mathbf{k}' \rangle = \int d^3 \mathbf{x} \, \langle \mathbf{k} | \mathbf{x} \rangle \langle \mathbf{x} | \mathbf{k}' \rangle = \int d^3 \mathbf{x} \, \frac{\exp[-i\,\mathbf{x} \cdot (\mathbf{k} - \mathbf{k}')]}{(2\pi)^3} = \delta^3(\mathbf{k} - \mathbf{k}'). \quad (10.15)$$

(See Section 2.6 and Exercise 10.4.)

Suppose that the scattering potential, $V(\mathbf{x})$, is non-zero only in some relatively localized region centered on the origin ($\mathbf{x} = \mathbf{0}$). Let us calculate the total wavefunction, $\psi(\mathbf{x})$, far from the scattering region. In other words, let us adopt the ordering $r \gg r'$, where $r = |\mathbf{x}|$ and $r' = |\mathbf{x}'|$. It is easily demonstrated that

$$|\mathbf{x} - \mathbf{x}'| \simeq r - \mathbf{e}_r \cdot \mathbf{x}' \qquad (10.16)$$

to first order in r'/r, where $\mathbf{e}_r = \mathbf{x}/r$ is a unit vector that is directed from the scattering region to the observation point. Let us define

$$\mathbf{k}' = k\,\mathbf{e}_r. \qquad (10.17)$$

Clearly, $\mathbf{k}'$ is the wavevector for particles that possess the same energy as the incoming particles (i.e., $k' = k$), but propagate from the scattering region to the observation point. Note that

$$\exp(\pm i\,k\,|\mathbf{x} - \mathbf{x}'|) \simeq \exp(\pm i\,k\,r)\exp(\mp i\,\mathbf{k}' \cdot \mathbf{x}'). \qquad (10.18)$$

In the large-r limit, Equations (10.13) and (10.14) reduce to

$$\psi^{\pm}(\mathbf{x}) \simeq \frac{\exp(i\,\mathbf{k} \cdot \mathbf{x})}{(2\pi)^{3/2}}$$

$$- \frac{m}{2\pi\,\hbar^2} \frac{\exp(\pm i\,k\,r)}{r} \int d^3 \mathbf{x}' \, \exp(\mp i\,\mathbf{k}' \cdot \mathbf{x}') \, V(\mathbf{x}') \, \psi^{\pm}(\mathbf{x}'). \qquad (10.19)$$

The first term on the right-hand side of the previous equation is the incident wave. The second term represents a spherical wave centered on the scattering region. The plus sign (on $\psi^{\pm}$) corresponds to a wave propagating away from the scattering region, whereas the minus sign corresponds to a wave propagating toward the scattering region. (See Exercise 10.5.) It is obvious that the former represents the physical solution. Thus, the wavefunction far from the scattering region can be written

$$\psi(\mathbf{x}) = \frac{1}{(2\pi)^{3/2}} \left[\exp(\mathrm{i}\,\mathbf{k}\cdot\mathbf{x}) + \frac{\exp(\mathrm{i}\,k\,r)}{r} f(\mathbf{k}',\mathbf{k}) \right], \tag{10.20}$$

where

$$f(\mathbf{k}',\mathbf{k}) = -\frac{(2\pi)^2 m}{\hbar^2} \int d^3\mathbf{x}' \, \frac{\exp(-\mathrm{i}\,\mathbf{k}'\cdot\mathbf{x}')}{(2\pi)^{3/2}} \, V(\mathbf{x}')\,\psi(\mathbf{x}')$$

$$= -\frac{(2\pi)^2 m}{\hbar^2} \langle \mathbf{k}' | H_1 | \psi \rangle. \tag{10.21}$$

[See Equations (10.11) and (10.14).]

Let us define the *differential scattering cross-section*, $d\sigma/d\Omega$, as the number of particles per unit time scattered into an element of solid angle $d\Omega$, divided by the incident particle flux. Recall, from Chapter 3, that the probability current (which is proportional to the particle flux) associated with a wavefunction ψ is

$$\mathbf{j} = \frac{\hbar}{m} \, \mathrm{Im}(\psi^* \, \nabla \psi). \tag{10.22}$$

Thus, the particle flux associated with the incident wavefunction,

$$\frac{\exp(\mathrm{i}\,\mathbf{k}\cdot\mathbf{x})}{(2\pi)^{3/2}}, \tag{10.23}$$

is proportional to

$$\mathbf{j}_{\text{incident}} = \frac{\hbar\,\mathbf{k}}{(2\pi)^3 \, m}. \tag{10.24}$$

Likewise, the particle flux associated with the scattered wavefunction,

$$\frac{\exp(\mathrm{i}\,k\,r)}{(2\pi)^{3/2}} \frac{f(\mathbf{k}',\mathbf{k})}{r}, \tag{10.25}$$

is proportional to

$$\mathbf{j}_{\text{scattered}} = \frac{\hbar\,\mathbf{k}'}{(2\pi)^3 \, m} \frac{|f(\mathbf{k}',\mathbf{k})|^2}{r^2}. \tag{10.26}$$

Now, by definition,

$$\frac{d\sigma}{d\Omega} \, d\Omega = \frac{r^2 \, d\Omega \, |\mathbf{j}_{\text{scattered}}|}{|\mathbf{j}_{\text{incident}}|}, \tag{10.27}$$

giving

$$\frac{d\sigma}{d\Omega} = |f(\mathbf{k}', \mathbf{k})|^2. \qquad (10.28)$$

Thus, $|f(\mathbf{k}', \mathbf{k})|^2$ is the differential cross-section for particles with incident momentum $\hbar\,\mathbf{k}$ to be scattered into states whose momentum vectors are directed in a range of solid angles $d\Omega$ about $\hbar\,\mathbf{k}'$. Note that the scattered particles possess the same energy as the incoming particles (i.e., $k' = k$). This is always the case for scattering Hamiltonians of the form specified in Equation (10.11).

10.3 Born Approximation

Equation (10.28) is not particularly useful, as it stands, because the quantity $f(\mathbf{k}', \mathbf{k})$ depends on the unknown ket $|\psi\rangle$. Recall that $\psi(\mathbf{x}) \equiv \langle \mathbf{x}|\psi\rangle$ is the solution of the integral equation

$$\psi(\mathbf{x}) = \phi(\mathbf{x}) - \frac{m}{2\pi\hbar^2} \int d^3\mathbf{x}'\, \frac{\exp(\mathrm{i}\,k\,|\mathbf{x} - \mathbf{x}'|)}{|\mathbf{x} - \mathbf{x}'|}\, V(\mathbf{x}')\,\psi(\mathbf{x}'), \qquad (10.29)$$

where $\phi(\mathbf{x})$ is the wavefunction of the incident state. [See Equation (10.13).] According to the previous equation, the total wavefunction is a superposition of the incident wavefunction and a great many spherical waves emitted from the scattering region. The strength of the spherical wave emitted at a given point in the scattering region is proportional to the local value of the scattering potential, $V(\mathbf{x})$, as well as the local value of the wavefunction, $\psi(\mathbf{x})$.

Suppose, however, that the scattering is not particularly intense. In this case, it is reasonable to suppose that the total wavefunction, $\psi(\mathbf{x})$, does not differ substantially from the incident wavefunction, $\phi(\mathbf{x})$. Thus, we can obtain an expression for $\psi(\mathbf{x})$ by making the substitution

$$\psi(\mathbf{x}) \to \phi(\mathbf{x}) = \frac{\exp(\mathrm{i}\,\mathbf{k} \cdot \mathbf{x})}{(2\pi)^{3/2}} \qquad (10.30)$$

on the right-hand side of Equation (10.29). [See Equation (10.14).] This simplification is known as the *Born approximation* [Born (1926)].

The Born approximation yields

$$f(\mathbf{k}', \mathbf{k}) \simeq -\frac{m}{2\pi\hbar^2} \int d^3\mathbf{x}'\, \exp\left[\mathrm{i}\,(\mathbf{k} - \mathbf{k}') \cdot \mathbf{x}'\right] V(\mathbf{x}'). \qquad (10.31)$$

[See Equation (10.21).] Thus, $f(\mathbf{k}', \mathbf{k})$ is proportional to the Fourier transform of the scattering potential, $V(\mathbf{x})$, with respect to the relative wavevector, $\mathbf{q} \equiv \mathbf{k} - \mathbf{k}'$.

For a spherically symmetric scattering potential,

$$f(\mathbf{k}', \mathbf{k}) \simeq -\frac{m}{2\pi\hbar^2} \int_0^\infty \int_0^\pi \int_0^{2\pi} dr'\, d\theta'\, d\varphi'\, r'^2\, \sin\theta'\, e^{\mathrm{i}\,q\,r'\cos\theta'}\, V(r'), \qquad (10.32)$$

giving

$$f(\mathbf{k'}, \mathbf{k}) \simeq -\frac{2\,m}{\hbar^2\,q} \int_0^\infty dr\, r\, V(r) \sin(q\,r).$$ (10.33)

Hence, it is clear that, for the special case of a spherically symmetric potential, $f(\mathbf{k'}, \mathbf{k})$ depends only on the magnitude of the relative wavevector, $\mathbf{q} = \mathbf{k} - \mathbf{k'}$, and is independent of its direction. Now, it is easily demonstrated that

$$q \equiv |\mathbf{k} - \mathbf{k'}| = 2\,k\,\sin(\theta/2),$$ (10.34)

where θ is the angle subtended between the vectors $\mathbf{k}$ and $\mathbf{k'}$. In other words, θ is the *angle of scattering*. Recall that the vectors $\mathbf{k}$ and $\mathbf{k'}$ have the same length, as a consequence of energy conservation. It follows that, according to the Born approximation, $f(\mathbf{k'}, \mathbf{k}) = f(\theta)$ for a spherically symmetric scattering potential, $V(r)$. Moreover, $f(\theta)$ is real. Finally, the differential scattering cross-section, $d\sigma/d\Omega = |f(\theta)|^2$, is invariant under the transformation $V \to -V$. In other words, the pattern of scattering is identical for attractive and repulsive scattering potentials of the same strength.

Consider scattering by a *Yukawa potential* [Yukawa (1935)],

$$V(r) = \frac{V_0\,\exp(-\mu\,r)}{\mu\,r},$$ (10.35)

where V_0 is a constant, and $1/\mu$ measures the "range" of the potential. It follows from Equation (10.33) that

$$f(\theta) = -\frac{2\,m\,V_0}{\hbar^2\,\mu}\,\frac{1}{q^2 + \mu^2},$$ (10.36)

because

$$\int_0^\infty dr\, \exp(-\mu\,r)\,\sin(q\,r) = \frac{q}{q^2 + \mu^2}.$$ (10.37)

(See Exercise 10.6.) Thus, the Born approximation yields a differential cross-section for scattering by a Yukawa potential of the form

$$\frac{d\sigma}{d\Omega} \simeq \left(\frac{2\,m\,V_0}{\hbar^2\,\mu}\right)^2 \frac{1}{\left[4\,k^2\,\sin^2(\theta/2) + \mu^2\right]^2}.$$ (10.38)

[See Equations (10.28) and (10.34).]

The Yukawa potential reduces to the familiar Coulomb potential in the limit $\mu \to 0$, provided that $V_0/\mu \to Z Z'\, e^2/4\pi\,\epsilon_0$. Here, $Z\,e$ and $Z'\,e$ are the electric charges of the two interacting particles. In the Coulomb limit, the previous Born differential cross-section transforms into

$$\frac{d\sigma}{d\Omega} \simeq \left(\frac{2\,m\,Z Z'\,e^2}{4\pi\,\epsilon_0\,\hbar^2}\right)^2 \frac{1}{16\,k^4\,\sin^4(\theta/2)}.$$ (10.39)

Recalling that $\hbar k$ is equivalent to $|\mathbf{p}|$, where $\mathbf{p}$ is the momentum of the incident particles, the preceding equation can be rewritten

$$\frac{d\sigma}{d\Omega} \simeq \left(\frac{Z Z' e^2}{16\pi \epsilon_0 E} \right)^2 \frac{1}{\sin^4(\theta/2)}, \tag{10.40}$$

where $E = p^2/(2m)$ is the kinetic energy of the incident particles. Equation (10.40) is identical to the well-known *Rutherford scattering cross-section* formula of classical physics [Jackson (1975)].

The Born approximation is valid provided $\psi(\mathbf{x})$ is not significantly different from $\phi(\mathbf{x})$ in the scattering region. It follows, from Equation (10.29), that the condition that must be satisfied in order that $\psi(\mathbf{x}) \simeq \phi(\mathbf{x})$ in the vicinity of $\mathbf{x} = \mathbf{0}$ is

$$\left| \frac{m}{2\pi \hbar^2} \int d^3\mathbf{x}' \, \frac{\exp(i k r')}{r'} V(\mathbf{x}') \right| \ll 1. \tag{10.41}$$

Consider the special case of the Yukawa potential, (10.35). At low energies (i.e., $k \ll \mu$), we can replace $\exp(i k r')$ by unity, giving

$$\frac{2m}{\hbar^2} \frac{|V_0|}{\mu^2} \ll 1 \tag{10.42}$$

as the condition for the validity of the Born approximation. Now, the criterion for the Yukawa potential to develop a bound state turns out to be

$$\frac{2m}{\hbar^2} \frac{|V_0|}{\mu^2} \geq 2.7, \tag{10.43}$$

provided V_0 is negative [Sakurai and Napolitano (2011)]. Thus, if the potential is strong enough to form a bound state then the Born approximation is likely to break down. In the high-k limit (i.e., $k \gg \mu$), Equation (10.41) yields

$$\frac{2m}{\hbar^2} \frac{|V_0|}{\mu k} \ll 1. \tag{10.44}$$

This inequality becomes progressively easier to satisfy as k increases, implying that the Born approximation becomes more accurate at high incident particle energies

10.4 Born Expansion

As we have seen, quantum scattering theory requires the solution of the integral equation (10.29),

$$\psi(\mathbf{x}) = \phi(\mathbf{x}) - \frac{m}{2\pi \hbar^2} \int d^3\mathbf{x}' \, \frac{\exp(i k |\mathbf{x} - \mathbf{x}'|)}{|\mathbf{x} - \mathbf{x}'|} V(\mathbf{x}') \psi(\mathbf{x}'), \tag{10.45}$$

where $\phi(\mathbf{x}) = \exp(\,i\,\mathbf{k} \cdot \mathbf{x})/(2\pi)^{3/2}$ is the incident wavefunction, and $V(\mathbf{x})$ the scattering potential. An obvious approach, in the weak-scattering limit, is to solve the preceding equation via a series of successive approximations. That is,

$$\psi^{(1)}(\mathbf{x}) = \phi(\mathbf{x}) - \frac{m}{2\pi\hbar^2} \int d^3\mathbf{x}' \, \frac{\exp(\,i\,k\,|\mathbf{x} - \mathbf{x}'|)}{|\mathbf{x} - \mathbf{x}'|} \, V(\mathbf{x}')\,\phi(\mathbf{x}'), \tag{10.46}$$

$$\psi^{(2)}(\mathbf{x}) = \phi(\mathbf{x}) - \frac{m}{2\pi\hbar^2} \int d^3\mathbf{x}' \, \frac{\exp(\,i\,k\,|\mathbf{x} - \mathbf{x}'|)}{|\mathbf{x} - \mathbf{x}'|} \, V(\mathbf{x}')\,\psi^{(1)}(\mathbf{x}'), \tag{10.47}$$

$$\psi^{(3)}(\mathbf{x}) = \phi(\mathbf{x}) - \frac{m}{2\pi\hbar^2} \int d^3\mathbf{x}' \, \frac{\exp(\,i\,k\,|\mathbf{x} - \mathbf{x}'|)}{|\mathbf{x} - \mathbf{x}'|} \, V(\mathbf{x}')\,\psi^{(2)}(\mathbf{x}'), \tag{10.48}$$

and so on. Assuming that $V(\mathbf{x})$ is only non-negligible relatively close to the origin, and taking the limit $|\mathbf{x}| \to \infty$, we find that

$$\psi(\mathbf{x}) = \frac{1}{(2\pi)^{3/2}} \left[\exp(\,i\,\mathbf{k} \cdot \mathbf{x}) + \frac{\exp(\,i\,k\,r)}{r} f(\mathbf{k}', \mathbf{k}) \right], \tag{10.49}$$

where

$$f(\mathbf{k}', \mathbf{k}) = f^{(1)}(\mathbf{k}', \mathbf{k}) + f^{(2)}(\mathbf{k}', \mathbf{k}) + f^{(3)}(\mathbf{k}', \mathbf{k}) + \cdots. \tag{10.50}$$

The first two terms in the previous series, which is generally known as the *Born expansion*, are

$$f^{(1)}(\mathbf{k}', \mathbf{k}) = -\frac{m}{2\pi\hbar^2} \int d^3\mathbf{x}' \, e^{\,i\,(\mathbf{k}-\mathbf{k}')\cdot\mathbf{x}'} \, V(\mathbf{x}'), \tag{10.51}$$

$$f^{(2)}(\mathbf{k}', \mathbf{k}) = \left(\frac{m}{2\pi\hbar^2}\right)^2 \int d^3\mathbf{x}' \int d^3\mathbf{x}'' \, e^{\,i\,(\mathbf{k}\cdot\mathbf{x}''-\mathbf{k}'\cdot\mathbf{x}')} \, \frac{e^{\,i\,k\,|\mathbf{x}'-\mathbf{x}''|}}{|\mathbf{x}' - \mathbf{x}''|} \, V(\mathbf{x}')\,V(\mathbf{x}''). \tag{10.52}$$

Of course, we recognize Expression (10.51) as that produced by the Born approximation discussed in the preceding section. In other words, the Born approximation essentially involves truncating the Born expansion after its first term. Incidentally, it can be proved that the Born expansion converges for all k (for a spherically symmetric scattering potential) provided; a) $\int_0^\infty dr\, r\,|V(r)| < \infty$; b) $\int_0^\infty dr\, r^2\,|V(r)| < \infty$; and; c) $-|V(r)|$ is too weak to form a bound state [Bushell (1972)]. Furthermore, the criterion for convergence becomes less stringent at high k [Kohn (1954)].

10.5   Partial Waves

We can assume, without loss of generality, that the incident wavefunction is characterized by a wavevector, $\mathbf{k}$, that is aligned parallel to the z-axis. The scattered wavefunction is characterized by a wavevector, $\mathbf{k}'$, that has the same magnitude as $\mathbf{k}$, but, in general, points in a different direction. The direction of $\mathbf{k}'$ is specified by the polar angle θ (i.e., the angle subtended between the two wavevectors), and an azimuthal angle φ measured about the z-axis. Equations (10.33) and (10.34) strongly suggest that for a spherically symmetric scattering potential [i.e., $V(\mathbf{x}) = V(r)$], the scattering amplitude is a function of θ only: that is,

$$f(\theta, \varphi) = f(\theta). \tag{10.53}$$

Let us assume that this is the case. It follows that neither the incident wavefunction,

$$\phi(\mathbf{x}) = \frac{\exp(i\,k\,z)}{(2\pi)^{3/2}} = \frac{\exp(i\,k\,r\cos\theta)}{(2\pi)^{3/2}} \tag{10.54}$$

[see Equation (10.14)], nor the total wavefunction far from the scattering region,

$$\psi(\mathbf{x}) = \frac{1}{(2\pi)^{3/2}} \left[\exp(i\,k\,r\cos\theta) + \frac{\exp(i\,k\,r)\,f(\theta)}{r} \right] \tag{10.55}$$

[see Equation (10.20)], depend on the azimuthal angle, φ.

Outside the range of the scattering potential, $\phi(\mathbf{x})$ and $\psi(\mathbf{x})$ both satisfy the free-space Schrödinger equation,

$$(\nabla^2 + k^2)\psi = 0. \tag{10.56}$$

Consider the most general solution to this equation that is independent of the azimuthal angle, φ. Separation of variables (in spherical coordinates) yields

$$\psi(r, \theta) = \sum_{l=0,\infty} R_l(r)\,P_l(\cos\theta) \tag{10.57}$$

(See Exercise 10.10.) The Legendre polynomials, $P_l(\cos\theta)$, are related to the associated Legendre functions, $P_{lm}(\cos\theta)$, as well as the spherical harmonics, $Y_{lm}(\theta, \varphi)$, introduced in Section 4.4, via $P_l(\cos\theta) = P_{l0}(\cos\theta)$, and

$$P_l(\cos\theta) = \sqrt{\frac{4\pi}{2\,l+1}}\,Y_{l0}(\theta, \varphi), \tag{10.58}$$

respectively. Equations (10.56) and (10.57) can be combined to give

$$r^2\frac{d^2R_l}{dr^2} + 2\,r\,\frac{dR_l}{dr} + [k^2\,r^2 - l(l+1)]\,R_l = 0. \tag{10.59}$$

(See Exercise 10.10.) The two independent solutions to this equation are the *spherical Bessel function*, $j_l(k\,r)$, and the *Neumann function*, $\eta_l(k\,r)$, where

$$j_l(y) = y^l\left(-\frac{1}{y}\frac{d}{dy}\right)^l\frac{\sin y}{y}, \tag{10.60}$$

$$\eta_l(y) = -y^l\left(-\frac{1}{y}\frac{d}{dy}\right)^l\frac{\cos y}{y} \tag{10.61}$$

[Abramowitz and Stegun (1965)]. Note that spherical Bessel functions are well behaved in the limit $y \to 0$, whereas Neumann functions become singular. The asymptotic behavior of these functions in the limit $y \to \infty$ is

$$j_l(y) \to \frac{\sin(y - l\pi/2)}{y}, \tag{10.62}$$

$$\eta_l(y) \to -\frac{\cos(y - l\pi/2)}{y} \tag{10.63}$$

[Abramowitz and Stegun (1965)].

We can write

$$\exp(\mathrm{i}\,k\,r\cos\theta) = \sum_{l=0,\infty} a_l\, j_l(k\,r)\, P_l(\cos\theta), \tag{10.64}$$

where the a_l are constants. Of course, there are no Neumann functions in this expansion because they are not well behaved as $r \to 0$ (whereas the function on the left-hand side is clearly finite at $r = 0$). As is well known, the Legendre polynomials are orthogonal functions,

$$\int_{-1}^{1} d\mu\, P_n(\mu)\, P_m(\mu) = \frac{\delta_{nm}}{n + 1/2} \tag{10.65}$$

[Abramowitz and Stegun (1965)], so we can invert the preceding expansion to give

$$a_l\, j_l(k\,r) = (l + 1/2) \int_{-1}^{1} d\mu\, \exp(\mathrm{i}\,k\,r\,\mu)\, P_l(\mu). \tag{10.66}$$

Now,

$$j_l(y) = \frac{(-\mathrm{i})^l}{2} \int_{-1}^{1} d\mu\, \exp(\mathrm{i}\,y\,\mu)\, P_l(\mu), \tag{10.67}$$

for $l = 0, \infty$ [Abramowitz and Stegun (1965)]. Thus, a comparison of the previous two equations yields

$$a_l = \mathrm{i}^l\,(2\,l + 1), \tag{10.68}$$

giving

$$\exp(\mathrm{i}\,k\,r\cos\theta) = \sum_{l=0,\infty} \mathrm{i}^l\,(2\,l + 1)\, j_l(k\,r)\, P_l(\cos\theta). \tag{10.69}$$

The preceding expression specifies how a plane wave can be decomposed into a series of spherical waves. The latter waves are usually referred to as *partial waves*.

The most general expression for the total wavefunction outside the scattering region is

$$\psi(\mathbf{x}) = \frac{1}{(2\pi)^{3/2}} \sum_{l=0,\infty} \left[A_l\, j_l(k\,r) + B_l\, \eta_l(k\,r) \right] P_l(\cos\theta), \tag{10.70}$$

where the A_l and B_l are constants. Note that the Neumann functions are allowed to appear in this expansion, because its region of validity does not include the origin. In the large-r limit, the total wavefunction reduces to

$$\psi(\mathbf{x}) \simeq \frac{1}{(2\pi)^{3/2}} \sum_{l=0,\infty} \left[A_l \frac{\sin(kr - l\pi/2)}{kr} - B_l \frac{\cos(kr - l\pi/2)}{kr} \right] P_l(\cos\theta), \quad (10.71)$$

where use has been made of Equations (10.62) and (10.63). The previous expression can also be written

$$\psi(\mathbf{x}) \simeq \frac{1}{(2\pi)^{3/2}} \sum_{l=0,\infty} C_l \frac{\sin(kr - l\pi/2 + \delta_l)}{kr} P_l(\cos\theta), \quad (10.72)$$

where

$$A_l = C_l \cos\delta_l, \quad (10.73)$$

$$B_l = -C_l \sin\delta_l. \quad (10.74)$$

Equation (10.72) yields

$$\psi(\mathbf{x}) \simeq \frac{1}{(2\pi)^{3/2}} \sum_{l=0,\infty} C_l \left[\frac{e^{i(kr-l\pi/2+\delta_l)} - e^{-i(kr-l\pi/2+\delta_l)}}{2ikr} \right] P_l(\cos\theta), \quad (10.75)$$

which contains both incoming and outgoing spherical waves. What is the source of the incoming waves? Obviously, they must form part of the large-r asymptotic expansion of the incident wavefunction. In fact, it is easily seen from Equations (10.54), (10.62), and (10.69) that

$$\phi(\mathbf{x}) \simeq \frac{1}{(2\pi)^{3/2}} \sum_{l=0,\infty} i^l (2l+1) \left[\frac{e^{i(kr-l\pi/2)} - e^{-i(kr-l\pi/2)}}{2ikr} \right] P_l(\cos\theta), \quad (10.76)$$

in the large-r limit. Now, Equations (10.54) and (10.55) give

$$(2\pi)^{3/2} [\psi(\mathbf{x}) - \phi(\mathbf{x})] = \frac{\exp(ikr)}{r} f(\theta). \quad (10.77)$$

Note that the right-hand side consists only of an outgoing spherical wave. This implies that the coefficients of the incoming spherical waves in the large-r expansions of $\psi(\mathbf{x})$ and $\phi(\mathbf{x})$ must be equal. It follows from Equations (10.75) and (10.76) that

$$C_l = (2l+1) \exp[i(\delta_l + l\pi/2)], \quad (10.78)$$

which leads to

$$\phi(\dot{\mathbf{x}}) = \frac{1}{(2\pi)^{3/2}} \sum_{l=0,\infty} i^l (2l+1) \frac{\sin(kr - l\pi/2)}{kr} P_l(\cos\theta), \quad (10.79)$$

$$\psi(\mathbf{x}) = \frac{1}{(2\pi)^{3/2}} \sum_{l=0,\infty} i^l (2l+1) e^{i\delta_l} \frac{\sin(kr - l\pi/2 + \delta_l)}{kr} P_l(\cos\theta). \quad (10.80)$$

Thus, it is apparent that the effect of the scattering is to introduce a phase-shift, δ_l, into the lth partial wave. Finally, Equation (10.77) yields

$$f(\theta) = \sum_{l=0,\infty} (2l+1) \frac{\exp(i\delta_l)}{k} \sin \delta_l P_l(\cos\theta) \qquad (10.81)$$

[Faxen and Holtzmark (1927)]. Clearly, determining the scattering amplitude, $f(\theta)$, via a decomposition into partial waves (i.e., spherical waves), is equivalent to determining the phase-shifts, δ_l.

It is helpful to write

$$\phi(\mathbf{r}) = \sum_{l=0,\infty} \left[\phi_l^+(r,\theta) + \phi_l^-(r,\theta) \right], \qquad (10.82)$$

$$\psi(\mathbf{r}) = \sum_{l=0,\infty} \left[S_l\, \phi_l^+(r,\theta) + \phi_l^-(r,\theta) \right], \qquad (10.83)$$

where

$$\phi_l^-(r,\theta) = -\frac{(2l+1)}{(2\pi)^{3/2}} \frac{e^{-i(kr-l\pi)}}{2ikr} P_l(\cos\theta) \qquad (10.84)$$

is an ingoing spherical wave, whereas

$$\phi_l^+(r,\theta) = \frac{(2l+1)}{(2\pi)^{3/2}} \frac{e^{ikr}}{2ikr} P_l(\cos\theta) \qquad (10.85)$$

is an outgoing spherical wave. Moreover,

$$S_l = e^{i2\delta_l}. \qquad (10.86)$$

[See Equations (10.79) and (10.80).] Note that $\phi_l^-(r,\theta)$ and $\phi_l^+(r,\theta)$ are both eigenstates of the magnitude of the total orbital angular momentum about the origin belonging to the eigenvalues $\sqrt{l(l+1)}\,\hbar$. (See Chapter 4.) Thus, in preforming a partial wave expansion, we have effectively separated the incoming and outgoing particles into streams possessing definite angular momenta about the origin. Moreover, the effect of the scattering is to introduce an angular-momentum-dependent phase-shift into the outgoing particle streams.

The net outward particle flux through a sphere of radius r, centered on the origin, is proportional to

$$\oint d\Omega\, r^2 j_r, \qquad (10.87)$$

where $\mathbf{j} = (\hbar/m)\,\mathrm{Im}(\psi^*\,\nabla\psi)$ is the probability current. It follows that

$$\oint d\Omega\, r^2 j_r = \frac{\hbar}{8\pi^2 km} \sum_{l=0,\infty} (2l+1)\,(|S_l|^2 - 1), \qquad (10.88)$$

where use has been made of Equation (10.65). Of course, the net particle flux must be zero, otherwise the number of particles would not be conserved. Particle conservation is ensured by the fact that $|S_l| = 1$ for all l. [See Equation (10.86).]

10.6 Optical Theorem

The differential scattering cross-section, $d\sigma/d\Omega$, is simply the modulus squared of the scattering amplitude, $f(\theta)$. [See Equation (10.28).] The *total scattering cross-section* is defined as

$$\sigma_{\text{total}} = \oint d\Omega \, \frac{d\sigma}{d\Omega} = \oint d\Omega \, |f(\theta)|^2$$

$$= \frac{1}{k^2} \oint d\varphi \int_{-1}^{1} d\mu \sum_{l=0,\infty} \sum_{l'=0,\infty} (2\,l+1)\,(2\,l'+1)\exp[\,\mathrm{i}\,(\delta_l - \delta_{l'})]$$

$$\sin \delta_l \, \sin \delta_{l'} \, P_l(\mu) \, P_{l'}(\mu), \quad (10.89)$$

where $\mu = \cos \theta$. It follows that

$$\sigma_{\text{total}} = \frac{4\pi}{k^2} \sum_{l=0,\infty} (2\,l+1)\,\sin^2 \delta_l, \quad (10.90)$$

where use has been made of Equation (10.65). A comparison of the preceding expression with Equation (10.81) reveals that

$$\sigma_{\text{total}} = \frac{4\pi}{k} \,\mathrm{Im}\,[f(0)] = \frac{4\pi}{k} \,\mathrm{Im}\,[f(\mathbf{k}, \mathbf{k})], \quad (10.91)$$

because $P_l(1) = 1$ [Abramowitz and Stegun (1965)]. This result is known as the *optical theorem* [Strutt (1871); Mie (1908)], and is a consequence of the fact that the very existence of scattering requires scattering in the forward ($\theta = 0$) direction, in order to interfere with the incident wave, and thereby reduce the probability current in that direction.

It is conventional to write

$$\sigma_{\text{total}} = \sum_{l=0,\infty} \sigma_l, \quad (10.92)$$

where

$$\sigma_l = \frac{4\pi}{k^2} \,(2\,l+1)\,\sin^2 \delta_l \quad (10.93)$$

is termed the lth *partial scattering cross-section*: that is, the contribution to the total scattering cross-section from the lth partial wave. Note that (at fixed k) the maximum value for the lth partial scattering cross-section occurs when the associated phase-shift, δ_l, takes the value $\pi/2$.

10.7   Determination of Phase-Shifts

Let us now consider how the partial wave phase-shifts, δ_l, can be evaluated. Consider a spherically symmetric potential, $V(r)$, that vanishes for $r > a$, where a is termed the range of the potential. In the region $r > a$, the wavefunction $\psi(\mathbf{x})$ satisfies the free-space Schrödinger equation, (10.56). According to Equations (10.70), (10.73), (10.74), and (10.78), the most general solution of this equation that is consistent with no incoming spherical waves, other than those contained in the incident wave, is

$$\psi(\mathbf{x}) = \frac{1}{(2\pi)^{3/2}} \sum_{l=0,\infty} i^l \,(2\,l+1)\, A_l(r)\, P_l(\cos\theta), \qquad (10.94)$$

where

$$A_l(r) = \exp(\mathrm{i}\,\delta_l)\,\left[\cos\delta_l\, j_l(k\,r) - \sin\delta_l\, \eta_l(k\,r)\right]. \qquad (10.95)$$

Note that Neumann functions are allowed to appear in the previous expression, because its region of validity does not include the origin (where $V \neq 0$). The logarithmic derivative of the lth radial wavefunction, $A_l(r)$, just outside the range of the potential is given by

$$\beta_{l+} = k\,a \left[\frac{\cos\delta_l\, j_l'(k\,a) - \sin\delta_l\, \eta_l'(k\,a)}{\cos\delta_l\, j_l(k\,a) - \sin\delta_l\, \eta_l(k\,a)} \right], \qquad (10.96)$$

where $j_l'(x)$ denotes $d\,j_l(x)/dx$, et cetera. The previous equation can be inverted to give

$$\tan\delta_l = \frac{k\,a\,j_l'(k\,a) - \beta_{l+}\, j_l(k\,a)}{k\,a\,\eta_l'(k\,a) - \beta_{l+}\, \eta_l(k\,a)}. \qquad (10.97)$$

Thus, the problem of determining the phase-shift, δ_l, is equivalent to that of determining β_{l+}.

The most general solution to Schrödinger's equation inside the range of the potential ($r < a$) that does not depend on the azimuthal angle, φ, is

$$\psi(\mathbf{x}) = \frac{1}{(2\pi)^{3/2}} \sum_{l=0,\infty} i^l \,(2\,l+1)\, A_l(r)\, P_l(\cos\theta), \qquad (10.98)$$

where

$$A_l(r) = \frac{u_l(r)}{r}, \qquad (10.99)$$

and

$$\frac{d^2 u_l}{dr^2} + \left[k^2 - \frac{2\,m}{\hbar^2}\, V - \frac{l(l+1)}{r^2} \right] u_l = 0. \qquad (10.100)$$

(See Exercise 10.12.) The boundary condition

$$u_l(0) = 0 \qquad (10.101)$$

ensures that the radial wavefunction is well behaved at the origin. We can launch a well-behaved solution of the previous equation from $r = 0$, integrate out to $r = a$, and form the logarithmic derivative [of $A_l(r)$]

$$\beta_{l-} = \frac{1}{(u_l/r)} \frac{d(u_l/r)}{dr}\bigg|_{r=a}. \qquad (10.102)$$

Because $\psi(\mathbf{x})$ and its first derivatives are necessarily continuous for physically acceptable wavefunctions, it follows that

$$\beta_{l+} = \beta_{l-}. \qquad (10.103)$$

The phase-shift, δ_l, is then obtained from Equation (10.97).

10.8 Hard-Sphere Scattering

Let us try out the scheme outlined in the previous section using a particularly simple example. Consider scattering by a hard sphere, for which the potential is infinite for $r < a$, and zero for $r > a$. It follows that $\psi(\mathbf{x})$ is zero in the region $r < a$, which implies that $u_l = 0$ for all l. Thus,

$$\beta_{l-} = \beta_{l+} = \infty \qquad (10.104)$$

for all l. Equation (10.97) yields

$$\tan \delta_l = \frac{j_l(ka)}{\eta_l(ka)}. \qquad (10.105)$$

In fact, this result is most easily obtained from the obvious requirement that $A_l(a) = 0$. [See Equation (10.95).]

Consider the $l = 0$ partial wave, which is usually referred to as the *S-wave*. Equation (10.105) gives

$$\tan \delta_0 = \frac{\sin(ka)/ka}{-\cos(ka)/ka} = -\tan(ka), \qquad (10.106)$$

where use has been made of Equations (10.60) and (10.61). It follows that

$$\delta_0 = -ka. \qquad (10.107)$$

The S-wave radial wave function is

$$A_0(r) = \exp(-ika)\left[\frac{\cos(ka)\sin(kr) - \sin(ka)\cos(kr)}{kr}\right]$$

$$= \exp(-ika)\frac{\sin[k(r-a)]}{kr}. \qquad (10.108)$$

[See Equation (10.95).] The corresponding radial wavefunction for the incident wave takes the form

$$\tilde{A}_0(r) = \frac{\sin(kr)}{kr}. \tag{10.109}$$

[See Equations (10.79), (10.80), (10.94), and (10.107).] It is clear that the actual $l = 0$ radial wavefunction is similar to the incident $l = 0$ wavefunction, except that it is phase-shifted by ka.

Let us consider the low- and high-energy asymptotic limits of $\tan \delta_l$. Low energy corresponds to $ka \ll 1$. In this limit, the spherical Bessel functions and Neumann functions reduce to

$$j_l(kr) \simeq \frac{(kr)^l}{(2l+1)!!}, \tag{10.110}$$

$$\eta_l(kr) \simeq -\frac{(2l-1)!!}{(kr)^{l+1}}, \tag{10.111}$$

where $n!! = n(n-2)(n-4)\cdots 1$ [Abramowitz and Stegun (1965)]. It follows that

$$\tan \delta_l = \frac{-(ka)^{2l+1}}{(2l+1)[(2l-1)!!]^2}. \tag{10.112}$$

It is clear that we can neglect δ_l, with $l > 0$, with respect to δ_0. In other words, at low energy, only S-wave scattering (i.e., spherically symmetric scattering) is important. It follows from Equations (10.28), (10.81), and (10.107) that

$$\frac{d\sigma}{d\Omega} = \frac{\sin^2(ka)}{k^2} \simeq a^2 \tag{10.113}$$

for $ka \ll 1$. Note that the total cross-section,

$$\sigma_{\text{total}} = \oint d\Omega \, \frac{d\sigma}{d\Omega} = 4\pi a^2, \tag{10.114}$$

is four times the geometric cross-section, πa^2 (i.e., the cross-section for classical particles bouncing off a hard sphere of radius a). However, low-energy scattering implies relatively long de Broglie wavelengths, so we would not expect to obtain the classical result in this limit.

Consider the high-energy limit, $ka \gg 1$. At high energies, by analogy with classical scattering, the scattered particles with the largest angular momenta about the origin have angular momenta $\hbar ka$ (i.e., the product of their incident momenta, $\hbar k$, and their maximum possible impact parameters, a). Given that particles in the lth partial wave have angular momenta $\sqrt{l(l+1)}\,\hbar$, we deduce that all partial waves up to $l_{\text{max}} \simeq ka$ contribute significantly to the scattering cross-section. It follows from Equation (10.90) that

$$\sigma_{\text{total}} = \frac{4\pi}{k^2} \sum_{l=0,l_{\text{max}}} (2l+1)\sin^2 \delta_l. \tag{10.115}$$

Making use of Equation (10.105), as well as the asymptotic expansions (10.62) and (10.63), we find that

$$\sin^2 \delta_l = \frac{\tan^2 \delta_l}{1 + \tan^2 \delta_l} = \frac{j_l^2(k\,a)}{j_l^2(k\,a) + \eta_l^2(k\,a)} = \sin^2(k\,a - l\,\pi/2). \tag{10.116}$$

In particular,

$$\sin^2 \delta_l + \sin^2 \delta_{l+1} = \sin^2(k\,a - l\,\pi/2) + \cos^2(k\,a - l\,\pi/2) = 1. \tag{10.117}$$

Hence, it is a good approximation to write

$$\sigma_{\text{total}} \simeq \frac{2\pi}{k^2} \sum_{l=0,l_{\max}} (2\,l+1) = \frac{2\pi}{k^2}\,(l_{\max} + 1)^2 \simeq 2\pi\,a^2 \tag{10.118}$$

[Gradshteyn *et al.* (1980)]. This is twice the classical result, which is somewhat surprising, because we might expect to obtain the classical result in the short-wavelength limit. In fact, for hard-sphere scattering, all incident particles with impact parameters less than a are deflected. However, in order to produce a shadow behind the sphere, there must be scattering in the forward direction (recall the optical theorem) to produce destructive interference with the incident plane wave. The effective cross-section associated with this forward scattering is $\pi\,a^2$, which, when combined with the cross-section for classical reflection, $\pi\,a^2$, gives the actual cross-section of $2\pi\,a^2$ [Sakurai and Napolitano (2011)].

10.9 Low-Energy Scattering

In general, at low energies (i.e., when $1/k$ is much larger than the range of the potential), partial waves with $l > 0$ make a negligible contribution to the scattering cross-section. It follows that, with a finite-range potential, only S-wave (i.e., spherically symmetric) scattering is important at such energies.

As a specific example, let us consider scattering by a finite potential well, characterized by $V = V_0$ for $r < a$, and $V = 0$ for $r \geq a$. Here, V_0 is a constant. The potential is repulsive for $V_0 > 0$, and attractive for $V_0 < 0$. The external wavefunction is given by [see Equation (10.95)]

$$A_0(r) = \exp(i\,\delta_0)\,[j_0(k\,r)\cos\delta_0 - \eta_0(k\,r)\sin\delta_0]$$

$$= \frac{\exp(i\,\delta_0)\,\sin(k\,r + \delta_0)}{k\,r}, \tag{10.119}$$

where use has been made of Equations (10.60) and (10.61). The internal wavefunction follows from Equation (10.100). We obtain

$$A_0(r) = B\,\frac{\sin(k'\,r)}{r}, \tag{10.120}$$

where use has been made of the boundary condition (10.101). Here, B is a constant, and

$$E - V_0 = \frac{\hbar^2 k'^2}{2m}. \tag{10.121}$$

Note that Equation (10.120) only applies when $E > V_0$. For $E < V_0$, we have

$$A_0(r) = B \frac{\sinh(\kappa r)}{r}, \tag{10.122}$$

where

$$V_0 - E = \frac{\hbar^2 \kappa^2}{2m}. \tag{10.123}$$

Matching $A_0(r)$, and its radial derivative, at $r = a$ yields

$$\tan(k a + \delta_0) = \frac{k}{k'} \tan(k' a) \tag{10.124}$$

for $E > V_0$, and

$$\tan(k a + \delta_0) = \frac{k}{\kappa} \tanh(\kappa a) \tag{10.125}$$

for $E < V_0$.

Consider an attractive potential, for which $E > V_0$. Suppose that $|V_0| \gg E$ (i.e., the depth of the potential well is much larger than the energy of the incident particles), so that $k' \gg k$. As can be seen from Equation (10.124), unless $\tan(k' a)$ becomes extremely large, the right-hand side of the equation is much less than unity, so replacing the tangent of a small quantity with the quantity itself, we obtain

$$k a + \delta_0 \simeq \frac{k}{k'} \tan(k' a). \tag{10.126}$$

This yields

$$\delta_0 \simeq k a \left[\frac{\tan(k' a)}{k' a} - 1 \right]. \tag{10.127}$$

According to Equation (10.115), the total scattering cross-section is given by

$$\sigma_{\text{total}} \simeq \frac{4\pi}{k^2} \sin^2 \delta_0 = 4\pi a^2 \left[\frac{\tan(k' a)}{k' a} - 1 \right]^2. \tag{10.128}$$

Now,

$$k' a = \sqrt{k^2 a^2 + \frac{2m |V_0| a^2}{\hbar^2}}, \tag{10.129}$$

so for sufficiently small values of $k a$,

$$k' a \simeq \sqrt{\frac{2m |V_0| a^2}{\hbar^2}}. \tag{10.130}$$

It follows that the total (S-wave) scattering cross-section is independent of the energy of the incident particles (provided that this energy is sufficiently small).

Note that there are values of $k'a$ (e.g., $k'a \simeq 4.493$) at which the scattering cross-section (10.128) vanishes, despite the very strong attraction of the potential. In reality, the cross-section is not exactly zero, because of contributions from $l > 0$ partial waves. But, at low incident energies, these contributions are small. It follows that there are certain values of $|V_0|$, a, and k that give rise to almost perfect transmission of the incident wave. This is called the *Ramsauer-Townsend effect*, and has been observed experimentally [Ramsauer (1921); Bailey and Townsend (1921)].

10.10   Resonant Scattering

There is a significant exception to the energy independence of the scattering cross-section at low incident energies described in the previous section. Suppose that the quantity $(2\,m\,|V_0|\,a^2/\hbar^2)^{1/2}$ is slightly less than $\pi/2$. As the incident energy increases, $k'a$, which is given by Equation (10.129), can reach the value $\pi/2$. In this case, $\tan(k'a)$ becomes infinite, so we can no longer assume that the right-hand side of Equation (10.124) is small. In fact, at the value of the incident energy at which $k'a = \pi/2$, it follows from Equation (10.124) that $ka + \delta_0 = \pi/2$, or $\delta_0 \simeq \pi/2$ (because we are assuming that $ka \ll 1$). This implies that

$$\sigma_{\text{total}} = \frac{4\pi}{k^2}\sin^2\delta_0 = 4\pi\,a^2\left(\frac{1}{k^2\,a^2}\right). \qquad (10.131)$$

Note that the total scattering cross-section now depends on the energy. Furthermore, the magnitude of the cross-section is much larger than that given in Equation (10.128) for $k'a \neq \pi/2$ (because $ka \ll 1$, whereas $k'a \sim 1$).

The origin of the rather strange behavior described in the previous paragraph is easily explained. The condition

$$\sqrt{\frac{2\,m\,|V_0|\,a^2}{\hbar^2}} = \frac{\pi}{2} \qquad (10.132)$$

is equivalent to the condition that a spherical well of depth $|V_0|$ possesses a bound state at zero energy. Thus, for a potential well that satisfies the preceding equation, the energy of the scattering system is essentially the same as the energy of the bound state. In this situation, an incident particle would like to form a bound state in the potential well. However, the bound state is not stable, because the system has a small positive energy. Nevertheless, this sort of *resonant scattering* is best understood as the capture of an incident particle to form a metastable bound state, followed by the decay of the bound state and release of the particle. The cross-section for resonant scattering is generally far higher than that for non-resonant scattering.

We have seen that there is a resonant effect when the phase-shift of the S-wave takes the value $\pi/2$. There is nothing special about the $l = 0$ partial wave, so it is reasonable to assume that there is a similar resonance when the phase-shift of the lth partial wave is $\pi/2$. Suppose that δ_l attains the value $\pi/2$ at the incident energy E_0, so that

$$\delta_l(E_0) = \frac{\pi}{2}. \tag{10.133}$$

Let us expand $\cot \delta_l$ in the vicinity of the resonant energy:

$$\cot \delta_l(E) = \cot \delta_l(E_0) + \left(\frac{d \cot \delta_l}{dE} \right)_{E=E_0} (E - E_0) + \cdots$$

$$= -\left(\frac{1}{\sin^2 \delta_l} \frac{d\delta_l}{dE} \right)_{E=E_0} (E - E_0) + \cdots. \tag{10.134}$$

Defining

$$\left[\frac{d\delta_l(E)}{dE} \right]_{E=E_0} = \frac{2}{\Gamma}, \tag{10.135}$$

we obtain

$$\cot \delta_l(E) = -\frac{2}{\Gamma} (E - E_0) + \cdots. \tag{10.136}$$

Recall, from Equation (10.93), that the contribution of the lth partial wave to the total scattering cross-section is

$$\sigma_l = \frac{4\pi}{k^2} (2l + 1) \sin^2 \delta_l = \frac{4\pi}{k^2} (2l + 1) \frac{1}{1 + \cot^2 \delta_l}. \tag{10.137}$$

Thus,

$$\sigma_l \simeq \frac{4\pi}{k^2} (2l + 1) \frac{\Gamma^2/4}{(E - E_0)^2 + \Gamma^2/4}, \tag{10.138}$$

which is known as the *Breit-Wigner formula* [Breit and Wigner (1936)]. The variation of the partial cross-section, σ_l, with the incident energy has the form of a classical resonance curve. The quantity Γ is the width of the resonance (in energy). We can interpret the Breit-Wigner formula as describing the absorption of an incident particle to form a metastable state, of energy E_0, and lifetime $\tau = \hbar/\Gamma$.

10.11 Elastic and Inelastic Scattering

According to the analysis of Section 10.5, for the case of a spherically symmetric scattering potential, the scattered wave is characterized by

$$f(\theta) = \sum_{l=0,\infty} (2\,l+1)\, f_l\, P_l(\cos\theta), \tag{10.139}$$

where

$$f_l = \frac{\exp(i\,\delta_l)}{k}\, \sin\delta_l = \frac{S_l - 1}{2\,i\,k} \tag{10.140}$$

is the amplitude of the *l*th partial wave, whereas δ_l is the associated phase-shift. Here,

$$S_l = e^{i\,2\,\delta_l}. \tag{10.141}$$

Moreover, the fact that $|S_l| = 1$ ensures that the scattering is *elastic* (i.e., that the number of particles is conserved). Finally, according to the analysis of Section 10.6, the net elastic scattering cross-section can be written

$$\sigma_{\text{elastic}} = \frac{4\pi}{k^2} \sum_{l=0,\infty} (2\,l+1)\, \sin^2\delta_l = 4\pi \sum_{l=0,\infty} (2\,l+1)\, |f_l|^2. \tag{10.142}$$

It turns out that many scattering experiments are characterized by the absorption of some of the incident particles. Such absorption may induce a change in the quantum state of the target, or, perhaps, the emergence of another particle. Note that scattering that does not conserve particle number is known as *inelastic scattering*. We can take inelastic scattering into account in our analysis by writing

$$S_l = \eta_l\, e^{i\,2\,\delta_l}, \tag{10.143}$$

where the real parameter η_l is such that

$$0 \leq \eta_l \leq 1. \tag{10.144}$$

It follows from Equation (10.140) that

$$f_l = \frac{\eta_l\, \sin(2\,\delta_l)}{2\,k} + i\left[\frac{1 - \eta_l\,\cos(2\,\delta_l)}{2\,k}\right]. \tag{10.145}$$

Hence, according to Equation (10.142), the net elastic scattering cross-section becomes

$$\sigma_{\text{elastic}} = 4\pi \sum_{l=0,\infty} (2\,l+1)\, |f_l|^2$$

$$= \frac{\pi}{k^2} \sum_{l=0,\infty} (2\,l+1) \left[1 + \eta_l^2 - 2\,\eta_l\,\cos(2\,\delta_l)\right]. \tag{10.146}$$

The net inelastic scattering (i.e., absorption) cross-section follows from Equations (10.24) and (10.88):

$$\sigma_{\text{inelastic}} = \frac{\oint d\Omega \, r^2 \, (-j_r)}{|\mathbf{j}_{\text{incident}}|} = \frac{\pi}{k^2} \sum_{l=0,\infty} (2\,l+1)\left(1 - |S_l|^2\right)$$

$$= \frac{\pi}{k^2} \sum_{l=0,\infty} (2\,l+1)\left(1 - \eta_l^2\right). \tag{10.147}$$

Thus, the total cross-section is

$$\sigma_{\text{total}} = \sigma_{\text{elastic}} + \sigma_{\text{inelastic}}$$

$$= \frac{2\pi}{k^2} \sum_{l=0,\infty} (2\,l+1)\left[1 - \eta_l \, \cos(2\,\delta_l)\right]. \tag{10.148}$$

Note, from Equations (10.139), (10.140), and (10.143) that

$$\text{Im}[f(0)] = \frac{1}{2\,k} \sum_{l=0,\infty} (2\,l+1)\left[1 - \eta_l \, \cos(2\,\delta_l)\right]. \tag{10.149}$$

In other words,

$$\sigma_{\text{total}} = \frac{4\pi}{k} \, \text{Im}[f(0)]. \tag{10.150}$$

Hence, we deduce that the optical theorem, described in Section 10.6, still applies in the presence of inelastic scattering.

If $\eta_l = 1$ then there is no absorption, and the lth partial wave is scattered in a completely elastic manner. On the other hand, if $\eta_l = 0$ then there is total absorption of the lth partial wave. However, such absorption is necessarily accompanied by some degree of elastic scattering. In order to illustrate this important point, let us investigate the special case of scattering by a *black sphere*. Such a sphere has a well-defined edge of radius a, and is completely absorbing. Consider short-wavelength scattering characterized by $k\,a \gg 1$. In this case, we expect all partial waves with $l \leq l_{\text{max}}$, where $l_{\text{max}} \simeq k\,a$, to be completely absorbed (because, by analogy with classical physics, the impact parameters of the associated particles are less than a—see Section 10.8), and all other partial waves to suffer neither absorption nor scattering. In other words, $\eta_l = 0$ for $0 \leq l_{\text{max}}$, and $\eta_l = 1$, $\delta_l = 0$ for $l > l_{\text{max}}$. It follows from Equations (10.146) and (10.147) that

$$\sigma_{\text{elastic}} = \frac{\pi}{k^2} \sum_{l=0,l_{\text{max}}} (2\,l+1) = \frac{\pi}{k^2} \, (1 + l_{\text{max}})^2 \simeq \pi\,a^2, \tag{10.151}$$

and

$$\sigma_{\text{inelastic}} = \frac{\pi}{k^2} \sum_{l=0,l_{\text{max}}} (2\,l+1) = \frac{\pi}{k^2} \, (1 + l_{\text{max}})^2 \simeq \pi\,a^2 \tag{10.152}$$

[Gradshteyn *et al.* (1980)]. Thus, the total scattering cross-section is

$$\sigma_{\text{total}} = \sigma_{\text{elastic}} + \sigma_{\text{inelastic}} = 2\pi a^2. \tag{10.153}$$

This result seems a little strange, at first, because, by analogy with classical physics, we would not expect the total cross-section to exceed the cross-section presented by the sphere. Nor would we expect a totally absorbing sphere to give rise to any elastic scattering. In fact, this reasoning is incorrect. The absorbing sphere removes flux proportional to πa^2 from the incident wave, which leads to the formation of a shadow behind the sphere. However, a long way from the sphere, the shadow gets filled in. In other words, the shadow is not visible infinitely far downstream of the sphere. The only way in which this can occur is via the diffraction of some of the incident wave around the edges of the sphere. Actually, the amount of the incident wave that must be diffracted is the same amount as was removed from the wave by absorption. Thus, the scattered flux is also proportional to πa^2 [Gasiorowicz (1996)].

Consider low-energy scattering by a hard-sphere potential. As we saw in Section 10.8, this process is dominated by S-wave (i.e., $l = 0$) scattering. Moreover, the phase-shift of the S-wave takes the form

$$\delta_0 = -k\,a, \tag{10.154}$$

where k is the wavenumber of the incident particles, and a is the radius of the sphere. Note that the low-energy limit corresponds to $k\,a \ll 1$. It follows that

$$S_0 = e^{i2\,\delta_0} \simeq 1 - 2\,i\,k\,a. \tag{10.155}$$

We can generalize the previous analysis to take absorption into account by writing

$$S_0 \simeq 1 - 2\,i\,k\,\alpha, \tag{10.156}$$

where α is complex, $k\,|\alpha| \ll 1$, and $\text{Im}(\alpha) < 0$. According to Equations (10.146) and (10.147),

$$\sigma_{\text{elastic}} \simeq \frac{\pi}{k^2}\,|S_0 - 1|^2 \simeq 4\pi\,|\alpha|^2, \tag{10.157}$$

$$\sigma_{\text{inelastic}} \simeq \frac{\pi}{k^2}\left(1 - |S_0|^2\right) \simeq \frac{4\pi\,\text{Im}(-\alpha)}{k}. \tag{10.158}$$

We conclude that the low-energy elastic scattering cross-section is again independent of the incident particle velocity (which is proportional to k), whereas the inelastic cross-section is inversely proportional to the particle velocity. Consequently, as the incident particle velocity decreases, inelastic scattering becomes more and more important in comparison with elastic scattering [Bethe (1935)].

10.12   Scattering of Identical Particles

Consider two identical particles that scatter off one another. In the center of mass frame, there is no way of distinguishing a deflection of a particle through an angle θ, and a deflection through an angle $\pi - \theta$, because momentum conservation demands that if one of the particles is scattered in the direction characterized by the angle θ then the other is scattered in the direction characterized by $\pi - \theta$ [Goldstein *et al.* (2002)]. (Here, for the sake of simplicity, we are assuming that the scattering potential is spherically symmetric, which implies that the motion of the two particles is confined to a fixed plane passing through the origin.)

In classical mechanics, the differential cross-section for scattering is affected by the identity of the particles because the number of particles counted by a detector located at angular position θ is the sum of the counts due to the two particles, which implies that

$$\frac{d\sigma_{\text{classical}}}{d\Omega} = \frac{d\sigma(\theta)}{d\Omega} + \frac{d\sigma(\pi - \theta)}{d\Omega} = |f(\theta)|^2 + |f(\pi - \theta)|^2 \,. \tag{10.159}$$

[See Equation (10.28).]

In quantum mechanics, the overall wavefunction must be either symmetric or antisymmetric under interchange of identical particles, depending on whether the particles in question are bosons or fermions, respectively. (See Section 9.3.) If the spatial wavefunction is symmetric then Equation (10.55) is replaced by

$$\psi(\mathbf{x}) = \frac{1}{(2\pi)^{3/2}} \left(e^{i k r \cos\theta} + e^{-i k r \cos\theta} + \frac{e^{ikr}}{r} \left[f(\theta) + f(\pi - \theta) \right] \right), \tag{10.160}$$

and the associated differential scattering cross-section becomes

$$\frac{d\sigma}{d\Omega} = |f(\theta) + f(\pi - \theta)|^2 \,. \tag{10.161}$$

[See Equation (10.28).] On the other hand, if the spatial wavefunction is antisymmetric then Equation (10.55) is replaced by

$$\psi(\mathbf{x}) = \frac{1}{(2\pi)^{3/2}} \left(e^{i k r \cos\theta} - e^{-i k r \cos\theta} + \frac{e^{ikr}}{r} \left[f(\theta) - f(\pi - \theta) \right] \right), \tag{10.162}$$

and the associated differential scattering cross-section is written

$$\frac{d\sigma}{d\Omega} = |f(\theta) - f(\pi - \theta)|^2 \,. \tag{10.163}$$

For the case of two identical spin-zero (i.e., boson) particles (e.g., α-particles), the spatial wavefunction is symmetric with respect to particle interchange, which implies that

$$\frac{d\sigma}{d\Omega} = |f(\theta) + f(\pi - \theta)|^2$$

$$= |f(\theta)|^2 + |f(\pi - \theta)|^2 + \left[f^*(\theta) f(\pi - \theta) + f(\theta) f^*(\pi - \theta) \right]. \tag{10.164}$$

The previous result differs from the classical one because of the interference term (i.e., the final term on the right-hand side), which leads to an enhancement of the differential scattering cross-section at $\theta = \pi/2$. In fact,

$$\left(\frac{d\sigma}{d\Omega}\right)_{\theta=\pi/2} = 4\left[f(\pi/2)\right]^2, \tag{10.165}$$

whereas

$$\left(\frac{d\sigma_{\text{classical}}}{d\Omega}\right)_{\theta=\pi/2} = 2\left[f(\pi/2)\right]^2. \tag{10.166}$$

For the case of two identical spin-$1/2$ (i.e., fermion) particles (e.g., electrons or protons), the overall wavefunction is antisymmetric under particle interchange. If the two particles are in the spin singlet state (see Section 7.9) then the spatial wavefunction is symmetric (because the spin wavefunction is antisymmetric), and

$$\frac{d\sigma_{\text{singlet}}}{d\Omega} = |f(\theta) + f(\pi - \theta)|^2$$

$$= |f(\theta)|^2 + |f(\pi - \theta)|^2 + \left[f^*(\theta)\,f(\pi-\theta) + f(\theta)\,f^*(\pi-\theta)\right]. \tag{10.167}$$

On the other hand, if the two particles are in the spin triplet state then the spatial wavefunction is antisymmetric (because the spin wavefunction is symmetric), which leads to

$$\frac{d\sigma_{\text{triplet}}}{d\Omega} = |f(\theta) - f(\pi - \theta)|^2$$

$$= |f(\theta)|^2 + |f(\pi - \theta)|^2 - \left[f^*(\theta)\,f(\pi-\theta) + f(\theta)\,f^*(\pi-\theta)\right]. \tag{10.168}$$

In the former case, the interference term leads to an enhancement (with respect to the classical case) of the differential scattering cross-section at $\theta = \pi/2$: that is,

$$\left(\frac{d\sigma_{\text{singlet}}}{d\Omega}\right)_{\theta=\pi/2} = 4\left[f(\pi/2)\right]^2. \tag{10.169}$$

In the latter case, the interference term leads to the complete suppression of scattering in the direction $\theta = \pi/2$: that is,

$$\left(\frac{d\sigma_{\text{triplet}}}{d\Omega}\right)_{\theta=\pi/2} = 0. \tag{10.170}$$

Consider the mutual scattering of two unpolarized beams of spin-$1/2$ particles. All spin states are equally likely, so the probability of finding a given pair of particles (one from each beam) in the triplet state is three times that of finding it in the singlet state (see Section 7.9), which implies that

$$\left(\frac{d\sigma_{\text{unpolarized}}}{d\Omega}\right) = \frac{1}{4}\left(\frac{d\sigma_{\text{singlet}}}{d\Omega}\right) + \frac{3}{4}\left(\frac{d\sigma_{\text{triplet}}}{d\Omega}\right) \tag{10.171}$$

$$= |f(\theta)|^2 + |f(\pi - \theta)|^2 - \frac{1}{2}\left[f^*(\theta)\,f(\pi-\theta) + f(\theta)\,f^*(\pi-\theta)\right].$$

In this case, the interference term leads to incomplete suppression (with respect to the classical case) of the differential scattering cross-section at $\theta = \pi/2$: that is,

$$\left(\frac{d\sigma_{\text{unpolarized}}}{d\Omega}\right)_{\theta=\pi/2} = [f(\pi/2)]^2. \tag{10.172}$$

10.13   Exercises

10.1 Justify Equation (10.5).

10.2 Justify Equation (10.7).

10.3 Consider the Green's function for the Helmholtz equation:

$$(\nabla^2 + k^2)\, G(\mathbf{x}, \mathbf{x}') = \delta^3(\mathbf{x} - \mathbf{x}').$$

(a) Assuming, as seems reasonable, that $G(\mathbf{x}, \mathbf{x}') = G(R)$, where $\mathbf{R} = \mathbf{x} - \mathbf{x}'$, demonstrate that the two independent solutions in the region $R > 0$ are

$$G(R) = -\frac{e^{\pm i k R}}{4\pi R}.$$

(b) Show that

$$\int_V dV\, (\nabla^2 + k^2)\, G(R) = 1,$$

where V is a sphere of vanishing radius centered on $R = 0$. Hence, deduce that $G(R)$ is the required Green's function.

10.4 (a) Demonstrate that

$$\frac{1}{2\pi} \int_{-\infty}^{\infty} dx\, \exp\left(-\frac{x^2}{2\sigma_x^2}\right) \cos(k\,x) = \frac{1}{(2\pi\sigma_k^2)^{1/2}} \exp\left(-\frac{k^2}{2\sigma_k^2}\right),$$

where $\sigma_k = 1/\sigma_x$.

(b) Show that

$$\frac{1}{(2\pi\sigma_k^2)^{1/2}} \int_{-\infty}^{\infty} dk\, \exp\left(-\frac{k^2}{2\sigma_k^2}\right) = 1.$$

(c) Finally, by taking the limit $\sigma_x \to \infty$, conclude that

$$\frac{1}{2\pi} \int_{-\infty}^{\infty} dx\, \cos(k\,x) = \delta(x),$$

where $\delta(x)$ is a Dirac delta function.

10.5 Demonstrate that the probability current associated with the spherical waves

$$\psi^{\pm}(\mathbf{x}) = \frac{\exp(\pm i\,k\,r)}{r},$$

where $r = |\mathbf{x}|$, is

$$\mathbf{j} = \pm \frac{\hbar\,k}{m}\,\frac{\nabla r}{r^2}.$$

Hence, deduce that $\psi^{+}(\mathbf{x})$ and $\psi^{-}(\mathbf{x})$ correspond to outgoing and ingoing spherical waves, respectively.

10.6 Justify Equation (10.37).

10.7 Show that, in the Born approximation, the total scattering cross-section associated with the Yukawa potential, (10.35), is

$$\sigma_{\text{total}} = \left(\frac{2\,m\,V_0}{\hbar^2}\right)^2 \frac{4\pi}{\mu^4\,(4\,k^2 + \mu^2)}.$$

10.8 Consider a scattering potential of the form

$$V(r) = V_0\,\exp\left(-\frac{r^2}{a^2}\right).$$

Demonstrate, using the Born approximation, that

$$\frac{d\sigma}{d\Omega} = \left(\frac{\sqrt{\pi}\,m\,V_0\,a^3}{2\,\hbar^2}\right)^2 \exp\left[-2\,(k\,a)^2\,\sin^2(\theta/2)\right],$$

and

$$\sigma_{\text{total}} = \left(\frac{\sqrt{\pi}\,m\,V_0\,a^3}{2\,\hbar^2}\right)^2 2\pi\left[\frac{1 - e^{-2(k\,a)^2}}{(k\,a)^2}\right].$$

10.9 Show that the differential cross-section for the elastic scattering of a fast electron by the ground state of a hydrogen atom is

$$\frac{d\sigma}{d\Omega} = \left(\frac{2\,m_e\,e^2}{4\pi\,\epsilon_0\,\hbar^2\,q^2}\right)^2 \left(1 - \frac{16}{[4 + (q\,a_0)^2]^2}\right)^2,$$

where $q = |\mathbf{k} - \mathbf{k}'|$, and a_0 is the Bohr radius.

10.10 Justify Equations (10.57) and (10.59).

10.11 Demonstrate that, for a spherically symmetric scattering potential,

$$\sigma_{\text{total}} \simeq \frac{m^2}{\pi\,\hbar^4} \int d^3\mathbf{x} \int d^3\mathbf{x}'\,\frac{\sin^2[k\,|\mathbf{x} - \mathbf{x}'|]}{k^2\,|\mathbf{x} - \mathbf{x}'|^2}\,V(r)\,V(r')$$

in each of the following ways:

(a) By directly integrating the differential scattering cross-section obtained from the Born approximation.

(b) By utilizing the optical theorem in combination with the first two terms in the Born expansion. [Note that the first term in this expansion is real, and, therefore, does not contribute to σ_{total}.]

10.12 Justify Equations (10.98) and (10.100).

10.13 Consider a scattering potential that takes the constant value V_0 for $r < R$, and is zero for $r > R$, where V_0 may be either positive or negative. Using the method of partial waves, show that for $|V_0| \ll E \equiv \hbar^2 k^2 / 2m$, and $kR \ll 1$,

$$\frac{d\sigma}{d\Omega} = \left(\frac{4}{9}\right)\left(\frac{m^2 V_0^2 R^6}{\hbar^4}\right)\left[1 + \frac{2}{5}(kR)^2 \cos\theta + O(kR)^4\right],$$

and

$$\sigma_{\text{total}} = \left(\frac{16\pi}{9}\right)\left(\frac{m^2 V_0^2 R^6}{\hbar^4}\right)\left[1 + O(kR)^4\right].$$

10.14 Consider scattering of particles of mass m and incident wavenumber k by a repulsive δ-shell potential:

$$V(r) = \left(\frac{\hbar^2}{2m}\right)\gamma\,\delta(r - a),$$

where $\gamma, a > 0$. Show that the S-wave phase-shift is given by

$$\delta_0 = -ka + \tan^{-1}\left[\frac{1}{\cot(ka) + \gamma/k}\right].$$

Assuming that $\gamma \gg k, a^{-1}$, demonstrate that if $\cot(ka) \sim O(1)$ then the solution of the previous equation takes the form

$$\delta_0 \simeq -ka + \frac{k}{\gamma} - \left(\frac{k}{\gamma}\right)^2 \cot(ka) + O\left(\frac{k}{\gamma}\right)^3.$$

Of course, in the limit $\gamma \to \infty$, the preceding equation yields $\delta_0 = -ka$, which is the same result obtained when particles are scattered by a hard sphere of radius a. (See Section 10.8.) This is not surprising, because a strong repulsive δ-shell potential is indistinguishable from hard sphere as far as external particles are concerned.

The previous solution breaks down when $ka \simeq n\pi$, where n is a positive integer. Suppose that

$$ka = n\pi - \frac{k}{\gamma} + \frac{k^2}{\gamma^2}\,y,$$

where $y \sim O(1)$. Demonstrate that the S-wave contribution to the total scattering cross-section takes the form

$$\sigma_0 \simeq \frac{4\pi}{k_n^2}\frac{1}{1 + y^2} = \frac{4\pi}{k_n^2}\frac{\Gamma_n^2/4}{(E - E_n)^2 + \Gamma_n^2/4}.$$

where

$$k_n \simeq \frac{n\pi}{a},$$

$$E_n \simeq \frac{n^2 \pi^2 \hbar^2}{2ma^2},$$

$$\Gamma_n \simeq \frac{4n\pi E_n}{(\gamma a)^2}.$$

Hence, deduce that the net S-wave contribution to the total scattering cross-section is

$$\sigma_0 \simeq \frac{4\pi}{k^2}\left(\sin^2(ka) + \sum_{n=1,\infty}\frac{\Gamma_n^2/4}{(E-E_n)^2 + \Gamma_n^2/4}\right).$$

Obviously, there are resonant contributions to the cross-section whenever $E \simeq E_n$. Note that the E_n are the possible energies of particles trapped within the δ-shell potential. Hence, the resonances are clearly associated with incident particles tunneling though the δ-shell and forming transient trapped states. However, the width of the resonances (in energy) decreases strongly as the strength, γ, of the shell increases.

10.15 Consider the mutual scattering of two, counter-propagating, unpolarized, proton beams in the center of mass frame. Making use of the Born approximation, demonstrate that the differential scattering cross-section is

$$\frac{d\sigma}{d\Omega} \simeq \left(\frac{m_p e^2}{16\pi \epsilon_0 \hbar^2 k^2}\right)^2 \left[\frac{1}{\sin^4(\theta/2)} + \frac{1}{\cos^4(\theta/2)} - \frac{1}{\sin^2(\theta/2)\cos^2(\theta/2)}\right],$$

where θ is the scattering angle, m_p the proton mass, and $\hbar k$ the magnitude of the incident momenta of protons in each beam.

Chapter 11

Relativistic Electron Theory

11.1 Introduction

The aim of this chapter is to develop a quantum mechanical theory of electron dynamics that is consistent with special relativity. Such a theory is needed to explain the origin of electron spin (which is essentially a relativistic effect). Relativistic electron theory is also required to fully understand the fine structure of the hydrogen atom energy levels. (Recall, from Section 7.7, and Exercises 7.11 and 7.13, that the modification to the energy levels due to spin-orbit coupling is of the same order of magnitude as the first-order correction due to the electron's relativistic mass increase.)

11.2 Preliminary Analysis

In the following, we shall employ x^1, x^2, x^3 to represent the Cartesian coordinates x, y, z, respectively, and x^0 to represent ct, where c is the velocity of light in vacuum. The time-dependent wavefunction then takes the form $\psi(x^0, x^1, x^2, x^3)$. Adopting standard relativistic notation, we write the four x's as x^μ, for $\mu = 0, 1, 2, 3$ [Rindler (1966)]. A space-time vector with four components that transform under Lorentz transformation in an analogous manner to the four space-time coordinates, x^μ, is termed a *4-vector* [Rindler (1966)], and its components are written like a^μ (i.e., with an upper Greek suffix). We can lower the suffix according to the rules

$$a_0 = +a^0, \tag{11.1}$$

$$a_1 = -a^1, \tag{11.2}$$

$$a_2 = -a^2, \tag{11.3}$$

$$a_3 = -a^3. \tag{11.4}$$

Here, the a^μ are called the *contravariant components* of the 4-vector a, whereas the a_μ are termed the *covariant components*. Two 4-vectors a^μ and b^μ have the Lorentz-invariant scalar product

$$a^0 b^0 - a^1 b^1 - a^2 a^2 - a^3 b^3 = a^\mu b_\mu = a_\mu b^\mu, \tag{11.5}$$

a summation being implied over a repeated letter suffix [Rindler (1966)]. The *metric tenor*, $g_{\mu\nu}$, is defined

$$g_{00} = +1, \tag{11.6}$$

$$g_{11} = -1, \tag{11.7}$$

$$g_{22} = -1, \tag{11.8}$$

$$g_{33} = -1, \tag{11.9}$$

with all other components zero [Rindler (1966)]. Thus,

$$a_\mu = g_{\mu\nu} a^\nu. \tag{11.10}$$

Likewise,

$$a^\mu = g^{\mu\nu} a_\nu, \tag{11.11}$$

where $g^{00} = 1$, $g^{11} = g^{22} = g^{33} = -1$, with all other components zero. Finally, $g_\nu^\mu = g^\mu_\nu = 1$ if $\mu = \nu$, and $g_\nu^\mu = g^\mu_\nu = 0$ otherwise.

In the Schrödinger representation (see Section 2.4), the momentum of a particle, whose Cartesian components are written p_x, p_y, p_z, or p^1, p^2, p^3, is represented by the operators

$$p^i = -i\hbar \frac{\partial}{\partial x^i}, \tag{11.12}$$

for $i = 1, 2, 3$. Now, the four operators $\partial/\partial x^\mu$ form the covariant components of a 4-vector whose contravariant components are written $\partial/\partial x_\mu$. (See Exercise 11.1.) So, to make expression (11.12) consistent with relativistic theory, we must first write it with its suffixes balanced,

$$p^i = i\hbar \frac{\partial}{\partial x_i}, \tag{11.13}$$

and then extend it to the complete 4-vector equation,

$$p^\mu = i\hbar \partial^\mu, \tag{11.14}$$

where $\partial^\mu \equiv \partial/\partial x_\mu$. According to standard relativistic theory, the new operator $p^0 = i\hbar \partial/\partial x_0 = i(\hbar/c) \partial/\partial t$, which forms a 4-vector when combined with the momenta p^i, is interpreted as the energy of the particle divided by c [Rindler (1966)].

11.3 Dirac Equation

Consider the motion of an electron in the absence of electromagnetic fields. In classical relativity, electron energy, E, is related to electron momentum, $\mathbf{p}$, according to the well-known formula

$$\frac{E}{c} = (p^2 + m_e^2 c^2)^{1/2}, \tag{11.15}$$

where m_e is the electron rest mass [Rindler (1966)]. The quantum mechanical equivalent of this expression is the wave equation

$$\left[p^0 - (p^1 p^1 + p^2 p^2 + p^3 p^3 + m_e^2 c^2)^{1/2} \right] \psi = 0, \tag{11.16}$$

where the p's are interpreted as differential operators according to Equation (11.14). The previous equation takes into account the correct relativistic relation between electron energy and momentum, but is nevertheless unsatisfactory from the point of view of relativistic theory, because it is highly asymmetric between p^0 and the other p's. This makes the equation difficult to generalize, in a manifestly Lorentz-invariant manner, in the presence of electromagnetic fields. We must therefore look for a new equation.

If we multiply the wave equation (11.16) by the operator

$$p^0 + (p^1 p^1 + p^2 p^2 + p^3 p^3 + m_e^2 c^2)^{1/2} \tag{11.17}$$

then we obtain

$$\left(p^0 p^0 - p^1 p^1 - p^2 p^2 - p^3 p^3 - m_e^2 c^2 \right) \psi \equiv \left(p^\mu p_\mu - m_e^2 c^2 \right) \psi = 0. \tag{11.18}$$

This equation is manifestly Lorentz invariant, and, therefore, forms a more convenient starting point for relativistic quantum mechanics. Note, however, that Equation (11.18) is not entirely equivalent to Equation (11.16), because, although each solution of Equation (11.16) is also a solution of Equation (11.18), the converse is not true. In fact, only those solutions of Equation (11.18) belonging to positive values of p^0 are also solutions of Equation (11.16).

The wave equation (11.18) is quadratic in p^0, and is, thus, not of the form required by the laws of quantum theory. (Recall that we demonstrated, from general principles, in Chapter 3, that the time evolution equation for the wavefunction should be linear in the operator $\partial/\partial t$, and, hence, in p^0.) We, therefore, seek a wave equation that is equivalent to Equation (11.18), but is linear in p^0. In order to ensure that this equation transforms in a simple way under a Lorentz transformation, we shall require it to be rational and linear in p^1, p^2, p^3, as well as p^0. We are, thus, lead to a wave equation of the form

$$\left(p^0 - \alpha_1 p^1 - \alpha_2 p^2 - \alpha_3 p^3 - \beta m_e c \right) \psi = 0, \tag{11.19}$$

where the α's and β are dimensionless, and independent of the p's. Moreover, according to standard relativity, because we are considering the case of no electromagnetic fields, all points in space-time must be equivalent. Hence, the α's and β must also be independent of the x's. This implies that the α's and β commute with the p's and the x's. We, therefore, deduce that the α's and β describe an internal degree of freedom that is independent of space-time coordinates. In fact, we shall demonstrate later on in this chapter that these operators are related to electron spin. Note that the previous equation can be written in the standard form

$$i\hbar \frac{\partial \psi}{\partial t} = H\psi, \tag{11.20}$$

where the effective Hamiltonian, H, is

$$H = -i\hbar c\left(\alpha_1 \frac{\partial}{\partial x^1} + \alpha_2 \frac{\partial}{\partial x^2} + \alpha_3 \frac{\partial}{\partial x^3}\right) + \beta m_e c^2. \tag{11.21}$$

Multiplying Equation (11.19) by the operator

$$p^0 + \alpha_1 p^1 + \alpha_2 p^2 + \alpha_3 p^3 + \beta m_e c, \tag{11.22}$$

we obtain

$$\left[p^0 p^0 - \frac{1}{2}\sum_{i,j=1,3}\{\alpha_i,\alpha_j\}\, p^i p^j - \sum_{i=1,3}\{\alpha_i,\beta\}\, p^i m_e c - \beta^2 m_e^2 c^2\right]\psi = 0, \tag{11.23}$$

where $\{a, b\} \equiv ab + ba$. This equation is equivalent to Equation (11.18) provided that

$$\{\alpha_i, \alpha_j\} = 2\,\delta_{ij}, \tag{11.24}$$

$$\{\alpha_i, \beta\} = 0, \tag{11.25}$$

$$\beta^2 = 1, \tag{11.26}$$

for $i, j = 1, 3$. It is helpful to define the γ^μ, for $\mu = 0, 3$, where

$$\beta = \gamma^0, \tag{11.27}$$

$$\alpha_i = \gamma^0 \gamma^i, \tag{11.28}$$

for $i = 1, 3$. Equations (11.24)–(11.26) can then be shown to reduce to

$$\{\gamma^\mu, \gamma^\nu\} = 2\, g^{\mu\nu}. \tag{11.29}$$

(See Exercise 11.2.) One way of satisfying the previous anti-commutation relations is to represent the operators γ^μ as matrices. It follows that the operators α_i and β are also matrices. [Incidentally, it is clear from Equation (11.21) that the α_i and β matrices must be Hermitian, otherwise our effective Hamiltonian would not be an Hermitian operator.] In fact, it is possible to prove that the α_i and β must be even-dimensional matrices. (See Exercise 11.3.) Unfortunately, it is not possible to find a system of four appropriate 2×2 matrices. (The three Pauli matrices have the correct properties, but there is no fourth matrix.) It turns out that the smallest dimension in which the γ^μ can be realized is four [Bjorken and Drell (1964)]. In fact, it is easily verified that the 4×4 matrices

$$\gamma^0 = \begin{pmatrix} 1, & 0 \\ 0, & -1 \end{pmatrix}, \tag{11.30}$$

$$\gamma^i = \begin{pmatrix} 0, & \sigma_i \\ -\sigma_i, & 0 \end{pmatrix}, \tag{11.31}$$

for $i = 1, 3$, satisfy the appropriate anti-commutation relations. (See Exercise 11.4) Here, 0 and 1 denote 2×2 null and identity matrices, respectively, whereas the σ_i represent the 2×2 Pauli matrices introduced in Section 5.7. It follows from Equations (11.27) and (11.28) that

$$\beta = \begin{pmatrix} 1, & 0 \\ 0, & -1 \end{pmatrix}, \tag{11.32}$$

$$\alpha_i = \begin{pmatrix} 0, & \sigma_i \\ \sigma_i, & 0 \end{pmatrix}. \tag{11.33}$$

Note that γ^0, β, and the α_i, are all Hermitian matrices, whereas the γ^μ, for $\mu = 1, 3$, are anti-Hermitian. However, the matrices $\gamma^0 \gamma^\mu$, for $\mu = 0, 3$, are Hermitian. Moreover, it is easily demonstrated that

$$\gamma^{\mu\dagger} = \gamma^0 \gamma^\mu \gamma^0, \tag{11.34}$$

for $\mu = 0, 3$.

Equation (11.19) can be written in the form

$$(\gamma^\mu p_\mu - m_e c) \psi \equiv (i \hbar \gamma^\mu \partial_\mu - m_e c) \psi = 0, \tag{11.35}$$

where $\partial_\mu \equiv \partial/\partial x^\mu$. Alternatively, we can write

$$i \hbar \frac{\partial \psi}{\partial t} = (c \, \boldsymbol{\alpha} \cdot \mathbf{p} + \beta \, m_e c^2) \psi, \tag{11.36}$$

where $\mathbf{p} = (p_x, p_y, p_z) = (p^1, p^2, p^3)$, and $\boldsymbol{\alpha}$ is the vector of the α_i matrices. The previous expression is known as the *Dirac equation* [Dirac (1928)]. Incidentally, it is clear that, corresponding to the four rows and columns of the γ^μ matrices, the wavefunction ψ must take the form of a 4×1 column matrix, each element of which is, in general, a function of the x^μ. We saw in Section 5.7 that the spin of the electron requires the wavefunction to have two components. The reason our present theory requires the wavefunction to have four components is because the wave equation (11.18) has twice as many solutions as it ought to have, half of them corresponding to negative energy states.

We can incorporate an electromagnetic field into the previous formalism by means of the standard prescription $E \to E + e\phi$, and $p^i \to p^i + e A^i$, where e is the magnitude of the electron charge, ϕ the scalar potential, and $\mathbf{A}$ the vector potential [Jackson (1975)]. This prescription can be expressed in the Lorentz-invariant form

$$p^\mu \to p^\mu + \frac{e}{c}\, \Phi^\mu, \qquad (11.37)$$

where $\Phi^\mu = (\phi, c\,\mathbf{A})$ is the *potential 4-vector* [Rindler (1966)]. Thus, Equation (11.35) becomes

$$\left[\gamma^\mu \left(p_\mu + \frac{e}{c}\, \Phi_\mu\right) - m_e\, c\right]\psi \equiv \left[\gamma^\mu \left(i\,\hbar\,\partial_\mu + \frac{e}{c}\, \Phi_\mu\right) - m_e\, c\right]\psi = 0, \qquad (11.38)$$

whereas Equation (11.36) generalizes to

$$i\,\hbar\, \frac{\partial\psi}{\partial t} = \left[-e\,\phi + c\,\boldsymbol{\alpha} \cdot (\mathbf{p} + e\,\mathbf{A}) + \beta\, m_e\, c^2\right]\psi = 0. \qquad (11.39)$$

If we write the wavefunction in the spinor form

$$\psi = \begin{pmatrix} \psi_0 \\ \psi_1 \\ \psi_2 \\ \psi_3 \end{pmatrix} \qquad (11.40)$$

then the Hermitian conjugate of Equation (11.39) becomes

$$-i\,\hbar\, \frac{\partial\psi^\dagger}{\partial t} = \psi^\dagger \left[-e\,\phi + c\,\boldsymbol{\alpha} \cdot (\mathbf{p} + e\,\mathbf{A}) + \beta\, m_e\, c^2\right] = 0, \qquad (11.41)$$

where

$$\psi^\dagger = \left(\psi_0^*,\, \psi_1^*,\, \psi_2^*,\, \psi_3^*\right). \qquad (11.42)$$

Here, use has been made of the fact that the α_i and the β are Hermitian matrices that commute with the p^i and the A^i.

It follows from $\psi^\dagger\, \gamma^0$ times Equation (11.38) that

$$\psi^\dagger \left[\gamma^0\, \gamma^\mu \left(i\,\hbar\,\partial_\mu - \frac{e}{c}\, \Phi_\mu\right) - \gamma^0\, m_e\, c\right]\psi = 0. \qquad (11.43)$$

The Hermitian conjugate of this expression is

$$\psi^\dagger \left[\left(-i\,\hbar\,\partial_\mu - \frac{e}{c}\, \Phi_\mu\right)\gamma^0\, \gamma^\mu - m_e\, c\, \gamma^0\right]\psi = 0, \qquad (11.44)$$

where ∂_μ now acts backward on $\psi^\dagger$, and use has been made of the fact that the matrices $\gamma^0\gamma^\mu$ and γ^0 are Hermitian. Taking the difference between the previous two equations, we obtain

$$\partial_\mu\, j^\mu = 0, \tag{11.45}$$

where

$$j^\mu = c\,\psi^\dagger\,\gamma^0\,\gamma^\mu\,\psi. \tag{11.46}$$

Writing $j^\mu = (c\rho, \mathbf{j})$, where

$$\rho = \psi^\dagger\,\psi, \tag{11.47}$$

$$j^i = c\,\psi^\dagger\,\gamma^0\,\gamma^i\,\psi = \psi^\dagger\,c\,\alpha_i\,\psi. \tag{11.48}$$

Equation (11.45) becomes

$$\frac{\partial\rho}{\partial t} + \nabla\cdot\mathbf{j} = 0. \tag{11.49}$$

The previous expression has the same form as the non-relativistic probability conservation equation, (3.65). This suggests that we can interpret the positive-definite real scalar field $\rho(\mathbf{x}, t) = |\psi|^2$ as the *relativistic probability density*, and the vector field $\mathbf{j}(\mathbf{x}, t)$ as the *relativistic probability current*. Integration of the preceding expression over all space, assuming that $|\psi(\mathbf{x}, t)| \to 0$ as $|\mathbf{x}| \to \infty$, yields

$$\frac{d}{dt}\int d^3\mathbf{x}\, \rho(\mathbf{x}, t) = 0. \tag{11.50}$$

This ensures that if the wavefunction is properly normalized at time $t = 0$, such that

$$\int d^3\mathbf{x}\, \rho(\mathbf{x}, 0) = 1, \tag{11.51}$$

then the wavefunction remains properly normalized at all subsequent times, as it evolves in accordance with the Dirac equation. In fact, if this were not the case then it would be impossible to interpret ρ as a probability density. Now, relativistic invariance demands that if the wavefunction is properly normalized in one particular inertial frame then it should be properly normalized in all inertial frames [Rindler (1966)]. This is the case provided that Equation (11.45) is Lorentz invariant (i.e., it has the property that if it holds in one inertial frame then it holds in all inertial frames), which is true as long as the j^μ transform as the contravariant components of a 4-vector under Lorentz transformation. (See the following section, and Exercise 11.6.)

11.4   Lorentz Invariance of Dirac Equation

Consider two inertial frames, S and S'. Let the x^μ and $x^{\mu'}$ be the space-time coordinates of a given event in each frame, respectively. These coordinates are related via a *Lorentz transformation*, which takes the general form

$$x^{\mu'} = a^\mu_{\ \nu} x^\nu, \tag{11.52}$$

where the $a^\mu_{\ \nu}$ are real numerical coefficients that are independent of the x^μ. We also have

$$x_{\mu'} = a_\mu^{\ \nu} x_\nu. \tag{11.53}$$

Now, because [see Equation (11.5)]

$$x^{\mu'} x_{\mu'} = x^\mu x_\mu, \tag{11.54}$$

it follows that

$$a^\mu_{\ \nu} a_\mu^{\ \lambda} = g_\nu^{\ \lambda}. \tag{11.55}$$

Moreover, it is easily shown that

$$x^\mu = a_\nu^{\ \mu} x^{\nu'}, \tag{11.56}$$

$$x_\mu = a^\nu_{\ \mu} x_{\nu'}. \tag{11.57}$$

By definition, a 4-vector, p^μ, has analogous transformation properties to the x^μ. Thus,

$$p^{\mu'} = a^\mu_{\ \nu} p^\nu, \tag{11.58}$$

$$p^\mu = a_\nu^{\ \mu} p^{\nu'}, \tag{11.59}$$

et cetera.

In frame S, the Dirac equation is written

$$\left[\gamma^\mu \left(p_\mu - \frac{e}{c} \Phi_\mu \right) - m_e c \right] \psi = 0. \tag{11.60}$$

Let ψ' be the wavefunction in frame S'. Suppose that

$$\psi' = A \psi, \tag{11.61}$$

where A is a 4×4 transformation matrix that is independent of the x^μ. (Hence, A commutes with the p_μ and the Φ_μ.) Multiplying Equation (11.60) by A, we obtain

$$\left[A \gamma^\mu A^{-1} \left(p_\mu - \frac{e}{c} \Phi_\mu \right) - m_e c \right] \psi' = 0. \tag{11.62}$$

Hence, given that the p_μ and Φ_μ are the covariant components of 4-vectors, we obtain

$$\left[A \gamma^\mu A^{-1} a^\nu_{\ \mu} \left(p_{\nu'} - \frac{e}{c} \Phi_{\nu'} \right) - m_e c \right] \psi' = 0. \tag{11.63}$$

Suppose that

$$A \gamma^{\mu} A^{-1} a^{\nu}_{\mu} = \gamma^{\nu}, \tag{11.64}$$

which is equivalent to

$$A^{-1} \gamma^{\nu} A = a^{\nu}_{\mu} \gamma^{\mu}. \tag{11.65}$$

Here, we have assumed that the a^{ν}_{μ} commute with A and the γ^{μ} (because they are just numbers). If Equation (11.64) holds then Equation (11.63) becomes

$$\left[\gamma^{\mu} \left(p_{\mu'} - \frac{e}{c} \Phi_{\mu'} \right) - m_e c \right] \psi' = 0. \tag{11.66}$$

A comparison of this equation with Equation (11.60) reveals that the Dirac equation takes the same form in frames S and S'. In other words, the Dirac equation is Lorentz invariant. Incidentally, it is clear from Equations (11.60) and (11.66) that the γ^{μ} matrices are the same in all inertial frames.

It remains to find a transformation matrix A that satisfies Equation (11.65). Consider an infinitesimal Lorentz transformation, for which

$$a^{\nu}_{\mu} = g^{\nu}_{\mu} + \delta \omega^{\nu}_{\mu}, \tag{11.67}$$

where the $\delta \omega^{\nu}_{\mu}$ are real numerical coefficients that are independent of the x^{μ}, and are also small compared to unity. To first order in small quantities, Equation (11.55) yields

$$\delta \omega^{\mu \nu} + \delta \omega^{\nu \mu} = 0. \tag{11.68}$$

Each of the six independent non-vanishing $\delta \omega^{\mu \nu}$ generates a particular infinitesimal Lorentz transformation. For instance,

$$\delta \omega^{01} = -\delta \omega^{10} = \delta \beta \tag{11.69}$$

for a transformation to a coordinate system moving with a velocity $c \, \delta \beta$ along the x^1-direction. Furthermore,

$$\delta \omega^{21} = -\delta \omega^{12} = \delta \varphi \tag{11.70}$$

for a rotation through an angle $\delta \varphi$ about the x^3-axis, and so on.

Let us write

$$A = 1 - \frac{i}{4} \sigma_{\mu \nu} \delta \omega^{\mu \nu}, \tag{11.71}$$

where the $\sigma_{\mu \nu}$ are $O(1)$ 4×4 matrices. To first order in small quantities,

$$A^{-1} = 1 + \frac{i}{4} \sigma_{\mu \nu} \delta \omega^{\mu \nu}. \tag{11.72}$$

Moreover, it follows from Equation (11.68) that

$$\sigma_{\mu \nu} = -\sigma_{\nu \mu}. \tag{11.73}$$

To first order in small quantities, Equations (11.65), (11.67), (11.71), and (11.72) yield

$$\delta\omega^{\nu}{}_{\beta}\,\gamma^{\beta} = -\frac{i}{4}\,\delta\omega^{\alpha\beta}\left(\gamma^{\nu}\,\sigma_{\alpha\beta} - \sigma_{\alpha\beta}\,\gamma^{\nu}\right). \tag{11.74}$$

Hence, making use of the symmetry property (11.68), we obtain

$$\delta\omega^{\alpha\beta}\left(g^{\nu}{}_{\alpha}\,\gamma_{\beta} - g^{\nu}{}_{\beta}\,\gamma_{\alpha}\right) = -\frac{i}{2}\,\delta\omega^{\alpha\beta}\left(\gamma^{\nu}\,\sigma_{\alpha\beta} - \sigma_{\alpha\beta}\,\gamma^{\nu}\right), \tag{11.75}$$

where $\gamma_{\mu} = g_{\mu\nu}\,\gamma^{\nu}$. Because this equation must hold for arbitrary $\delta\omega^{\alpha\beta}$, we deduce that

$$2\,i\left(g^{\nu}{}_{\alpha}\,\gamma_{\beta} - g^{\nu}{}_{\beta}\,\gamma_{\alpha}\right) = [\gamma^{\nu}, \sigma_{\alpha\beta}]. \tag{11.76}$$

Making use of the anti-commutation relations (11.29), it can be shown that a suitable solution of the previous equation is

$$\bullet\;\sigma_{\mu\nu} = \frac{i}{2}\,[\gamma_{\mu}, \gamma_{\nu}]. \tag{11.77}$$

(See Exercise 11.7.) Hence,

$$A = 1 + \frac{1}{8}\,[\gamma_{\mu}, \gamma_{\nu}]\,\delta\omega^{\mu\nu}, \tag{11.78}$$

$$A^{-1} = 1 - \frac{1}{8}\,[\gamma_{\mu}, \gamma_{\nu}]\,\delta\omega^{\mu\nu}. \tag{11.79}$$

Now that we have found the correct transformation rules for an infinitesimal Lorentz transformation, we can easily find those for a finite transformation by building it up from a large number of successive infinitesimal transforms [Bjorken and Drell (1964)]. (See Exercises 11.8 and 11.9.)

Making use of Equation (11.34), as well as $\gamma^{0}\,\gamma^{0} = 1$, the Hermitian conjugate of Equation (11.78) can be shown to take the form

$$A^{\dagger} = 1 - \frac{1}{8}\,\gamma^{0}\,[\gamma_{\mu}, \gamma_{\nu}]\,\gamma^{0}\,\delta\omega^{\mu\nu} = \gamma^{0}\,A^{-1}\,\gamma^{0}. \tag{11.80}$$

Hence, Equation (11.65) yields

$$A^{\dagger}\,\gamma^{0}\,\gamma^{\mu}\,A = a^{\mu}{}_{\nu}\,\gamma^{0}\,\gamma^{\nu}. \tag{11.81}$$

It follows that

$$\psi^{\dagger}\,A^{\dagger}\,\gamma^{0}\,\gamma^{\mu}\,A\,\psi = a^{\mu}{}_{\nu}\,\psi^{\dagger}\,\gamma^{0}\,\gamma^{\nu}\,\psi, \tag{11.82}$$

or

$$\psi^{\dagger\prime}\,\gamma^{0}\,\gamma^{\mu}\,\psi' = a^{\mu}{}_{\nu}\,\psi^{\dagger}\,\gamma^{0}\,\gamma^{\nu}\,\psi, \tag{11.83}$$

which implies that

$$j^{\mu\prime} = a^{\mu}{}_{\nu}\,j^{\nu}, \tag{11.84}$$

where the j^{μ} are defined in Equation (11.46). This proves that the j^{μ} transform as the contravariant components of a 4-vector.

11.5 Free Electron Motion

According to Equation (11.36), the relativistic Hamiltonian of a free electron takes the form

$$H = c\,\boldsymbol{\alpha} \cdot \mathbf{p} + \beta\, m_e\, c^2. \tag{11.85}$$

Let us use the Heisenberg picture to investigate the motion of such an electron. For the sake of brevity, we shall omit the suffix t that should be appended to dynamical variables that vary in time, according to the formalism of Section 3.3.

The previous Hamiltonian is independent of $\mathbf{x}$. Hence, the momentum, $\mathbf{p}$, commutes with the Hamiltonian, and is therefore a constant of the motion. The x component of the velocity is

$$\dot{x} = \frac{[x, H]}{i\,\hbar} = c\,\alpha_1, \tag{11.86}$$

where use has been made of the standard commutation relations between position and momentum operators. (See Section 2.2.) This result is rather surprising, because it implies a relationship between velocity and momentum that is quite different from that in classical mechanics. This relationship, however, is clearly connected to the expression $j_x = \psi^\dagger c\,\alpha_1\,\psi$ for the x-component of the probability current. The operator $\dot{x}$, specified in the previous equation, has the eigenvalues $\pm c$, corresponding to the eigenvalues ± 1 of α_1. Because $\dot{y}$ and $\dot{z}$ are similar to $\dot{x}$, we conclude that a measurement of a velocity component of a free electron is certain to yield the result $\pm c$. As is easily demonstrated, this conclusion also holds in the presence of an electromagnetic field.

Of course, electrons are often observed to have velocities considerably less than that of light. Hence, the previous conclusion seems to be in conflict with experimental observations. The conflict is not real, however, because the theoretical velocity discussed previously is the velocity at one instance in time, whereas observed velocities are always averages over a finite time interval. We shall find, on further examination of the equations of motion, that the velocity of a free electron is not constant, but oscillates rapidly about a mean value that agrees with the experimentally observed value.

In order to understand why a measurement of a velocity component must lead to the result $\pm c$ in a relativistic theory, consider the following argument. To measure the velocity we must measure the position at two slightly different times, and then divide the change in position by the time interval. (We cannot just measure the momentum and then apply a formula, because the ordinary connection between velocity and momentum is no longer valid.) In order that our measured velocity may approximate to the instantaneous velocity, the time interval between the two measurements of position must be very short, and the measurements themselves very accurate. However, the great accuracy with which the position of the electron is known during the time interval leads to an almost complete indeterminacy in its momentum, according to the Heisenberg uncertainty principle. This means that almost all values of the momentum are equally likely, so that the momentum is almost certain to be infinite. But, an infinite value of a momentum component corresponds to the values $\pm c$ for the corresponding velocity component.

Let us now examine how the election velocity varies in time. We have

$$i\hbar \dot{\alpha}_1 = \alpha_1 H - H \alpha_1. \tag{11.87}$$

Now, α_1 anti-commutes with all terms in H except $c\,\alpha_1\,p^1$, so

$$\alpha_1 H + H \alpha_1 = \alpha_1 c \alpha_1 p^1 + c \alpha_1 p^1 \alpha_1 = 2 c p^1. \tag{11.88}$$

Here, use has been made of the fact that α_1 commutes with p^1, and also that $\alpha_1^2 = 1$. Hence, we get

$$i\hbar \dot{\alpha}_1 = 2 \alpha_1 H - 2 c p^1. \tag{11.89}$$

Because H and p^1 are constants of the motion, this equation yields

$$i\hbar \ddot{\alpha}_1 = 2 \dot{\alpha}_1 H, \tag{11.90}$$

which can be integrated to give

$$\dot{\alpha}_1(t) = \dot{\alpha}_1(0) \exp\left(\frac{-2\,i\,H\,t}{\hbar}\right). \tag{11.91}$$

It follows from Equation (11.86) and (11.89) that

$$\dot{x}(t) = c\,\alpha_1(t) = \frac{i\,\hbar\,c}{2}\,\dot{\alpha}_1(0)\,\exp\left(\frac{-2\,i\,H\,t}{\hbar}\right) H^{-1} + c^2\,p_x\,H^{-1}, \tag{11.92}$$

and

$$x(t) = x(0) - \frac{\hbar^2 c}{4}\,\dot{\alpha}_1(0)\,\exp\left(\frac{-2\,i\,H\,t}{\hbar}\right) H^{-2} + c^2\,p_x\,H^{-1}\,t. \tag{11.93}$$

We can see that the x-component of velocity consists of two parts, a constant part, $c^2\,p_x\,H^{-1}$, connected with the momentum according to the classical relativistic formula, and an oscillatory part whose frequency, $2\,H/h$, is high (being at least $2\,m_e\,c^2/h$). Only the constant part would be observed in a practical measurement of velocity (i.e., an average over a short time interval that is still much longer than $h/2\,m_e\,c^2$). The oscillatory part ensures that the instantaneous value of $\dot{x}$ has the eigenvalues $\pm c$. Note, finally, that the oscillatory part of x is small, being of order $\hbar/m_e\,c$. The rapid oscillatory motion of a relativistic electron is known as *zitterbewegung* (which is german for "trembling motion") [Schrödinger (1930)].

11.6 Electron Spin

According to Equation (11.39), the relativistic Hamiltonian of an electron in an electromagnetic field is

$$H = -e\,\phi + c\,\alpha \cdot (\mathbf{p} + e\,\mathbf{A}) + \beta\,m_e\,c^2. \tag{11.94}$$

Hence,

$$\left(\frac{H}{c} + \frac{e}{c}\,\phi\right)^2 = [\alpha \cdot (\mathbf{p} + e\,\mathbf{A}) + \beta\,m_e\,c]^2 = [\alpha \cdot (\mathbf{p} + e\,\mathbf{A})]^2 + m_e^2\,c^2, \tag{11.95}$$

where use has been made of Equations (11.25) and (11.26). Now, we can write

$$\alpha_i = \gamma^5\,\Sigma_i, \tag{11.96}$$

for $i = 1, 3$, where

$$\gamma^5 = \begin{pmatrix} 0, & 1 \\ 1, & 0 \end{pmatrix}, \tag{11.97}$$

and

$$\Sigma_i = \begin{pmatrix} \sigma_i, & 0 \\ 0, & \sigma_i \end{pmatrix}. \tag{11.98}$$

Here, 0 and 1 denote 2×2 null and identity matrices, respectively, whereas the σ_i are conventional 2×2 Pauli matrices. Note that $\gamma^5\,\gamma^5 = 1$, and

$$[\gamma^5, \Sigma_i] = 0. \tag{11.99}$$

It follows from Equation (11.95) that

$$\left(\frac{H}{c} + \frac{e}{c}\,\phi\right)^2 = [\Sigma \cdot (\mathbf{p} + e\,\mathbf{A})]^2 + m_e^2\,c^2. \tag{11.100}$$

Now, a straightforward generalization of Equation (5.93) gives

$$(\Sigma \cdot \mathbf{a})(\Sigma \cdot \mathbf{b}) = \mathbf{a} \cdot \mathbf{b} + i\,\Sigma \cdot (\mathbf{a} \times \mathbf{b}), \tag{11.101}$$

where $\mathbf{a}$ and $\mathbf{b}$ are any two three-dimensional vectors that commute with Σ. It follows that

$$[\Sigma \cdot (\mathbf{p} + e\,\mathbf{A})]^2 = (\mathbf{p} + e\,\mathbf{A})^2 + i\,\Sigma \cdot (\mathbf{p} + e\,\mathbf{A}) \times (\mathbf{p} + e\,\mathbf{A}). \tag{11.102}$$

However,

$$(\mathbf{p} + e\,\mathbf{A}) \times (\mathbf{p} + e\,\mathbf{A}) = e\,\mathbf{p} \times \mathbf{A} + e\,\mathbf{A} \times \mathbf{p} = -i\,e\,\hbar\,\nabla \times \mathbf{A} - i\,e\,\hbar\,\mathbf{A} \times \nabla$$

$$= -i\,e\,\hbar\,\mathbf{B}, \tag{11.103}$$

where $\mathbf{B} = \nabla \times \mathbf{A}$ is the magnetic field-strength. Hence, we obtain

$$\left(\frac{H}{c} + \frac{e}{c}\phi\right)^2 = (\mathbf{p} + e\,\mathbf{A})^2 + m_e^2\,c^2 + e\,\hbar\,\mathbf{\Sigma} \cdot \mathbf{B}. \tag{11.104}$$

Consider the non-relativistic limit. In this case, we can write

$$H = m_e\,c^2 + \delta H, \tag{11.105}$$

where δH is small compared to $m_e\,c^2$. Substituting into Equation (11.104), and neglecting δH^2, and other terms involving c^{-2}, we get

$$\delta H \simeq -e\,\phi + \frac{1}{2\,m_e}\,(\mathbf{p} + e\,\mathbf{A})^2 + \frac{e\,\hbar}{2\,m_e}\,\mathbf{\Sigma} \cdot \mathbf{B}. \tag{11.106}$$

This Hamiltonian is the same as the classical Hamiltonian of a non-relativistic electron, except for the final term. (See Section 3.6.) This term may be interpreted as arising from the electron having an intrinsic magnetic moment

$$\boldsymbol{\mu} = -\frac{e\,\hbar}{2\,m_e}\,\mathbf{\Sigma}. \tag{11.107}$$

(See Section 5.6.)

In order to demonstrate that the electron's intrinsic magnetic moment is associated with an intrinsic angular momentum, consider the motion of an electron in a central electrostatic potential: that is, $\phi = \phi(r)$ and $\mathbf{A} = \mathbf{0}$. In this case, the Hamiltonian (11.94) becomes

$$H = -e\,\phi(r) + c\,\gamma^5\,\mathbf{\Sigma} \cdot \mathbf{p} + \beta\,m_e\,c^2. \tag{11.108}$$

Consider the x-component of the electron's orbital angular momentum,

$$L_x = y\,p_z - z\,p_y = i\,\hbar\left(z\,\frac{\partial}{\partial y} - y\,\frac{\partial}{\partial z}\right). \tag{11.109}$$

The Heisenberg equation of motion for this quantity is

$$i\,\hbar\,\dot{L}_x = [L_x, H]. \tag{11.110}$$

However, it is easily demonstrated that

$$[L_x, r] = 0, \tag{11.111}$$

$$[L_x, p_x] = 0, \tag{11.112}$$

$$[L_x, p_y] = i\,\hbar\,p_z, \tag{11.113}$$

$$[L_x, p_z] = -i\,\hbar\,p_y. \tag{11.114}$$

Hence, we obtain

$$[L_x, H] = i\,\hbar\,c\,\gamma^5\,(\Sigma_2\,p_z - \Sigma_3\,p_y), \tag{11.115}$$

which implies that

$$\dot{L}_x = c\,\gamma^5\,(\Sigma_2\,p_z - \Sigma_3\,p_y). \tag{11.116}$$

It can be seen that L_x is not a constant of the motion. However, the x-component of the total angular momentum of the system must be a constant of the motion (because a central electrostatic potential exerts zero torque on the system). Hence, we deduce that the electron possesses additional angular momentum that is not connected with its motion through space. Now,

$$i\,\hbar\,\dot{\Sigma}_1 = [\Sigma_1, H]. \tag{11.117}$$

However,

$$[\Sigma_1, \beta] = 0, \tag{11.118}$$

$$[\Sigma_1, \gamma^5] = 0, \tag{11.119}$$

$$[\Sigma_1, \Sigma_1] = 0, \tag{11.120}$$

$$[\Sigma_1, \Sigma_2] = 2\,i\,\Sigma_3, \tag{11.121}$$

$$[\Sigma_1, \Sigma_3] = -2\,i\,\Sigma_2, \tag{11.122}$$

so

$$[\Sigma_1, H] = 2\,i\,c\,\gamma^5\,(\Sigma_3\,p_y - \Sigma_2\,p_z), \tag{11.123}$$

which implies that

$$\frac{\hbar}{2}\,\dot{\Sigma}_1 = -c\,\gamma^5\,(\Sigma_2\,p_z - \Sigma_3\,p_y). \tag{11.124}$$

Hence, we deduce that

$$\dot{L}_x + \frac{\hbar}{2}\,\dot{\Sigma}_1 = 0. \tag{11.125}$$

Because there is nothing special about the x-direction, we conclude that the vector $\mathbf{L} + (\hbar/2)\,\Sigma$ is a constant of the motion. We can interpret this result by saying that the electron has a spin angular momentum $\mathbf{S} = (\hbar/2)\,\Sigma$, which must be added to its orbital angular momentum in order to obtain a constant of the motion. According to Equation (11.107), the relationship between the electron's spin angular momentum and its intrinsic (i.e., non-orbital) magnetic moment is

$$\mu = -\frac{e\,g}{2\,m_e}\,\mathbf{S}, \tag{11.126}$$

where the gyromagnetic ratio, g, takes the value

$$g = 2. \tag{11.127}$$

As explained in Section 5.5, this is twice the value one would naively predict by analogy with classical physics.

11.7   Motion in Central Field

To further study the motion of an electron in a central field, whose Hamiltonian is

$$H = -e\,\phi(r) + c\,\boldsymbol{\alpha} \cdot \mathbf{p} + \beta\, m_e\, c^2, \tag{11.128}$$

it is convenient to transform to polar coordinates. Let

$$r = \left(x^2 + y^2 + z^2\right)^{1/2}, \tag{11.129}$$

and

$$r\, p_r = \mathbf{x} \cdot \mathbf{p}. \tag{11.130}$$

It is easily demonstrated that

$$[r, p_r] = \mathrm{i}\,\hbar, \tag{11.131}$$

which implies that in the Schrödinger representation

$$p_r = -\mathrm{i}\,\hbar\, \frac{\partial}{\partial r}. \tag{11.132}$$

Now, by symmetry, an energy eigenstate in a central field is a simultaneous eigenstate of the total angular momentum

$$\mathbf{J} = \mathbf{L} + \frac{\hbar}{2}\, \boldsymbol{\Sigma}. \tag{11.133}$$

Furthermore, we know from general principles that the eigenvalues of J^2 are $j\,(j+1)\,\hbar^2$, where j is a positive half-integer (because $j = |l + 1/2|$, where l is the standard non-negative integer quantum number associated with orbital angular momentum). (See Chapter 5.11.)

It follows from Equation (11.101) that

$$(\boldsymbol{\Sigma} \cdot \mathbf{L})\,(\boldsymbol{\Sigma} \cdot \mathbf{L}) = L^2 + \mathrm{i}\,\boldsymbol{\Sigma} \times (\mathbf{L} \times \mathbf{L}). \tag{11.134}$$

However, because $\mathbf{L}$ is an angular momentum, its components satisfy the standard commutation relations

$$\mathbf{L} \times \mathbf{L} = \mathrm{i}\,\hbar\, \mathbf{L}. \tag{11.135}$$

(See Section 4.1.) Thus, we obtain

$$(\boldsymbol{\Sigma} \cdot \mathbf{L})\,(\boldsymbol{\Sigma} \cdot \mathbf{L}) = L^2 - \hbar\,\boldsymbol{\Sigma} \cdot \mathbf{L} = J^2 - 2\,\hbar\,\boldsymbol{\Sigma} \cdot \mathbf{L} - \frac{\hbar^2}{4}\, \Sigma^2. \tag{11.136}$$

However, $\Sigma^2 = 3$, so

$$(\boldsymbol{\Sigma} \cdot \mathbf{L} + \hbar)^2 = J^2 + \frac{1}{4}\, \hbar^2. \tag{11.137}$$

Further application of Equation (11.101) yields

$$(\boldsymbol{\Sigma} \cdot \mathbf{L})(\boldsymbol{\Sigma} \cdot \mathbf{p}) = \mathbf{L} \cdot \mathbf{p} + i\,\boldsymbol{\Sigma} \cdot \mathbf{L} \times \mathbf{p} = i\,\boldsymbol{\Sigma} \cdot \mathbf{L} \times \mathbf{p}, \tag{11.138}$$

$$(\boldsymbol{\Sigma} \cdot \mathbf{p})(\boldsymbol{\Sigma} \cdot \mathbf{L}) = \mathbf{p} \cdot \mathbf{L} + i\,\boldsymbol{\Sigma} \cdot \mathbf{p} \times \mathbf{L} = i\,\boldsymbol{\Sigma} \cdot \mathbf{p} \times \mathbf{L}. \tag{11.139}$$

However, it is easily demonstrated from the fundamental commutation relations between position and momentum operators that

$$\mathbf{L} \times \mathbf{p} + \mathbf{p} \times \mathbf{L} = 2\,i\,\hbar\,\mathbf{p}. \tag{11.140}$$

(See Section 2.2.) Thus,

$$(\boldsymbol{\Sigma} \cdot \mathbf{L})(\boldsymbol{\Sigma} \cdot \mathbf{p}) + (\boldsymbol{\Sigma} \cdot \mathbf{p})(\boldsymbol{\Sigma} \cdot \mathbf{L}) = -2\,\hbar\,\boldsymbol{\Sigma} \cdot \mathbf{p}, \tag{11.141}$$

which implies that

$$\{\boldsymbol{\Sigma} \cdot \mathbf{L} + \hbar, \boldsymbol{\Sigma} \cdot \mathbf{p}\} = 0. \tag{11.142}$$

Now, $\boldsymbol{\Sigma} = \gamma^5 \boldsymbol{\alpha}$. Moreover, γ^5 commutes with $\mathbf{p}$, $\mathbf{L}$, and $\boldsymbol{\Sigma}$. Hence, we conclude that

$$\{\boldsymbol{\Sigma} \cdot \mathbf{L} + \hbar, \boldsymbol{\alpha} \cdot \mathbf{p}\} = 0. \tag{11.143}$$

Finally, because β commutes with $\mathbf{p}$ and $\mathbf{L}$, but anti-commutes with the components of $\boldsymbol{\alpha}$, we obtain

$$[\zeta, \boldsymbol{\alpha} \cdot \mathbf{p}] = 0, \tag{11.144}$$

where

$$\zeta = \beta \left(\boldsymbol{\Sigma} \cdot \mathbf{L} + \hbar\right). \tag{11.145}$$

If we repeat the previous analysis, starting at Equation (11.138), but substituting $\mathbf{x}$ for $\mathbf{p}$, and making use of the easily demonstrated result

$$\mathbf{L} \times \mathbf{x} + \mathbf{x} \times \mathbf{L} = 2\,i\,\hbar\,\mathbf{x}, \tag{11.146}$$

we find that

$$[\zeta, \boldsymbol{\alpha} \cdot \mathbf{x}] = 0. \tag{11.147}$$

Now, r commutes with β, as well as the components of $\boldsymbol{\Sigma}$ and $\mathbf{L}$. Hence,

$$[\zeta, r] = 0. \tag{11.148}$$

Moreover, β commutes with the components of $\mathbf{L}$, and can easily be shown to commute with all of the components of $\boldsymbol{\Sigma}$. It follows that

$$[\zeta, \beta] = 0. \tag{11.149}$$

Hence, Equations (11.128), (11.144), (11.148), and (11.149) imply that

$$[\zeta, H] = 0. \tag{11.150}$$

In other words, an eigenstate of the Hamiltonian is a simultaneous eigenstate of ζ. Now,

$$\zeta^2 = [\beta(\mathbf{\Sigma} \cdot \mathbf{L} + \hbar)]^2 = (\mathbf{\Sigma} \cdot \mathbf{L} + \hbar)^2 = J^2 + \frac{1}{4}\hbar^2, \tag{11.151}$$

where use has been made of Equation (11.137), as well as $\beta^2 = 1$. It follows that the eigenvalues of ζ^2 are $j(j+1)\hbar^2 + (1/4)\hbar^2 = (j + 1/2)^2\hbar^2$. Thus, the eigenvalues of ζ can be written $k\hbar$, where $k = \pm(j + 1/2)$ is a non-zero integer.

Equation (11.101) implies that

$$(\mathbf{\Sigma} \cdot \mathbf{x})(\mathbf{\Sigma} \cdot \mathbf{p}) = \mathbf{x} \cdot \mathbf{p} + i\mathbf{\Sigma} \cdot \mathbf{x} \times \mathbf{p} = r\,p_r + i\mathbf{\Sigma} \cdot \mathbf{L}$$

$$= r\,p_r + i(\beta\zeta - \hbar), \tag{11.152}$$

where use has been made of Equations (11.130) and (11.145).

It is helpful to define the dimensionless operator ϵ, where

$$r\epsilon = \boldsymbol{\alpha} \cdot \mathbf{x}. \tag{11.153}$$

Moreover, it is evident that

$$[\epsilon, r] = 0. \tag{11.154}$$

Hence,

$$r^2\,\epsilon^2 = (\boldsymbol{\alpha} \cdot \mathbf{x})^2 = \frac{1}{2}\sum_{i,j=1,3}\{\alpha_i, \alpha_j\}\,x^i\,x^j = \sum_{i=1,3}x^i\,x^i = r^2, \tag{11.155}$$

where use has been made of Equation (11.24). It follows that

$$\epsilon^2 = 1. \tag{11.156}$$

We have already seen that ζ commutes with $\boldsymbol{\alpha} \cdot \mathbf{x}$ and r. Thus,

$$[\zeta, \epsilon] = 0. \tag{11.157}$$

Because $\mathbf{\Sigma}$ commutes with $\mathbf{x}$ and $\mathbf{p}$, and $\mathbf{x} \cdot \mathbf{p} = -i\hbar\,r\,\partial/\partial r$, as well as $r\,\partial\mathbf{x}/\partial r = \mathbf{x}$, we obtain

$$(\mathbf{\Sigma} \cdot \mathbf{x})(\mathbf{x} \cdot \mathbf{p}) - (\mathbf{x} \cdot \mathbf{p})(\mathbf{\Sigma} \cdot \mathbf{x}) = \mathbf{\Sigma} \cdot [\mathbf{x}(\mathbf{x} \cdot \mathbf{p}) - (\mathbf{x} \cdot \mathbf{p})\mathbf{x}] = i\hbar\mathbf{\Sigma} \cdot \mathbf{x}. \tag{11.158}$$

However, $\mathbf{x} \cdot \mathbf{p} = r\,p_r$ and $\mathbf{\Sigma} \cdot \mathbf{x} = \gamma^5\,r\,\epsilon$, so, multiplying through by γ^5, we get

$$r^2\,\epsilon\,p_r - r\,p_r\,r\,\epsilon = i\hbar\,r\,\epsilon. \tag{11.159}$$

Equation (11.131) then yields

$$[\epsilon, p_r] = 0. \tag{11.160}$$

Equation (11.152) implies that

$$(\alpha \cdot \mathbf{x})(\alpha \cdot \mathbf{p}) = r\,p_r + \mathrm{i}\,(\beta\zeta - \hbar). \qquad (11.161)$$

Making use of Equations (11.148), (11.153), (11.154), and (11.156), we get

$$\alpha \cdot \mathbf{p} = \epsilon\,(p_r - \mathrm{i}\,\hbar/r) + \mathrm{i}\,\epsilon\,\beta\,\zeta/r. \qquad (11.162)$$

Hence, the Hamiltonian (11.128) becomes

$$H = -e\,\phi(r) + c\,\epsilon\,(p_r - \mathrm{i}\,\hbar/r) + \mathrm{i}\,c\,\epsilon\,\beta\,\zeta/r + \beta\,m_e\,c^2. \qquad (11.163)$$

Now, we wish to solve the energy eigenvalue problem

$$H\,\psi = E\,\psi, \qquad (11.164)$$

where E is the energy eigenvalue. However, we have already shown that an eigenstate of the Hamiltonian is a simultaneous eigenstate of the ζ operator belonging to the eigenvalue $k\,\hbar$, where k is a non-zero integer. Hence, the eigenvalue problem reduces to

$$\left[-e\,\phi(r) + c\,\epsilon\,(p_r - \mathrm{i}\,\hbar/r) + \mathrm{i}\,c\,\hbar\,k\,\epsilon\,\beta/r + \beta\,m_e\,c^2\right]\psi = E\,\psi, \qquad (11.165)$$

which only involves the radial coordinate, r. It is easily demonstrated that ϵ anti-commutes with β. Hence, given that β takes the form (11.32), and that $\epsilon^2 = 1$, we can represent ϵ as the matrix

$$\epsilon = \begin{pmatrix} 0, & -\mathrm{i} \\ \mathrm{i}, & 0 \end{pmatrix}. \qquad (11.166)$$

Thus, writing ψ in the spinor form

$$\psi = \begin{pmatrix} \psi_a(r) \\ \psi_b(r) \end{pmatrix}, \qquad (11.167)$$

and making use of Equation (11.132), the energy eigenvalue problem for an electron in a central field reduces to the following two coupled radial differential equations:

$$\hbar\,c\left(\frac{d}{dr} + \frac{k+1}{r}\right)\psi_b + (E - m_e\,c^2 + e\,\phi)\,\psi_a = 0, \qquad (11.168)$$

$$\hbar\,c\left(\frac{d}{dr} - \frac{k-1}{r}\right)\psi_a - (E + m_e\,c^2 + e\,\phi)\,\psi_b = 0. \qquad (11.169)$$

11.8   Fine Structure of Hydrogen Energy Levels

For the case of a hydrogen atom,

$$\phi(r) = \frac{e}{4\pi\,\epsilon_0\,r}. \tag{11.170}$$

Hence, Equations (11.168) and (11.169) yield

$$\left(\frac{1}{a_1} - \frac{\alpha}{y}\right)\psi_a - \left(\frac{d}{dy} + \frac{k+1}{y}\right)\psi_b = 0, \tag{11.171}$$

$$\left(\frac{1}{a_2} + \frac{\alpha}{y}\right)\psi_b - \left(\frac{d}{dy} - \frac{k-1}{y}\right)\psi_a = 0, \tag{11.172}$$

where $y = r/a_0$, and

$$a_1 = \frac{\alpha}{1 - \mathcal{E}}, \tag{11.173}$$

$$a_2 = \frac{\alpha}{1 + \mathcal{E}}, \tag{11.174}$$

with $\mathcal{E} = E/(m_e\,c^2)$. Here, $a_0 = 4\pi\,\epsilon_0\,\hbar^2/(m_e\,e^2)$ is the Bohr radius, and $\alpha = e^2/(4\pi\,\epsilon_0\,\hbar\,c)$ the fine structure constant. Writing

$$\psi_a(y) = \frac{e^{-y/a}}{y}\,f(y), \tag{11.175}$$

$$\psi_b(y) = \frac{e^{-y/a}}{y}\,g(y), \tag{11.176}$$

where

$$a = (a_1\,a_2)^{1/2} = \frac{\alpha}{\sqrt{1 - \mathcal{E}^2}}, \tag{11.177}$$

we obtain

$$\left(\frac{1}{a_1} - \frac{\alpha}{y}\right)f - \left(\frac{d}{dy} - \frac{1}{a} + \frac{k}{y}\right)g = 0, \tag{11.178}$$

$$\left(\frac{1}{a_2} + \frac{\alpha}{y}\right)g - \left(\frac{d}{dy} - \frac{1}{a} - \frac{k}{y}\right)f = 0. \tag{11.179}$$

Let us search for power-law solutions of the form

$$f(y) = \sum_s c_s\,y^s, \tag{11.180}$$

$$g(y) = \sum_s c_s'\,y^s, \tag{11.181}$$

where successive values of s differ by unity. Substitution of these solutions into Equations (11.178) and (11.179) leads to the recursion relations

$$\frac{c_{s-1}}{a_1} - \alpha c_s - (s+k) c'_s + \frac{c'_{s-1}}{a} = 0, \tag{11.182}$$

$$\frac{c'_{s-1}}{a_2} + \alpha c'_s - (s-k) c_s + \frac{c_{s-1}}{a} = 0. \tag{11.183}$$

Multiplying the first of these equations by a, and the second by a_2, and then subtracting, we eliminate both c_{s-1} and c'_{s-1}, because $a/a_1 = a_2/a$. We are left with

$$[a\,\alpha - a_2\,(s-k)]\,c_s + [a_2\,\alpha + a\,(s+k)]\,c'_s = 0. \tag{11.184}$$

The physical boundary conditions at $y = 0$ require that $y\,\psi_a \to 0$ and $y\,\psi_b \to 0$ as $y \to 0$. Thus, it follows from Equations (11.175) and (11.176) that $f \to 0$ and $g \to 0$ as $y \to 0$. Consequently, the series (11.180) and (11.181) must terminate at small positive s. If s_0 is the minimum value of s for which c_s and c'_s do not both vanish then it follows from Equations (11.182) and (11.183), putting $s = s_0$ and $c_{s_0-1} = c'_{s_0-1} = 0$, that

$$\alpha\,c_{s_0} + (s_0 + k)\,c'_{s_0} = 0, \tag{11.185}$$

$$\alpha\,c'_{s_0} - (s_0 - k)\,c_{s_0} = 0, \tag{11.186}$$

which implies that

$$\alpha^2 = -s_0^2 + k^2. \tag{11.187}$$

Because the boundary condition requires that the minimum value of s_0 be greater than zero, we must take

$$s_0 = (k^2 - \alpha^2)^{1/2}. \tag{11.188}$$

To investigate the convergence of the series (11.180) and (11.181) at large y, we shall determine the ratio c_s/c_{s-1} for large s. In the limit of large s, Equations (11.183) and (11.184) yield

$$s\,c_s \simeq \frac{c_{s-1}}{a} + \frac{c'_{s-1}}{a_2}, \tag{11.189}$$

$$a_2\,c_s \simeq a\,c'_s, \tag{11.190}$$

because $\alpha \simeq 1/137 \ll 1$. Thus,

$$\frac{c_s}{c_{s-1}} \simeq \frac{2}{a\,s}. \tag{11.191}$$

However, this is the ratio of coefficients in the series expansion of $\exp(2\,y/a)$. Hence, we deduce that the series (11.180) and (11.181) diverge unphysically at large y unless they terminate at large s.

Suppose that the series (11.180) and (11.181) terminate with the terms c_s and c'_s, so that $c_{s+1} = c'_{s+1} = 0$. It follows from Equations (11.182) and (11.183), with $s + 1$ substituted for s, that

$$\frac{c_s}{a_1} + \frac{c'_s}{a} = 0, \tag{11.192}$$

$$\frac{c'_s}{a_2} + \frac{c_s}{a} = 0. \tag{11.193}$$

These two expressions are equivalent, because $a^2 = a_1 a_2$. When combined with Equation (11.184) they give

$$a_1 \left[a\,\alpha - a_2\,(s - k) \right] = a\left[a_2\,\alpha + a\,(s + k) \right], \tag{11.194}$$

which reduces to

$$2\,a_1\,a_2\,s = a\,(a_1 - a_2)\,\alpha, \tag{11.195}$$

or

$$\mathcal{E} = \left(1 + \frac{\alpha^2}{s^2} \right)^{-1/2}. \tag{11.196}$$

Here, s, which specifies the last term in the series, must be greater than s_0 by some non-negative integer i. Thus,

$$s = i + (k^2 - \alpha^2)^{1/2} = i + \left[(j + 1/2)^2 - \alpha^2 \right]^{1/2}, \tag{11.197}$$

where $j\,(j + 1)\,\hbar^2$ is the eigenvalue of J^2. Hence, the energy eigenvalues of the hydrogen atom become

$$\frac{E}{m_e\,c^2} = \left\{ 1 + \frac{\alpha^2}{\left(i + [(j + 1/2)^2 - \alpha^2]^{1/2} \right)^2} \right\}^{-1/2} \tag{11.198}$$

[Darwin (1928); Gordon (1928)]. Given that $\alpha \simeq 1/137$, we can expand the previous expression in α^2 to give

$$\frac{E}{m_e\,c^2} = 1 - \frac{\alpha^2}{2\,n^2} - \frac{\alpha^4}{2\,n^4} \left(\frac{n}{j + 1/2} - \frac{3}{4} \right) + O(\alpha^6), \tag{11.199}$$

where $n = i + j + 1/2$ is a positive integer. Of course, the first term in the previous expression corresponds to the electron's rest mass energy. The second term corresponds to the standard non-relativistic expression for the hydrogen energy levels, with n playing the role of the radial quantum number. (See Section 4.6.) Finally, the third term corresponds to the fine structure correction to these energy levels. (See Exercise 7.14). Note that this correction only depends on the quantum numbers n and j. Now, we showed in Exercise 7.14 that the fine structure correction to the energy levels of the hydrogen atom is a combined effect of spin-orbit coupling, the electron's relativistic mass increase, and the Darwin term. Hence, it is evident that all of these effects are automatically taken into account in the Dirac equation.

11.9 Positron Theory

We have already mentioned that the Dirac equation admits twice as many solutions as it ought to, half of them belonging to states with negative values for the kinetic energy $c\,p^0 + e\,\Phi^0$. This difficulty was introduced when we passed from Equation (11.16) to Equation (11.18), and is inherent in any relativistic theory.

Let us examine the negative energy solutions of the equation

$$\left[\left(p^0 + \frac{e}{c}\Phi^0\right) - \alpha_1\left(p^1 + \frac{e}{c}\Phi^1\right) - \alpha_2\left(p^2 + \frac{e}{c}\Phi^2\right)\right.$$
$$\left. -\alpha_3\left(p^3 + \frac{e}{c}\Phi^3\right) - \beta m_e c\right]\psi = 0 \quad (11.200)$$

a little more closely. For this purpose, it is convenient to use a representation of the α's and β in which all the elements of the matrices α_1, α_2, and α_3 are real, and all of those of the matrix representing β are imaginary or zero. Such a representation can be obtained from the standard representation by interchanging the expressions for α_2 and β. If Equation (11.200) is expressed as a matrix equation in this representation, and we then substitute $-i$ for i, we get [remembering the factor i in Equation (11.14)]

$$\left[\left(p^0 - \frac{e}{c}\Phi^0\right) - \alpha_1\left(p^1 - \frac{e}{c}\Phi^1\right) - \alpha_2\left(p^2 - \frac{e}{c}\Phi^2\right)\right.$$
$$\left. -\alpha_3\left(p^3 - \frac{e}{c}\Phi^3\right) - \beta m_e c\right]\psi^* = 0. \quad (11.201)$$

Thus, each solution, ψ, of the wave equation (11.200) has for its complex conjugate, ψ^*, a solution of the wave equation (11.201). Furthermore, if the solution, ψ, of Equation (11.200) belongs to a negative value for $c\,p^0 + e\,\Phi^0$ then the corresponding solution, ψ^*, of Equation (11.201) will belong to a positive value for $c\,p^0 - e\,\Phi^0$. But, the operator in Equation (11.201) is just what we would get if we substituted $-e$ for e in the operator in Equation (11.200). It follows that each negative energy solution of Equation (11.200) is the complex conjugate of a positive energy solution of the wave equation obtained from Equation (11.200) by the substitution of $-e$ for e. The latter solution represents an electron of charge $+e$ (instead of $-e$, as we have had up to now) moving through the given electromagnetic field.

We conclude that the negative energy solutions of Equation (11.200) refer to the motion of a new type of particle having the mass of an electron, but the opposite charge [Dirac (1930)]. Such particles have been observed experimentally, and are called *positrons* [Anderson (1933)]. Note that we cannot simply assert that the negative energy solutions represent positrons, because this would make the dynamical relations all wrong. For instance, it is certainly not true that a positron has a negative kinetic energy. Instead, we assume that nearly all of the negative energy states are occupied, with one electron in each state, in accordance with the Pauli exclusion principle. An unoccupied negative energy state will now appear as a particle with a positive energy, because to make it disappear we would have to add an electron with a negative energy to the system. We assume that these unoccupied negative energy states correspond to positrons.

The previous assumptions require there to be a distribution of electrons of infinite density everywhere in space. A perfect vacuum is a region of space in which all states of positive energy are unoccupied, and all of those of negative energy are occupied. In such a vacuum, the Maxwell equation

$$\nabla \cdot \mathbf{E} = 0 \tag{11.202}$$

must be valid. This implies that the infinite distribution of negative energy electrons does not contribute to the electric field. Thus, only departures from the vacuum distribution contribute to the electric charge density ρ in the Maxwell equation

$$\nabla \cdot \mathbf{E} = \frac{\rho}{\epsilon_0}. \tag{11.203}$$

In other words, there is a contribution $-e$ for each occupied state of positive energy, and a contribution $+e$ for each unoccupied state of negative energy.

The exclusion principle ordinarily prevents a positive energy electron from making transitions to states of negative energy. However, it is still possible for such an electron to drop into an unoccupied state of negative energy. In this case, we would observe an electron and a positron simultaneously disappearing, their energy being emitted in the form of radiation. The converse process would consist in the creation of an electron-positron pair from electromagnetic radiation.

11.10   Exercises

11.1 Demonstrate that the four operators $\partial_\mu \equiv \partial/\partial x^\mu$ transform under Lorentz transformation as the covariant components of 4-vector, whereas the four operators $\partial^\mu \equiv \partial/\partial x_\mu$ transform as the contravariant components of the same vector.

11.2 Demonstrate that Equation (11.29) is equivalent to Equations (11.24)–(11.26).

11.3 Noting that $\alpha_i = -\beta \alpha_i \beta$, prove that the α_i and β matrices all have zero trace. Hence, deduce that each of these matrices has n eigenvalues $+1$, and n eigenvalues -1, where $2n$ is the dimension of the matrices.

11.4 Verify that the matrices (11.30) and (11.31) satisfy the anti-commutation relations (11.29).

11.5 Verify that the matrices (11.32) and (11.33) satisfy Equations (11.24)–(11.26).

11.6 Verify that if

$$\partial_\mu j^\mu = 0,$$

where j^μ is a 4-vector field, then

$$\int d^3\mathbf{x}\, j^0$$

is Lorentz invariant, where the integral is over all space, and it is assumed that $j^\mu \to 0$ as $|\mathbf{x}| \to \infty$.

11.7 Verify that Equation (11.77) is a solution of Equations (11.76).

11.8 A Lorentz transformation between frames S and S' takes the form

$$x^{\mu'} = a^\mu_\nu x^\nu.$$

If

$$a^\mu_\nu = g^\mu_\nu + \delta\omega\, I^\mu_\nu,$$

where $I^0_1 = I^1_0 = -1$, and $I^\mu_\nu = 0$ otherwise, then the transformation corresponds to an infinitesimal velocity boost, $\delta v = c\,\delta\omega$, parallel to the x_1-axis. Show that if a finite boost is built up from a great many such boosts then the transformation matrix becomes

$$a^\mu_\nu = \begin{pmatrix} \cosh\omega, & -\sinh\omega, & 0, & 0 \\ -\sinh\omega, & \cosh\omega, & 0, & 0 \\ 0, & 0, & 1, & 0 \\ 0, & 0, & 0, & 1 \end{pmatrix},$$

where $v = c\tanh\omega$ is the velocity of frame S' relative to frame S. Show that the corresponding transformation rule for spinor wavefunctions is $\psi' = A\,\psi$, where

$$A = \exp\left(-\frac{\omega}{2}\alpha_1\right) = \cosh\left(\frac{\omega}{2}\right) - \sinh\left(\frac{\omega}{2}\right)\alpha_1.$$

11.9 Show that the transformation rule for spinor wavefunctions associated with a Lorentz transformation from frame S to some frame S' moving with velocity $\mathbf{v} = v\mathbf{n}$ with respect to S is $\psi' = A\,\psi$, where

$$A = \exp\left[-\frac{\omega}{2}(\mathbf{n}\cdot\boldsymbol{\alpha})\right] = \cosh\left(\frac{\omega}{2}\right) - \sinh\left(\frac{\omega}{2}\right)(\mathbf{n}\cdot\boldsymbol{\alpha}),$$

and $\omega = \tanh^{-1}(v/c)$.

11.10 Consider the spinors

$$\psi^r = e^{-i\,\epsilon_r\,(m_e\,c^2/\hbar)\,t}\,w^r(0),$$

for $r = 1, 2, 3, 4$. Here, $\epsilon_r = +1$ for $r = 1, 2$, and $\epsilon_r = -1$ for $r = 3, 4$. Moreover,

$$w^1(0) = \begin{pmatrix} 1 \\ 0 \\ 0 \\ 0 \end{pmatrix}, \quad w^2(0) = \begin{pmatrix} 0 \\ 1 \\ 0 \\ 0 \end{pmatrix}, \quad w^3(0) = \begin{pmatrix} 0 \\ 0 \\ 1 \\ 0 \end{pmatrix}, \quad w^4(0) = \begin{pmatrix} 0 \\ 0 \\ 0 \\ 1 \end{pmatrix}.$$

Verify that the ψ^r are solutions of the Dirac equation in free space corresponding to electrons of energy $\epsilon_r\,m_e\,c^2$, momentum $\mathbf{p} = \mathbf{0}$, and spin angular momentum parallel to the x_3-axis $S_3 = \zeta_r\,\hbar/2$, where $\zeta_r = +1$ for $r = 1, 3$, and $\zeta_r = -1$ for $r = 2, 4$.

11.11 Show that the four solutions of the Dirac equation corresponding to an electron of energy E and momentum $\mathbf{p}$ moving in free space take the form

$$\psi^r = e^{-i\,\epsilon_r\,(p_\mu\,x^\mu/\hbar)}\,w^r(\mathbf{p}),$$

where

$$w^1(\mathbf{p}) = \sqrt{\frac{E + m_e\,c^2}{2\,m_e\,c^2}} \begin{pmatrix} 1 \\ 0 \\ \frac{p_z\,c}{E + m_e\,c^2} \\ \frac{p_+\,c}{E + m_e\,c^2} \end{pmatrix}, \quad w^2(\mathbf{p}) = \sqrt{\frac{E + m_e\,c^2}{2\,m_e\,c^2}} \begin{pmatrix} 0 \\ 1 \\ \frac{p_-\,c}{E + m_e\,c^2} \\ \frac{-p_z\,c}{E + m_e\,c^2} \end{pmatrix},$$

$$w^3(\mathbf{p}) = \sqrt{\frac{E + m_e\,c^2}{2\,m_e\,c^2}} \begin{pmatrix} \frac{p_z\,c}{E + m_e\,c^2} \\ \frac{p_+\,c}{E + m_e\,c^2} \\ 1 \\ 0 \end{pmatrix}, \quad w^4(\mathbf{p}) = \sqrt{\frac{E + m_e\,c^2}{2\,m_e\,c^2}} \begin{pmatrix} \frac{p_-\,c}{E + m_e\,c^2} \\ \frac{-p_z\,c}{E + m_e\,c^2} \\ 0 \\ 1 \end{pmatrix}.$$

Here, $p_\pm = p_x \pm i\,p_y$. Demonstrate that these spinors become identical to the ψ^r of the previous exercise in the limit that $\mathbf{p} \to \mathbf{0}$.

11.12 Verify that the 4×4 matrices Σ_i, defined in Equation (11.98), satisfy the standard anti-commutation relations for Pauli matrices: that is,

$$\{\Sigma_i, \Sigma_j\} = 2\,\delta_{ij}.$$

Appendix A

Physical Constants

Planck's constant : $h = 6.62606957 \times 10^{-34}\,\mathrm{J\,s}$

Reduced Planck's constant : $\hbar = 1.05457173 \times 10^{-34}\,\mathrm{J\,s}$

Electron mass : $m_e = 9.10938291 \times 10^{-31}\,\mathrm{kg}$

Electron charge (magnitude) : $e = 1.60217657 \times 10^{-19}\,\mathrm{C}$

Proton mass : $m_p = 1.67262178 \times 10^{-27}\,\mathrm{kg}$

Permittivity of free space : $\epsilon_0 = 8.85418782 \times 10^{-12}\,\mathrm{F\,m^{-1}}$

Permeability of free space : $\mu_0 = 4\pi \times 10^{-7}\,\mathrm{H\,m^{-1}}$

Speed of light in vacuum : $c = 2.99792458 \times 10^{8}\,\mathrm{m\,s^{-1}}$

Fine structure constant : $\alpha = \dfrac{e^2}{4\pi\,\epsilon_0\,\hbar\,c} = \dfrac{1}{137.035999}$

Bohr radius : $a_0 = \dfrac{\hbar}{\alpha\,m_e\,c} = 5.2917721092 \times 10^{-11}\,\mathrm{m}$

Hydrogen ground-state energy : $E_0 = -\dfrac{1}{2}\,\alpha^2\,m_e\,c^2 = -13.6057\,\mathrm{eV}$

Bohr magnetron : $\mu_B = \dfrac{e\,\hbar}{2\,m_e} = 9.27400968 \times 10^{-24}\,\mathrm{J\,T^{-1}}$

Nuclear magnetron : $\mu_N = \dfrac{e\,\hbar}{2\,m_p} = 5.05078324 \times 10^{-27}\,\mathrm{J\,T^{-1}}$

Classical electron radius : $r_e = \dfrac{e^2}{4\pi\,\epsilon_0\,m_e\,c^2} = 2.8179403267 \times 10^{-15}\,\mathrm{m}$

Molar ideal gas constant : $\qquad R = 8.3144621\,\mathrm{J\,K^{-1}}$

Boltzmann constant : $\qquad k_B = 1.3806488 \times 10^{-23}\,\mathrm{J\,K^{-1}}$

Appendix B

Solutions to Exercises

B.1 Chapter 1

1.1 The velocity of an electron in a classical circular orbit of radius r about a proton satisfies

$$\frac{v^2}{c^2} = \frac{e^2}{4\pi \epsilon_0 m_e c^2 r} = \alpha^2 \frac{a_0}{r}, \tag{B.1.1}$$

where $a_0 = 4\pi \epsilon_0 \hbar^2/(m_e e^2)$ is the Bohr radius, and $\alpha = e^2/(4\pi \epsilon_0 \hbar c)$ the fine structure constant. The energy of the orbit is

$$\mathcal{E} = \frac{1}{2} m_e v^2 - \frac{e^2}{4\pi \epsilon_0 r} = -\frac{1}{2} m_e c^2 \alpha^2 \frac{a_0}{r}. \tag{B.1.2}$$

The orbital acceleration, a, satisfies

$$a \frac{a_0}{c^2} = \frac{v^2}{c^2} \frac{a_0}{r} = \alpha^2 \left(\frac{a_0}{r}\right)^2. \tag{B.1.3}$$

Finally, the radiated power is

$$P = \frac{e^2 a^2}{6\pi \epsilon_0 c^3} = \frac{2}{3} \frac{c}{a_0} m_e c^2 \alpha^6 \left(\frac{a_0}{r}\right)^4. \tag{B.1.4}$$

Conservation of energy yields

$$\frac{d\mathcal{E}}{dt} = -P, \tag{B.1.5}$$

which implies that

$$\frac{d}{dt} \left(\frac{r}{a_0}\right)^3 = -\frac{4c\alpha^4}{a_0}. \tag{B.1.6}$$

Hence, assuming that $r = a_0$ at $t = 0$, which is the case for an electron initially (i.e., at $t = 0$) in the ground state, we obtain

$$r = a_0 (1 - t/\tau)^{1/3}, \tag{B.1.7}$$

299

where $\tau = a_0/(4\,\alpha^4\,c) = 1.6 \times 10^{-11}$ s is the classical lifetime of the hydrogen atom: that is, the time required for the ground-state orbital radius to decay to zero.

1.2 Let

$$X = \sum_{i=1,N} |i\rangle\langle i| - 1.$$ (B.1.8)

It follows that

$$X|j\rangle = \sum_{i=1,N} |i\rangle\langle i|j\rangle - |j\rangle = \sum_{i=1,N} \delta_{ij}\,|i\rangle - |j\rangle = |j\rangle - |j\rangle = |0\rangle$$ (B.1.9)

for $j = 1, N$. Hence, because the $|j\rangle$, for $j = 1, N$, span the ket space, we deduce that

$$X|A\rangle = |0\rangle,$$ (B.1.10)

where $|A\rangle$ is a general ket. Hence, X is the null operator. Thus, $X + 1 = 1$, where 1 is the identity operator, which implies that

$$\sum_{i=1,N} |i\rangle\langle i| = 1.$$ (B.1.11)

1.3 Let the ket space be N-dimensional, and let the $|i\rangle$ be a set of orthonormal kets that span the space. It follows that

$$\sum_{i=1,N} |i\rangle\langle i| = 1.$$ (B.1.12)

Thus,

$$|A\rangle = 1\,|A\rangle = \sum_{i=1,N} |i\rangle\langle i|A\rangle = \sum_{i=1,N} \alpha_i\,|i\rangle,$$ (B.1.13)

where

$$\alpha_i = \langle i|A\rangle.$$ (B.1.14)

Likewise,

$$\langle A| = \langle A|\,1 = \sum_{i=1,N} \langle A|i\rangle\langle i| = \sum_{i=1,N} \beta_i\,\langle i|,$$ (B.1.15)

where

$$\beta_i = \langle A|i\rangle.$$ (B.1.16)

However,

$$|A\rangle \overset{\text{DC}}{\longleftrightarrow} \langle A|,$$ (B.1.17)

and

$$\alpha_i\,|i\rangle \overset{\text{DC}}{\longleftrightarrow} \alpha_i^*\,\langle i|.$$ (B.1.18)

Hence,

$$\beta_i = \alpha_i^*,$$ (B.1.19)

or

$$\langle A|i\rangle = \langle i|A\rangle^*.$$ (B.1.20)

Now,

$$\langle A|B \rangle = \langle A| \, 1 \, |B \rangle = \sum_{n=1,N} \langle A|i \rangle \langle i|B \rangle, \tag{B.1.21}$$

and

$$\langle B|A \rangle = \langle B| \, 1 \, |A \rangle = \sum_{n=1,N} \langle B|i \rangle \langle i|A \rangle. \tag{B.1.22}$$

Thus, it follows from Equation (B.1.20) that

$$\langle B|A \rangle = \sum_{n=1,N} \langle i|B \rangle^* \langle A|i \rangle^* = \left(\sum_{n=1,N} \langle A|i \rangle \langle i|B \rangle \right)^* = \langle A|B \rangle^*. \tag{B.1.23}$$

1.4 Let the ket space be N-dimensional, and let the $|i\rangle$ be a set of orthonormal kets that span the space. It follows that $\sum_{i=1,N} |i\rangle\langle i| = 1$.

(a) Now,

$$\langle A|i \rangle = \langle i|A \rangle^*. \tag{B.1.24}$$

Let

$$|A\rangle = X \, |j\rangle. \tag{B.1.25}$$

It follows that

$$\langle A| = \langle j| \, X^\dagger, \tag{B.1.26}$$

and hence that

$$\langle j| \, X^\dagger \, |i\rangle = \langle i| \, X \, |j\rangle^*. \tag{B.1.27}$$

Now,

$$\langle A| X \, |B\rangle = \sum_{i,j=1,N} \langle A|i \rangle \langle i| X \, |j \rangle \langle j|B \rangle, \tag{B.1.28}$$

and so

$$\langle B| X^\dagger \, |A\rangle = \sum_{i,j=1,N} \langle B|j \rangle \langle j| X^\dagger \, |i \rangle \langle i|A \rangle = \sum_{i,j=1,N} \langle j|B \rangle^* \langle i| X \, |j \rangle^* \langle A|i \rangle^*$$

$$= \left(\sum_{i,j=1,N} \langle A|i \rangle \langle i| X \, |j \rangle \langle j|B \rangle \right)^* = \langle A| X \, |B\rangle^*. \tag{B.1.29}$$

(b) From Part (a), we have

$$\langle A| X \, |B\rangle = \langle B| X^\dagger \, |A\rangle^*. \tag{B.1.30}$$

Making the substitutions $A \leftrightarrow B$ and $X \to X^\dagger$, we also have

$$\langle B| X^\dagger \, |A\rangle = \langle A| (X^\dagger)^\dagger \, |B\rangle^*. \tag{B.1.31}$$

Hence,

$$\langle A| X \, |B\rangle = \left(\langle A| (X^\dagger)^\dagger \, |B\rangle^* \right)^* = \langle A| (X^\dagger)^\dagger \, |B\rangle. \tag{B.1.32}$$

Because $|A\rangle$ and $|B\rangle$ are general kets, we conclude that

$$(X^\dagger)^\dagger = X. \tag{B.1.33}$$

(c) We have

$$\langle A|B\rangle = \langle B|A\rangle^*. \tag{B.1.34}$$

Let

$$|A\rangle = X^\dagger |C\rangle, \tag{B.1.35}$$
$$|B\rangle = Y|D\rangle. \tag{B.1.36}$$

It follows that

$$\langle A| = \langle C|(X^\dagger)^\dagger = \langle C|X, \tag{B.1.37}$$
$$\langle B| = \langle D|Y^\dagger, \tag{B.1.38}$$

and

$$\langle C|XY|D\rangle = \langle D|Y^\dagger X^\dagger |C\rangle^*. \tag{B.1.39}$$

However, from Part (a),

$$\langle C|XY|D\rangle = \langle D|(XY)^\dagger |C\rangle^*, \tag{B.1.40}$$

which implies that

$$\langle D|(XY)^\dagger |C\rangle = \langle D|Y^\dagger X^\dagger |C\rangle. \tag{B.1.41}$$

Because $|C\rangle$ and $|D\rangle$ are general kets, we deduce that

$$(XY)^\dagger = Y^\dagger X^\dagger. \tag{B.1.42}$$

(d) Now, from Part (c),

$$(XA)^\dagger = A^\dagger X^\dagger. \tag{B.1.43}$$

Let

$$A = YZ. \tag{B.1.44}$$

It follows that

$$A^\dagger = Z^\dagger Y^\dagger. \tag{B.1.45}$$

Hence,

$$(XYZ)^\dagger = Z^\dagger Y^\dagger X^\dagger. \tag{B.1.46}$$

(e) Let

$$X = |A\rangle\langle B|. \tag{B.1.47}$$

Thus,

$$\langle C|X|D\rangle = \langle C|A\rangle\langle B|D\rangle. \tag{B.1.48}$$

Furthermore, from Part (a),

$$\langle D| X^\dagger |C\rangle = \langle C| X |D\rangle^* = \langle C|A\rangle^* \langle B|D\rangle^* = \langle D|B\rangle \langle A|C\rangle. \tag{B.1.49}$$

It follows that

$$\langle D| (|A\rangle\langle B|)^\dagger |C\rangle = \langle D| (|B\rangle\langle A|) |C\rangle. \tag{B.1.50}$$

Because $|C\rangle$ and $|D\rangle$ are general kets, we conclude that

$$(|A\rangle\langle B|)^\dagger = |B\rangle\langle A|. \tag{B.1.51}$$

1.5 Given that A and B are Hermitian, we have

$$A^\dagger = A, \tag{B.1.52}$$

$$B^\dagger = B. \tag{B.1.53}$$

Consider $C = A\,B$. It follows that

$$C^\dagger = B^\dagger A^\dagger = B\,A. \tag{B.1.54}$$

Thus,

$$C^\dagger = C \tag{B.1.55}$$

only if

$$B\,A = A\,B. \tag{B.1.56}$$

According to the *binomial theorem* [Abramowitz and Stegun (1965)],

$$D = (A + B)^n = \sum_{k=0,n} \binom{n}{k} A^{n-k} B^k, \tag{B.1.57}$$

assuming that A and B commute. It follows that

$$D^\dagger = \sum_{k=0,n} \binom{n}{k} (B^\dagger)^k (A^\dagger)^{n-k} = \sum_{k=0,n} \binom{n}{k} B^k A^{n-k}. \tag{B.1.58}$$

Thus, $D^\dagger = D$ provided that A and B commute.

1.6 Let $B = A + A^\dagger$. Thus,

$$B^\dagger = A^\dagger + (A^\dagger)^\dagger = A^\dagger + A = A + A^\dagger = B. \tag{B.1.59}$$

Let $C = i(A - A^\dagger)$. Thus,

$$C^\dagger = -i[A^\dagger - (A^\dagger)^\dagger] = -i(A^\dagger - A) = i(A - A^\dagger) = C. \tag{B.1.60}$$

Let $D = A\,A^\dagger$. Thus,

$$D^\dagger = (A^\dagger)^\dagger A^\dagger = A\,A^\dagger = D. \tag{B.1.61}$$

Hence, B, C, and D are all Hermitian.

1.7 Let

$$X = \exp(iH) \equiv \sum_{n=0,\infty} \frac{(iH)^n}{n!}, \tag{B.1.62}$$

where H is an Hermitian operator. It follows that

$$X^\dagger = \sum_{n=0,\infty} \frac{[(iH)^n]^\dagger}{n!} = \sum_{n=0,\infty} \frac{[(iH)^\dagger]^n}{n!} = \sum_{n=0,\infty} \frac{(-iH^\dagger)^n}{n!} = \sum_{n=0,\infty} \frac{(-iH)^n}{n!}$$

$$\equiv \exp(-iH), \tag{B.1.63}$$

or

$$\exp(iH)^\dagger = \exp(-iH). \tag{B.1.64}$$

1.8 We have

$$\exp(A) = \sum_{n=0,\infty} \frac{A^n}{n!}, \tag{B.1.65}$$

$$\exp(B) = \sum_{n=0,\infty} \frac{B^n}{n!}, \tag{B.1.66}$$

$$\exp(A + B) = \sum_{n=0,\infty} \frac{(A + B)^n}{n!}. \tag{B.1.67}$$

However, according to the binomial theorem [Abramowitz and Stegun (1965)],

$$(A + B)^n = \sum_{k=0,n} \frac{n!}{(n - k)!\, k!} A^{n-k} B^k. \tag{B.1.68}$$

Note that the previous expression is only valid provided that A and B commute. Thus, it follows that

$$\exp(A + B) = \sum_{n=0,\infty} \sum_{k=0,n} \frac{A^{n-k} B^k}{(n - k)!\, k!}. \tag{B.1.69}$$

Now,

$$\exp(A)\exp(B) = \sum_{m=0,\infty} \sum_{k=0,\infty} \frac{A^m B^k}{m!\, k!}. \tag{B.1.70}$$

Let $m = n - k$. We obtain

$$\exp(A)\exp(B) = \sum_{k=0,\infty} \sum_{n=k,\infty} \frac{A^{n-k} B^k}{(n - k)!\, k!}. \tag{B.1.71}$$

However, if we reverse the order of the summation in k and n then we get

$$\exp(A)\exp(B) = \sum_{n=0,\infty} \sum_{k=0,n} \frac{A^{n-k} B^k}{(n - k)!\, k!}, \tag{B.1.72}$$

which is identical to Equation (B.1.69). Hence, we conclude that

$$\exp(A + B) = \exp(A)\,\exp(B),\tag{B.1.73}$$

provided that A and B commute.

1.9 Given that ξ is an observable with discrete eigenvalues, ξ', we can write

$$\langle \xi''|\xi'\rangle = \delta_{\xi''\xi'}.\tag{B.1.74}$$

Furthermore, any general ket, $|A\rangle$, can be expressed as a linear combination of the $|\xi'\rangle$: that is,

$$|A\rangle = \sum_{\xi'} \alpha_{\xi'}\,|\xi'\rangle.\tag{B.1.75}$$

It follows that

$$\langle \xi''|A\rangle = \sum_{\xi'} \alpha_{\xi'}\,\langle \xi''|\xi'\rangle = \sum_{\xi'} \alpha_{\xi'}\,\delta_{\xi''\xi'} = \alpha_{\xi''}.\tag{B.1.76}$$

Hence,

$$|A\rangle = \sum_{\xi'} \langle \xi'|A\rangle\,|\xi'\rangle.\tag{B.1.77}$$

Now, if

$$\sum_{\xi'} |\xi'\rangle\langle \xi'| = 1\tag{B.1.78}$$

then, operating on $|A\rangle$, we obtain

$$\sum_{\xi'} \langle \xi'|A\rangle\,|\xi'\rangle = |A\rangle.\tag{B.1.79}$$

However, Equations (B.1.77) and (B.1.79) are identical. Hence, we deduce the validity of Equation (B.1.78).

1.10 Let the $|\xi''_j\rangle$ for $j = 1, i - 1$ be a set of mutually orthogonal kets: that is,

$$\langle \xi''_k|\xi''_j\rangle = \langle \xi''_j|\xi''_j\rangle\,\delta_{jk}\tag{B.1.80}$$

for j, k in the range 1 to $i - 1$. Consider

$$|\xi''_i\rangle = |\xi'_i\rangle - \sum_{j=1,i-1} \frac{\langle \xi''_j|\xi'_i\rangle}{\langle \xi''_j|\xi''_j\rangle}\,|\xi''_j\rangle,\tag{B.1.81}$$

where the $|\xi'_i\rangle$ are a set of non-orthogonal kets. It follows from Equations (B.1.80) and (B.1.81) that

$$\langle \xi''_k|\xi''_i\rangle = \langle \xi''_k|\xi'_i\rangle - \langle \xi''_k|\xi'_i\rangle = 0,\tag{B.1.82}$$

for $k = 1, i - 1$. In other words, $|\xi''_i\rangle$ is orthogonal to all of the $|\xi''_j\rangle$, for $j = 1, i - 1$. Hence, we can recursively construct a set of orthogonal kets, the $|\xi''_i\rangle$, as linear combinations of the $|\xi'_i\rangle$. Moreover, there are as many $|\xi''_i\rangle$ as there are $|\xi'_i\rangle$. If the $|\xi'_i\rangle$ are eigenkets of an Hermitian operator belonging to the same eigenvalue then linear combinations of the $|\xi_i'\rangle$ are also eigenkets belonging to the same eigenvalue. Hence, the $|\xi_i''\rangle$ are a set of mutually orthogonal eigenkets belonging to this eigenvalue.

1.11 Let

$$H^\dagger = H, \tag{B.1.83}$$

and

$$c = \langle A|H|A\rangle. \tag{B.1.84}$$

In other words, H is an Hermitian operator, and c its expectation value. It follows that

$$c^* = \langle A|H^\dagger|A\rangle = \langle A|H|A\rangle = c. \tag{B.1.85}$$

Hence, c is real.

Let

$$H^\dagger = -H, \tag{B.1.86}$$

and

$$c = \langle A|H|A\rangle. \tag{B.1.87}$$

In other words, H is an anti-Hermitian operator, and c is its expectation value. It follows that

$$c^* = \langle A|H^\dagger|A\rangle = -\langle A|H|A\rangle = -c. \tag{B.1.88}$$

Hence, c is imaginary.

1.12 Given that H is Hermitian, we have

$$H^\dagger = H. \tag{B.1.89}$$

Let

$$c = \langle A|H^2|A\rangle = \langle A|H^\dagger H|A\rangle = \langle B|B\rangle, \tag{B.1.90}$$

where

$$|B\rangle = H|A\rangle. \tag{B.1.91}$$

However, $\langle B|B\rangle$ is real and non-negative (because it is the norm of a ket). Hence, c, which is the expectation value of H^2, is real and non-negative.

1.13 Suppose that

$$H^4 = 1. \tag{B.1.92}$$

Let

$$H|\lambda\rangle = \lambda|\lambda\rangle, \tag{B.1.93}$$

where $|\lambda\rangle$ is an eigenket corresponding to the eigenvalue λ. It follows that

$$H^2|\lambda\rangle = HH|\lambda\rangle = H\lambda|\lambda\rangle = \lambda H|\lambda\rangle = \lambda^2|\lambda\rangle. \tag{B.1.94}$$

Hence, by analogy,

$$H^4|\lambda\rangle = \lambda^4|\lambda\rangle. \tag{B.1.95}$$

Equations (B.1.92) and (B.1.95) imply that

$$\left(\lambda^4 - 1\right)|\lambda\rangle = |0\rangle. \tag{B.1.96}$$

Assuming that $|\lambda\rangle$ is not the null ket, we deduce that

$$\lambda^4 = 1. \tag{B.1.97}$$

The roots of this equation are ± 1, $\pm i$. However, an Hermitian operator can only have real eigenvalues. So, in this case, the possible eigenvalues are $\lambda = \pm 1$. In the general case, the possible eigenvalues are $\lambda = \pm 1$, $\pm i$.

1.14 Suppose that

$$U^\dagger U = 1. \tag{B.1.98}$$

Let

$$|B\rangle = U|A\rangle. \tag{B.1.99}$$

It follows that

$$\langle B| = \langle A| U^\dagger, \tag{B.1.100}$$

and, hence, that

$$\langle B|B\rangle = \langle A| U^\dagger U |A\rangle = \langle A|A\rangle. \tag{B.1.101}$$

Thus, if $\langle A|A\rangle = 1$ then $\langle B|B\rangle = 1$.

1.15 Let

$$U = \exp(i H). \tag{B.1.102}$$

It follows from Exercise 1.7 that if H is Hermitian then

$$U^\dagger = \exp(-i H). \tag{B.1.103}$$

However, according to Exercise 1.8,

$$U U^\dagger = \exp(i H) \exp(-i H) = \exp[i (H - H)] = \exp(0) = 1, \tag{B.1.104}$$

which demonstrates that U is unitary.

1.16 If

$$|v_i\rangle = U |u_i\rangle \tag{B.1.105}$$

then

$$\langle v_i| = \langle u_i| U^\dagger. \tag{B.1.106}$$

Hence,

$$\langle v_i|v_j\rangle = \langle u_i| U^\dagger U |u_j\rangle = \langle u_i|u_j\rangle = \delta_{ij}. \tag{B.1.107}$$

Here, we have made use of the fact that U is unitary, as well as the given result $\langle u_i|u_j\rangle = \delta_{ij}$. It follows that the $|v_i\rangle$ are orthonormal.

1.17 We have

$$A\,|j\rangle = a_j\,|j\rangle. \qquad (B.1.108)$$

It follows that

$$\langle i|\,A^\dagger = a_i\,\langle i|, \qquad (B.1.109)$$

where we have made use of the fact that the a_i are real. Now, given that the $|i\rangle$ are orthonormal, we can write $\langle i|j\rangle = \delta_{ij}$. Hence, Equation (B.1.108) yields

$$\langle i|\,A\,|j\rangle = a_j\,\langle i|j\rangle = a_i\,\delta_{ij}. \qquad (B.1.110)$$

Likewise, Equation (B.1.109) gives

$$\langle i|\,A^\dagger\,|j\rangle = a_i\,\langle i|j\rangle = a_i\,\delta_{ij}. \qquad (B.1.111)$$

Thus,

$$\langle i|\,A - A^\dagger\,|j\rangle = 0. \qquad (B.1.112)$$

However, given that the $|i\rangle$ form a complete set, we deduce that

$$\langle B|\,A - A^\dagger\,|C\rangle = 0, \qquad (B.1.113)$$

where $|B\rangle$ and $|C\rangle$ are general kets. This implies that

$$A - A^\dagger = 0, \qquad (B.1.114)$$

or

$$A = A^\dagger. \qquad (B.1.115)$$

Thus, the operator A is Hermitian.

1.18 We are told that

$$\langle \xi'|\xi''\rangle = \delta(\xi' - \xi''). \qquad (B.1.116)$$

Given that ξ is an observable, any general ket, $|A\rangle$, can be expressed as a linear combination of the $|\xi'\rangle$: that is,

$$|A\rangle = \int_{\xi'} d\xi'\,\alpha(\xi')\,|\xi'\rangle. \qquad (B.1.117)$$

It follows that

$$\langle \xi'|A\rangle = \int_{\xi''} d\xi'\,\alpha(\xi'')\,\langle \xi'|\xi''\rangle = \int_{\xi''} d\xi'\,\alpha(\xi'')\,\delta(\xi' - \xi'') = \alpha(\xi'). \qquad (B.1.118)$$

Hence,

$$|A\rangle = \int_{\xi'} d\xi'\,\langle \xi'|A\rangle\,|\xi'\rangle. \qquad (B.1.119)$$

Now, if

$$\int_{\xi'} d\xi'\,|\xi'\rangle\langle \xi'| = 1 \qquad (B.1.120)$$

then, operating on $|A\rangle$, we obtain

$$\int_{\xi'} d\xi' \, \langle \xi'|A\rangle \, |\xi'\rangle = |A\rangle. \tag{B.1.121}$$

However, Equations (B.1.119) and (B.1.121) are identical. Hence, we deduce the validity of Equation (B.1.120).

1.19 We have

$$\delta(x) = \lim_{\eta \to 0} \frac{1}{\pi} \frac{\eta}{x^2 + \eta^2}. \tag{B.1.122}$$

Thus,

$$\delta(-x) = \lim_{\eta \to 0} \frac{1}{\pi} \frac{\eta}{(-x)^2 + \eta^2} = \lim_{\eta \to 0} \frac{1}{\pi} \frac{\eta}{x^2 + \eta^2} = \delta(x). \tag{B.1.123}$$

Also,

$$\delta(a\,x) = \lim_{\eta \to 0} \frac{1}{\pi} \frac{\eta}{a^2 \, x^2 + \eta^2} = \lim_{\mu \to 0} \frac{1}{a\pi} \frac{\mu}{x^2 + \mu^2} = \frac{1}{a} \delta(x), \tag{B.1.124}$$

where $\mu = \eta/a$.

B.2 Chapter 2

2.1 Consider an N degree of freedom, classical, dynamical system whose generalized coordinates are the q_i (for $i = 1, N$), and whose corresponding conjugate momenta are the p_i. It follows from standard calculus, as well as the fact that the p_i are independent of the q_i, that

$$\frac{\partial q_i}{\partial q_j} = \delta_{ij}, \qquad \frac{\partial p_i}{\partial p_j} = \delta_{ij}, \qquad \frac{\partial q_i}{\partial p_j} = 0, \qquad \frac{\partial p_i}{\partial q_j} = 0. \tag{B.2.1}$$

The classical Poisson bracket is defined

$$[u, v]_{cl} = \sum_{k=1,N} \left(\frac{\partial u}{\partial q_k} \frac{\partial v}{\partial p_k} - \frac{\partial u}{\partial p_k} \frac{\partial v}{\partial q_k} \right). \tag{B.2.2}$$

Thus:

(a)

$$[q_i, q_j]_{cl} = \sum_{k=1,N} \left(\frac{\partial q_i}{\partial q_k} \frac{\partial q_j}{\partial p_k} - \frac{\partial q_i}{\partial p_k} \frac{\partial q_j}{\partial q_k} \right) = \sum_{k=1,N} \left(\delta_{ik} \, 0 - 0 \, \delta_{jk} \right) = 0. \tag{B.2.3}$$

(b)

$$[p_i, p_j]_{cl} = \sum_{k=1,N} \left(\frac{\partial p_i}{\partial q_k} \frac{\partial p_j}{\partial p_k} - \frac{\partial p_i}{\partial p_k} \frac{\partial p_j}{\partial q_k} \right) = \sum_{k=1,N} \left(0 \, \delta_{jk} - \delta_{ik} \, 0 \right) = 0. \tag{B.2.4}$$

(c)

$$[q_i, p_j]_{cl} = \sum_{k=1,N} \left(\frac{\partial q_i}{\partial q_k} \frac{\partial p_j}{\partial p_k} - \frac{\partial q_i}{\partial p_k} \frac{\partial p_j}{\partial q_k} \right) = \sum_{k=1,N} \left(\delta_{ik} \delta_{jk} - 0\,0 \right) = \delta_{ij}. \qquad \text{(B.2.5)}$$

2.2 The classical Poisson bracket is defined

$$[u, v]_{cl} = \sum_{k=1,N} \left(\frac{\partial u}{\partial q_k} \frac{\partial v}{\partial p_k} - \frac{\partial u}{\partial p_k} \frac{\partial v}{\partial q_k} \right). \qquad \text{(B.2.6)}$$

The quantum mechanical Poisson bracket is defined

$$[u, v]_{qm} = \frac{uv - vu}{i\,\hbar}. \qquad \text{(B.2.7)}$$

Thus:

(a)

$$[v, u]_{cl} = \sum_{k=1,N} \left(\frac{\partial v}{\partial q_k} \frac{\partial u}{\partial p_k} - \frac{\partial v}{\partial p_k} \frac{\partial u}{\partial q_k} \right) = -[u, v]_{cl}, \qquad \text{(B.2.8)}$$

and

$$i\,\hbar\,[v, u]_{qm} = vu - uv = -i\,\hbar\,[u, v]_{qm}. \qquad \text{(B.2.9)}$$

(b)

$$[u, c]_{cl} = \sum_{k=1,N} \left(\frac{\partial u}{\partial q_k} \frac{\partial c}{\partial p_k} - \frac{\partial u}{\partial p_k} \frac{\partial c}{\partial q_k} \right) = 0, \qquad \text{(B.2.10)}$$

and

$$i\,\hbar\,[u, c]_{qm} = uc - cu = 0, \qquad \text{(B.2.11)}$$

because $\partial c/\partial q_k = \partial c/\partial p_k = 0$, and $uc = cu$.

(c)

$$[u_1 + u_2, v]_{cl} = \sum_{k=1,N} \left(\frac{\partial(u_1 + u_2)}{\partial q_k} \frac{\partial v}{\partial p_k} - \frac{\partial(u_1 + u_2)}{\partial p_k} \frac{\partial v}{\partial q_k} \right)$$

$$= \sum_{k=1,N} \left(\frac{\partial u_1}{\partial q_k} \frac{\partial v}{\partial p_k} - \frac{\partial u_1}{\partial p_k} \frac{\partial v}{\partial q_k} \right) + \sum_{k=1,N} \left(\frac{\partial u_2}{\partial q_k} \frac{\partial v}{\partial p_k} - \frac{\partial u_2}{\partial p_k} \frac{\partial v}{\partial q_k} \right)$$

$$= [u_1, v]_{cl} + [u_2, v]_{cl}, \qquad \text{(B.2.12)}$$

and

$$i\,\hbar\,[u_1 + u_2, v]_{qm} = (u_1 + u_2)v - v(u_1 + u_2) = u_1 v - v u_1 + u_2 v - v u_2$$

$$= i\,\hbar\,[u_1, v]_{qm} + i\,\hbar\,[u_2, v]_{qm}. \qquad \text{(B.2.13)}$$

(d)

$$[u, v_1 + v_2]_{cl} = -[v_1 + v_2, u]_{cl} = -[v_1, u]_{cl} - [v_2, u]_{cl}$$
$$= [u, v_1]_{cl} + [u, v_2]_{cl}, \qquad (B.2.14)$$

and

$$[u, v_1 + v_2]_{qm} = -[v_1 + v_2, u]_{qm} = -[v_1, u]_{qm} - [v_2, u]_{qm}$$
$$= [u, v_1]_{qm} + [u, v_2]_{qm}. \qquad (B.2.15)$$

(e)

$$[u_1 u_2, v]_{cl} = \sum_{k=1,N} \left(\frac{\partial(u_1 u_2)}{\partial q_k} \frac{\partial v}{\partial p_k} - \frac{\partial(u_1 u_2)}{\partial p_k} \frac{\partial v}{\partial q_k} \right)$$
$$= \sum_{k=1,N} \left(\frac{\partial u_1}{\partial q_k} \frac{\partial v}{\partial p_k} - \frac{\partial u_1}{\partial p_k} \frac{\partial v}{\partial q_k} \right) u_2 + u_1 \sum_{k=1,N} \left(\frac{\partial u_2}{\partial q_k} \frac{\partial v}{\partial p_k} - \frac{\partial u_2}{\partial p_k} \frac{\partial v}{\partial q_k} \right)$$
$$= [u_1, v]_{cl} u_2 + u_1 [u_2, v]_{cl}, \qquad (B.2.16)$$

and

$$i\,\hbar\,[u_1 u_2, v]_{qm} = u_1 u_2 v - v u_1 u_2 = (u_1 v - v u_1) u_2 + u_1 (u_2 v - v u_2)$$
$$= [u_1, v]_{qm} u_2 + u_1 [u_2, v]_{qm}. \qquad (B.2.17)$$

(f)

$$[u, v_1 v_2]_{cl} = -[v_1 v_2, u]_{cl} = -[v_1, u]_{cl} v_2 - v_1 [v_2, u]_{cl}$$
$$= [u, v_1]_{cl} v_2 + v_1 [u, v_2]_{cl}, \qquad (B.2.18)$$

and

$$[u, v_1 v_2]_{qm} = -[v_1 v_2, u]_{qm} = -[v_1, u]_{qm} v_2 - v_1 [v_2, u]_{qm}$$
$$= [u, v_1]_{qm} v_2 + v_1 [u, v_2]_{qm}. \qquad (B.2.19)$$

(g) For the sake of simplicity, we shall only consider a one degree of freedom (i.e., $N = 1$) classical system. The generalization to a many degree of freedom system is straightforward. Let $q_1 = x$ and $p_1 = y$. We shall adopt the convenient shorthand $\partial u/\partial x = u_x$, $\partial^2 u/(\partial x\,\partial y) = u_{xy}$, et cetera. Thus,

$$[u, v]_{cl} = u_x v_y - u_y v_x, \qquad (B.2.20)$$

and

$$[u, [v, w]_{cl}]_{cl} = u_x (v_x w_y - v_y w_x)_y - u_y (v_x w_y - v_y w_x)_x \qquad (B.2.21)$$

It follows that

$$[u,[v,w]_{cl}]_{cl} = u_x(v_{xy}w_y + v_x w_{yy} - v_{yy}w_x - v_y w_{xy})$$
$$- u_y(v_{xx}w_y + v_x w_{xy} - v_{xy}w_x - v_y w_{xx}), \tag{B.2.22}$$

$$[v,[w,u]_{cl}]_{cl} = v_x(w_{xy}u_y + w_x u_{yy} - w_{yy}u_x - w_y u_{xy})$$
$$- v_y(w_{xx}u_y + w_x u_{xy} - w_{xy}u_x - w_y u_{xx}), \tag{B.2.23}$$

$$[w,[u,v]_{cl}]_{cl} = w_x(u_{xy}v_y + u_x v_{yy} - u_{yy}v_x - u_y v_{xy})$$
$$- w_y(u_{xx}v_y + u_x v_{xy} - u_{xy}v_x - u_y v_{xx}). \tag{B.2.24}$$

By the symmetry of the problem, it is sufficient to show that all of the terms multiplying u_x sum to zero. The sum of the terms multiplying u_x in the second and third equations is

$$-v_x w_{yy} + v_y w_{xy} + w_x v_{yy} - w_y v_{xy} = -(v_{xy}w_y + v_x w_{yy} - v_{yy}w_x - v_y w_{xy}). \tag{B.2.25}$$

But, this is minus the sum of the term multiplying u_x in the first equation. Thus, summing all three equations leads to the cancellation of all terms multiplying u_x. It is easily shown that the terms multiplying u_y, v_x, v_y, w_x, and w_y similarly sum to zero. Thus,

$$[u,[v,w]_{cl}]_{cl} + [v,[w,u]_{cl}]_{cl} + [w,[u,v]_{cl}]_{cl} = 0. \tag{B.2.26}$$

Now,

$$(i\hbar)^2[u,[v,w]_{qm}]_{qm} = u(vw - wv) - (vw - wv)u$$
$$= uvw - uwv - vwu + wvu. \tag{B.2.27}$$

Similarly,

$$(i\hbar)^2[v,[w,u]_{qm}]_{qm} = vwu - vuw - wuv + uwv, \tag{B.2.28}$$

$$(i\hbar)^2[w,[u,v]_{qm}]_{qm} = wuv - wvu - uvw + vuw. \tag{B.2.29}$$

It is evident that the right-hand sides of the previous three equations sum to zero. Thus,

$$[u,[v,w]_{qm}]_{qm} + [v,[w,u]_{qm}]_{qm} + [w,[u,v]_{qm}]_{qm} = 0. \tag{B.2.30}$$

2.3 We have

$$[\xi^m,\xi] = \xi^m\xi - \xi\xi^m = \xi^{m+1} - \xi^{m+1} = 0. \tag{B.2.31}$$

Thus, if

$$f(\xi) = \sum_{m=0,\infty} c_m \xi^m, \tag{B.2.32}$$

where the c_m are numbers, then

$$[f(\xi),\xi] = \sum_{m=0,\infty} c_m[\xi^m,\xi] = 0. \tag{B.2.33}$$

Also,

$$\xi^m |\xi'\rangle = \xi^{m-1} \xi' |\xi'\rangle = \xi' \xi^{m-1} |\xi'\rangle. \tag{B.2.34}$$

Hence, by induction,

$$\xi^m |\xi'\rangle = \xi'^m |\xi'\rangle. \tag{B.2.35}$$

Thus,

$$f(\xi)|\xi'\rangle = \sum_{m=0,\infty} c_m \xi^m |\xi'\rangle = \left(\sum_{m=0,\infty} c_m \xi'^m \right) |\xi'\rangle = f(\xi')|\xi'\rangle. \tag{B.2.36}$$

It is obvious that

$$[\xi^m, \eta^n] = 0, \tag{B.2.37}$$

because all m factors of ξ commute with all n factors of η. Let

$$g(\eta) = \sum_{n=0,\infty} d_n \eta^n, \tag{B.2.38}$$

where the c_n are numbers. It follows that

$$[f(\xi), g(\eta)] = \sum_{n,m=0,\infty} c_m d_n [\xi^m, \eta^n] = 0. \tag{B.2.39}$$

2.4 We are told that

$$\psi(x') = \psi_0 \exp\left[-\frac{(x'-x_0)^2}{4\sigma^2}\right]. \tag{B.2.40}$$

The normalization constraint

$$\int_{-\infty}^{\infty} dx' \, |\psi(x')|^2 = 1, \tag{B.2.41}$$

leads to

$$|\psi_0|^2 \int_{-\infty}^{\infty} dx' \, \exp\left[-\frac{(x'-x_0)^2}{2\sigma^2}\right] = 1, \tag{B.2.42}$$

or

$$|\psi_0|^2 (2\sigma^2)^{1/2} \int_{-\infty}^{\infty} dy \, e^{-y^2} = 1. \tag{B.2.43}$$

Hence, given that [Reif (1965)]

$$\int_{-\infty}^{\infty} dy \, e^{-y^2} = \sqrt{\pi}, \tag{B.2.44}$$

we deduce that

$$\psi_0 = \frac{1}{(2\pi\sigma^2)^{1/4}}. \tag{B.2.45}$$

Note that we can chose ψ_0 to be real without loss of generality.

(a) Now,

$$\langle x \rangle = \int_{-\infty}^{\infty} dx'\, \psi^*(x')\, x'\, \psi(x'), \tag{B.2.46}$$

giving

$$\langle x \rangle = \frac{1}{(2\pi\sigma^2)^{1/2}} \int_{-\infty}^{\infty} dx'\, x'\, \exp\left[-\frac{(x'-x_0)^2}{2\sigma^2}\right]$$

$$= \frac{1}{\sqrt{\pi}} \int_{-\infty}^{\infty} dy \left(\sqrt{2\sigma^2}\, y + x_0\right) e^{-y^2} = \frac{x_0}{\sqrt{\pi}} \int_{-\infty}^{\infty} dy\, e^{-y^2} = x_0, \tag{B.2.47}$$

because

$$\int_{-\infty}^{\infty} y\, e^{-y^2}\, dy = 0 \tag{B.2.48}$$

by symmetry.

(b) Now,

$$\langle (\Delta x)^2 \rangle = \langle x^2 \rangle - \langle x \rangle^2 = \langle x^2 \rangle - x_0^2. \tag{B.2.49}$$

Moreover,

$$\langle x^2 \rangle = \int_{-\infty}^{\infty} dx'\, \psi^*(x')\, x'^2\, \psi(x'), \tag{B.2.50}$$

giving

$$\langle x^2 \rangle = \frac{1}{(2\pi\sigma^2)^{1/2}} \int_{-\infty}^{\infty} dx'\, x'^2\, \exp\left[-\frac{(x'-x_0)^2}{2\sigma^2}\right]$$

$$= \frac{1}{\sqrt{\pi}} \int_{-\infty}^{\infty} dy \left(\sqrt{2\sigma^2}\, y + x_0\right)^2 e^{-y^2}$$

$$= \frac{2\sigma^2}{\sqrt{\pi}} \int_{-\infty}^{\infty} dy\, y^2\, e^{-y^2} + \frac{2\sqrt{2\sigma^2}\, x_0}{\sqrt{\pi}} \int_{-\infty}^{\infty} dy\, y\, e^{-y^2}$$

$$+ \frac{x_0^2}{\sqrt{\pi}} \int_{-\infty}^{\infty} dy\, e^{-y^2} = \sigma^2 + x_0^2, \tag{B.2.51}$$

because [Reif (1965)]

$$\int_{-\infty}^{\infty} dy\, y^2\, e^{-y^2} = \frac{\sqrt{\pi}}{2}. \tag{B.2.52}$$

Hence,

$$\langle (\Delta x)^2 \rangle = \sigma^2. \tag{B.2.53}$$

(c) Making use of the Schrödinger representation, we can write

$$p_x = -i\hbar \frac{d}{dx}. \tag{B.2.54}$$

Hence,

$$\langle p_x \rangle = -\mathrm{i}\,\hbar \int_{-\infty}^{\infty} dx'\, \psi^*(x')\, \frac{d}{dx'}\, \psi(x') = -\frac{\mathrm{i}\,\hbar}{2} \left[\psi^2(x') \right]_{-\infty}^{\infty} = 0, \qquad \text{(B.2.55)}$$

because $\psi(x')$ is real.

(d) Now,

$$\langle (\Delta p_x)^2 \rangle = \langle p_x^2 \rangle - \langle p_x \rangle^2 = \langle p_x^2 \rangle. \qquad \text{(B.2.56)}$$

Moreover,

$$\langle p_x^2 \rangle = -\hbar^2 \int_{-\infty}^{\infty} \psi^*(x')\, \frac{d^2}{dx'^2}\, \psi(x')\, dx', \qquad \text{(B.2.57)}$$

giving

$$\langle p^2 \rangle = -\hbar^2 \left[\psi\, \frac{d\psi}{dx'} \right]_{-\infty}^{\infty} + \hbar^2 \int_{-\infty}^{\infty} dx' \left(\frac{d\psi}{dx'} \right)^2 = \hbar^2 \int_{-\infty}^{\infty} dx' \left(\frac{d\psi}{dx'} \right)^2, \qquad \text{(B.2.58)}$$

because $\psi(x')$ is real. Thus,

$$\langle p_x^2 \rangle = \frac{\hbar^2}{(2\pi\sigma^2)^{1/2}} \int_{-\infty}^{\infty} dx'\, \frac{(x'-x_0)^2}{4\sigma^2}\, \exp\left[-\frac{(x'-x_0)^2}{2\sigma^2} \right]$$

$$= \frac{\hbar^2}{2\sigma^2} \int_{-\infty}^{\infty} dy\, y^2\, \mathrm{e}^{-y^2} = \frac{\hbar^2}{4\sigma^2}. \qquad \text{(B.2.59)}$$

Hence,

$$\langle (\Delta p_x)^2 \rangle = \frac{\hbar^2}{4\sigma^2}. \qquad \text{(B.2.60)}$$

2.5 We know that

$$D_x(\Delta x) = \exp\left(-\frac{\mathrm{i}\, p_x\, \Delta x}{\hbar} \right), \qquad \text{(B.2.61)}$$

$$D_y(\Delta y) = \exp\left(-\frac{\mathrm{i}\, p_y\, \Delta y}{\hbar} \right). \qquad \text{(B.2.62)}$$

Thus, from Exercise 1.8,

$$D_x(\Delta x_2)\, D_x(\Delta x_1) = \exp\left(-\frac{\mathrm{i}\, p_x\, \Delta x_2}{\hbar} \right) \exp\left(-\frac{\mathrm{i}\, p_x\, \Delta x_1}{\hbar} \right)$$

$$= \exp\left[-\frac{\mathrm{i}\, p_x\, (\Delta x_2 + \Delta x_1)}{\hbar} \right] = D_x(\Delta x_2 + \Delta x_1)$$

$$= \exp\left[-\frac{\mathrm{i}\, p_x\, (\Delta x_1 + \Delta x_2)}{\hbar} \right] = D_x(\Delta x_1)\, D_x(\Delta x_2). \qquad \text{(B.2.63)}$$

This signifies that a displacement of the system a distance Δx_1 along the x-axis, followed by another displacement in the same direction a distance Δx_2, is equivalent to a single displacement a distance $\Delta x_2 + \Delta x_1$ along the x-axis, which, in turn, is equivalent to the first two displacements taken in reverse order. We also have

$$[D_x(\Delta x), D_y(\Delta y)] = 0, \tag{B.2.64}$$

because $D_x(\Delta x)$ and $D_y(\Delta y)$ are functions of two commuting operators (i.e., p_x and p_y). (See Exercises 2.3.) Thus,

$$D_x(\Delta x) D_y(\Delta y) = D_y(\Delta y) D_x(\Delta x). \tag{B.2.65}$$

This signifies that a displacement a distance Δy along the y-axis, followed by a displacement a distance Δx along the x-axis, is equivalent to the two displacements taken in reverse order.

2.6 We are told that

$$D_x(a) = \exp\left(-\frac{i\, p_x a}{\hbar}\right) \equiv \sum_{n=0,\infty} \frac{1}{n!} \left(-\frac{i\, p_x a}{\hbar}\right)^n. \tag{B.2.66}$$

It follows that

$$D_x(a)^\dagger = \exp\left(\frac{i\, p_x^\dagger a}{\hbar}\right) = \exp\left(\frac{i\, p_x a}{\hbar}\right), \tag{B.2.67}$$

because p_x is Hermitian. Note, from Exercises 1.15, that $D_x(a) D_x(a)^\dagger = 1$. Now,

$$D_x(a)\, x\, D_x(a)^\dagger = \exp\left(-\frac{i\, p_x a}{\hbar}\right) x \,\exp\left(\frac{i\, p_x a}{\hbar}\right). \tag{B.2.68}$$

Adopting the momentum representation, we can write

$$x = i\hbar \frac{d}{dp_x}. \tag{B.2.69}$$

Thus,

$$D_x(a)\, x\, D_x(a)^\dagger = i\hbar \,\exp\left(-\frac{i\, p_x a}{\hbar}\right) \frac{d}{dp_x} \exp\left(\frac{i\, p_x a}{\hbar}\right)$$

$$= i\hbar \,\exp\left(-\frac{i\, p_x a}{\hbar}\right) \left[\frac{i\, a}{\hbar} \exp\left(\frac{i\, p_x a}{\hbar}\right) + \exp\left(\frac{i\, p_x a}{\hbar}\right) \frac{d}{dp_x}\right]$$

$$= -a + i\hbar \frac{d}{dp_x} = x - a. \tag{B.2.70}$$

Left-multiplication by $D_x(a)^\dagger$ yields

$$x\, D_x(a)^\dagger = D_x(a)^\dagger\, (x - a). \tag{B.2.71}$$

Hence,

$$D_x(a)\, x^m\, D_x(a)^\dagger = D_x(a)\, x^{m-1}\, x\, D_x(a)^\dagger = D_x(a)\, x^{m-1}\, D_x(a)^\dagger (x-a). \qquad \text{(B.2.72)}$$

Thus, it follows by induction that

$$D_x(a)\, x^m\, D_x(a)^\dagger = (x-a)^m. \qquad \text{(B.2.73)}$$

Suppose that $V(x)$ can be expanded as a power series:

$$V(x) = \sum_{m=0,\infty} \alpha_m\, x^m, \qquad \text{(B.2.74)}$$

where the α_m are numbers. It follows that

$$D_x(a)\, V(x)\, D_x(a)^\dagger = \sum_{m=0,\infty} \alpha_m\, D_x(a)\, x^m\, D_x(a)^\dagger = \sum_{m=0,\infty} \alpha_m\, (x-a)^m$$

$$= V(x-a). \qquad \text{(B.2.75)}$$

Consider

$$|A\rangle = \sum_{n=-\infty,\infty} c_n\, |k' + n\,k_a\rangle, \qquad \text{(B.2.76)}$$

where $k = p_x/\hbar$, the c_n are numbers, and $k_a = 2\pi/a$. Now,

$$p_x\, |k' + n\,k_a\rangle = \hbar\, (k' + n\,k_a)\, |k' + n\,k_a\rangle. \qquad \text{(B.2.77)}$$

Thus, from Exercise 2.3,

$$F(p_x)\, |k' + n\,k_a\rangle = F(\hbar\, [k' + n\,k_a])\, |k' + n\,k_a\rangle. \qquad \text{(B.2.78)}$$

It follows that

$$D_x(a)\, |k' + n\,k_a\rangle = \exp\!\left(-\frac{i\,p_x\,a}{\hbar}\right) |k' + n\,k_a\rangle = \exp\left[-i\,(k'\,a + n\,2\pi)\right] |k' + n\,k_a\rangle$$

$$= \exp\left(-i\,k'\,a\right) |k' + n\,k_a\rangle. \qquad \text{(B.2.79)}$$

Thus,

$$D_x(a)\, |A\rangle = \sum_{n=-\infty,\infty} c_n\, D_x(a)\, |k' + n\,k_a\rangle$$

$$= \sum_{n=-\infty,\infty} c_n\, \exp(-i\,k'\,a)\, |k' + n\,k_a\rangle = \exp(-i\,k'\,a)\, |A\rangle, \qquad \text{(B.2.80)}$$

which implies that $|A\rangle$ is an eigenket of the $D_x(a)$ operator belonging to the eigenvalue $\exp(-i\,k'\,a)$.

Now,

$$\psi_A(x') = \langle x'|A \rangle = \sum_{n=-\infty,\infty} c_n \langle x'|k' + n\,k_a \rangle. \tag{B.2.81}$$

However, $|k' + n\,k_a\rangle$ is an eigenket of the p_x operator belonging to the eigenvalue $\hbar(k' + n\,k_a)$, and

$$\langle x'|p_x' \rangle = \frac{1}{h^{1/2}} \exp\left(\frac{i\,p_x'\,x'}{\hbar}\right). \tag{B.2.82}$$

Hence,

$$\psi_A(x') = \sum_{n=-\infty,\infty} c_n \frac{1}{h^{1/2}} \exp\left[i\left(k' + \frac{n\,2\pi}{a}\right)x'\right] = \exp(i\,k'\,x')\,u(x'), \tag{B.2.83}$$

where

$$u(x') = \frac{1}{h^{1/2}} \sum_{n=-\infty,\infty} c_n \exp\left(i\,n\,2\pi\,\frac{x'}{a}\right) \tag{B.2.84}$$

is a periodic function that satisfies $u(x' + a) = u(x')$ for all x'.

B.3   Chapter 3

3.1 In the momentum representation,

$$x_i = i\,\hbar\,\frac{\partial}{\partial p_i}. \tag{B.3.1}$$

Thus,

$$[x_i, F(\mathbf{x}, \mathbf{p})] = i\,\hbar\,\frac{\partial}{\partial p_i}\,F(\mathbf{x}, \mathbf{p}) - F(\mathbf{x}, \mathbf{p})\,i\,\hbar\,\frac{\partial}{\partial p_i} = i\,\hbar\,\frac{\partial F(\mathbf{x}, \mathbf{p})}{\partial p_i}. \tag{B.3.2}$$

In the Schrödinger representation,

$$p_i = -i\,\hbar\,\frac{\partial}{\partial x_i}. \tag{B.3.3}$$

Thus,

$$[p_i, G(\mathbf{x}, \mathbf{p})] = -i\,\hbar\,\frac{\partial}{\partial x_i}\,G(\mathbf{x}, \mathbf{p}) + G(\mathbf{x}, \mathbf{p})\,i\,\hbar\,\frac{\partial}{\partial x_i} = -i\,\hbar\,\frac{\partial G(\mathbf{x}, \mathbf{p})}{\partial x_i}. \tag{B.3.4}$$

The previous results can also be obtained by expanding $F(\mathbf{x}, \mathbf{p})$ and $G(\mathbf{x}, \mathbf{p})$ as power series in $\mathbf{x}$ and $\mathbf{p}$, and then making use of the fundamental commutation relations between position and momentum.

3.2 Let us work in one dimension, for the sake of clarity. Multiplying the Schrödinger time-dependent wave equation, (3.55), by ψ^*, we obtain

$$\mathrm{i}\,\hbar\,\psi^* \, \frac{\partial \psi}{\partial t} + \frac{\hbar^2}{2\,m}\,\psi^* \, \frac{\partial^2 \psi}{\partial x'^2} = V\,\psi^*\,\psi. \tag{B.3.5}$$

The complex conjugate of this equation yields

$$-\mathrm{i}\,\hbar\,\psi \, \frac{\partial \psi^*}{\partial t} + \frac{\hbar^2}{2\,m}\,\psi \, \frac{\partial^2 \psi^*}{\partial x'^2} = V^*\,\psi^*\,\psi. \tag{B.3.6}$$

Taking the difference of the previous two equations, and dividing by $\mathrm{i}\,\hbar$, we get

$$\left(\psi^* \, \frac{\partial \psi}{\partial t} + \psi \, \frac{\partial \psi^*}{\partial t}\right) - \frac{\mathrm{i}\,\hbar}{2\,m}\left(\psi^* \, \frac{\partial^2 \psi}{\partial x'^2} - \psi \, \frac{\partial^2 \psi^*}{\partial x'^2}\right) = -\frac{\mathrm{i}}{\hbar}\,(V - V^*)\,|\psi|^2, \tag{B.3.7}$$

or

$$\frac{\partial |\psi|^2}{\partial t} + \frac{\hbar}{m} \, \frac{\partial}{\partial x'}\left[\mathrm{Im}\left(\psi^* \, \frac{\partial \psi}{\partial x'}\right)\right] = 2\,\frac{\mathrm{Im}(V)}{\hbar}\,|\psi|^2. \tag{B.3.8}$$

In three dimensions, the previous equation generalizes to give

$$\frac{\partial \rho}{\partial t} + \nabla' \cdot \mathbf{j} = 2\,\frac{\mathrm{Im}(V)}{\hbar}\,\rho, \tag{B.3.9}$$

where

$$\rho(\mathbf{x}', t) = |\psi(\mathbf{x}', t)|^2, \tag{B.3.10}$$

and

$$\mathbf{j}(\mathbf{x}', t) = \left(\frac{\hbar}{m}\right)\mathrm{Im}(\psi^* \, \nabla' \psi). \tag{B.3.11}$$

3.3 (a) We can write

$$\langle H \rangle = \langle A|A \rangle + \langle B|B \rangle \geq 0, \tag{B.3.12}$$

where $|B\rangle = (2\,m)^{-1/2}\,p_x\,|\,\rangle$ and $|A\rangle = (m/2)^{1/2}\,\omega\,x\,|\,\rangle$. This follows because p_x and x are Hermitian operators.

(b) Let

$$A = \alpha\,x + \beta\,p_x, \tag{B.3.13}$$

where

$$\alpha = \sqrt{\frac{m\,\omega}{2\,\hbar}}, \tag{B.3.14}$$

$$\beta = \frac{\mathrm{i}}{\sqrt{2\,m\,\omega\,\hbar}}. \tag{B.3.15}$$

Thus,

$$[A, A^\dagger] = (\alpha x + \beta p_x)(\alpha^* x + \beta^* p_x) - (\alpha^* x + \beta^* p_x)(\alpha x + \beta p_x)$$

$$= (\alpha \beta^* - \alpha^* \beta)[x, p_x] = 1, \qquad (B.3.16)$$

because x and p_x are Hermitian operators, $[x, p_x] = i\hbar$, and $\alpha \beta^* = -i/(2\hbar)$. Now,

$$A^\dagger A = (\alpha^* x + \beta^* p_x)(\alpha x + \beta p_x) = |\alpha|^2 x^2 + \alpha^* \beta[x, p_x] + |\beta|^2 p_x^2, \quad (B.3.17)$$

because $\alpha \beta^* = -\alpha^* \beta$. Hence,

$$A^\dagger A = |\alpha|^2 x^2 + \alpha^* \beta i\hbar + |\beta|^2 p_x^2 = \frac{m\omega}{2\hbar} x^2 - \frac{1}{2} + \frac{p_x^2}{2m\omega\hbar}, \qquad (B.3.18)$$

which implies that

$$\hbar\omega \left(\frac{1}{2} + A^\dagger A \right) = \frac{p_x^2}{2m} + \frac{1}{2} m\omega^2 x^2 = H. \qquad (B.3.19)$$

Now,

$$[H, A] = \hbar\omega [A^\dagger A, A] = \hbar\omega [A^\dagger, A] A = -\hbar\omega A. \qquad (B.3.20)$$

Likewise,

$$[H, A^\dagger] = \hbar\omega [A^\dagger A, A^\dagger] = \hbar\omega A^\dagger [A, A^\dagger] = \hbar\omega A^\dagger. \qquad (B.3.21)$$

(c) Let

$$H|E\rangle = E|E\rangle. \qquad (B.3.22)$$

It follows that

$$H A |E\rangle = (A H + [H, A])|E\rangle = (E - \hbar\omega) A |E\rangle. \qquad (B.3.23)$$

Likewise,

$$H A^\dagger |E\rangle = (A^\dagger H + [H, A^\dagger])|E\rangle = (E + \hbar\omega) A^\dagger |E\rangle. \qquad (B.3.24)$$

Now,

$$\langle E| H |E\rangle = E = \frac{1}{2m} \langle E| p_x^\dagger p_x |E\rangle + \frac{1}{2} m\omega^2 \langle E| x^\dagger x |E\rangle, \qquad (B.3.25)$$

because x and p_x are Hermitian operators. But, $\langle E| p_x^\dagger p |E\rangle$ and $\langle E| x^\dagger x |E\rangle$ are the norms of the ket vectors $p_x |E\rangle$ and $x|E\rangle$, respectively, and are thus positive definite. Hence, we deduce that $E > 0$: that is, there are no negative energy states. Thus, there must exist a minimum energy state, $|E_{min}\rangle$. Moreover,

$$A |E_{min}\rangle = 0, \qquad (B.3.26)$$

otherwise $A|E_{\min}\rangle$ would represent an eigenket with energy $E_{\min} - \hbar\omega$, and $|E_{\min}\rangle$ would thus not be the minimum energy state. Thus,

$$H|E_{\min}\rangle = E_{\min}|E_{\min}\rangle = \frac{1}{2}\hbar\omega|E_{\min}\rangle + \hbar\omega A^\dagger A|E_{\min}\rangle = \frac{1}{2}\hbar\omega|E_{\min}\rangle. \quad \text{(B.3.27)}$$

Hence,

$$E_{\min} = \frac{1}{2}\hbar\omega. \quad \text{(B.3.28)}$$

It follows from Equation (B.3.24) that $A^\dagger|E_{\min}\rangle$ is an energy eigenket corresponding to the energy $E_{\min} + \hbar\omega$. Likewise, $A^{\dagger 2}|E_{\min}\rangle$ is an energy eigenket corresponding to the energy $E_{\min} + 2\hbar\omega$. Hence, the allowed energy eigenvalues are

$$E_n = (n + 1/2)\hbar\omega, \quad \text{(B.3.29)}$$

where n is a non-negative integer.

(d) Suppose that $H|E_n\rangle = E_n|E_n\rangle$ and $\langle E_n|E_n\rangle = 1$. It is clear from Equations (B.3.23) and (B.3.24) that

$$A|E_n\rangle = c_n^-|E_{n-1}\rangle, \quad \text{(B.3.30)}$$

$$A^\dagger|E_n\rangle = c_n^+|E_{n+1}\rangle, \quad \text{(B.3.31)}$$

where the c_n^- and the c_n^+ are complex numbers. It follows that

$$\langle E_n|A^\dagger A|E_n\rangle = |c_n^-|^2 = \langle E_n|H/\hbar\omega - 1/2|E_n\rangle = \frac{E_n}{\hbar\omega} - \frac{1}{2} = n. \quad \text{(B.3.32)}$$

Likewise,

$$\langle E_n|A A^\dagger|E_n\rangle = |c_n^+|^2 = \langle E_n|A^\dagger A + 1|E_n\rangle = n + 1. \quad \text{(B.3.33)}$$

Without loss of generality, we can choose c_n^- and c_n^+ to be real and positive. Thus,

$$c_n^- = \sqrt{n}, \quad \text{(B.3.34)}$$

$$c_n^+ = \sqrt{n+1}. \quad \text{(B.3.35)}$$

It follows that

$$|E_1\rangle = \frac{1}{\sqrt{1}} A^\dagger|E_0\rangle, \quad \text{(B.3.36)}$$

$$|E_2\rangle = \frac{1}{\sqrt{2}} A^\dagger|E_1\rangle = \frac{1}{\sqrt{2\times 1}} A^{\dagger 2}|E_0\rangle, \quad \text{(B.3.37)}$$

$$|E_3\rangle = \frac{1}{\sqrt{3}} A^\dagger|E_2\rangle = \frac{1}{\sqrt{3\times 2\times 1}} A^{\dagger 3}|E_0\rangle. \quad \text{(B.3.38)}$$

Hence, we deduce that

$$|E_n\rangle = \frac{1}{\sqrt{n!}} (A^\dagger)^n |E_0\rangle. \tag{B.3.39}$$

(e) We know that

$$A |E_0\rangle = |0\rangle. \tag{B.3.40}$$

Hence,

$$\langle x' | A | E_0 \rangle = \langle x' | \left(\frac{x}{\sqrt{2}\, x_0} + \frac{x_0}{\sqrt{2}} \frac{d}{dx} \right) |E_0\rangle$$

$$= \left(\frac{x'}{\sqrt{2}\, x_0} + \frac{x_0}{\sqrt{2}} \frac{d}{dx'} \right) \langle x' | E_0 \rangle = 0, \tag{B.3.41}$$

where $x_0 = (\hbar / m\,\omega)^{1/2}$, and use has been made of the Schrödinger representation. Thus,

$$\left(\frac{x'}{x_0} + x_0 \frac{d}{dx'} \right) \psi_0(x') = 0. \tag{B.3.42}$$

The solution is

$$\psi_0(x') = A \, \exp\left[-\frac{1}{2} \left(\frac{x'}{x_0} \right)^2 \right]. \tag{B.3.43}$$

The constant A is determined from the normalization constraint

$$1 = \int_{-\infty}^{\infty} dx' \, |\psi_0|^2 = A^2 \, x_0 \int_{-\infty}^{\infty} dy \, e^{-y^2} = A^2 \, x_0 \, \pi^{1/2} \tag{B.3.44}$$

[Reif (1965)]. Hence,

$$\psi_0(x') = \frac{1}{\pi^{1/4} \, x_0^{1/2}} \exp\left[-\frac{1}{2} \left(\frac{x'}{x_0} \right)^2 \right]. \tag{B.3.45}$$

Finally, according to Equation (B.3.39)

$$\psi_n(x') = \frac{1}{\sqrt{n!}} \langle x' | (A^\dagger)^n | E_0 \rangle$$

$$= \frac{1}{\sqrt{n!}} \langle x' | \left(\frac{x}{\sqrt{2}\, x_0} - \frac{x_0}{\sqrt{2}} \frac{d}{dx} \right)^n |E_0\rangle$$

$$= \frac{1}{\sqrt{n!}} \left(\frac{x'}{\sqrt{2}\, x_0} - \frac{x_0}{\sqrt{2}} \frac{d}{dx'} \right)^n \psi_0(x')$$

$$= \frac{1}{\pi^{1/4} \, (2^n \, n!)^{1/2} \, x_0^{n+1/2}} \left(x' - x_0^2 \frac{d}{dx'} \right)^n \exp\left[-\frac{1}{2} \left(\frac{x'}{x_0} \right)^2 \right]. \tag{B.3.46}$$

3.4 We have

$$\langle n' | n \rangle = \delta_{n' n}, \tag{B.3.47}$$

as well as

$$A |n\rangle = \sqrt{n} |n - 1\rangle, \tag{B.3.48}$$

$$A^\dagger |n\rangle = \sqrt{n + 1} |n + 1\rangle. \tag{B.3.49}$$

It follows that

$$A^2 |n\rangle = \sqrt{n(n-1)} |n - 2\rangle, \tag{B.3.50}$$

$$A^{\dagger 2} |n\rangle = \sqrt{(n+1)(n+2)} |n + 2\rangle, \tag{B.3.51}$$

$$A^\dagger A |n\rangle = n |n\rangle, \tag{B.3.52}$$

$$A A^\dagger |n\rangle = (n+1) n\rangle. \tag{B.3.53}$$

Hence,

$$\langle n' | A |n\rangle = \sqrt{n} \, \delta_{n' \, n-1}, \tag{B.3.54}$$

$$\langle n' | A^\dagger |n\rangle = \sqrt{n+1} \, \delta_{n' \, n+1}, \tag{B.3.55}$$

$$\langle n' | A^\dagger A |n\rangle = n \, \delta_{n' \, n}, \tag{B.3.56}$$

$$\langle n' | A A^\dagger |n\rangle = (n+1) \, \delta_{n' \, n}, \tag{B.3.57}$$

$$\langle n' | A^2 |n\rangle = \sqrt{n(n-1)} \, \delta_{n' \, n-2}, \tag{B.3.58}$$

$$\langle n' | A^{\dagger 2} |n\rangle = \sqrt{(n+1)(n+2)} \, \delta_{n' \, n+2}. \tag{B.3.59}$$

(a) Now, making use of the previous results,

$$x = \sqrt{\frac{\hbar}{2 m \omega}} \, (A + A^\dagger), \tag{B.3.60}$$

so

$$\langle n' | x |n\rangle = \sqrt{\frac{\hbar}{2 m \omega}} \left(\sqrt{n} \, \delta_{n' \, n-1} + \sqrt{n+1} \, \delta_{n' \, n+1} \right). \tag{B.3.61}$$

(b) Now,

$$p_x = i \sqrt{\frac{m \hbar \omega}{2}} \, (-A + A^\dagger), \tag{B.3.62}$$

so

$$\langle n' | p_x |n\rangle = \sqrt{\frac{m \hbar \omega}{2}} \left(-\sqrt{n} \, \delta_{n' \, n-1} + \sqrt{n+1} \, \delta_{n' \, n+1} \right). \tag{B.3.63}$$

(c) Now,

$$x^2 = \left(\frac{\hbar}{2 m \omega} \right) \left(A^2 + A A^\dagger + A^\dagger A + A^{\dagger 2} \right), \tag{B.3.64}$$

so

$$\langle n' | x^2 | n \rangle = \left(\frac{\hbar}{2 m \omega} \right) \Big[\sqrt{n (n-1)} \, \delta_{n' \, n-2} + \sqrt{(n+1)(n+2)} \, \delta_{n' \, n+2}$$

$$+ (2n+1) \, \delta_{n' \, n} \Big]. \tag{B.3.65}$$

(d) Now,

$$p_x^2 = \left(\frac{m \hbar \omega}{2} \right) \left(-A^2 - A A^\dagger + A^\dagger A + A^{\dagger 2} \right), \tag{B.3.66}$$

so

$$\langle n' | p_x^2 | n \rangle = \left(\frac{m \hbar \omega}{2} \right) \Big[- \sqrt{n (n-1)} \, \delta_{n' \, n-2} - \sqrt{(n+1)(n+2)} \, \delta_{n' \, n+2}$$

$$+ (2n+1) \, \delta_{n' \, n} \Big]. \tag{B.3.67}$$

(e) Making use of the previous results,

$$\langle x \rangle = \langle n | x | n \rangle = 0, \tag{B.3.68}$$

$$\langle p_x \rangle = \langle n | p_x | n \rangle = 0, \tag{B.3.69}$$

$$\langle x^2 \rangle = \langle n | x^2 | n \rangle = \left(\frac{\hbar}{2 m \omega} \right) (2n+1), \tag{B.3.70}$$

$$\langle p_x^2 \rangle = \langle n | p_x^2 | n \rangle = \left(\frac{m \hbar \omega}{2} \right) (2n+1), \tag{B.3.71}$$

so

$$\langle (\Delta x)^2 \rangle = \langle x^2 \rangle - \langle x \rangle^2 = \left(\frac{\hbar}{2 m \omega} \right) (2n+1), \tag{B.3.72}$$

$$\langle (\Delta p_x)^2 \rangle = \langle p_x^2 \rangle - \langle p_x \rangle^2 = \left(\frac{m \hbar \omega}{2} \right) (2n+1). \tag{B.3.73}$$

Thus, we deduce that

$$\langle (\Delta x)^2 \rangle \langle (\Delta p_x)^2 \rangle = (n + 1/2)^2 \, \hbar^2. \tag{B.3.74}$$

3.5 We have

$$A | \alpha \rangle = \alpha | \alpha \rangle, \tag{B.3.75}$$

$$\langle \alpha | A^\dagger = \alpha^* \langle \alpha |, \tag{B.3.76}$$

$$\langle \alpha | \alpha \rangle = 1. \tag{B.3.77}$$

Moreover, from the previous two exercises,

$$AA^\dagger = 1 + A^\dagger A, \tag{B.3.78}$$

$$x = \sqrt{\frac{\hbar}{2m\omega}}\,(A + A^\dagger), \tag{B.3.79}$$

$$p_x = i\sqrt{\frac{m\hbar\omega}{2}}\,(-A + A^\dagger). \tag{B.3.80}$$

(a) Now, making use of the previous six equations,

$$\langle x \rangle = \sqrt{\frac{\hbar}{2m\omega}}\,\langle \alpha | A + A^\dagger | \alpha \rangle = \sqrt{\frac{\hbar}{2m\omega}}\,(\alpha + \alpha^*) = \sqrt{\frac{2\hbar}{m\omega}}\,\mathrm{Re}(\alpha), \tag{B.3.81}$$

and

$$\langle p_x \rangle = i\sqrt{\frac{m\hbar\omega}{2}}\,\langle \alpha | -A + A^\dagger | \alpha \rangle = i\sqrt{\frac{m\hbar\omega}{2}}\,(-\alpha + \alpha^*)$$

$$= \sqrt{2m\hbar\omega}\,\mathrm{Im}(\alpha). \tag{B.3.82}$$

Moreover,

$$x^2 = \frac{\hbar}{2m\omega}\,(AA + 2A^\dagger A + A^\dagger A^\dagger + 1), \tag{B.3.83}$$

so

$$\langle x^2 \rangle = \frac{\hbar}{2m\omega}\,(\alpha^2 + 2\alpha\alpha^* + \alpha^{*2} + 1) = \frac{2\hbar}{m\omega}\left(\frac{1}{4} + [\mathrm{Re}(\alpha)]^2\right). \tag{B.3.84}$$

Furthermore,

$$p_x^2 = -\frac{m\hbar\omega}{2}\,(AA - 2A^\dagger A + A^\dagger A^\dagger - 1), \tag{B.3.85}$$

so

$$\langle p_x^2 \rangle = -\frac{m\hbar\omega}{2}\,(\alpha^2 - 2\alpha\alpha^* + \alpha^{*2} - 1) = 2m\hbar\omega\left(\frac{1}{4} + [\mathrm{Im}(\alpha)]^2\right). \tag{B.3.86}$$

It follows that

$$\langle (\Delta x)^2 \rangle = \langle x^2 \rangle - \langle x \rangle^2 = \frac{\hbar}{2m\omega}, \tag{B.3.87}$$

and

$$\langle (\Delta p_x)^2 \rangle = \langle p_x^2 \rangle - \langle p_x \rangle^2 = \frac{m\hbar\omega}{2}. \tag{B.3.88}$$

Hence,

$$\langle (\Delta x)^2 \rangle \langle (\Delta p_x)^2 \rangle = \frac{\hbar^2}{4}. \tag{B.3.89}$$

(b) Let

$$|\alpha\rangle = \sum_{n=0,\infty} c_n |n\rangle, \tag{B.3.90}$$

where

$$\langle n|n'\rangle = \delta_{nn'}. \tag{B.3.91}$$

We have

$$A |\alpha\rangle = \alpha |\alpha\rangle, \tag{B.3.92}$$

so

$$\sum_{n'=0,\infty} c_{n'} A |n'\rangle = \alpha \sum_{n'=0,\infty} c_{n'} |\alpha\rangle. \tag{B.3.93}$$

It follows from Equation (B.3.91) that

$$\sum_{n'=0,\infty} c_{n'} \langle n| A |n'\rangle = \alpha c_n. \tag{B.3.94}$$

However, from Exercise 3.3,

$$A |n'\rangle = \sqrt{n'} |n'-1\rangle, \tag{B.3.95}$$

which implies that

$$\langle n| A |n'\rangle = \sqrt{n'} \, \delta_{nn'-1}. \tag{B.3.96}$$

Hence,

$$\sum_{n'=0,\infty} c_{n'} \sqrt{n'} \, \delta_{nn'-1} = \alpha c_n, \tag{B.3.97}$$

which yields

$$\alpha c_n = \sqrt{n+1} \, c_{n+1}. \tag{B.3.98}$$

Thus, we deduce that

$$c_n = \frac{\alpha^n}{\sqrt{n!}} c_0. \tag{B.3.99}$$

However, we know that

$$\langle \alpha|\alpha\rangle = 1, \tag{B.3.100}$$

which leads to

$$\sum_{n=0,\infty} |c_n|^2 = 1, \tag{B.3.101}$$

where use has been made of Equation (B.3.91). Hence,

$$|c_0|^2 \sum_{n=0,\infty} \frac{|\alpha|^{2n}}{n!} = 1. \tag{B.3.102}$$

However,

$$\sum_{n=0,\infty} \frac{|\alpha|^{2n}}{n!} \equiv e^{|\alpha|^2}, \tag{B.3.103}$$

which implies that

$$c_0 = \exp\left(-\frac{|\alpha|^2}{2}\right). \tag{B.3.104}$$

Thus, we obtain

$$c_n = \frac{\alpha^2}{\sqrt{n!}} \exp\left(-\frac{|\alpha|^2}{2}\right). \tag{B.3.105}$$

(c) From Exercise 3.3, the Hamiltonian can be written

$$H = \left(A^\dagger A + \frac{1}{2}\right) \hbar \omega. \tag{B.3.106}$$

Thus,

$$\langle H \rangle = \hbar \omega \langle \alpha| \left(A^\dagger A + \frac{1}{2}\right) |\alpha\rangle = \left(\alpha^* \alpha + \frac{1}{2}\right) \hbar \omega = \left(|\alpha|^2 + \frac{1}{2}\right) \hbar \omega. \tag{B.3.107}$$

(d) The energy eigenstates evolve in time as

$$|n, t\rangle = e^{-i E_n t / \hbar} |n\rangle, \tag{B.3.108}$$

where

$$E_n = (n + 1/2) \hbar \omega. \tag{B.3.109}$$

Thus,

$$|\alpha, t\rangle = \sum_{n=0,\infty} \frac{\alpha^n}{\sqrt{n!}} \exp\left(-\frac{|\alpha|^2}{2}\right) e^{-i(n+1/2)\omega t} |n\rangle. \tag{B.3.110}$$

It follows that

$$A |\alpha, t\rangle = \sum_{n=0,\infty} \frac{\alpha^n}{\sqrt{n!}} \exp\left(-\frac{|\alpha|^2}{2}\right) e^{-i(n+1/2)\omega t} A |n\rangle, \tag{B.3.111}$$

or

$$A |\alpha, t\rangle = \sum_{n=0,\infty} \frac{\alpha^n}{\sqrt{n!}} \exp\left(-\frac{|\alpha|^2}{2}\right) e^{-i(n+1/2)\omega t} \sqrt{n} |n - 1\rangle, \tag{B.3.112}$$

where use has been made of Equation (B.3.95). If $n' = n - 1$ then

$$A |\alpha, t\rangle = \alpha e^{-i\omega t} \sum_{n'=0,\infty} \frac{\alpha^{n'}}{\sqrt{n'!}} \exp\left(-\frac{|\alpha|^2}{2}\right) e^{-i(n'+1/2)\omega t} |n'\rangle$$

$$= \alpha e^{-i\omega t} |\alpha, t\rangle. \tag{B.3.113}$$

Thus, we conclude that $|\alpha, t\rangle$ is an eigenket of A corresponding to the eigenvalue

$$\alpha(t) = \alpha e^{-i\omega t}. \tag{B.3.114}$$

Furthermore, Equation (B.3.110) yields

$$|\alpha, t\rangle = e^{-i\omega t/2} |\alpha\rangle. \tag{B.3.115}$$

(e) From Equations (B.3.81) and (B.3.82), if $\alpha = (m\omega/2\hbar)^{1/2} a$, where a is real, then $\alpha(t) = (m\omega/2\hbar)^{1/2} a\,e^{-i\omega t}$, and

$$\langle x \rangle = a\cos(\omega t), \tag{B.3.116}$$

$$\langle p_x \rangle = -m\omega a \sin(\omega t). \tag{B.3.117}$$

(f) In the Schrödinger representation,

$$A = \sqrt{\frac{m\omega}{2\hbar}}\,x + i\,\frac{p_x}{\sqrt{2m\omega h}} \rightarrow \sqrt{\frac{m\omega}{2\hbar}}\,x' + \sqrt{\frac{\hbar}{2m\omega}}\,\frac{d}{dx'}. \tag{B.3.118}$$

Now, the wavefunction associated with $|\alpha, t\rangle$ satisfies $A\psi = \alpha(t)\psi$, or

$$\left(\sqrt{\frac{m\omega}{2\hbar}}\,x' + \sqrt{\frac{\hbar}{2m\omega}}\,\frac{d}{dx'} \right)\psi = \alpha(t)\psi. \tag{B.3.119}$$

Writing $x_0 = \sqrt{\hbar/m\omega}$ and $\alpha(t) = (m\omega/2\hbar)^{1/2} a(t)$, we obtain

$$\left(\frac{x'}{x_0} + x_0 \frac{d}{dx'} \right)\psi = \frac{a}{x_0}\,\psi, \tag{B.3.120}$$

which can be solved to give

$$\psi(x', t) = B \exp\left[-\frac{(x'-a)^2}{2x_0^2} \right], \tag{B.3.121}$$

where B is a constant. From Exercise 3.3, B takes the value $1/(\pi^{1/4} x_0^{1/2})$, so

$$\psi(x', t) = \frac{1}{\pi^{1/4} x_0^{1/2}} \exp\left[-\frac{(x'-a)^2}{2x_0^2} \right] \equiv \psi_0\left(x' - \sqrt{\frac{2\hbar}{m\omega}}\,\alpha(t) \right), \tag{B.3.122}$$

where $\psi_0(x')$ is the properly normalized, stationary, ground-state wavefunction.

3.6 The Hamiltonian is

$$H = \frac{p_x^2}{2m} + V(x). \tag{B.3.123}$$

It is easily seen that

$$[V(x), x] = 0. \tag{B.3.124}$$

Moreover,

$$[p_x^2, x] = p_x [p_x, x] - [x, p_x] p_x = -2i\hbar p_x. \tag{B.3.125}$$

Thus,

$$[H, x] = -\frac{i\hbar p_x}{m}, \tag{B.3.126}$$

and

$$[[H, x], x] = -\frac{i\hbar}{m}[p_x, x] = -\frac{\hbar^2}{m}. \tag{B.3.127}$$

Let

$$H|n\rangle = E_n|n\rangle, \tag{B.3.128}$$

where

$$\langle n|n'\rangle = \delta_{nn'}. \tag{B.3.129}$$

It follows that

$$\langle n|[[H, x], x]|n\rangle = -\langle n|\frac{\hbar^2}{m}|n\rangle = -\frac{\hbar^2}{m}. \tag{B.3.130}$$

However,

$$[[H, x], x] = Hx^2 - 2xHx + x^2H, \tag{B.3.131}$$

so

$$\langle n|[[H, x], x]|n\rangle = \langle n|E_n x^2 - 2xHx + x^2 E_n|n\rangle$$
$$= 2E_n\langle n|x^2|n\rangle - 2\langle n|xHx|n\rangle. \tag{B.3.132}$$

It follows that

$$\langle n|x(H - E_n)x|n\rangle = \frac{\hbar^2}{2m}. \tag{B.3.133}$$

Now, because the $|n\rangle$ must span ket space, we have

$$\sum_{n'}|n'\rangle\langle n'| = 1. \tag{B.3.134}$$

Thus,

$$\langle n|x(H - E_n)x|n\rangle = \sum_{n'}\langle n|x|n'\rangle\langle n'|(H - E_n)x|n\rangle$$
$$= \sum_{n'}(E_{n'} - E_n)\langle n|x|n'\rangle\langle n'|x|n\rangle$$
$$= \sum_{n'}(E_{n'} - E_n)|\langle n|x|n'\rangle|^2. \tag{B.3.135}$$

Hence,

$$\sum_{n'}(E_{n'} - E_n)|\langle n|x|n'\rangle|^2 = \frac{\hbar^2}{2m}. \tag{B.3.136}$$

3.7 From Exercise 2.6, we know that

$$D_x(a)V(x)D_x(a)^\dagger = V(x - a), \tag{B.3.137}$$

where the operator $D_x(a)$ displaces the system a distance a parallel to the x-axis. If

$$V(x - a) = V(x) \tag{B.3.138}$$

for all x then

$$D_x(a) V(x) - V(x) D_x(a) = 0, \tag{B.3.139}$$

because $D_x(a)$ is unitary [i.e, $D_x(a) D_x(a)^\dagger = 1$]. Now, $D_x(a)$ is a function of the p_x operator. Thus, it obviously commutes with $p_x^2/2\,m$. Hence, we deduce that $D_x(a)$ commutes with a Hamiltonian of the form

$$H = \frac{p_x^2}{2\,m} + V(x), \tag{B.3.140}$$

where $V(x - a) = V(x)$ for all x. This implies that eigenstates of the Hamiltonian are also eigenstates of $D_x(a)$. However, according to Exercise 2.6, the most general wavefunction of an eigenstate of $D_x(a)$ takes the form

$$\psi(x') = e^{i k' x} u(x'), \tag{B.3.141}$$

where k' is a real parameter, and $u(x'-a) = u(x')$ for all x. Hence, an energy eigenstate must also have a wavefunction of this form.

3.8   The Heisenberg equation of motion for the ladder operator A is

$$\frac{dA}{dt} = -\frac{i}{\hbar} [A, H]. \tag{B.3.142}$$

However, according to Exercise 3.3,

$$[A, H] = \hbar \omega A. \tag{B.3.143}$$

Hence,

$$\frac{dA}{dt} = -i \omega A. \tag{B.3.144}$$

The conjugate to the previous equation yields

$$\frac{dA^\dagger}{dt} = i \omega A^\dagger. \tag{B.3.145}$$

Thus, it follows that

$$A(t) = e^{-i \omega t} A(0), \tag{B.3.146}$$

$$A^\dagger(t) = e^{i \omega t} A^\dagger(0). \tag{B.3.147}$$

Now, according to Exercise 3.3,

$$x(t) \sqrt{\frac{\hbar}{2m\omega}} \left[A(t) + A^\dagger(t) \right], \tag{B.3.148}$$

$$p_x(t) = \mathrm{i} \sqrt{\frac{m\hbar\omega}{2}} \left[-A(t) + A^\dagger(t) \right]. \tag{B.3.149}$$

Furthermore,

$$A(0) = \sqrt{\frac{m\omega}{2\hbar}}\, x(0) + \mathrm{i}\, \frac{p_x(0)}{\sqrt{2m\omega\hbar}}, \tag{B.3.150}$$

$$A^\dagger(0) = \sqrt{\frac{m\omega}{2\hbar}}\, x(0) - \mathrm{i}\, \frac{p_x(0)}{\sqrt{2m\omega\hbar}}. \tag{B.3.151}$$

Hence, we obtain

$$x(t) = \frac{1}{2} \left(\mathrm{e}^{-\mathrm{i}\omega t} \left[x(0) + \mathrm{i}\, \frac{p_x(0)}{m\omega} \right] + \mathrm{e}^{\mathrm{i}\omega t} \left[x(0) - \mathrm{i}\, \frac{p_x(0)}{m\omega} \right] \right), \tag{B.3.152}$$

$$p_x(t) = \mathrm{i}\, \frac{m\omega}{2} \left(-\mathrm{e}^{-\mathrm{i}\omega t} \left[x(0) + \mathrm{i}\, \frac{p_x(0)}{m\omega} \right] + \mathrm{e}^{\mathrm{i}\omega t} \left[x(0) - \mathrm{i}\, \frac{p_x(0)}{m\omega} \right] \right). \tag{B.3.153}$$

The previous two equations yield

$$x(t) = \cos(\omega t)\, x(0) + \frac{\sin(\omega t)}{m\omega}\, p_x(0), \tag{B.3.154}$$

$$p_x(t) = \cos(\omega t)\, p_x(0) - m\omega\, \sin(\omega t)\, x(0). \tag{B.3.155}$$

3.9 The time-independent Schrödinger equation has the form

$$\left[-\frac{\hbar^2}{2m}\, \frac{d^2}{dx'^2} + V(x') \right] \psi(x') = E\, \psi(x'). \tag{B.3.156}$$

It follows that

$$\int_{-\infty}^{\infty} dx'\, \psi^*(x') \left[-\frac{\hbar^2}{2m}\, \frac{d^2}{dx'^2} + V(x') \right] \psi(x') = E \int_{-\infty}^{\infty} |\psi|^2\, dx', \tag{B.3.157}$$

which implies that

$$\left\langle \frac{p_x^2}{2m} \right\rangle + \langle V \rangle = E, \tag{B.3.158}$$

assuming that $\psi(x')$ is properly normalized. Multiplying Equation (B.3.156) by $x'\, d\psi/dx'$ and integrating, we obtain

$$-\frac{\hbar^2}{2m} \int_{-\infty}^{\infty} dx'\, x'\, \frac{d}{dx'} \left(\frac{d\psi}{dx'} \right)^2 + \int_{-\infty}^{\infty} dx'\, x'\, V\, \frac{d\psi^2}{dx'} = E \int_{-\infty}^{\infty} dx'\, x'\, \frac{d\psi^2}{dx'}. \tag{B.3.159}$$

Integrating by parts, making use of the fact that $|\psi| \to 0$ as $|x'| \to \infty$ in a bound state, we get

$$\frac{\hbar^2}{2m} \int_{-\infty}^{\infty} dx' \left(\frac{d\psi}{dx'}\right)^2 - \int_{-\infty}^{\infty} dx' \frac{d(x'\,V)}{dx'} \psi^2 = -E \int_{\infty}^{\infty} dx' \,\psi^2. \qquad \text{(B.3.160)}$$

Integrating by parts again, we obtain

$$-\frac{\hbar^2}{2m} \int_{-\infty}^{\infty} dx' \,\psi \frac{d^2}{dx'^2} \psi - \int_{-\infty}^{\infty} dx' \frac{d(x'\,V)}{dx'} \psi^2 = -E \int_{\infty}^{\infty} dx' \,\psi^2, \qquad \text{(B.3.161)}$$

or, because $\psi(x)$ can be taken as real,

$$\int_{-\infty}^{\infty} dx' \,\psi^*(x') \left[-\frac{\hbar^2}{2m} \frac{d^2}{dx'^2} - x' \frac{dV}{dx'} - V \right] \psi(x') = -E \int_{-\infty}^{\infty} dx' \,|\psi|^2. \qquad \text{(B.3.162)}$$

Hence, we get

$$\left\langle \frac{p_x^2}{2m} \right\rangle - \left\langle x \frac{dV}{dx} \right\rangle - \langle V \rangle = -E. \qquad \text{(B.3.163)}$$

Finally, Equations (B.3.158) and (B.3.163) can be combined to give

$$\left\langle \frac{p_x^2}{2m} \right\rangle = \frac{1}{2} \left\langle x \frac{dV}{dx} \right\rangle. \qquad \text{(B.3.164)}$$

3.10 The Hamiltonian of the system is

$$H = \frac{\Pi_x^2 + \Pi_y^2}{2m}. \qquad \text{(B.3.165)}$$

Note that $\Pi_z = 0$, because the particle is moving in the x-y plane (and, thus, has no z-directed mechanical momentum). We also know that

$$[\Pi_x, \Pi_y] = i\,\hbar\,q\,B_z. \qquad \text{(B.3.166)}$$

Making use of this commutation relation, we can write

$$H = \hbar\,\omega \left(\frac{1}{2} + A^\dagger A \right), \qquad \text{(B.3.167)}$$

where $\omega = q\,B_z/m$, and

$$A = \frac{\Pi_x + i\,\Pi_y}{\sqrt{2\,\hbar\,q\,B_z}}. \qquad \text{(B.3.168)}$$

Now,

$$[A, A^\dagger] = -\frac{i}{\hbar\,q\,B_z}\,[\Pi_x, \Pi_y] = 1. \qquad \text{(B.3.169)}$$

Our system is now formally equivalent to the harmonic oscillator system considered in Exercise 3.3. Hence, we can immediately deduce that

$$E_n = (n + 1/2)\hbar\omega. \tag{B.3.170}$$

3.11 Schrödinger's equation is written

$$i\hbar\frac{\partial\psi}{\partial t} = \frac{1}{2m}(-i\hbar\nabla' - q\mathbf{A})\cdot(-i\hbar\nabla' - q\mathbf{A})\psi + q\phi\psi, \tag{B.3.171}$$

which becomes

$$i\hbar\frac{\partial\psi}{\partial t} = \frac{1}{2m}(-\hbar^2\nabla'^2 + 2i\hbar q\mathbf{A}\cdot\nabla' + i\hbar q\nabla'\cdot\mathbf{A} + q^2 A^2)\psi + q\phi\psi. \tag{B.3.172}$$

If $\nabla'\cdot\mathbf{A} = 0$ then the previous equation simplifies to give

$$i\hbar\frac{\partial\psi}{\partial t} = -\frac{\hbar^2}{2m}\nabla'^2\psi + i\frac{\hbar q}{m}\mathbf{A}\cdot\nabla'\psi + \frac{q^2}{2m}A^2\psi + q\phi\psi. \tag{B.3.173}$$

Multiplying the previous equation by ψ^*, we obtain

$$i\hbar\psi^*\frac{\partial\psi}{\partial t} = -\frac{\hbar^2}{2m}\psi^*\nabla'^2\psi + i\frac{\hbar q}{m}\psi^*\mathbf{A}\cdot\nabla'\psi + \frac{q^2}{2m}A^2|\psi|^2 + q\phi|\psi|^2. \tag{B.3.174}$$

The complex conjugate of this expression is

$$-i\hbar\psi\frac{\partial\psi^*}{\partial t} = -\frac{\hbar^2}{2m}\psi\nabla'^2\psi^* - i\frac{\hbar q}{m}\psi\mathbf{A}\cdot\nabla'\psi^* + \frac{q^2}{2m}A^2|\psi|^2 + q\phi|\psi|^2. \tag{B.3.175}$$

Taking the difference between the previous two equations, and dividing by $i\hbar$, we get

$$\psi^*\frac{\partial\psi}{\partial t} + \psi\frac{\partial\psi^*}{\partial t} = \frac{i\hbar}{2m}(\psi^*\nabla'^2\psi - \psi\nabla'^2\psi^*) + \frac{q}{m}(\psi^*\mathbf{A}\cdot\nabla'\psi + \psi\mathbf{A}\cdot\nabla'\psi^*), \tag{B.3.176}$$

or

$$\frac{\partial|\psi|^2}{\partial t} = \frac{i\hbar}{2m}\nabla'\cdot(\psi^*\nabla'\psi - \psi\nabla'\psi^*) + \frac{q}{m}\mathbf{A}\cdot\nabla'|\psi|^2, \tag{B.3.177}$$

which can be written

$$\frac{\partial\rho}{\partial t} + \nabla'\cdot\mathbf{j} = 0, \tag{B.3.178}$$

where

$$\rho = |\psi|^2, \tag{B.3.179}$$

and

$$\mathbf{j} = -\left(\frac{i\hbar}{2m}\right)[\psi^*\nabla'\psi - (\nabla'\psi)^*\psi] - \frac{q}{m}\rho\mathbf{A}. \tag{B.3.180}$$

Here, use has been made of the fact that $\nabla' \cdot \mathbf{A} = 0$. In a gauge transformation,

$$\mathbf{A} \to \mathbf{A} - \nabla f, \qquad\qquad\qquad (B.3.181)$$

$$\psi \to e^{i\Lambda} \psi, \qquad\qquad\qquad (B.3.182)$$

where

$$\Lambda = -\frac{q}{\hbar} f. \qquad\qquad\qquad (B.3.183)$$

It follows that

$$\rho \to \left(e^{-i\Lambda} \psi^*\right)\left(e^{i\Lambda} \psi\right) = |\psi|^2 = \rho. \qquad\qquad (B.3.184)$$

Moreover,

$$\psi^* \nabla' \psi - (\nabla' \psi)^* \psi \to \psi^* \nabla' \psi - (\nabla' \psi)^* \psi + 2i \nabla\Lambda |\psi|^2. \qquad (B.3.185)$$

Hence,

$$\mathbf{j} \to \mathbf{j}. \qquad\qquad\qquad (B.3.186)$$

B.4   Chapter 4

4.1 (a) We have

$$[L^2, L_z] = [L_x^2, L_z] + [L_y^2, L_z] + [L_z^2, L_z]. \qquad\qquad (B.4.1)$$

However, making use of the fundamental commutation relations $\mathbf{L} \times \mathbf{L} = i\hbar \mathbf{L}$, we obtain

$$[L_x^2, L_z] = L_x [L_x, L_z] + [L_x, L_z] L_x = -i\hbar L_x L_y - i\hbar L_y L_x, \qquad (B.4.2)$$

$$[L_y^2, L_z] = L_y [L_y, L_z] + [L_y, L_z] L_y = i\hbar L_y L_x + i\hbar L_x L_y, \qquad (B.4.3)$$

$$[L_z^2, L_z] = L_z [L_z, L_z] + [L_z, L_z] L_z = 0. \qquad\qquad (B.4.4)$$

Hence,

$$[L^2, L_z] = 0. \qquad\qquad\qquad (B.4.5)$$

(b) We have

$$[L^{\pm}, L_z] = [L_x \pm i L_y, L_z] = [L_x, L_z] \pm i [L_y, L_z] = -i\hbar L_y \mp \hbar L_x$$
$$= \mp\hbar \left(L_x \pm i L_y\right) = \mp\hbar L^{\mp}. \qquad\qquad (B.4.6)$$

(c) We have

$$[L^+, L^-] = [L_x + i L_y, L_x - i L_y]$$
$$= [L_x, L_x] + i [L_y, L_x] - i [L_x, L_y] + [L_y, L_y] = 2\hbar L_z. \qquad (B.4.7)$$

4.2 We have

$$x = r \sin\theta \cos\varphi, \tag{B.4.8}$$

$$y = r \sin\theta \sin\varphi, \tag{B.4.9}$$

$$z = r \cos\theta, \tag{B.4.10}$$

which can be inverted to give

$$r = \left(x^2 + y^2 + z^2\right)^{1/2}, \tag{B.4.11}$$

$$\theta = \tan^{-1}\left(\frac{[x^2 + y^2]^{1/2}}{z}\right), \tag{B.4.12}$$

$$\varphi = \tan^{-1}\left(\frac{y}{x}\right). \tag{B.4.13}$$

It follows that

$$\frac{\partial r}{\partial x} = \sin\theta \cos\varphi, \qquad \frac{\partial r}{\partial y} = \sin\theta \sin\varphi, \qquad \frac{\partial r}{\partial z} = \cos\theta, \tag{B.4.14}$$

$$\frac{\partial \theta}{\partial x} = \frac{\cos\theta \cos\varphi}{r}, \qquad \frac{\partial \theta}{\partial y} = \frac{\cos\theta \sin\varphi}{r}, \qquad \frac{\partial \theta}{\partial z} = -\frac{\sin\theta}{r}, \tag{B.4.15}$$

$$\frac{\partial \varphi}{\partial x} = -\frac{\sin\varphi}{r \sin\theta}, \qquad \frac{\partial \varphi}{\partial y} = \frac{\cos\varphi}{r \sin\theta}, \qquad \frac{\partial \varphi}{\partial z} = 0. \tag{B.4.16}$$

Hence,

$$\frac{\partial}{\partial x} = \frac{\partial r}{\partial x}\frac{\partial}{\partial r} + \frac{\partial \theta}{\partial x}\frac{\partial}{\partial \theta} + \frac{\partial \varphi}{\partial x}\frac{\partial}{\partial \varphi}$$

$$= \sin\theta \cos\varphi \frac{\partial}{\partial r} + \frac{\cos\theta \cos\varphi}{r}\frac{\partial}{\partial \theta} - \frac{\sin\varphi}{r \sin\theta}\frac{\partial}{\partial \varphi}. \tag{B.4.17}$$

Similarly,

$$\frac{\partial}{\partial y} = \sin\theta \sin\varphi \frac{\partial}{\partial r} + \frac{\cos\theta \sin\varphi}{r}\frac{\partial}{\partial \theta} + \frac{\cos\varphi}{r \sin\theta}\frac{\partial}{\partial \varphi}, \tag{B.4.18}$$

$$\frac{\partial}{\partial z} = \cos\theta \frac{\partial}{\partial r} - \frac{\sin\theta}{r}\frac{\partial}{\partial \theta}. \tag{B.4.19}$$

Thus,

$$y \frac{\partial}{\partial z} - z \frac{\partial}{\partial y} = -\sin \varphi \frac{\partial}{\partial \theta} - \cot \theta \cos \varphi \frac{\partial}{\partial \varphi}, \tag{B.4.20}$$

$$z \frac{\partial}{\partial x} - x \frac{\partial}{\partial z} = \cos \varphi \frac{\partial}{\partial \theta} - \cot \theta \sin \varphi \frac{\partial}{\partial \varphi}, \tag{B.4.21}$$

$$x \frac{\partial}{\partial y} - y \frac{\partial}{\partial x} = \frac{\partial}{\partial \varphi}. \tag{B.4.22}$$

Hence,

$$L_x = -i \hbar \left(y \frac{\partial}{\partial z} - z \frac{\partial}{\partial y} \right) = i \hbar \left(\sin \varphi \frac{\partial}{\partial \theta} + \cot \theta \cos \varphi \frac{\partial}{\partial \varphi} \right), \tag{B.4.23}$$

$$L_y = -i \hbar \left(z \frac{\partial}{\partial x} - x \frac{\partial}{\partial z} \right) = -i \hbar \left(\cos \varphi \frac{\partial}{\partial \theta} - \cot \theta \sin \varphi \frac{\partial}{\partial \varphi} \right), \tag{B.4.24}$$

$$L_z = -i \hbar \left(x \frac{\partial}{\partial y} - y \frac{\partial}{\partial x} \right) = -i \hbar \frac{\partial}{\partial \varphi}. \tag{B.4.25}$$

We have

$$L_x \pm i L_y = \hbar \left[\pm (\cos \varphi \pm i \sin \varphi) \frac{\partial}{\partial \theta} + i \cot \theta (\cos \varphi \pm i \sin \varphi) \frac{\partial}{\partial \varphi} \right], \tag{B.4.26}$$

or

$$L^\pm = \pm \hbar \, e^{\pm i \varphi} \left(\frac{\partial}{\partial \theta} \pm i \cot \theta \frac{\partial}{\partial \varphi} \right). \tag{B.4.27}$$

We have

$$L^2 = L_z^2 + \frac{1}{2} (L^+ L^- + L^- L^+). \tag{B.4.28}$$

Now,

$$\begin{aligned}
\frac{L^+ L^-}{-\hbar^2} &= e^{i\varphi} \left(\frac{\partial}{\partial \theta} + i \cot \theta \frac{\partial}{\partial \varphi} \right) e^{-i\varphi} \left(\frac{\partial}{\partial \theta} - i \cot \theta \frac{\partial}{\partial \varphi} \right) \\
&= \frac{\partial}{\partial \theta} \left(\frac{\partial}{\partial \theta} - i \cot \theta \frac{\partial}{\partial \varphi} \right) + e^{i\varphi} i \cot \theta \frac{\partial}{\partial \varphi} e^{-i\varphi} \left(\frac{\partial}{\partial \theta} - i \cot \theta \frac{\partial}{\partial \varphi} \right) \\
&= \frac{\partial^2}{\partial \theta^2} + \frac{i}{\sin^2 \theta} \frac{\partial}{\partial \varphi} - i \cot \theta \frac{\partial^2}{\partial \theta \partial \varphi} + \cot \theta \left(\frac{\partial}{\partial \theta} - i \cot \theta \frac{\partial}{\partial \varphi} \right) \\
&\quad + i \cot \theta \frac{\partial^2}{\partial \theta \partial \varphi} + \cot^2 \varphi \frac{\partial^2}{\partial \varphi^2} \\
&= \frac{\partial^2}{\partial \theta^2} + i \frac{\partial}{\partial \varphi} + \cot \theta \frac{\partial}{\partial \theta} + \cot^2 \theta \frac{\partial^2}{\partial \varphi^2},
\end{aligned} \tag{B.4.29}$$

and

$$\frac{L^- L^+}{-\hbar^2} = \frac{(L^+ L^-)^*}{-\hbar^2} = \frac{\partial^2}{\partial\theta^2} - i\,\frac{\partial}{\partial\varphi} + \cot\theta\,\frac{\partial}{\partial\theta} + \cot^2\theta\,\frac{\partial^2}{\partial\varphi^2}. \tag{B.4.30}$$

Also,

$$\frac{L_z^2}{-\hbar^2} = \frac{\partial^2}{\partial\varphi^2}. \tag{B.4.31}$$

Hence,

$$
\begin{aligned}
L^2 &= -\hbar^2 \left[\frac{\partial^2}{\partial\theta^2} + \cot\theta\,\frac{\partial}{\partial\theta} + \left(1 + \cot^2\theta\right)\frac{\partial^2}{\partial\varphi^2} \right] \\
&= -\hbar^2 \left[\frac{1}{\sin\theta}\,\frac{\partial}{\partial\theta}\,\sin\theta\,\frac{\partial}{\partial\theta} + \frac{1}{\sin^2\theta}\,\frac{\partial^2}{\partial\varphi^2} \right].
\end{aligned}
\tag{B.4.32}
$$

4.3 A system whose wavefunction is $\psi(\theta,\varphi) = Y_{lm}(\theta,\varphi)$ is in a simultaneous eigenstate of L^2 and L_z. The corresponding state ket is $|l,m\rangle$ where

$$L^2 |l,m\rangle = l(l+1)\hbar^2 |l,m\rangle, \tag{B.4.33}$$

$$L_z |l,m\rangle = m\hbar |l,m\rangle, \tag{B.4.34}$$

$$\langle l,m|l'm'\rangle = \delta_{ll'}\,\delta_{mm'}. \tag{B.4.35}$$

We know that

$$L^\pm = L_x \pm i L_y, \tag{B.4.36}$$

$$L^\pm |l,m\rangle = c_{lm}^\pm |l,m\pm 1\rangle, \tag{B.4.37}$$

$$c_{lm}^\pm = [l(l+1) - m(m\pm 1)]^{1/2}\hbar. \tag{B.4.38}$$

Thus,

$$
\begin{aligned}
\langle L_x\rangle &= \langle l,m|\, L_x\, |l,m\rangle = \frac{1}{2}\,\langle l,m|\, L^+ + L^- \,|l,m\rangle \\
&= \frac{1}{2}\,c_{lm}^+ \langle l,m|l,m+1\rangle + \frac{1}{2}\,c_{lm}^- \langle l,m|l,m-1\rangle = 0.
\end{aligned}
\tag{B.4.39}
$$

Likewise,

$$
\begin{aligned}
\langle L_y\rangle &= \langle l,m|\, L_y\, |l,m\rangle = \frac{1}{2i}\,\langle l,m|\, L^+ - L^- \,|l,m\rangle \\
&= \frac{1}{2i}\,c_{lm}^+ \langle l,m|l,m+1\rangle - \frac{1}{2i}\,c_{lm}^- \langle l,m|l,m-1\rangle = 0.
\end{aligned}
\tag{B.4.40}
$$

Similarly,

$$\langle L_x^2 \rangle = \frac{1}{4} \langle l, m | L^{+2} + L^+ L^- + L^- L^+ + L^{-2} | l, m \rangle$$

$$= \frac{1}{4} c_{lm+1}^+ c_{lm}^+ \langle l, m | l, m+2 \rangle + \frac{1}{4} c_{lm}^- c_{lm-1}^+ \langle l, m | l, m \rangle$$

$$+ \frac{1}{4} c_{lm}^+ c_{lm+1}^- \langle l, m | l, m \rangle + \frac{1}{4} c_{lm-1}^- c_{lm}^- \langle l, m | l, m-2 \rangle$$

$$= \frac{1}{4} \left(c_{lm}^- c_{lm-1}^+ + c_{lm}^+ c_{lm+1}^- \right) = \frac{\hbar^2}{2} \left[l(l+1) - m^2 \right]. \tag{B.4.41}$$

Finally,

$$\langle L_y^2 \rangle = \langle L^2 \rangle - \langle L_x^2 \rangle - \langle L_z^2 \rangle$$

$$= \hbar^2 l(l+1) - \frac{\hbar^2}{2} \left[l(l+1) - m^2 \right] - \hbar^2 m^2 = \frac{\hbar^2}{2} \left[l(l+1) - m^2 \right]. \tag{B.4.42}$$

4.4 Now,

$$L^{\pm} = \pm \hbar\, \mathrm{e}^{\pm \mathrm{i}\varphi} \left(\frac{\partial}{\partial \theta} \pm \mathrm{i} \cot \theta \frac{\partial}{\partial \varphi} \right), \tag{B.4.43}$$

and

$$Y_{lm}(\theta, \varphi) = \sqrt{\frac{2l+1}{4\pi} \frac{(l-m)!}{(l+m)!}} \, P_{lm}(\cos \theta)\, \mathrm{e}^{\mathrm{i}m\varphi}. \tag{B.4.44}$$

Thus,

$$L^+ Y_{lm} = \hbar \sqrt{\frac{2l+1}{4\pi} \frac{(l-m)!}{(l+m)!}} \left[-P'_{lm}(\cos \theta) \sin \theta - m \cot \theta\, P_{lm}(\cos \theta) \right] \mathrm{e}^{\mathrm{i}(m+1)\varphi}. \tag{B.4.45}$$

But,

$$P'_{lm}(\cos \theta) = -\frac{P_{lm+1}(\cos \theta)}{\sin \theta} - \frac{m \cot \theta}{\sin \theta} P_{lm}(\cos \theta), \tag{B.4.46}$$

so

$$L^+ Y_{lm} = \hbar \sqrt{\frac{2l+1}{4\pi} \frac{(l-m)!}{(l+m)!}} \, P_{lm+1}(\cos \theta)\, \mathrm{e}^{\mathrm{i}(m+1)\varphi}$$

$$= \hbar \sqrt{l(l+1) - m(m+1)} \, Y_{lm+1}. \tag{B.4.47}$$

Furthermore,

$$L^- Y_{lm} = -\hbar \sqrt{\frac{2l+1}{4\pi} \frac{(l-m)!}{(l+m)!}} \left[-P'_{lm}(\cos \theta) \sin \theta + m \cot \theta\, P_{lm}(\cos \theta) \right] \mathrm{e}^{\mathrm{i}(m-1)\varphi}. \tag{B.4.48}$$

But,

$$P'_{lm}(\cos\theta) = \frac{(l+m)(l-m+1)}{\sin\theta} P_{lm-1}(\cos\theta) + \frac{m\cot\theta}{\sin\theta} P_{lm}(\cos\theta), \qquad (B.4.49)$$

so

$$L^- Y_{lm} = \hbar \sqrt{\frac{2l+1}{4\pi} \frac{(l-m)!}{(l+m)!}} (l+m)(l-m+1) P_{lm-1}(\cos\theta)\, e^{i(m-1)\varphi}$$

$$= \hbar \sqrt{l(l+1) - m(m-1)}\, Y_{lm-1}. \qquad (B.4.50)$$

4.5 The eigenvalues and eigenfunctions of L_z are $m\hbar$ and $Y_{lm}(\theta,\varphi)$, respectively, where θ, φ are spherical angles defined with respect to the z-axis. Here, l is a non-negative integer, and m is an integer lying in the range $-l \leq m \leq l$. By analogy (because there is nothing special about the z-axis), the eigenvalues and eigenfunctions of L_x are $m\hbar$ and $Y_{lm}(\theta',\varphi')$, respectively, where θ', φ' are spherical angles defined with respect to the x-axis. We know that

$$\mathbf{e}_r = (\sin\theta\cos\varphi,\ \sin\theta\sin\varphi,\ \cos\theta). \qquad (B.4.51)$$

By analogy,

$$\mathbf{e}_r = (\cos\theta',\ \sin\theta'\cos\varphi',\ \sin\theta'\sin\varphi'). \qquad (B.4.52)$$

Hence,

$$\cos\theta' = \sin\theta\cos\varphi, \qquad (B.4.53)$$

$$\sin\theta'\cos\varphi' = \sin\theta\sin\varphi, \qquad (B.4.54)$$

$$\sin\theta'\sin\varphi' = \cos\theta. \qquad (B.4.55)$$

It follows that the eigenfunction of L_x can be written

$$Y'_{lm}(\theta,\varphi) = Y_{lm}(\theta',\varphi')$$

$$= Y_{lm}(\cos^{-1}(\sin\theta\cos\varphi), \tan^{-1}(\cot\theta\,\mathrm{cosec}\,\varphi)). \qquad (B.4.56)$$

In particular, consider $l = 1$. We have

$$Y_{1-1} = \frac{1}{2}\sqrt{\frac{3}{2\pi}} (\sin\theta\cos\varphi - i\sin\theta\sin\varphi), \qquad (B.4.57)$$

$$Y_{10} = \frac{1}{2}\sqrt{\frac{3}{\pi}}\cos\theta, \qquad (B.4.58)$$

$$Y_{11} = -\frac{1}{2}\sqrt{\frac{3}{2\pi}} (\sin\theta\cos\varphi + i\sin\theta\sin\varphi). \qquad (B.4.59)$$

Hence,

$$Y'_{1-1} = \frac{1}{2} \sqrt{\frac{3}{2\pi}} (\sin\theta' \cos\varphi' - i \sin\theta' \sin\varphi'),$$ (B.4.60)

$$Y'_{10} = \frac{1}{2} \sqrt{\frac{3}{\pi}} \cos\theta',$$ (B.4.61)

$$Y'_{11} = -\frac{1}{2} \sqrt{\frac{3}{2\pi}} (\sin\theta' \cos\varphi' + i \sin\theta' \sin\varphi').$$ (B.4.62)

It follows from Equations (B.4.53)–(B.4.55) that

$$Y'_{1-1} = \frac{1}{2} \sqrt{\frac{3}{2\pi}} (\sin\theta \sin\varphi - i \cos\theta),$$ (B.4.63)

$$Y'_{10} = \frac{1}{2} \sqrt{\frac{3}{\pi}} \sin\theta \cos\varphi,$$ (B.4.64)

$$Y'_{11} = -\frac{1}{2} \sqrt{\frac{3}{2\pi}} (\sin\theta \sin\varphi + i \cos\theta).$$ (B.4.65)

Hence,

$$Y'_{1-1} = \frac{i}{2} (Y_{1-1} + Y_{11}) - \frac{i}{\sqrt{2}} Y_{10},$$ (B.4.66)

$$Y'_{10} = \frac{1}{\sqrt{2}} (Y_{1-1} - Y_{11}),$$ (B.4.67)

$$Y'_{11} = -\frac{i}{2} (Y_{1-1} + Y_{11}) - \frac{i}{\sqrt{2}} Y_{10}.$$ (B.4.68)

4.6 If a measurement of L_x yields the result $\hbar$ then the system is left in the state Y'_{11}. From Equation (B.4.68), it is clear that a subsequent measurement of L_z will yield the results $-\hbar$, 0, $+\hbar$ with the probabilities 1/4, 1/2, 1/4, respectively. If a measurement of L_x yields the result 0 then the system is left in the state Y'_{10}. From Equation (B.4.67), it is clear that a subsequent measurement of L_z will yield the results $-\hbar$, 0, $+\hbar$ with the probabilities 1/2, 0, 1/2, respectively. If a measurement of L_x yields the result $-\hbar$ then the system is left in the state Y'_{1-1}. From Equation (B.4.66), it is clear that a subsequent measurement of L_z will yield the results $-\hbar$, 0, $+\hbar$ with the probabilities 1/4, 1/2, 1/4, respectively.

4.7 The Hamiltonian of the system is

$$H = \frac{L_x^2 + L_y^2}{2 I_\perp} + \frac{L_z^2}{2 I_\parallel},$$ (B.4.69)

which can also be written

$$H = \frac{L^2}{2 I_\perp} + \left(\frac{1}{2 I_\parallel} - \frac{1}{2 I_\perp} \right) L_z^2. \tag{B.4.70}$$

It is clear that H commutes with both L^2 and L_z. Hence, the energy eigenstates are also eigenstates of L^2 and L_z. Let $|l, m\rangle$ represent a simultaneous eigenstate of L^2 and L_z. Thus,

$$L^2 |l, m\rangle = l(l+1)\hbar^2 |l, m\rangle, \tag{B.4.71}$$

$$L_z |l, m\rangle = m\hbar |l, m\rangle, \tag{B.4.72}$$

where l is a non-negative integer, and m is an integer lying in the range $-l \le m \le m$. The energy eigenvalue problem is written

$$H |l, m\rangle = E |l, m\rangle. \tag{B.4.73}$$

It follows that

$$E = \hbar^2 \left[\frac{l(l+1)}{2 I_\perp} + \left(\frac{1}{2 I_\parallel} - \frac{1}{2 I_\perp} \right) m^2 \right]. \tag{B.4.74}$$

4.8 We have

$$\mathbf{x} \cdot \mathbf{p} = x p_x + y p_y + z p_z, \tag{B.4.75}$$

and

$$H = \frac{p^2}{2m} + V(r). \tag{B.4.76}$$

Now,

$$[V(r), x p_x] = -x [p_x, V(r)] = i\hbar x \frac{\partial V}{\partial x}, \tag{B.4.77}$$

where use has been made of Equation (3.34). By analogy,

$$[V(r), \mathbf{x} \cdot \mathbf{p}] = i\hbar \left(x \frac{\partial V}{\partial x} + y \frac{\partial V}{\partial y} + z \frac{\partial V}{\partial z} \right) = i\hbar r \frac{dV}{dr}. \tag{B.4.78}$$

Now,

$$[p^2, x p_x] = -[x, p^2] p_x = -i\hbar \frac{\partial p^2}{\partial p_x} p_x = -2 i\hbar p_x^2, \tag{B.4.79}$$

where use has been made of Equation (3.33). By analogy,

$$[p^2, \mathbf{x} \cdot \mathbf{p}] = -2 i\hbar p^2. \tag{B.4.80}$$

Hence,

$$0 = \frac{i}{\hbar} \langle [H, \mathbf{x} \cdot \mathbf{p}] \rangle = \left\langle \frac{p^2}{m} \right\rangle - \left\langle r \frac{dV}{dr} \right\rangle, \tag{B.4.81}$$

which implies that

$$\left\langle \frac{p^2}{2m} \right\rangle = \frac{1}{2} \left\langle r \frac{dV}{dr} \right\rangle. \tag{B.4.82}$$

4.9 The hydrogen atom Hamiltonian is written

$$H = \frac{p^2}{2m_e} + V(r), \tag{B.4.83}$$

where

$$V = -\frac{e^2}{4\pi \epsilon_0 r}. \tag{B.4.84}$$

Now, for an energy eigenstate corresponding to the radial quantum number n,

$$\langle H \rangle = \left\langle \frac{p^2}{2m_e} \right\rangle + \langle V \rangle = \frac{E_0}{n^2}, \tag{B.4.85}$$

where

$$E_0 = -\frac{m_e e^4}{32\pi^2 \epsilon_0^2 \hbar^2} \tag{B.4.86}$$

is the ground-state energy. The virial theorem of the previous exercise implies that

$$\left\langle \frac{p^2}{2m} \right\rangle = \frac{1}{2} \left\langle r \frac{dV}{dr} \right\rangle = -\frac{1}{2} \langle V \rangle. \tag{B.4.87}$$

Hence,

$$\frac{E_0}{n^2} = \frac{1}{2} \langle V \rangle, \tag{B.4.88}$$

or

$$\frac{m_e e^4}{32\pi^2 \epsilon_0^2 \hbar^2 n^2} = \frac{e^2}{8\pi \epsilon_0} \left\langle \frac{1}{r} \right\rangle, \tag{B.4.89}$$

which implies that

$$\left\langle \frac{1}{r} \right\rangle = \frac{1}{n^2 a_0}, \tag{B.4.90}$$

where $a_0 = 4\pi \epsilon_0 \hbar^2 / (m_e e^2)$.

4.10 The Hamiltonian is

$$H = \frac{p^2}{2m} + V(\mathbf{x}). \tag{B.4.91}$$

The angular momentum operator takes the form

$$\mathbf{L} = \mathbf{x} \times \mathbf{p}. \tag{B.4.92}$$

In the Schrödinger representation,

$$\mathbf{p} = -i\hbar \nabla. \tag{B.4.93}$$

Hence,

$$L_i = -\mathrm{i}\,\hbar\,\epsilon_{ijk}\,x_j\,\frac{\partial}{\partial x_k}, \tag{B.4.94}$$

and

$$p^2 = -\hbar^2\,\frac{\partial}{\partial x_l}\,\frac{\partial}{\partial x_l}. \tag{B.4.95}$$

So,

$$
\begin{aligned}
[L_i, p^2] &= \mathrm{i}\,\hbar^3\,\epsilon_{ijk}\,x_j\,\frac{\partial}{\partial x_k}\,\frac{\partial}{\partial x_l}\,\frac{\partial}{\partial x_l} - \mathrm{i}\,\hbar^3\,\epsilon_{ijk}\,\frac{\partial}{\partial x_l}\,\frac{\partial}{\partial x_l}\,x_j\,\frac{\partial}{\partial x_k} \\
&= \mathrm{i}\,\hbar^3\,\epsilon_{ijk}\,x_j\,\frac{\partial}{\partial x_k}\,\frac{\partial}{\partial x_l}\,\frac{\partial}{\partial r_l} - \mathrm{i}\,\hbar^3\,\epsilon_{ijk}\,\frac{\partial}{\partial x_l}\left(x_j\,\frac{\partial}{\partial x_l} + \delta_{jl}\right)\frac{\partial}{\partial x_k} \\
&= \mathrm{i}\,\hbar^3\,\epsilon_{ijk}\,x_j\,\frac{\partial}{\partial x_k}\,\frac{\partial}{\partial x_l}\,\frac{\partial}{\partial x_l} - \mathrm{i}\,\hbar^3\,\epsilon_{ijk}\left(x_j\,\frac{\partial}{\partial r_l} + \delta_{jl}\right)\frac{\partial}{\partial x_l}\,\frac{\partial}{\partial x_k} \\
&= -\mathrm{i}\,\hbar^3\,\epsilon_{ijk}\,\frac{\partial}{\partial x_j}\,\frac{\partial}{\partial x_k} = 0, \tag{B.4.96}
\end{aligned}
$$

and

$$
\begin{aligned}
[L_i, V] &= -\mathrm{i}\,\hbar\,\epsilon_{ijk}\,x_j\,\frac{\partial}{\partial x_k}\,V + \mathrm{i}\,\hbar\,\epsilon_{ijk}\,V\,x_j\,\frac{\partial}{\partial r_k} \\
&= -\mathrm{i}\,\hbar\,\epsilon_{ijk}\,x_j\left(\frac{\partial V}{\partial x_k} + \frac{\partial}{\partial x_k}\,V\right) + \mathrm{i}\,\hbar\,\epsilon_{ijk}\,V\,x_j\,\frac{\partial}{\partial x_k} \\
&= -\mathrm{i}\,\hbar\,\epsilon_{ijk}\,x_j\,\frac{\partial V}{\partial x_k}. \tag{B.4.97}
\end{aligned}
$$

It follows that

$$[\mathbf{L}, p^2] = \mathbf{0}, \tag{B.4.98}$$

$$[\mathbf{L}, V] = -\mathrm{i}\,\hbar\,\mathbf{x} \times \nabla V, \tag{B.4.99}$$

so that

$$[\mathbf{L}, H] = -\mathrm{i}\,\hbar\,\mathbf{x} \times \nabla V. \tag{B.4.100}$$

In the Heisenberg picture,

$$\mathrm{i}\,\hbar\,\frac{d\mathbf{L}}{dt} = [\mathbf{L}, H] = -\mathrm{i}\,\hbar\,\mathbf{x} \times \nabla V. \tag{B.4.101}$$

Hence,

$$\frac{d\langle\mathbf{L}\rangle}{dt} = -\langle\mathbf{x} \times \nabla V\rangle. \tag{B.4.102}$$

Finally, if $V = V(r)$ then $\nabla V \propto \mathbf{x}$. It follows that

$$\frac{d\langle\mathbf{L}\rangle}{dt} = \mathbf{0}. \tag{B.4.103}$$

4.11 Let

$$I_n = \int_0^\infty y^n \, e^{-y} \, dy. \tag{B.4.104}$$

Integration by parts yields

$$I_n = n \, I_{n-1}. \tag{B.4.105}$$

However,

$$I_0 = \int_0^\infty e^{-y} \, dy = [e^{-y}]_\infty^0 = 1. \tag{B.4.106}$$

Hence, we obtain

$$I_n = n!. \tag{B.4.107}$$

4.12 We have

$$R(r) = \frac{f(r/a)}{r} \, e^{-r/a}, \tag{B.4.108}$$

$$a = n \, a_0, \tag{B.4.109}$$

$$f(y) = \sum_{k=l+1,n} c_k \, y^k, \tag{B.4.110}$$

$$c_k \, [k \, (k-1) - l \, (l+1)] = c_{k-1} \, [2 \, (k-1) - 2 \, n], \tag{B.4.111}$$

$$\int_0^\infty dr \, r^2 \, [R(r)]^2 = 1. \tag{B.4.112}$$

(a) For $n = 1$ and $l = 0$,

$$f(y) = c_1 \, y. \tag{B.4.113}$$

Thus,

$$R_{10}(r) = \frac{c_1}{a_0} \exp\left(-\frac{r}{a_0}\right), \tag{B.4.114}$$

and

$$\left(\frac{c_1}{a_0}\right)^2 \int_0^\infty dr \, r^2 \, \exp\left(-\frac{2\,r}{a_0}\right) = 1. \tag{B.4.115}$$

The previous equation becomes

$$\frac{c_1^2 \, a_0}{8} \int_0^\infty d\zeta \, \zeta^2 \, e^{-\zeta} = 1. \tag{B.4.116}$$

The integral evaluates to 2 (see Exercise 4.11), so we obtain $c_1 = 2/a_0^{1/2}$. Hence,

$$R_{10}(r) = \frac{2}{a_0^{3/2}} \exp\left(-\frac{r}{a_0}\right). \tag{B.4.117}$$

(b) For $n = 2$ and $l = 0$, we have

$$f(y) = c_1 y + c_2 y^2,$$ (B.4.118)

and

$$c_2 = -c_1.$$ (B.4.119)

Hence,

$$R_{20}(r) = \frac{c_1}{2 a_0} \left(1 - \frac{r}{2 a_0}\right) \exp\left(-\frac{r}{2 a_0}\right),$$ (B.4.120)

and

$$\left(\frac{c_1}{2 a_0}\right)^2 \int_0^\infty dr\, r^2 \left(1 - \frac{r}{2 a_0}\right)^2 \exp\left(-\frac{r}{a_0}\right) = 1.$$ (B.4.121)

The previous equation becomes

$$\frac{c_1^2 a_0}{4} \int_0^\infty d\zeta \left(\zeta^2 - \zeta^3 + \frac{\zeta^4}{4}\right) e^{-\zeta} = 1.$$ (B.4.122)

The integral evaluates to 2 (see Exercise 4.11), so we obtain $c_1 = (2/a_0)^{1/2}$. Hence,

$$R_{20}(r) = \frac{2}{(2 a_0)^{3/2}} \left(1 - \frac{r}{2 a_0}\right) \exp\left(-\frac{r}{2 a_0}\right).$$ (B.4.123)

(c) For $n = 2$ and $l = 1$,

$$f(y) = c_2 y^2.$$ (B.4.124)

Thus,

$$R_{21}(r) = \frac{c_2}{4 a_0} \frac{r}{a_0} \exp\left(-\frac{r}{2 a_0}\right),$$ (B.4.125)

and

$$\left(\frac{c_2}{4 a_0}\right)^2 \int_0^\infty dr\, r^2 \left(\frac{r}{a_0}\right)^2 \exp\left(-\frac{r}{a_0}\right) = 1.$$ (B.4.126)

The previous equation becomes

$$\frac{c_2^2 a_0}{16} \int_0^\infty d\zeta\, \zeta^4\, e^{-\zeta} = 1.$$ (B.4.127)

The integral evaluates to 24 (see Exercise 4.11), so we obtain $c_2 = (2/3 a_0)^{1/2}$. Hence,

$$R_{21}(r) = \frac{1}{\sqrt{2}\,(2 a_0)^{3/2}} \frac{r}{a_0} \exp\left(-\frac{r}{2 a_0}\right).$$ (B.4.128)

(d) For $n = 3$ and $l = 0$,

$$f(y) = c_1 y + c_2 y^2 + c_3 y^3,$$ (B.4.129)

where

$$c_2 \doteq -2\,c_1, \tag{B.4.130}$$

$$c_3 = -\frac{1}{3}\,c_2 = \frac{2}{3}\,c_1. \tag{B.4.131}$$

Thus,

$$R_{30}(r) = \frac{c_1}{3\,a_0}\left(1 - \frac{2\,r}{3\,a_0} + \frac{2\,r^2}{27\,a_0^2}\right)\exp\left(-\frac{r}{3\,a_0}\right), \tag{B.4.132}$$

and

$$\left(\frac{c_1}{3\,a_0}\right)^2 \int_0^\infty dr\,r^2\left(1 - \frac{2\,r}{3\,a_0} + \frac{2\,r^2}{27\,a_0^2}\right)^2 \exp\left(-\frac{2\,r}{3\,a_0}\right) = 1. \tag{B.4.133}$$

The previous equation becomes

$$\frac{3\,c_1^2\,a_0}{8} \int_0^\infty d\zeta\left(\zeta^2 - 2\,\zeta^3 + \frac{4}{3}\,\zeta^4 - \frac{\zeta^5}{5} + \frac{\zeta^6}{36}\right)e^{-\zeta} = 1. \tag{B.4.134}$$

The integral evaluates to 2 (see Exercise 4.11), so we obtain $c_1 = 2/(3\,a_0)^{1/2}$. Hence,

$$R_{30}(r) = \frac{2}{(3\,a_0)^{3/2}}\left(1 - \frac{2\,r}{3\,a_0} + \frac{2\,r^2}{27\,a_0^2}\right)\exp\left(-\frac{r}{3\,a_0}\right). \tag{B.4.135}$$

(e) For $n = 3$ and $l = 1$, we have

$$f(y) = c_2\,y^2 + c_3\,y^3, \tag{B.4.136}$$

and

$$c_3 = -\frac{1}{2}\,c_2. \tag{B.4.137}$$

Hence,

$$R_{31}(r) = \frac{c_2}{9\,a_0}\,\frac{r}{a_0}\left(1 - \frac{r}{6\,a_0}\right)\exp\left(-\frac{r}{3\,a_0}\right), \tag{B.4.138}$$

and

$$\left(\frac{c_2}{9\,a_0}\right)^2 \int_0^\infty dr\,r^2\left(\frac{r}{a_0}\right)^2\left(1 - \frac{r}{6\,a_0}\right)^2 \exp\left(-\frac{2\,r}{3\,a_0}\right) = 1. \tag{B.4.139}$$

The previous equation becomes

$$\frac{3\,c_2^2\,a_0}{32} \int_0^\infty d\zeta\left(\zeta^4 - \frac{\zeta^5}{5} + \frac{\zeta^6}{16}\right)e^{-\zeta} = 1. \tag{B.4.140}$$

The integral evaluates to 9 (see Exercise 4.11), so we obtain $c_2 = 4\sqrt{2}/3\sqrt{3}\,a_0$. Hence,

$$R_{31}(r) = \frac{4\sqrt{2}}{9\,(3\,a_0)^{3/2}} \frac{r}{a_0}\left(1 - \frac{r}{6\,a_0}\right)\exp\left(-\frac{r}{3\,a_0}\right). \tag{B.4.141}$$

(f) For $n = 3$ and $l = 2$,

$$f(y) = c_3\,y^3. \tag{B.4.142}$$

Thus,

$$R_{32}(r) = \frac{c_3}{27\,a_0}\left(\frac{r}{a_0}\right)^2\exp\left(-\frac{r}{3\,a_0}\right), \tag{B.4.143}$$

and

$$\left(\frac{c_3}{27\,a_0}\right)^2 \int_0^\infty dr\, r^2\left(\frac{r}{a_0}\right)^4\exp\left(-\frac{2\,r}{3\,a_0}\right) = 1. \tag{B.4.144}$$

The previous equation becomes

$$\frac{3\,c_3^2\,a_0}{128}\int_0^\infty d\zeta\,\zeta^6\,e^{-\zeta} = 1. \tag{B.4.145}$$

The integral evaluates to 720 (see Exercise 4.11), so we obtain $c_3 = (8/135\,a_0)^{1/2}$. Hence,

$$R_{32}(r) = \frac{2\sqrt{2}}{27\sqrt{5}\,(5\,a_0)^{3/2}}\left(\frac{r}{a_0}\right)^2\exp\left(-\frac{r}{3\,a_0}\right). \tag{B.4.146}$$

4.13 For the hydrogen ground state, we can write

$$\langle r^k\rangle = \int_0^\infty r^{2+k}\,[R_{10}(r)]^2\,dr. \tag{B.4.147}$$

Thus, it follows from the previous exercise that

$$\langle r^k\rangle = \int_0^\infty \frac{4\,r^{2+k}}{a_0^3}\exp\left(-\frac{2\,r}{a_0}\right)dr = \frac{I_{2+k}}{2^{1+k}}\,a_0^k, \tag{B.4.148}$$

where $I_k = k!$ is defined in Exercise 4.11. Hence,

$$\langle r^k\rangle = \frac{(2+k)!}{2^{1+k}}\,a_0^k. \tag{B.4.149}$$

Now,

$$\langle x^2\rangle + \langle y^2\rangle + \langle z^2\rangle = \langle r^2\rangle. \tag{B.4.150}$$

However, the hydrogen ground state is spherically symmetric, which implies that

$$\langle x^2\rangle = \langle y^2\rangle = \langle z^2\rangle. \tag{B.4.151}$$

Hence,

$$\langle x^2\rangle = \langle y^2\rangle = \langle z^2\rangle = \frac{1}{3}\,\langle r^2\rangle. \tag{B.4.152}$$

But, from Equation (B.4.149), $\langle r^2 \rangle = 3\,a_0^2$. Thus, we obtain

$$\langle x^2 \rangle = \langle y^2 \rangle = \langle z^2 \rangle = a_0^2. \tag{B.4.153}$$

4.14 The probability of an observation of r yielding a result in the range r to $r + dr$ is $P(r)\,dr$, where

$$P(r) = \oint d\Omega\, r^2\, |\psi|^2. \tag{B.4.154}$$

For the hydrogen ground state, this reduces to

$$P(r) = r^2\, |R_{10}|^2 = \frac{4\,r^2}{a_0^3}\, \exp\!\left(-\frac{2\,r}{a_0}\right). \tag{B.4.155}$$

The most probable value of r is that which maximizes $P(r)$. It is easily seen that $dP(a_0)/dr = 0$ and $d^2 P(a_0)/dr^2 < 0$. Hence, we deduce that the most probable value of r is a_0.

4.15 According to Equation (4.96), $z = r\cos\theta = r\,(4\pi/3)^{1/2}\, Y_{10}(\theta,\varphi)$. Moreover,

$$\psi_{200} = R_{20}(r)\, Y_{00}(\theta,\varphi), \tag{B.4.156}$$

$$\psi_{210} = R_{21}(r)\, Y_{10}(\theta,\varphi). \tag{B.4.157}$$

Hence,

$$\langle 2,0,0|\, z\, |2,1,0 \rangle = \int_0^\infty dr\, r^2 \oint d\Omega\, \psi_{200}^*\, z\, \psi_{210} \tag{B.4.158}$$

$$= \sqrt{\frac{4\pi}{3}} \int_0^\infty dr\, r^3\, R_{20}(r)\, R_{21}(r) \oint d\Omega\, Y_{00}(\theta,\varphi)\, Y_{10}(\theta,\varphi)\, Y_{1,0}(\theta,\varphi).$$

. Here, we have made use of the fact that $R_{20}(r)$ and $Y_{00}(\theta,\varphi)$ are both real. Now, $Y_{00} = 1/\sqrt{4\pi}$, and

$$\oint d\Omega\, Y_{10}(\theta,\varphi)\, Y_{10}(\theta,\varphi) = \oint d\Omega\, Y_{10}^*(\theta,\varphi)\, Y_{10}(\theta,\varphi) = 1, \tag{B.4.159}$$

where we have made use of the fact that Y_{10} is real. Hence, we obtain

$$\langle 2,0,0|\, z\, |2,1,0 \rangle = \frac{1}{\sqrt{3}} \int_0^\infty dr\, r^3\, R_{20}(r)\, R_{21}(r). \tag{B.4.160}$$

Making use of the expressions for $R_{20}(r)$ and $R_{21}(r)$ derived in Exercise 4.12, the previous expression reduces to

$$\langle 2,0,0|\, z\, |2,1,0 \rangle = \frac{a_0}{12}\left(I_4 - \frac{I_5}{2}\right), \tag{B.4.161}$$

where $I_n = n!$ is defined in Exercise 4.11. Hence,

$$\langle 2,0,0|z|2,1,0\rangle = \frac{a_0}{12}\left(4! - \frac{5!}{2}\right) = -3\,a_0. \tag{B.4.162}$$

4.16 (a) From Section 4.6,

$$\langle r^k\rangle = \int_0^\infty dr\,r^{k+2}\,[R_{nl}(r)]^2. \tag{B.4.163}$$

Writing $x = r/(n\,a_0)$ and $R_{nl}(r) = u(x)/x$, we obtain

$$\langle r^k\rangle = (n\,a_0)^{k+3}\,J_k, \tag{B.4.164}$$

where

$$J_k = \int_0^\infty dx\,x^k\,[u(x)]^2, \tag{B.4.165}$$

and $u(x)$ is a well-behaved solution of the differential equation

$$u'' = \left[\frac{l(l+1)}{x^2} - \frac{2n}{x} + 1\right]u. \tag{B.4.166}$$

Here, $' \equiv d/dx$.

(b) We have

$$\int_0^\infty dx\,x^k\,u\,u' = \frac{1}{2}\int_0^\infty dx\,x^k\,\frac{du^2}{dx} = \frac{1}{2}\left[x^k u^2\right]_0^\infty - \frac{k}{2}\int_0^\infty dx\,x^{k-1}\,u^2, \tag{B.4.167}$$

which implies that

$$\int_0^\infty dx\,x^k\,u\,u' = -\frac{k}{2}\,J_k. \tag{B.4.168}$$

Here, we have made use of the fact that $u \to 0$ as $x \to 0$ and $x \to \infty$. We can also write

$$\int_0^\infty dx\,x^k\,(u')^2 = \left[\frac{x^{k+1}}{k+1}\,(u')^2\right]_0^\infty - \frac{2}{k+1}\int_0^\infty dx\,x^{k+1}\,u'\,u'', \tag{B.4.169}$$

which yields

$$\int_0^\infty dx\,x^k\,(u')^2 = -\frac{2}{k+1}\int_0^\infty dx\,x^{k+1}\,u'\,u''. \tag{B.4.170}$$

Finally, we have

$$\int_0^\infty dx\,x^k\,u\,u'' = \left[x^k u\,u'\right]_0^\infty - \int_0^\infty dx\,(x^k u)'\,u'$$

$$= -k\int_0^\infty dx\,x^{k-1}\,u\,u' - \int_0^\infty dx\,x^k\,(u')^2. \tag{B.4.171}$$

Thus, it follows from Equations (B.4.168) and (B.4.170) that

$$\int_0^\infty dx\, x^k\, u\, u'' = \frac{k\,(k-1)}{2}\, J_{k-2} + \frac{2}{k+1} \int_0^\infty dx\, x^{k+1}\, u'\, u''. \tag{B.4.172}$$

(c) Multiplying Equation (B.4.166) by $u\, x^k$, and integrating, we obtain

$$\int_0^\infty dx\, x^k\, u\, u'' = l\,(l+1)\, J_{k-2} - 2\, n\, J_{k-1} + J_k. \tag{B.4.173}$$

(d) Equations (B.4.166) and (B.4.172) can be combined to give

$$\int_0^\infty dx\, x^k\, u\, u'' = \frac{k\,(k-1)}{2}\, J_{k-2}$$
$$+ \frac{2}{k+1} \int_0^\infty dx\, x^{k+1}\, u' \left[\frac{l\,(l+1)}{x^2} - \frac{2\,n}{x} + 1 \right] u. \tag{B.4.174}$$

Making use of Equation (B.4.168), this reduces to

$$\int_0^\infty dx\, x^k\, u\, u'' = \frac{k\,(k-1)}{2}\, J_{k-2}$$
$$- \frac{l\,(l+1)\,(k-1)}{k+1}\, J_{k-2} + \frac{2\,n\,k}{k+1}\, J_{k-1} - J_k. \tag{B.4.175}$$

(e) Equating the right-hand sides of Equations (B.4.173) and (B.4.175), we obtain

$$\frac{k}{4}\, [(2\,l+1)^2 - k^2]\, J_{k-2} - n\,(2\,k+1)\, J_{k-1} + (k+1)\, J_k = 0. \tag{B.4.176}$$

Hence, it follows from Equation (B.4.164) that

$$\frac{k\, a_0^2}{4} [(2\,l+1)^2 - k^2] \langle r^{k-2} \rangle - a_0\,(2\,k+1)\,\langle r^{k-1} \rangle + \frac{(k+1)}{n^2}\, \langle r^k \rangle = 0. \tag{B.4.177}$$

(f) Of course,

$$\langle 1 \rangle = 1. \tag{B.4.178}$$

Making use of Equation (B.4.177), with $k = 0$, we get

$$-a_0\, \langle r^{-1} \rangle + \frac{1}{n^2}\, \langle 1 \rangle = 0, \tag{B.4.179}$$

or

$$\left\langle \frac{1}{r} \right\rangle = \frac{1}{n^2\, a_0}. \tag{B.4.180}$$

Making use of Equation (B.4.177), with $k = 1$, we get

$$\frac{a_0^2}{4} [(2\,l+1)^2 - 1]\, \langle r^{-1} \rangle - 3\, a_0\, \langle 1 \rangle + \frac{2}{n^2}\, \langle r \rangle = 0, \tag{B.4.181}$$

which reduces to

$$\langle r \rangle = \frac{a_0}{2}\,[3\,n^2 - l\,(l+1)].\tag{B.4.182}$$

Finally, making use of Equation (B.4.177), with $k = 2$, we get

$$\frac{2\,a_0^2}{4}\,[(2\,l+1)^2 - 4]\,\langle 1 \rangle - 5\,a_0\,\langle r \rangle + \frac{3}{n^2}\,\langle r^2 \rangle = 0,\tag{B.4.183}$$

which reduces to

$$\langle r^2 \rangle = \frac{a_0^2\,n^2}{2}\,[5\,n^2 + 1 - 3\,l\,(l+1)].\tag{B.4.184}$$

4.17. (a) We can write $x = y/n$ and $u_{nl}(x) = v_{nl}(y)/n$, where $x = r/(n\,a_0)$ and $R_{nl}(r) = u_{nl}(x)/x = v_{nl}(y)/y$. It follows from Equation (4.138) that

$$\frac{d^2 v_{nl}}{dy^2} = \left[\frac{l(l+1)}{y^2} - \frac{2}{y} + \frac{1}{n^2}\right] v_{nl}.\tag{B.4.185}$$

(b) Let $v_{nl} = c\,y^\alpha$. At small $|y|$, the previous differential equation yields $\alpha\,(\alpha - 1) = l\,(l+1)$. The solutions are $\alpha = 1 + l$ and $\alpha = -l$. However, the latter solution is unphysical, because it would cause $R_{nl}(r)$ to diverge as $r \to 0$. Hence, $v_{nl} \sim y^{1+l}$ as $|y| \to 0$.

(c) Equation (B.4.185) gives

$$v_{ml}\,\frac{d^2 v_{nl}}{dy^2} = \left[\frac{l(l+1)}{y^2} - \frac{2}{y} + \frac{1}{n^2}\right] v_{nl}\,v_{ml}.\tag{B.4.186}$$

Likewise,

$$v_{nl}\,\frac{d^2 v_{ml}}{dy^2} = \left[\frac{l(l+1)}{y^2} - \frac{2}{y} + \frac{1}{m^2}\right] v_{nl}\,v_{ml}.\tag{B.4.187}$$

Taking the difference between the previous two equations, we obtain

$$\frac{d}{dy}\left[v_{ml}\,\frac{dv_{nl}}{dy} - v_{nl}\,\frac{dv_{ml}}{dy}\right] = \left(\frac{1}{n^2} - \frac{1}{m^2}\right) v_{nl}\,v_{ml}.\tag{B.4.188}$$

Hence,

$$\left[v_{ml}\,\frac{dv_{nl}}{dy} - v_{nl}\,\frac{dv_{ml}}{dy}\right]_0^\infty = \left(\frac{1}{n^2} - \frac{1}{m^2}\right)\int_0^\infty dy\, v_{nl}\,v_{ml}.\tag{B.4.189}$$

However, given that $v_{nl} \sim y^{1+l}$ at small y, and $v_{nl} \to 0$ as $y \to \infty$, the left-hand side of the previous equation is obviously zero. Thus, we deduce that

$$\left(\frac{1}{n^2} - \frac{1}{m^2}\right)\int_0^\infty dy\, v_{nl}(y)\, v_{ml}(y) = 0.\tag{B.4.190}$$

(d) Given that $v_{nl}(y)/y = R_{nl}(r)$, and $y = r/a_0$, the previous equation yields

$$\left(\frac{1}{n^2} - \frac{1}{m^2}\right) \int_0^\infty r^2 R_{nl}(r) R_{ml}(r) = 0.$$ (B.4.191)

It follows that

$$\int_0^\infty r^2 R_{nl}(r) R_{ml}(r) = 0$$ (B.4.192)

if $n \neq m$.

B.5   Chapter 5

5.1 Now,

$$S_x = \frac{\hbar}{2} \left(|+\rangle\langle-| + |-\rangle\langle+|\right),$$ (B.5.1)

so

$$S_x^\dagger = \frac{\hbar}{2} \left(|-\rangle\langle+| + |+\rangle\langle-|\right) = S_x.$$ (B.5.2)

Now,

$$S_y = \frac{i\hbar}{2} \left(-|+\rangle\langle-| + |-\rangle\langle+|\right),$$ (B.5.3)

so

$$S_y^\dagger = -\frac{i\hbar}{2} \left(-|-\rangle\langle+| + |+\rangle\langle-|\right) = S_y.$$ (B.5.4)

Now,

$$S_z = \frac{\hbar}{2} \left(|+\rangle\langle+| - |-\rangle\langle-|\right),$$ (B.5.5)

so

$$S_z^\dagger = -\frac{\hbar}{2} \left(|+\rangle\langle+| - |-\rangle\langle-|\right) = S_z.$$ (B.5.6)

Now,

$$S_x S_y = \frac{i\hbar^2}{4} \left(|+\rangle\langle-| + |-\rangle\langle+|\right) \left(-|+\rangle\langle-| + |-\rangle\langle+|\right)$$

$$= \frac{i\hbar^2}{2} \left(|+\rangle\langle+| - |-\rangle\langle-|\right) = \frac{i\hbar}{2} S_z,$$ (B.5.7)

where use has been made of

$$\langle+|-\rangle = 0,$$ (B.5.8)

$$\langle+|+\rangle = \langle-|-\rangle = 1.$$ (B.5.9)

Also,

$$S_y S_x = \frac{i\hbar^2}{4} \left(-|+\rangle\langle-| + |-\rangle\langle+|\right) \left(|+\rangle\langle-| + |-\rangle\langle+|\right)$$

$$= \frac{i\hbar^2}{2} \left(-|+\rangle\langle+| + |-\rangle\langle-|\right) = -\frac{i\hbar}{2} S_z. \tag{B.5.10}$$

Thus,

$$[S_x, S_y] = S_x S_y - S_y S_x = i\hbar S_z. \tag{B.5.11}$$

Now,

$$S_y S_z = \frac{i\hbar^2}{4} \left(-|+\rangle\langle-| + |-\rangle\langle+|\right) \left(|+\rangle\langle+| - |-\rangle\langle-|\right)$$

$$= \frac{i\hbar^2}{2} \left(|+\rangle\langle-| + |-\rangle\langle+|\right) = \frac{i\hbar}{2} S_x, \tag{B.5.12}$$

and

$$S_z S_y = \frac{i\hbar^2}{4} \left(|+\rangle\langle+| - |-\rangle\langle-|\right) \left(-|+\rangle\langle-| + |-\rangle\langle+|\right)$$

$$= \frac{i\hbar^2}{2} \left(-|+\rangle\langle-| - |-\rangle\langle+|\right) = -\frac{i\hbar}{2} S_x. \tag{B.5.13}$$

Thus,

$$[S_y, S_z] = S_y S_z - S_z S_y = i\hbar S_x. \tag{B.5.14}$$

Now,

$$S_z S_x = \frac{\hbar^2}{4} \left(|+\rangle\langle+| - |-\rangle\langle-|\right) \left(|+\rangle\langle-| + |-\rangle\langle+|\right)$$

$$= \frac{\hbar^2}{2} \left(|+\rangle\langle-| - |-\rangle\langle+|\right) = \frac{i\hbar}{2} S_y, \tag{B.5.15}$$

and

$$S_x S_z = \frac{\hbar^2}{4} \left(|+\rangle\langle-| + |-\rangle\langle+|\right) \left(|+\rangle\langle+| - |-\rangle\langle-|\right)$$

$$= \frac{\hbar^2}{2} \left(-|+\rangle\langle-| + |-\rangle\langle+|\right) = -\frac{i\hbar}{2} S_y. \tag{B.5.16}$$

Thus,

$$[S_z, S_x] = S_z S_x - S_x S_z = i\hbar S_y. \tag{B.5.17}$$

5.2 Given that

$$\exp(G t) \equiv \sum_{n=0,\infty} \frac{G^n t^n}{n!}, \tag{B.5.18}$$

where G is an operator, and t a number, we can write

$$\frac{d}{dt} e^{Gt} = \sum_{n=0,\infty} \frac{n\,G^n\,t^{n-1}}{n!} = G \sum_{n=1,\infty} \frac{G^{n-1}\,t^{n-1}}{(n-1)!} = G \sum_{k=0,\infty} \frac{G^k\,t^k}{k!} = G\,e^{Gt}. \qquad \text{(B.5.19)}$$

Similarly,

$$\frac{d}{dt} e^{-Gt} = -G\,e^{-Gt}. \qquad \text{(B.5.20)}$$

Hence, if

$$A(t) = e^{Gt}\,A\,e^{-Gt}, \qquad \text{(B.5.21)}$$

where A is another operator, then

$$\frac{dA(t)}{dt} = G\,e^{Gt}\,A\,e^{-Gt} + e^{Gt}\,A\,(-G)\,e^{-Gt}$$

$$= G\,A(t) - A(t)\,G = [G, A(t)]. \qquad \text{(B.5.22)}$$

Thus,

$$\frac{d^2 A(t)}{dt^2} = \frac{d}{dt}\frac{dA(t)}{dt} = [G, \frac{dA(t)}{dt}] = [G, [G, A(t)]]. \qquad \text{(B.5.23)}$$

Similarly,

$$\frac{d^3 A(t)}{dt^3} = [G, [G, [G, A(t)]]]. \qquad \text{(B.5.24)}$$

By Taylor expansion,

$$A(t) = A(0) + t\,\frac{dA(0)}{dt} + \frac{t^2}{2!}\frac{d^2 A(0))}{dt^2} + \frac{t^3}{3!}\frac{d^3 A(0)}{dt^3} + \cdots. \qquad \text{(B.5.25)}$$

Hence, given that $A(0) = A$, we obtain

$$e^{Gt}\,A\,e^{-Gt} = A + t\,[G, A] + \frac{t^2}{2!}\,[G, [G, A]] + \frac{t^3}{3!}\,[G, [G, [G, A]]] + \cdots. \qquad \text{(B.5.26)}$$

Finally, writing $t = i\,\lambda$, we get

$$e^{iG\lambda}\,A\,e^{-iG\lambda} = A + i\,\lambda\,[G, A] + \frac{i^2\,\lambda^2}{2!}\,[G, [G, A]] + \frac{i^3\,\lambda^3}{3!}\,[G, [G, [G, A]]] + \cdots. \qquad \text{(B.5.27)}$$

5.3 We have

$$\sigma_1 \sigma_1 = \begin{pmatrix} 0, & 1 \\ 1, & 0 \end{pmatrix} \begin{pmatrix} 0, & 1 \\ 1, & 0 \end{pmatrix} = \begin{pmatrix} 1, & 0 \\ 0, & 1 \end{pmatrix} = 1, \tag{B.5.28}$$

$$\sigma_2 \sigma_2 = \begin{pmatrix} 0, & -i \\ i, & 0 \end{pmatrix} \begin{pmatrix} 0, & -i \\ i, & 0 \end{pmatrix} = \begin{pmatrix} 1, & 0 \\ 0, & 1 \end{pmatrix} = 1, \tag{B.5.29}$$

$$\sigma_3 \sigma_3 = \begin{pmatrix} 1, & 0 \\ 0, & -1 \end{pmatrix} \begin{pmatrix} 1, & 0 \\ 0, & -1 \end{pmatrix} = \begin{pmatrix} 1, & 0 \\ 0, & 1 \end{pmatrix} = 1, \tag{B.5.30}$$

$$\sigma_1 \sigma_2 = \begin{pmatrix} 0, & 1 \\ 1, & 0 \end{pmatrix} \begin{pmatrix} 0, & -i \\ i, & 0 \end{pmatrix} = \begin{pmatrix} i, & 0 \\ 0, & -i \end{pmatrix} = i \sigma_3, \tag{B.5.31}$$

$$\sigma_2 \sigma_1 = \begin{pmatrix} 0, & -i \\ i, & 0 \end{pmatrix} \begin{pmatrix} 0, & 1 \\ 1, & 0 \end{pmatrix} = \begin{pmatrix} -i, & 0 \\ 0, & i \end{pmatrix} = -i \sigma_3, \tag{B.5.32}$$

$$\sigma_1 \sigma_3 = \begin{pmatrix} 0, & 1 \\ 1, & 0 \end{pmatrix} \begin{pmatrix} 1, & 0 \\ 0, & -1 \end{pmatrix} = \begin{pmatrix} 0, & -1 \\ 1, & 0 \end{pmatrix} = -i \sigma_2, \tag{B.5.33}$$

$$\sigma_3 \sigma_1 = \begin{pmatrix} 1, & 0 \\ 0, & -1 \end{pmatrix} \begin{pmatrix} 0, & 1 \\ 1, & 0 \end{pmatrix} = \begin{pmatrix} 0, & 1 \\ -1, & 0 \end{pmatrix} = i \sigma_2, \tag{B.5.34}$$

$$\sigma_2 \sigma_3 = \begin{pmatrix} 0, & -i \\ i, & 0 \end{pmatrix} \begin{pmatrix} 1, & 0 \\ 0, & -1 \end{pmatrix} = \begin{pmatrix} 0, & i \\ i, & 0 \end{pmatrix} = i \sigma_1, \tag{B.5.35}$$

$$\sigma_3 \sigma_2 = \begin{pmatrix} 1, & 0 \\ 0, & -1 \end{pmatrix} \begin{pmatrix} 0, & -i \\ i, & 0 \end{pmatrix} = \begin{pmatrix} 0, & -i \\ -i, & 0 \end{pmatrix} = -i \sigma_1. \tag{B.5.36}$$

The previous nine formulae can be summed up by writing

$$\sigma_j \sigma_k = \delta_{jk} + i \epsilon_{jkl} \sigma_l. \tag{B.5.37}$$

Furthermore,

$$[\sigma_1, \sigma_2] = \sigma_1 \sigma_2 - \sigma_2 \sigma_1 = 2 i \sigma_3, \tag{B.5.38}$$

$$[\sigma_2, \sigma_3] = \sigma_2 \sigma_3 - \sigma_3 \sigma_2 = 2 i \sigma_1, \tag{B.5.39}$$

$$[\sigma_3, \sigma_1] = \sigma_3 \sigma_1 - \sigma_1 \sigma_3 = 2 i \sigma_2, \tag{B.5.40}$$

$$\{\sigma_1, \sigma_1\} = 2 \sigma_1 \sigma_1 = 2, \tag{B.5.41}$$

$$\{\sigma_2, \sigma_2\} = 2 \sigma_2 \sigma_2 = 2, \tag{B.5.42}$$

$$\{\sigma_3, \sigma_3\} = 2 \sigma_3 \sigma_3 = 2, \tag{B.5.43}$$

$$\{\sigma_1, \sigma_2\} = \sigma_1 \sigma_2 + \sigma_2 \sigma_1 = 0, \tag{B.5.44}$$

$$\{\sigma_2, \sigma_3\} = \sigma_2 \sigma_3 + \sigma_3 \sigma_2 = 0, \tag{B.5.45}$$

$$\{\sigma_3, \sigma_1\} = \sigma_3 \sigma_1 + \sigma_1 \sigma_3 = 0. \tag{B.5.46}$$

5.4 (a) In the Pauli representation

$$S_x = \frac{\hbar}{2}\sigma_2 = \frac{\hbar}{2}\begin{pmatrix} 0, & 1 \\ 1, & 0 \end{pmatrix}. \tag{B.5.47}$$

The eigenvalues of S_x are $+\hbar/2$ and $-\hbar/2$. As is clear by inspection, the corresponding properly normalized eigenstates are

$$\chi_+ = \frac{1}{\sqrt{2}}\begin{pmatrix} 1 \\ 1 \end{pmatrix}, \tag{B.5.48}$$

and

$$\chi_- = \frac{1}{\sqrt{2}}\begin{pmatrix} 1 \\ -1 \end{pmatrix}, \tag{B.5.49}$$

respectively.

(b) In the Pauli representation,

$$S_y = \frac{\hbar}{2}\sigma_2 = \frac{\hbar}{2}\begin{pmatrix} 0, & -i \\ i, & 0 \end{pmatrix}. \tag{B.5.50}$$

The eigenvalues of S_y are $+\hbar/2$ and $-\hbar/2$. As is clear by inspection, the corresponding properly normalized eigenstates are

$$\chi_+ = \frac{1}{\sqrt{2}}\begin{pmatrix} 1 \\ i \end{pmatrix}, \tag{B.5.51}$$

and

$$\chi_- = \frac{1}{\sqrt{2}}\begin{pmatrix} 1 \\ -i \end{pmatrix}, \tag{B.5.52}$$

respectively.

5.5 The requisite state is obtained by operating on an eigenstate of S_z corresponding to the eigenvalue $\hbar/2$ with a spin-space operator that rotates through an angle θ about the y-axis. The initial spinor is

$$\begin{pmatrix} 1 \\ 0 \end{pmatrix}. \tag{B.5.53}$$

According to Equation (5.100), the rotation matrix takes the form

$$\begin{pmatrix} \cos(\theta/2), & -\sin(\theta/2) \\ \sin(\theta/2), & \cos(\theta/2) \end{pmatrix}. \tag{B.5.54}$$

Hence, the final spinor is

$$\chi = \begin{pmatrix} \cos(\theta/2), & -\sin(\theta/2) \\ \sin(\theta/2), & \cos(\theta/2) \end{pmatrix}\begin{pmatrix} 1 \\ 0 \end{pmatrix} = \begin{pmatrix} \cos(\theta/2) \\ \sin(\theta/2) \end{pmatrix}. \tag{B.5.55}$$

In other words,

$$\chi = \cos(\theta/2)\chi_+ + \sin(\theta/2)\chi_-, \tag{B.5.56}$$

where $\chi_\pm$ are properly normalized eigenstates of S_z corresponding to the eigenvalues $\pm\hbar/2$. It follows that an observation of S_z yields the result $+\hbar/2$ with probability $\cos^2(\theta/2)$ and the result $-\hbar/2$ with probability $\sin^2(\theta/2)$.

5.6 We are told that

$$\chi = A\begin{pmatrix} 1 - 2\,\mathrm{i} \\ 2 \end{pmatrix}. \tag{B.5.57}$$

(a) The normalization constraint is

$$\chi^\dagger \chi = 1, \tag{B.5.58}$$

which implies that

$$|A|^2 (1 + 2\,\mathrm{i}, 2)\begin{pmatrix} 1 - 2\,\mathrm{i} \\ 2 \end{pmatrix} = 9|A|^2 = 1. \tag{B.5.59}$$

Without loss of generality, we can assume that A is real and positive. Hence, $A = 1/3$, and

$$\chi = \frac{1}{3}\begin{pmatrix} 1 - 2\,\mathrm{i} \\ 2 \end{pmatrix}. \tag{B.5.60}$$

(b) The eigenvalues of S_z are $+\hbar/2$ and $-\hbar/2$. The properly normalized eigenstates are

$$\chi_+ = \begin{pmatrix} 1 \\ 0 \end{pmatrix} \tag{B.5.61}$$

and

$$\chi_- = \begin{pmatrix} 0 \\ 1 \end{pmatrix}, \tag{B.5.62}$$

respectively. It follows that

$$\chi = c_+ |+\rangle + c_- |-\rangle, \tag{B.5.63}$$

where $c_+ = \chi_+^\dagger \chi = (1/3)(1 - 2\,\mathrm{i})$, and $c_- = \chi_-^\dagger \chi = 2/3$. A measurement of S_z yields the result $+\hbar/2$ with probability $P_+ = |c_+|^2 = 5/9$, and the result $-\hbar/2$ with probability $P_- = |c_-|^2 = 4/9$. The expectation value of S_z is $\langle S_z \rangle = P_+ (\hbar/2) - P_- (\hbar/2) = \hbar/18$.

(c) The eigenvalues of S_x are $+\hbar/2$ and $-\hbar/2$. The properly normalized eigenstates are

$$\chi_+ = \frac{1}{\sqrt{2}}\begin{pmatrix} 1 \\ 1 \end{pmatrix} \tag{B.5.64}$$

and

$$\chi_- = \frac{1}{\sqrt{2}}\begin{pmatrix} 1 \\ -1 \end{pmatrix}, \tag{B.5.65}$$

respectively. It follows that

$$\chi = c_+ |+\rangle + c_- |-\rangle, \tag{B.5.66}$$

where $c_+ = \chi_+^\dagger \chi = (1/\sqrt{18})(3 - 2\,\mathrm{i})$, and $c_- = \chi_-^\dagger \chi = -(1/\sqrt{18})(1 + 2\,\mathrm{i})$. A measurement of S_x yields the result $+\hbar/2$ with probability $P_+ = |c_+|^2 = 13/18$, and the result $-\hbar/2$ with probability $P_- = |c_-|^2 = 5/18$. The expectation value of S_x is $\langle S_x \rangle = P_+ (\hbar/2) - P_- (\hbar/2) = 2\,\hbar/9$.

(d) The eigenvalues of S_y are $+\hbar/2$ and $-\hbar/2$. The properly normalized eigenstates are

$$\chi_+ = \frac{1}{\sqrt{2}} \begin{pmatrix} 1 \\ \mathrm{i} \end{pmatrix} \tag{B.5.67}$$

and

$$\chi_- = \frac{1}{\sqrt{2}} \begin{pmatrix} 1 \\ -\mathrm{i} \end{pmatrix}, \tag{B.5.68}$$

respectively. It follows that

$$\chi = c_+ |+\rangle + c_- |-\rangle, \tag{B.5.69}$$

where $c_+ = \chi_+^\dagger \chi = -(1/\sqrt{18})(1 - 4\,\mathrm{i})$, and $c_- = \chi_-^\dagger \chi = 1/\sqrt{18}$. A measurement of S_y yields the result $+\hbar/2$ with probability $P_- = |c_+|^2 = 17/18$, and the result $-\hbar/2$ with probability $P_- = |c_-|^2 = 1/18$. The expectation value of S_y is $\langle S_y \rangle = P_+ (\hbar/2) - P_- (\hbar/2) = 4\,\hbar/9$.

5.7 The properly normalized eigenstates of S_y are

$$\chi_+ = \frac{1}{\sqrt{2}} \begin{pmatrix} 1 \\ \mathrm{i} \end{pmatrix}, \tag{B.5.70}$$

and

$$\chi_- = \frac{1}{\sqrt{2}} \begin{pmatrix} 1 \\ -\mathrm{i} \end{pmatrix}, \tag{B.5.71}$$

respectively. We have

$$\chi = \begin{pmatrix} \cos \alpha \\ \sin \alpha\, \mathrm{e}^{\mathrm{i}\beta} \end{pmatrix}. \tag{B.5.72}$$

However,

$$\chi = c_+ \chi_+ + c_- \chi_-, \tag{B.5.73}$$

where

$$c_+ = \chi_+^\dagger \chi = \frac{1}{\sqrt{2}} \left(-\mathrm{i} \cos \alpha + \sin \alpha\, \mathrm{e}^{\mathrm{i}\beta} \right), \tag{B.5.74}$$

$$c_- = \chi_-^\dagger \chi = \frac{1}{\sqrt{2}} \left(+\mathrm{i} \cos \alpha + \sin \alpha\, \mathrm{e}^{\mathrm{i}\beta} \right). \tag{B.5.75}$$

Thus, the probability of a measurement of S_y yielding the result $-\hbar/2$ is

$$|c_-|^2 = \frac{1}{2}\left[1 - \sin(2\alpha)\sin\beta\right]. \tag{B.5.76}$$

5.8 (a) The Hamiltonian is

$$H = -\boldsymbol{\mu}\cdot\mathbf{B} = \frac{e}{m_e}\mathbf{S}\cdot\mathbf{B} = \Omega\cos(\omega t)S_z, \tag{B.5.77}$$

where $\Omega = eB_0/m_e$.
(b) The properly normalized eigenstates of S_x are

$$\chi_+ = \frac{1}{\sqrt{2}}\begin{pmatrix}1\\1\end{pmatrix}, \tag{B.5.78}$$

and

$$\chi_- = \frac{1}{\sqrt{2}}\begin{pmatrix}1\\-1\end{pmatrix}, \tag{B.5.79}$$

respectively. Let us write

$$\chi(t) = \begin{pmatrix}c_1(t)\\c_2(t)\end{pmatrix}. \tag{B.5.80}$$

The time evolution of the system is governed by

$$i\hbar\frac{\partial\chi}{\partial t} = H\chi = \Omega\cos(\omega t)\frac{\hbar}{2}\sigma_3\chi, \tag{B.5.81}$$

which yields

$$\frac{dc_1}{dt} = -i\frac{\Omega}{2}\cos(\omega t)c_1, \tag{B.5.82}$$

$$\frac{dc_2}{dt} = +i\frac{\Omega}{2}\cos(\omega t)c_2. \tag{B.5.83}$$

The initial condition is that $\chi(0) = \chi_+$, which implies that $c_1(0) = c_2(0) = 1/\sqrt{2}$. Hence,

$$c_1(t) = \frac{1}{\sqrt{2}}\exp\left[-i\frac{\Omega}{2\omega}\sin(\omega t)\right], \tag{B.5.84}$$

$$c_2(t) = \frac{1}{\sqrt{2}}\exp\left[+i\frac{\Omega}{2\omega}\sin(\omega t)\right], \tag{B.5.85}$$

and

$$\chi(t) = \frac{1}{\sqrt{2}}\begin{pmatrix}\exp[-i(\Omega/2\omega)\sin(\omega t)]\\\exp[+i(\Omega/2\omega)\sin(\omega t)]\end{pmatrix}. \tag{B.5.86}$$

(c) We can write

$$\chi(t) = d_+ \chi_+ + d_- \chi_-. \tag{B.5.87}$$

Now,

$$d_- = \chi_-^\dagger \chi = -\mathrm{i}\,\sin\left[(\Omega/2\,\omega)\,\sin(\omega\,t)\right]. \tag{B.5.88}$$

So, the probability of a measurement of S_x yielding the result $-\hbar/2$ is

$$|d_-|^2 = \sin^2\left[(\Omega/2\,\omega)\,\sin(\omega\,t)\right]. \tag{B.5.89}$$

(d) A complete flip in S_x implies that $|d_-|^2 = 1$. This is only possible if $\Omega/(2\,\omega) > \pi/2$. Hence,

$$B_{0\,\mathrm{min}} = \frac{\pi\,m_e\,\omega}{e}. \tag{B.5.90}$$

5.9 We have

$$\mathrm{i}\,\hbar\,\frac{\partial\chi}{\partial t} = \frac{1}{2\,m_e}\left[(-\mathrm{i}\,\hbar\,\nabla + e\,\mathbf{A})^2 + e\,\hbar\,\boldsymbol{\sigma}\cdot\mathbf{B}\right]\chi - e\,\phi\,\chi. \tag{B.5.91}$$

However, according to Equation (5.93),

$$(\boldsymbol{\sigma}\cdot\mathbf{a})(\boldsymbol{\sigma}\cdot\mathbf{b}) = \mathbf{a}\cdot\mathbf{b} + \mathrm{i}\,\boldsymbol{\sigma}\cdot(\mathbf{a}\times\mathbf{b}). \tag{B.5.92}$$

Hence,

$$\begin{aligned}
[\boldsymbol{\sigma}\cdot(-\mathrm{i}\,\hbar\,\nabla + e\,\mathbf{A})]^2 &= (-\mathrm{i}\,\hbar\,\nabla + e\,\mathbf{A})^2 \\
&\quad + \mathrm{i}\,\boldsymbol{\sigma}\cdot(-\mathrm{i}\,\hbar\,\nabla + e\,\mathbf{A})\times(-\mathrm{i}\,\hbar\,\nabla + e\,\mathbf{A}).
\end{aligned} \tag{B.5.93}$$

However,

$$(-\mathrm{i}\,\hbar\,\nabla + e\,\mathbf{A})\times(-\mathrm{i}\,\hbar\,\nabla + e\,\mathbf{A}) = -\mathrm{i}\,\hbar\,e\,\nabla\times\mathbf{A} = -\mathrm{i}\,\hbar\,e\,\mathbf{B}, \tag{B.5.94}$$

so

$$[\boldsymbol{\sigma}\cdot(-\mathrm{i}\,\hbar\,\nabla + e\,\mathbf{A})]^2 = (-\mathrm{i}\,\hbar\,\nabla + e\,\mathbf{A})^2 + e\,\hbar\,\boldsymbol{\sigma}\cdot\mathbf{B}. \tag{B.5.95}$$

Thus, we obtain

$$\mathrm{i}\,\hbar\,\frac{\partial\chi}{\partial t} = \frac{1}{2\,m_e}\,[\boldsymbol{\sigma}\cdot(-\mathrm{i}\,\hbar\,\nabla + e\,\mathbf{A})]^2\,\chi - e\,\phi\,\chi. \tag{B.5.96}$$

5.10 (a) We have

$$\sigma_1^2 = \frac{1}{2}\begin{pmatrix} 0, 1, 0 \\ 1, 0, 1 \\ 0, 1, 0 \end{pmatrix}\begin{pmatrix} 0, 1, 0 \\ 1, 0, 1 \\ 0, 1, 0 \end{pmatrix} = \begin{pmatrix} 1/2, 0, 1/2 \\ 0, 1, 0 \\ 1/2, 0, 1/2 \end{pmatrix} = \sigma_1', \tag{B.5.97}$$

and

$$\sigma_1^3 = \frac{1}{\sqrt{2}}\begin{pmatrix} 0, 1, 0 \\ 1, 0, 1 \\ 0, 1, 0 \end{pmatrix}\begin{pmatrix} 1/2, 0, 1/2 \\ 0, 1, 0 \\ 1/2, 0, 1/2 \end{pmatrix} = \frac{1}{\sqrt{2}}\begin{pmatrix} 0, 1, 0 \\ 1, 0, 1 \\ 0, 1, 0 \end{pmatrix} = \sigma_1. \tag{B.5.98}$$

It follows, by induction, that

$$(\sigma_1)^k = \sigma_1' \qquad \text{for } k \text{ even,} \qquad (\text{B.5.99})$$

$$(\sigma_1)^k = \sigma_1 \qquad \text{for } k \text{ odd.} \qquad (\text{B.5.100})$$

(b) We have

$$\sigma_1^2 = \frac{1}{2} \begin{pmatrix} 0, & -i, & 0, \\ i, & 0, & -i \\ 0 & i, & 0 \end{pmatrix} \begin{pmatrix} 0, & -i, & 0 \\ i, & 0, & -i \\ 0 & i, & 0 \end{pmatrix} = \begin{pmatrix} 1/2, & 0, & -1/2 \\ 0, & 1, & 0 \\ -1/2 & 0, & 1/2 \end{pmatrix} = \sigma_2', \qquad (\text{B.5.101})$$

and

$$\sigma_2^3 = \frac{1}{\sqrt{2}} \begin{pmatrix} 0, & -i, & 0 \\ i, & 0, & -i \\ 0 & i, & 0 \end{pmatrix} \begin{pmatrix} 1/2, & 0, & -1/2 \\ 0, & 1, & 0 \\ -1/2 & 0, & 1/2 \end{pmatrix} = \frac{1}{\sqrt{2}} \begin{pmatrix} 0, & -i, & 0 \\ i, & 0, & -i \\ 0 & i, & 0 \end{pmatrix} = \sigma_2. \qquad (\text{B.5.102})$$

It follows, by induction, that

$$(\sigma_2)^k = \sigma_2' \qquad \text{for } k \text{ even,} \qquad (\text{B.5.103})$$

$$(\sigma_2)^k = \sigma_2 \qquad \text{for } k \text{ odd.} \qquad (\text{B.5.104})$$

(c) We have

$$\sigma_3^2 = \begin{pmatrix} 1, & 0, & 0 \\ 0, & 0, & 0 \\ 0 & 0, & -1 \end{pmatrix} \begin{pmatrix} 1, & 0, & 0 \\ 0, & 0, & 0 \\ 0 & 0, & -1 \end{pmatrix} = \begin{pmatrix} 1 & 0, & 0 \\ 0 & 0, & 0 \\ 0 & 0, & 1 \end{pmatrix} = \sigma_3', \qquad (\text{B.5.105})$$

and

$$\sigma_3^3 = \begin{pmatrix} 1, & 0, & 0 \\ 0, & 0, & 0 \\ 0 & 0, & -1 \end{pmatrix} \begin{pmatrix} 1, & 0, & 0 \\ 0, & 0, & 0 \\ 0 & 0, & 1 \end{pmatrix} = \begin{pmatrix} 1, & 0, & 0 \\ 0, & 0, & 0 \\ 0 & 0, & -1 \end{pmatrix} = \sigma_3. \qquad (\text{B.5.106})$$

It follows, by induction, that

$$(\sigma_3)^k = \sigma_3' \qquad \text{for } k \text{ even,} \qquad (\text{B.5.107})$$

$$(\sigma_3)^k = \sigma_3 \qquad \text{for } k \text{ odd.} \qquad (\text{B.5.108})$$

(d) The spinor matrix for a rotation through an angle $\Delta\varphi$ about the x-axis is

$$T_1(\Delta\varphi) = \exp(-i\,\sigma_1\,\Delta\varphi) \equiv \sum_{n=0,\infty} \frac{(i\,\sigma_1\,\Delta\varphi)^n}{n!}. \qquad (\text{B.5.109})$$

Making use of Equations (B.5.99) and (B.5.100), we obtain

$$T_1(\Delta\varphi) = 1 - \sigma_1' + \cos(\Delta\varphi)\,\sigma_1' - i\,\sin(\Delta\varphi)\,\sigma_1, \qquad (\text{B.5.110})$$

which evaluates to give

$$T_1(\Delta\varphi) = \begin{pmatrix} c^2, & -i\sqrt{2}\,sc, & -s^2 \\ -i\sqrt{2}\,sc, & c^2-s^2, & -i\sqrt{2}\,sc \\ -s^2, & -i\sqrt{2}\,sc, & c^2 \end{pmatrix}, \qquad (B.5.111)$$

where $s = \sin(\Delta\varphi/2)$ and $c = \cos(\Delta\varphi/2)$.

The spinor matrix for a rotation through an angle $\Delta\varphi$ about the y-axis is

$$T_2(\Delta\varphi) = \exp(-i\,\sigma_2\,\Delta\varphi) \equiv \sum_{n=0,\infty} \frac{(i\,\sigma_2\,\Delta\varphi)^n}{n!}. \qquad (B.5.112)$$

Making use of Equations (B.5.103) and (B.5.104), we obtain

$$T_2(\Delta\varphi) = 1 - \sigma_2' + \cos(\Delta\varphi)\,\sigma_2' - i\,\sin(\Delta\varphi)\,\sigma_2, \qquad (B.5.113)$$

which evaluates to give

$$T_2(\Delta\varphi) = \begin{pmatrix} c^2, & -\sqrt{2}\,sc, & s^2 \\ \sqrt{2}\,sc, & c^2-s^2, & -\sqrt{2}\,sc \\ s^2, & \sqrt{2}\,sc, & c^2 \end{pmatrix}. \qquad (B.5.114)$$

Finally, the spinor matrix for a rotation through an angle $\Delta\varphi$ about the z-axis is

$$T_3(\Delta\varphi) = \exp(-i\,\sigma_3\,\Delta\varphi) \equiv \sum_{n=0,\infty} \frac{(i\,\sigma_3\,\Delta\varphi)^n}{n!}. \qquad (B.5.115)$$

Making use of Equations (B.5.107) and (B.5.108), we obtain

$$T_3(\Delta\varphi) = 1 - \sigma_3' + \cos(\Delta\varphi)\,\sigma_3' - i\,\sin(\Delta\varphi)\,\sigma_3, \qquad (B.5.116)$$

which evaluates to give

$$T_3(\Delta\varphi) = \begin{pmatrix} \exp(-i\,\Delta\varphi), & 0, & 0 \\ 0, & 1, & 0 \\ 0, & 0, & \exp(i\,\Delta\varphi) \end{pmatrix}. \qquad (B.5.117)$$

(e) The spinors that represent the eigenstates of S_z corresponding to the eigenvalues $+\hbar$, 0, and $-\hbar$ are

$$\chi_1 = \begin{pmatrix} 1 \\ 0 \\ 0 \end{pmatrix}, \qquad \chi_0 = \begin{pmatrix} 0 \\ 1 \\ 0 \end{pmatrix}, \qquad \chi_{-1} = \begin{pmatrix} 0 \\ 0 \\ 1 \end{pmatrix}, \qquad (B.5.118)$$

respectively. The spinor that represents the state of the system is obtained by rotating χ_1 through an angle θ about the y-axis. In other words,

$$\chi = T_2(\theta)\,\chi_1. \qquad (B.5.119)$$

It follows from Equations (B.5.114) and (B.5.118) that

$$\chi = \cos^2(\theta/2)\chi_1 + \sqrt{2}\,\sin(\theta/2)\,\cos(\theta/2)\chi_0 + \sin^2(\theta/2)\chi_{-1}. \qquad \text{(B.5.120)}$$

Hence, the probabilities of a measurement of S_z yielding the results $+\hbar$, 0, and $-\hbar$ are $\cos^4(\theta/2)$, $2\sin^2(\theta/2)\cos^2(\theta/2)$, and $\sin^2(\theta/2)$, respectively.

B.6 Chapter 6

6.1 We have $j_1 = 1$ and $j_2 = 1/2$. Thus, the allowed values of m_1 are 1, 0, and -1, whereas the allowed values of m_2 are $1/2$ and $-1/2$. Given that $|j_1 - j_2| \le j \le j_1 + j_2$, the allowed values of j are $1/2$ and $3/2$. In the first case, the allowed values of m are $3/2$, $1/2$, $-1/2$, and $-3/2$. In the second case, the allowed values of m are $1/2$ and $-1/2$. Given that the Clebsch-Gordon coefficients are zero unless $m_1 + m_2 = m$, and also that each row and column of the Clebsch-Gordon table must have a unit norm, we deduce that the table takes the form:

m_1	m_2						
1	1/2	1	0	0	0	0	0
1	−1/2	0	c_1	0	0	d_1	0
0	1/2	0	c_2	0	0	d_2	0
0	−1/2	0	0	c_3	0	0	d_3
−1	1/2	0	0	c_4	0	0	d_4
−1	−1/2	0	0	0	± 1	0	0
$j_1=1$	j	3/2	3/2	3/2	3/2	1/2	1/2
$j_2=1/2$	m	3/2	1/2	−1/2	−3/2	1/2	−1/2

Here, we have given the coefficient $\langle 1, 1/2; 3/2, 3/2 \rangle$ a conventional positive sign.

Making use of the recursion relation (6.32), choosing the upper sign, with $j = 3/2$, $m = 1/2$, $m_1 = 1$, and $m_2 = 1/2$, we obtain

$$\sqrt{3} = \sqrt{2}\,c_2 + c_1. \qquad \text{(B.6.1)}$$

However, the fact that the second column of the table must have a unit norm implies that

$$c_1^2 + c_2^2 = 1. \qquad \text{(B.6.2)}$$

It follows that

$$c_1 = \sqrt{\frac{1}{3}}, \qquad\qquad c_2 = \sqrt{\frac{2}{3}}. \qquad \text{(B.6.3)}$$

Making use of the recursion relation, choosing the upper sign, with $j = 3/2$, $m =$

$-1/2$, $m_1 = 0$, and $m_2 = 1/2$, we get

$$2\,c_2 = 2\sqrt{\frac{2}{3}} = \sqrt{2}\,c_4 + c_3.\tag{B.6.4}$$

However, the fact that the third column of the table must have a unit norm implies that

$$c_3^2 + c_4^2 = 1.\tag{B.6.5}$$

It follows that

$$c_3 = \sqrt{\frac{2}{3}},\qquad\qquad c_4 = \sqrt{\frac{1}{3}}.\tag{B.6.6}$$

Making use of the recursion relation, choosing the lower sign, with $j = 3/2$, $m = -1/2$, $m_1 = -1$, and $m_2 = -1/2$, we get

$$\sqrt{3}\,\langle -1, -1/2; 3/2, -3/2\rangle = \sqrt{2}\,c_3 + c_4 = \sqrt{3},\tag{B.6.7}$$

which implies that $\langle -1, -1/2; 3/2, -3/2\rangle = +1$.

Making use of the recursion relation, choosing the upper sign, with $j = 1/2$, $m = -1/2$, $m_1 = 0$, and $m_2 = 1/2$, we obtain

$$d_2 = \sqrt{2}\,d_4 + d_3.\tag{B.6.8}$$

However, the fact that all rows and columns of the table must have unit norms, and also be mutually perpendicular, implies that

$$c_1^2 + d_1^2 = 1,\tag{B.6.9}$$
$$c_2^2 + d_2^2 = 1,\tag{B.6.10}$$
$$c_3^2 + d_3^2 = 1,\tag{B.6.11}$$
$$c_4^2 + d_4^2 = 1,\tag{B.6.12}$$
$$c_1\,c_2 + d_1\,d_2 = 0,\tag{B.6.13}$$
$$c_3\,c_4 \dotplus d_3\,d_4 = 0.\tag{B.6.14}$$

It follows that

$$d_1 = \sqrt{\frac{2}{3}},\qquad d_2 = -\sqrt{\frac{1}{3}},\qquad d_3 = \sqrt{\frac{1}{3}},\qquad d_4 = -\sqrt{\frac{2}{3}},\tag{B.6.15}$$

where we have given d_1 a conventional positive sign. Hence, the complete Clebsch-

Gordon table for adding spin one-half to spin one takes the form:

m_1	m_2						
1	1/2	1	0	0	0	0	0
1	−1/2	0	$\sqrt{1/3}$	0	0	$\sqrt{2/3}$	0
0	1/2	0	$\sqrt{2/3}$	0	0	$-\sqrt{1/3}$	0
0	−1/2	0	0	$\sqrt{2/3}$	0	0	$\sqrt{1/3}$
−1	1/2	0	0	$\sqrt{1/3}$	0	0	$-\sqrt{2/3}$
−1	−1/2	0	0	0	1	0	0
$j_1=1$	j	3/2	3/2	3/2	3/2	1/2	1/2
$j_2=1/2$	m	3/2	1/2	−1/2	−3/2	1/2	−1/2

6.2 Let us, first, consider the angular momentum state of the electron. The electron is in an $l = 1$ orbital state (because such a state has the angular eigenfunction $Y_{1m}(\theta,\varphi)$, where $m = 1, 0,$ or -1), and has spin one-half. Thus, we are adding spin one to spin one-half. In other words, $j_1 = 1$ and $j_2 = 1/2$, where the subscript 1 refers to the orbital angular momentum, and the subscript 2 to the spin angular momentum. It is clear that the angular momentum state of the electron is

$$|\rangle = \sqrt{\frac{1}{3}}\,|1,1/2;0,1/2\rangle + \sqrt{\frac{2}{3}}\,|1,1/2;1,-1/2\rangle, \qquad (B.6.16)$$

where the kets on the right-hand side are $|j_1,j_2;m_1,m_2\rangle$ kets. However, from the Clebsch-Gordon table in the previous exercise,

$$|1,1/2;0,1/2\rangle = \sqrt{\frac{2}{3}}\,|1,1/2;3/2,1/2\rangle - \sqrt{\frac{1}{3}}\,|1,1/2;1/2,1/2\rangle, \qquad (B.6.17)$$

$$|1,1/2;1,-1/2\rangle = \sqrt{\frac{1}{3}}\,|1,1/2;3/2,1/2\rangle + \sqrt{\frac{2}{3}}\,|1,1/2;1/2,1/2\rangle, \qquad (B.6.18)$$

where the kets on the left-hand side are $|j_1,j_2;m_1,m_2\rangle$ kets, whereas those on the right-hand side are $|j_1,j_2;j,m\rangle$ kets. Thus, it follows that

$$|\rangle = \sqrt{\frac{8}{9}}\,|1,1/2;3/2,1/2\rangle + \sqrt{\frac{1}{9}}\,|1,1/2;1/2,1/2\rangle, \qquad (B.6.19)$$

where the kets on the right-hand side are $|j_1,j_2;j,m\rangle$ kets.

(a) A measurement of L^2 yields $j_1(j_1+1)\hbar^2$. Given that $j_1 = 1$, such a measurement is bound to give the result $2\hbar^2$.
(b) A measurement of L_z yields $m_1\hbar$. It follows from Equation (B.6.16) that such a measurement yields 0 with probability 1/3, and $\hbar$ with probability 2/3.
(c) A measurement of S^2 yields $j_2(j_2+1)\hbar^2$. Given that $j_2 = 1/2$, such a measurement is bound to give the result $(3/4)\hbar^2$.
(d) A measurement of S_z yields $m_2\hbar$. It follows from Equation (B.6.16) that such a measurement yields $\hbar/2$ with probability 1/3, and $-\hbar/2$ with probability 2/3.

(e) A measurement of J^2 yields $j(j+1)\hbar^2$. It follows from Equation (B.6.19) that such a measurement yields $(15/4)\hbar^2$ with probability 8/9, and $(3/4)\hbar^2$ with probability 1/9.

(f) A measurement of J_z yields $m\hbar$. It follows from Equation (B.6.19) that such a measurement is bound to give the result $\hbar/2$.

(g) The probability density for finding the electron at position r, θ, φ is

$$P(r,\theta,\varphi) = \chi^\dagger \chi = |R_{21}|^2 \left[\sqrt{\frac{1}{3}}\, Y_{10}^* \, \chi_+^\dagger + \sqrt{\frac{2}{3}}\, Y_{11}^* \, \chi_-^\dagger\right]$$

$$\left[\sqrt{\frac{1}{3}}\, Y_{10}\, \chi_+ + \sqrt{\frac{2}{3}}\, Y_{11}\, \chi_+\right], \quad \text{(B.6.20)}$$

which yields

$$P(r,\theta,\varphi) = |R_{21}|^2 \left(\frac{1}{3}\,|Y_{10}|^2 + \frac{2}{3}\,|Y_{11}|^2\right), \quad \text{(B.6.21)}$$

where use has been made of the orthonormality of the $\chi_\pm$. It follows from Equations (4.96) and (4.97) that

$$P(r,\theta,\varphi) = \frac{|R_{21}(r)|^2}{4\pi}. \quad \text{(B.6.22)}$$

Finally, Exercise 4.12 gives

$$P(r,\theta,\varphi) = \frac{1}{96\pi\, a_0^3}\left(\frac{r}{a_0}\right)^2 \exp\left(-\frac{r}{a_0}\right), \quad \text{(B.6.23)}$$

where a_0 is the Bohr radius.

(h) The probability density for finding the electron in the spin-up state at position r, θ, φ is

$$P(r,\theta,\varphi) = |R_{21}|^2 \, \frac{1}{3}\,|Y_{10}|^2. \quad \text{(B.6.24)}$$

It follows from Equation (4.96) that

$$P(r,\theta,\varphi) = \frac{|R_{21}|^2}{4\pi}\cos^2\theta. \quad \text{(B.6.25)}$$

The probability density for finding the electron at radius r is

$$P(r) = r^2 \oint d\Omega\, P(r,\theta,\varphi) = \frac{r^2\,|R_{21}|^2}{2}\int_0^\pi d\theta\,\cos^2\theta\,\sin\theta \;=\; \frac{r^2\,|R_{21}|^2}{3}. \quad \text{(B.6.26)}$$

Thus, we obtain

$$P(r) = \frac{1}{72\,a_0}\left(\frac{r}{a_0}\right)^4 \exp\left(-\frac{r}{a_0}\right). \quad \text{(B.6.27)}$$

6.3 The potential energy can be written

$$V(r) = V_1(r) + V_2(r)\left[\left(3\frac{x^2}{r^2} - 1\right)\langle\sigma_1\sigma_1\rangle + \left(3\frac{y^2}{r^2} - 1\right)\langle\sigma_2\sigma_2\rangle + \left(3\frac{z^2}{r^2} - 1\right)\langle\sigma_3\sigma_3\rangle\right.$$

$$\left. + 6\frac{xy}{r^2}\langle\sigma_1\sigma_2\rangle + 6\frac{yz}{r^2}\langle\sigma_2\sigma_3\rangle + 6\frac{xz}{r^2}\langle\sigma_1\sigma_3\rangle\right]$$

$$+ V_3(r)\left(\langle\sigma_1\sigma_1\rangle + \langle\sigma_2\sigma_2\rangle + \langle\sigma_3\sigma_3\rangle\right). \tag{B.6.28}$$

Here, $\sigma_i\sigma_j$ denotes the ith neutron Pauli matrix paired with the jth proton Pauli matrix. The expectation values are taken with respect to overall spin state.

(a) For the singlet state,

$$\chi = \frac{1}{\sqrt{2}}\left(|+\rangle|-\rangle - |-\rangle|+\rangle\right), \tag{B.6.29}$$

where $|+\rangle|-\rangle$ denotes a spin state in which the neutron is spin-up, and the proton spin-down, et cetera. Thus,

$$\langle\sigma_i\sigma_j\rangle = \chi^\dagger\sigma_i\sigma_j\chi$$

$$= \frac{1}{2}\left[(\sigma_i)_{11}(\sigma_j)_{22} + (\sigma_i)_{22}(\sigma_j)_{11} - (\sigma_i)_{12}(\sigma_j)_{21} - (\sigma_i)_{21}(\sigma_j)_{12}\right]. \tag{B.6.30}$$

It follows that

$$\langle\sigma_1\sigma_1\rangle = \langle\sigma_2\sigma_2\rangle = \langle\sigma_3\sigma_3\rangle = -1, \tag{B.6.31}$$

and

$$\langle\sigma_1\sigma_2\rangle = \langle\sigma_2\sigma_3\rangle = \langle\sigma_1\sigma_3\rangle = 0. \tag{B.6.32}$$

Hence,

$$V(r) = V_1(r) - 3V_3(r). \tag{B.6.33}$$

(b) For the $m = 0$ triplet state,

$$\chi = \frac{1}{\sqrt{2}}\left(|+\rangle|-\rangle + |-\rangle|+\rangle\right). \tag{B.6.34}$$

Thus,

$$\langle\sigma_i\sigma_j\rangle = \frac{1}{2}\left[(\sigma_i)_{11}(\sigma_j)_{22} + (\sigma_i)_{22}(\sigma_j)_{11} + (\sigma_i)_{12}(\sigma_j)_{21} + (\sigma_i)_{21}(\sigma_j)_{12}\right]. \tag{B.6.35}$$

It follows that

$$\langle\sigma_1\sigma_1\rangle = \langle\sigma_2\sigma_2\rangle = -\langle\sigma_3\sigma_3\rangle = 1, \tag{B.6.36}$$

and

$$\langle\sigma_1\sigma_2\rangle = \langle\sigma_2\sigma_3\rangle = \langle\sigma_1\sigma_3\rangle = 0. \tag{B.6.37}$$

Hence,

$$V(r) = V_1(r) - 2V_2(r)(3\cos^2\theta - 1) + V_3(r), \tag{B.6.38}$$

where $\cos\theta = z/r$.

For the $m = 1$ triplet state,

$$\chi = |+\rangle|+\rangle. \tag{B.6.39}$$

Thus,

$$\langle\sigma_i\,\sigma_j\rangle = (\sigma_i)_{11}\,(\sigma_j)_{11}. \tag{B.6.40}$$

It follows that

$$\langle\sigma_1\,\sigma_1\rangle = \langle\sigma_2\,\sigma_2\rangle = 0, \tag{B.6.41}$$

and

$$\langle\sigma_3\,\sigma_3\rangle = 1, \tag{B.6.42}$$

and

$$\langle\sigma_1\,\sigma_2\rangle = \langle\sigma_2\,\sigma_3\rangle = \langle\sigma_1\,\sigma_3\rangle = 0. \tag{B.6.43}$$

Hence,

$$V(r) = V_1(r) + V_2(r)\,(3\,\cos^2\theta - 1) + V_3(r). \tag{B.6.44}$$

By symmetry, the potential is the same in the $m = -1$ triplet state.

6.4 We are effectively adding spin one-half to spin one-half. In other words, $j_1 = 1/2$ and $j_2 = 1/2$, where the subscript 1 refers to the spin angular momentum of the first electron, and the subscript 2 to the spin angular momentum of the second electron. The electron pair is in a state of overall angular momentum characterized by the quantum numbers $j = 0$ and $m = 0$. In other words, the electron pair is in a state of zero overall angular momentum. Thus, we can represent the pair's angular momentum state by the ket

$$|\rangle = |1/2, 1/2; 0, 0\rangle, \tag{B.6.45}$$

where the ket on the right-hand side is a $|j_1, j_2; j, m\rangle$ ket. However, from the completed table in Section 6.4, we can also write

$$|\rangle = \frac{1}{\sqrt{2}}|1/2, 1/2; 1/2, -1/2\rangle - \frac{1}{\sqrt{2}}|1/2, 1/2; -1/2, 1/2\rangle, \tag{B.6.46}$$

where the kets on the right-hand side are $|j_1.j_2; m_1, m_2\rangle$ kets.

(a) It is evident from Equation (B.6.46) that if $m_1 = 1/2$ then $m_2 = -1/2$, and vice versa. Thus, the probability that a measurement of S_z for one electron giving the result $\hbar/2$, and a subsequent measurement of the spin of the other electron giving the same result, is zero.

(b) We can represent the ket (B.6.46) as the spinor

$$\chi = \frac{1}{\sqrt{2}}\,(\chi_{z+}\chi_{z-} - \chi_{z-}\chi_{z+}), \tag{B.6.47}$$

where the $\chi_{z\pm}$ are the eigenstates of S_z corresponding to the eigenvalues $\pm\hbar/2$, respectively. Moreover, the first spinor in the products refers to the spin state of the first electron, whereas the second spinor refers to the spin state of the second electron. Now, if the two electron are in a state of zero angular momentum then the projection of the overall angular momentum vector is zero along the y-axis, as well as along the z-axis. It follows, by analogy with the previous equation, that

$$\chi = \frac{1}{\sqrt{2}} \left(\chi_{y+} \chi_{y-} - \chi_{y-} \chi_{y+} \right),$$

(B.6.48)

where the $\chi_{y\pm}$ are the eigenstates of S_y corresponding to the eigenvalues $\pm\hbar/2$, respectively. Thus, if one electron is in the χ_{y+} state then the other electron is in the χ_{y-} state. Now, from the answer to Exercise 5.4,

$$\chi_{x\pm} = \frac{1}{\sqrt{2}} \left(\chi_{z+} \pm \chi_{z-} \right),$$

(B.6.49)

$$\chi_{y\pm} = \frac{1}{\sqrt{2}} \left(\chi_{z+} \pm i \chi_{z-} \right),$$

(B.6.50)

where the $\chi_{x\pm}$ are the eigenstates of S_x corresponding to the eigenvalues $\pm\hbar/2$, respectively. Hence,

$$\chi_{y-} = \frac{1}{\sqrt{2}} \left(e^{-i\pi/4} \chi_{x+} + e^{i\pi/4} \chi_{x-} \right).$$

(B.6.51)

It follows that the probability of a measurement of the other electron's x-component of spin yielding the result $-\hbar/2$ is $1/2$.

(c) We already know that the singlet (i.e., spin zero) state can be represented by the spinor

$$\chi_{00} = \frac{1}{\sqrt{2}} \left(\chi_{z+} \chi_{z-} - \chi_{z-} \chi_{z+} \right).$$

(B.6.52)

Here, the two subscripts on the left-hand side refer to the quantum numbers j and m, respectively. It is apparent from the completed table in Section 6.4 that the three triplet (i.e., spin one) states take the form

$$\chi_{11} = \chi_{z+} \chi_{z+},$$

(B.6.53)

$$\chi_{10} = \frac{1}{\sqrt{2}} \left(\chi_{z+} \chi_{z-} + \chi_{z-} \chi_{z+} \right),$$

(B.6.54)

$$\chi_{1-1} = \chi_{z-} \chi_{z-}.$$

(B.6.55)

Thus,

$$\chi_{z+} \chi_{z+} = \chi_{11},$$

(B.6.56)

$$\chi_{z+} \chi_{z-} = \frac{1}{\sqrt{2}} \left(\chi_{10} + \chi_{00} \right),$$

(B.6.57)

$$\chi_{z-} \chi_{z+} = \frac{1}{\sqrt{2}} \left(\chi_{10} - \chi_{00} \right),$$

(B.6.58)

$$\chi_{z-} \chi_{z-} = \chi_{1-1}.$$

(B.6.59)

We are told that

$$\chi = \left(\cos\alpha_1\,\chi_{z+} + \sin\alpha_1\,e^{i\beta_1}\chi_{z-}\right)\left(\cos\alpha_2\,\chi_{z+} + \sin\alpha_2\,e^{i\beta_2}\chi_{z-}\right)$$

$$= \cos\alpha_1\,\cos\alpha_2\,\chi_{z+}\chi_{z+} + \sin\alpha_1\,\sin\alpha_2\,e^{i(\beta_1+\beta_2)}\,\chi_{z-}\chi_{z-}$$

$$+ \cos\alpha_1\,\sin\alpha_2\,e^{i\beta_2}\,\chi_{z+}\chi_{z-} + \sin\alpha_1\,\cos\alpha_2\,e^{i\beta_1}\,\chi_{z-}\chi_{z+}$$

$$= \cos\alpha_1\,\cos\alpha_2\,\chi_{11} + \sin\alpha_1\,\sin\alpha_2\,e^{i(\beta_1+\beta_2)}\,\chi_{1-1}$$

$$+ \frac{1}{\sqrt{2}}\left(\cos\alpha_1\,\sin\alpha_2\,e^{i\beta_2} + \sin\alpha_1\,\cos\alpha_2\,e^{i\beta_1}\right)\chi_{10}$$

$$+ \frac{1}{\sqrt{2}}\left(\cos\alpha_1\,\sin\alpha_2\,e^{i\beta_2} - \sin\alpha_1\,\cos\alpha_2\,e^{i\beta_1}\right)\chi_{00}. \tag{B.6.60}$$

Thus, the probabilities of finding the system in the various overall spin states are

$$P_{11} = \cos^2\alpha_1\,\cos^2\alpha_2, \tag{B.6.61}$$

$$P_{10} = \frac{1}{2}\Big[\cos^2\alpha_1\,\sin^2\alpha_2 + \sin^2\alpha_1\,\cos^2\alpha_2$$

$$+ 2\cos\alpha_1\,\sin\alpha_1\,\cos\alpha_2\,\sin\alpha_2\,\cos(\beta_1-\beta_2)\Big], \tag{B.6.62}$$

$$P_{1-1} = \sin^2\alpha_1\,\sin^2\alpha_2, \tag{B.6.63}$$

$$P_{00} = \frac{1}{2}\Big[\cos^2\alpha_1\,\sin^2\alpha_2 + \sin^2\alpha_1\,\cos^2\alpha_2$$

$$- 2\cos\alpha_1\,\sin\alpha_1\,\cos\alpha_2\,\sin\alpha_2\,\cos(\beta_1-\beta_2)\Big]. \tag{B.6.64}$$

Finally, the probability of finding the system in one of the triplet states is

$$P_1 = P_{11} + P_{10} + P_{1-1}$$

$$= \frac{3}{4} + \frac{1}{4}\left[\cos(2\,\alpha_1)\,\cos(2\,\alpha_2) + \sin(2\,\alpha_1)\,\sin(2\,\alpha_2)\,\cos(\beta_1-\beta_2)\right]. \tag{B.6.65}$$

B.7   Chapter 7

7.1 From Equation (7.6), we have

$$\begin{pmatrix} \tilde{E}_1 - E, & e_{12} \\ e_{12}^*, & \tilde{E}_2 - E \end{pmatrix}\begin{pmatrix} \langle 1|E\rangle \\ \langle 2|E\rangle \end{pmatrix} = \begin{pmatrix} 0 \\ 0 \end{pmatrix}, \tag{B.7.1}$$

where $\tilde{E}_1 = E_1 + e_{11}$ and $\tilde{E}_2 = E_2 + e_{22}$. Setting the determinant of the matrix to zero yields the quadratic equation

$$E^2 - (\tilde{E}_1 + \tilde{E}_2)\,E + \tilde{E}_1\,\tilde{E}_2 - |e_{12}|^2 = 0, \tag{B.7.2}$$

The matrix equation itself gives

$$\frac{\langle 1|E\rangle}{\langle 2|E\rangle} = -\frac{e_{12}}{\tilde{E}_1 - E} = -\frac{\tilde{E}_2 - E}{e_{12}^*}. \tag{B.7.3}$$

We also have

$$|\langle 1|E\rangle|^2 + |\langle 2|E\rangle|^2 = 1, \tag{B.7.4}$$

and

$$|E\rangle = \langle 1|E\rangle|1\rangle + \langle 2|E\rangle|2\rangle. \tag{B.7.5}$$

Equation (B.7.2) yields

$$E = \frac{\tilde{E}_1 + \tilde{E}_2 \pm \sqrt{(\tilde{E}_1 - \tilde{E}_2)^2 + 4|e_{12}|^2}}{2}$$

$$= \frac{1}{2}(\tilde{E}_1 + \tilde{E}_2) \pm \frac{1}{2}(\tilde{E}_1 - \tilde{E}_2)(1 + 2\epsilon^2 + \cdots), \tag{B.7.6}$$

where

$$\epsilon = \frac{|e_{12}|}{\tilde{E}_1 - \tilde{E}_2} = \frac{|e_{12}|}{E_1 - E_2} + O(\epsilon). \tag{B.7.7}$$

The two roots of Equation (B.7.6) are

$$E_1' = E_1 + e_{11} + \frac{|e_{12}|^2}{E_1 - E_2} + O(\epsilon^3), \tag{B.7.8}$$

$$E_2' = E_2 + e_{22} - \frac{|e_{12}|^2}{E_1 - E_2} + O(\epsilon^3). \tag{B.7.9}$$

Substitution into Equations (B.7.3)–(B.7.5) gives

$$|1\rangle' = |1\rangle + \frac{e_{12}^*}{E_1 - E_2}|2\rangle + O(\epsilon^2), \tag{B.7.10}$$

$$|2\rangle' = |2\rangle - \frac{e_{12}}{E_1 - E_2}|1\rangle + O(\epsilon^2). \tag{B.7.11}$$

7.2 If the unperturbed energy eigenstates are also eigenstates of the perturbing Hamiltonian then $e_{12} = 0$. In this case, Equation (B.7.2) reduces to

$$(E - \tilde{E}_1)(E - \tilde{E}_2) = 0, \tag{B.7.12}$$

which has the exact solutions

$$E_1' = \tilde{E}_1 = E_1 + e_{11}, \tag{B.7.13}$$

$$E_2' = \tilde{E}_2 = E_2 + e_{22}. \tag{B.7.14}$$

Moreover, Equations (B.7.3)–(B.7.5) have the exact solutions

$$|1\rangle' = |1\rangle, \tag{B.7.15}$$

$$|2\rangle' = |2\rangle. \tag{B.7.16}$$

7.3 If $E_1 = E_2 = E_{12}$ then Equation (B.7.6) yields

$$E^\pm = E_{12} + \lambda^\pm, \tag{B.7.17}$$

where

$$\lambda^\pm = \frac{1}{2}\,(e_{11} + e_{22}) \pm \frac{1}{2}\left[(e_{11} - e_{22})^2 + 4\,|e_{12}|^2\right]^{1/2}. \tag{B.7.18}$$

Likewise, Equation (B.7.3) gives

$$\frac{\langle 1|E^\pm\rangle}{\langle 2|E^\pm\rangle} = -\frac{e_{12}}{e_{11} - \lambda^\pm}. \tag{B.7.19}$$

Note that the $|E^\pm\rangle$ are automatically eigenstates of H_0, because they are linear combinations of eigenstates corresponding to the same eigenvalue.

Let us search for the eigenstates of H_1:

$$H_1\,|\mu\rangle = \mu\,|\mu\rangle. \tag{B.7.20}$$

We can write

$$|\mu\rangle = \langle 1|\mu\rangle|1\rangle + \langle 2|\mu\rangle|2\rangle. \tag{B.7.21}$$

It follows that

$$\begin{pmatrix} e_{11} - \mu, & e_{12} \\ e_{12}^*, & e_{22} - \mu \end{pmatrix}\begin{pmatrix} \langle 1|\mu\rangle \\ \langle 2|\mu\rangle \end{pmatrix} = \begin{pmatrix} 0 \\ 0 \end{pmatrix}. \tag{B.7.22}$$

Setting the determinant of the matrix to zero yields

$$\mu^\pm = \frac{1}{2}\,(e_{11} + e_{22}) \pm \frac{1}{2}\left[(e_{11} - e_{22})^2 + 4\,|e_{12}|^2\right]^{1/2}. \tag{B.7.23}$$

The matrix equation itself gives

$$\frac{\langle 1|\mu^\pm\rangle}{\langle 2|\mu^\pm\rangle} = -\frac{e_{12}}{e_{11} - \mu^\pm}. \tag{B.7.24}$$

Thus, the $|E^\pm\rangle$ are simultaneous eigenstates of H_0 and H_1. Moreover, $\lambda^\pm$ is the eigenvalue of H_1.

7.4 The properly normalized ground-state wavefunction is

$$\psi_0(x) = \left(\frac{m\,\omega}{\pi\,\hbar}\right)^{1/4} \exp\left(-\frac{m\,\omega\,x^2}{2\,\hbar}\right). \tag{B.7.25}$$

(See Exercise 3.3.) The lowest-order energy-shift due to the perturbation $V(x) = \lambda x^4$ is

$$\Delta E = \int_{-\infty}^{\infty} dx \, \psi_0^*(x) \, V(x) \, \psi_0(x), \tag{B.7.26}$$

which yields

$$\Delta E = \frac{\lambda q}{\pi^{1/2}} \int_{-\infty}^{\infty} dx \, x^4 \, e^{-q^2 x^2}, \tag{B.7.27}$$

where $q = (m\omega/\hbar)^{1/2}$. Now [Reif (1965)],

$$\int_{-\infty}^{\infty} dx \, x^4 \, e^{-\alpha x^2} = \frac{d^2}{d\alpha^2} \int_{-\infty}^{\infty} dx \, e^{-\alpha x^2} = \frac{d^2}{d\alpha^2} \left(\frac{\pi^{1/2}}{\alpha^{1/2}} \right) = \frac{3}{4} \frac{\pi^{1/2}}{\alpha^{5/2}}, \tag{B.7.28}$$

so

$$\int_{-\infty}^{\infty} dx \, x^4 \, e^{-q^2 x^2} = \frac{3}{4} \frac{\pi^{1/2}}{q^5}. \tag{B.7.29}$$

Hence,

$$\Delta E = \frac{3}{4} \frac{\lambda}{q^4} = \frac{3}{4} \lambda \left(\frac{\hbar}{m\omega} \right)^2. \tag{B.7.30}$$

7.5 According to Section 4.6, the nine allowed $n = 3$ states are $|3,0,0\rangle$, $|3,1,0\rangle$, $|3,1,\pm1\rangle$, $|3,2,0\rangle$, $|3,2,\pm1\rangle$, and $|3,2,\pm2\rangle$. Let

$$c_{l',m',l,m} = \langle 3, l', m' | z | 3, l, m \rangle. \tag{B.7.31}$$

According to the selection rules derived in Section 7.4, $c_{l',m',l,m} = 0$ unless $l' = l\pm1$ and $m' = m$. It follows that the only non-zero matrix elements are $c_{0,0,1,0}$, $c_{1,0,0,0}$, $c_{1,0,2,0}$, $c_{2,0,1,0}$, $c_{1,\pm1,2,\pm1}$, and $c_{2,\pm1,1,\pm1}$. According to Section 4.6, the wavefunction associated with the $|3,l,m\rangle$ state takes the form

$$\psi_{3lm}(r,\theta,\varphi) = R_{3l}(r) \, Y_{lm}(\theta,\varphi). \tag{B.7.32}$$

Thus, given that $z = r\cos\theta$, we obtain

$$c_{l',m,l,m} = \int d^3\mathbf{x} \, \psi_{3l'm}^*(\mathbf{x}) \, r\cos\theta \, \psi_{3lm}(\mathbf{x}) = A_{3,l',3,l} \, B_{l',m,l,m}, \tag{B.7.33}$$

where

$$A_{3,l',3,l} = \int_0^r dr \, r^3 \, R_{3l'}(r) \, R_{3l}(r), \tag{B.7.34}$$

and

$$B_{l',m,l,m} = 2\pi \int_0^\pi d\theta \, \cos\theta \, \sin\theta \, Y_{l'm}^* \, Y_{lm}. \tag{B.7.35}$$

It follows from the radial wavefunctions given in Exercise 4.12 that

$$A_{3,1,3,0} = A_{3,0,3,1} = \frac{a_0}{2\sqrt{2}} \int_0^\infty dy\, y^4 \left(1 - y + \frac{y^2}{6}\right)\left(1 - \frac{y}{4}\right) e^{-y} = -9\sqrt{2}\, a_0, \quad \text{(B.7.36)}$$

$$A_{3,2,3,1} = A_{3,1,3,2} = \frac{a_0}{24\sqrt{5}} \int_0^\infty y^6 \left(1 - \frac{y}{4}\right) e^{-y} = -\frac{9\sqrt{5}}{2}\, a_0, \quad \text{(B.7.37)}$$

where use has been made of Exercise 4.11. It follows from the expressions for the spherical harmonics given in Section 4.4 that

$$B_{1,0,0,0} = B_{0,0,1,0} = \frac{1}{\sqrt{3}}, \quad \text{(B.7.38)}$$

$$B_{1,0,2,0} = B_{2,0,1,0} = \frac{2}{\sqrt{15}}, \quad \text{(B.7.39)}$$

$$B_{1,\pm1,2,\pm1} = B_{2,\pm1,1,\pm1} = \frac{1}{\sqrt{5}}. \quad \text{(B.7.40)}$$

Thus, the non-zero matrix elements are

$$\langle 3,1,0|z|3,0,0\rangle = \langle 3,0,0|z|3,1,0\rangle = -3\sqrt{6}\, a_0, \quad \text{(B.7.41)}$$

$$\langle 3,1,0|z|3,2,0\rangle = \langle 3,2,0|z|3,1,0\rangle = -3\sqrt{3}\, a_0, \quad \text{(B.7.42)}$$

$$\langle 3,1,\pm1|z|3,2,\pm1\rangle = \langle 3,2,\pm1|z|3,1,\pm1\rangle = -\frac{9}{2}\, a_0. \quad \text{(B.7.43)}$$

7.6 The perturbing Hamiltonian is

$$H_1 = e\,|\mathbf{E}|\,z. \quad \text{(B.7.44)}$$

It is clear from the previous exercise that this Hamiltonian couples the $|3,0,0\rangle$, $|3,1,0\rangle$, and $|3,2,0\rangle$ states together, separately couples the $|3,1,1\rangle$ and $|3,2,1\rangle$ states together, separately couples the $|3,1,-1\rangle$ and $|3,2,-1\rangle$ states together, and leaves the $|3,2,2\rangle$ and $|3,2,-2\rangle$ states unaffected.

We can determine the energy-shifts by associated with the $|3,0,0\rangle$, $|3,1,0\rangle$, and $|3,2,0\rangle$ states by solving an eigenvalue problem of the form

$$\mathbf{U}\mathbf{x} = \lambda\mathbf{x}, \quad \text{(B.7.45)}$$

where $\mathbf{x}$ is the column vector of the $|3,0,0\rangle$, $|3,1,0\rangle$, and $|3,2,0\rangle$ kets, and

$$\mathbf{U} = -3\sqrt{3}\, e\,|\mathbf{E}|\, a_0 \begin{pmatrix} 0, & \sqrt{2}, & 0 \\ \sqrt{2}, & 0, & 1 \\ 0, & 1, & 0 \end{pmatrix}. \quad \text{(B.7.46)}$$

Here, use has been made of the matrix elements calculated in the previous exercise. The eigenvalues of the previous matrix are $\lambda_1 = 9\,e\,|\mathbf{E}|\, a_0$, $\lambda_2 = 0$, and $\lambda_3 = -9\,e\,|\mathbf{E}|\, a_0$.

The corresponding properly normalized eigenvectors are

$$\mathbf{x}_1 = \begin{pmatrix} \sqrt{1/3} \\ -\sqrt{1/2} \\ \sqrt{1/6} \end{pmatrix}, \qquad \mathbf{x}_2 = \begin{pmatrix} \sqrt{1/3} \\ 0 \\ -\sqrt{2/3} \end{pmatrix}, \qquad \mathbf{x}_3 = \begin{pmatrix} \sqrt{1/3} \\ \sqrt{1/2} \\ \sqrt{1/6} \end{pmatrix}. \qquad \text{(B.7.47)}$$

We can determine the energy-shifts by associated with the $|3, 1, 1\rangle$ and $|3, 2, 1\rangle$ states by solving an eigenvalue problem of the form (B.7.45), where $\mathbf{x}$ is the column vector of the $|3, 1, 1\rangle$ and $|3, 2, 1\rangle$ kets, and

$$\mathbf{U} = -\frac{9}{2} e |\mathbf{E}| a_0 \begin{pmatrix} 0, 1 \\ 1, 0 \end{pmatrix}. \qquad \text{(B.7.48)}$$

The eigenvalues of the previous matrix are $\lambda_1 = (9/2) e |\mathbf{E}| a_0$ and $\lambda_2 = -(9/2) e |\mathbf{E}| a_0$. The corresponding properly normalized eigenvectors are

$$\mathbf{x}_1 = \begin{pmatrix} \sqrt{1/2} \\ 0 \\ -\sqrt{1/2} \end{pmatrix}, \qquad \mathbf{x}_2 = \begin{pmatrix} \sqrt{1/2} \\ 0 \\ \sqrt{1/2} \end{pmatrix}. \qquad \text{(B.7.49)}$$

We can determine the energy-shifts by associated with the $|3, 1, -1\rangle$ and $|3, 2, -1\rangle$ states by solving an eigenvalue problem of the form (B.7.45), where $\mathbf{x}$ is the column vector of the $|3, 1, -1\rangle$ and $|3, 2, -1\rangle$ kets, and

$$\mathbf{U} = -\frac{9}{2} e |\mathbf{E}| a_0 \begin{pmatrix} 0, 1 \\ 1, 0 \end{pmatrix}. \qquad \text{(B.7.50)}$$

The eigenvalues of the previous matrix are $\lambda_1 = (9/2) e |\mathbf{E}| a_0$ and $\lambda_2 = -(9/2) e |\mathbf{E}| a_0$. The corresponding properly normalized eigenvectors are

$$\mathbf{x}_1 = \begin{pmatrix} \sqrt{1/2} \\ 0 \\ -\sqrt{1/2} \end{pmatrix}, \qquad \mathbf{x}_2 = \begin{pmatrix} \sqrt{1/2} \\ 0 \\ \sqrt{1/2} \end{pmatrix}. \qquad \text{(B.7.51)}$$

In summary, in the presence of an external electric field, the nine degenerate $n = 3$ states of a hydrogen atom are split into five groups. There is one state with energy-shift

$$\Delta E_1 = 9 e |\mathbf{E}| a_0, \qquad \text{(B.7.52)}$$

two states with energy-shift

$$\Delta E_2 = \frac{9}{2} e |\mathbf{E}| a_0, \qquad \text{(B.7.53)}$$

three states with energy-shift

$$\Delta E_3 = 0 \qquad \text{(B.7.54)}$$

two states with energy-shift

$$\Delta E_4 = -\frac{9}{2} e \, |\mathbf{E}| \, a_0,$$ (B.7.55)

and one state with energy-shift

$$\Delta E_5 = -9 \, e \, |\mathbf{E}| \, a_0.$$ (B.7.56)

The eigenket of the state with energy-shift ΔE_1 is

$$|1\rangle = \frac{1}{\sqrt{3}} |3, 0, 0\rangle - \frac{1}{\sqrt{2}} |3, 1, 0\rangle + \frac{1}{\sqrt{6}} |3, 2, 0\rangle.$$ (B.7.57)

The eigenkets of the states with energy-shift ΔE_2 are

$$|2\rangle = \frac{1}{\sqrt{2}} |3, 1, 1\rangle - \frac{1}{\sqrt{2}} |3, 2, 1\rangle,$$ (B.7.58)

$$|3\rangle = \frac{1}{\sqrt{2}} |3, 1, -1\rangle - \frac{1}{\sqrt{2}} |3, 2, -1\rangle.$$ (B.7.59)

The eigenkets of the states with energy-shift ΔE_3 are

$$|4\rangle = \frac{1}{\sqrt{3}} |3, 0, 0\rangle - \sqrt{\frac{2}{3}} |3, 2, 0\rangle,$$ (B.7.60)

$$|5\rangle = |3, 2, 2\rangle,$$ (B.7.61)

$$|6\rangle = |3, 2, -2\rangle.$$ (B.7.62)

The eigenkets of the states with energy-shift ΔE_4 are

$$|7\rangle = \frac{1}{\sqrt{2}} |3, 1, 1\rangle + \frac{1}{\sqrt{2}} |3, 2, 1\rangle,$$ (B.7.63)

$$|8\rangle = \frac{1}{\sqrt{2}} |3, 1, -1\rangle + \frac{1}{\sqrt{2}} |3, 2, -1\rangle.$$ (B.7.64)

Finally, the eigenket of the state with energy-shift ΔE_5 is

$$|9\rangle = \frac{1}{\sqrt{3}} |3, 0, 0\rangle + \frac{1}{\sqrt{2}} |3, 1, 0\rangle + \frac{1}{\sqrt{6}} |3, 2, 0\rangle.$$ (B.7.65)

7.7 We can write

$$H(\lambda + \delta\lambda) = H(\lambda) + \delta H,$$ (B.7.66)

where

$$\delta H = \delta\lambda \, \frac{\partial H}{\partial \lambda}.$$ (B.7.67)

Let

$$E_n(\lambda + \delta\lambda) = E_n(\lambda) + \delta E_n.$$ (B.7.68)

According to first-order perturbation theory,

$$\delta E_n = \langle n(\lambda) | \, \delta H \, | n(\lambda) \rangle = \delta \lambda \left\langle n \left| \frac{\partial H}{\partial \lambda} \right| n \right\rangle, \tag{B.7.69}$$

which implies that

$$\frac{\delta E_n}{\delta \lambda} = \left\langle n \left| \frac{\partial H}{\partial \lambda} \right| n \right\rangle. \tag{B.7.70}$$

Thus, in the limit $\delta \lambda \to 0$, we obtain

$$\frac{\partial E_n}{\partial \lambda} = \left\langle n \left| \frac{\partial H}{\partial \lambda} \right| n \right\rangle. \tag{B.7.71}$$

Note that if the E_n are degenerate then the $|n\rangle$ must be chosen such as to be simultaneous eigenstates of H and $\partial H / \partial \lambda$ in order to avoid singular terms in the perturbation expansion to second order.

7.8 We have

$$\frac{\partial H}{\partial l} = \frac{\hbar^2}{2\, m_e} \frac{2\, l + 1}{r^2}, \tag{B.7.72}$$

and

$$\frac{\partial E_n}{\partial l} = -\frac{2\, E_0}{(k+l)^3}. \tag{B.7.73}$$

The Feynman-Hellmann theorem implies that

$$\left. \frac{\partial E_n}{\partial l} \right|_{k=n-l} = \left\langle \frac{\partial H}{\partial l} \right\rangle, \tag{B.7.74}$$

or

$$-\frac{2\, E_0}{n^3} = \frac{\hbar^2}{2\, m_e} (2\, l + 1) \left\langle \frac{1}{r^2} \right\rangle. \tag{B.7.75}$$

However,

$$E_0 = -\frac{\hbar^2}{2\, m_e\, a_0^2}, \tag{B.7.76}$$

so we obtain

$$\left\langle \frac{1}{r^2} \right\rangle = \frac{1}{(l+1/2)\, n^3\, a_0^2}. \tag{B.7.77}$$

7.9 Kramer's relation evaluated for $k = -1$ yields

$$\left\langle \frac{1}{r^3} \right\rangle = \frac{1}{l\,(l+1)\, a_0} \left\langle \frac{1}{r^2} \right\rangle. \tag{B.7.78}$$

(See Exercise 4.16.) Combining the previous two equations, we obtain

$$\left\langle \frac{1}{r^3} \right\rangle = \frac{1}{l\,(l+1/2)\,(l+1)\, n^3\, a_0^3}. \tag{B.7.79}$$

7.10 We can write

$$E = \left[p^2 c^2 + m^2 c^4\right]^{1/2} = m c^2 \left[1 + \left(\frac{p}{m c}\right)^2\right]^{1/2}. \tag{B.7.80}$$

It follows that, in the limit $p \ll m c$,

$$E \simeq m c^2 \left[1 + \frac{1}{2}\left(\frac{p}{m c}\right)^2 - \frac{1}{8}\left(\frac{p}{m c}\right)^4 + \cdots\right]. \tag{B.7.81}$$

Hence,

$$K = E - m c^2 \simeq \frac{1}{2}\frac{p^2}{m} - \frac{1}{8}\frac{p^4}{m^3 c^2}. \tag{B.7.82}$$

7.11 According to first-order perturbation theory, the lowest-order relativistic correction to the energy of an energy eigenstate of a hydrogen-like atom, characterized by the standard quantum numbers n, l, and m, is given by

$$\Delta E_{nlm} = \langle n, l, m| H_R |n, l, m\rangle = -\frac{1}{8 m_e^3 c^2} \langle n, l, m| p^4 |n, l, m\rangle$$

$$= -\frac{1}{8 m_e^3 c^2} \langle n, l, m| p^2 \, p^2 |n, l, m\rangle. \tag{B.7.83}$$

However, Schrödinger's time-independent equation for the unperturbed eigenstate can be written

$$p^2 |n, l, m\rangle = 2 m_e \left(E_n - V\right)|n, l, m\rangle, \tag{B.7.84}$$

where $V(r)$ is the potential energy, and E_n the energy eigenvalue. Because p^2 is an Hermitian operator, it follows that

$$\Delta E_{nlm} = -\frac{1}{2 m_e c^2} \langle n, l, m| (E_n - V)^2 |n, l, m\rangle$$

$$= -\frac{1}{2 m_e c^2} \left(E_n^2 - 2 E_n \langle n, l, m| V |n, l, m\rangle + \langle n, l, m| V^2 |n, l, m\rangle\right). \tag{B.7.85}$$

Note, incidentally, that the matrix elements of V and V^2 do not couple degenerate eigenstates with the same values of n, but different values of l and m, because V is a function of r only. Hence, there is no danger of singular terms arising to second order in the perturbation expansion.

For a hydrogen atom,

$$V(r) = -\frac{e^2}{4\pi \epsilon_0 \, r}. \tag{B.7.86}$$

Hence, we obtain

$$\Delta E_{nlm} = -\frac{1}{2 m_e c^2}\left(E_n^2 + 2 E_n \frac{e^2}{4\pi \epsilon_0}\left\langle\frac{1}{r}\right\rangle + \left(\frac{e^2}{4\pi \epsilon_0}\right)^2 \left\langle\frac{1}{r^2}\right\rangle\right), \tag{B.7.87}$$

where the expectation value is taken using the $|n, l, m\rangle$ eigenket. However,

$$\frac{E_n}{m_e c^2} = -\frac{\alpha^2}{2 n^2}, \tag{B.7.88}$$

$$\frac{e^2}{4\pi \epsilon_0 E_n} = -2 n^2 a_0, \tag{B.7.89}$$

where a_0 is the Bohr radius. Thus,

$$\Delta E_{nlm} = \frac{\alpha^2 E_n}{n^2} \left[\frac{1}{2} - n^2 \left\langle \frac{a_0}{r} \right\rangle + n^4 \left\langle \frac{a_0^2}{r^2} \right\rangle \right]. \tag{B.7.90}$$

However, according to Exercises 4.16 and 7.8,

$$\left\langle \frac{a_0}{r} \right\rangle = \frac{1}{n^2}, \tag{B.7.91}$$

$$\left\langle \frac{a_0^2}{r^2} \right\rangle = \frac{1}{(l + 1/2) n^3}. \tag{B.7.92}$$

It follows that

$$\Delta E_{nlm} = \frac{\alpha^2 E_n}{n^2} \left(\frac{n}{l + 1/2} - \frac{3}{4} \right). \tag{B.7.93}$$

7.12 According to first-order perturbation theory, the modification to the energy of an energy eigenstate of a hydrogen atom, characterized by the standard quantum numbers n, l, and m, caused by the Darwin term is

$$\Delta E_{nlm} = \langle n, l, m| H_D |n, l, m\rangle = \frac{e^2 \hbar^2}{8 \epsilon_0 m_e^2 c^2} |\psi_{nlm}(0)|^2. \tag{B.7.94}$$

However,

$$|\psi_{nlm}(0)| = \frac{1}{\sqrt{\pi} (n a_0)^{3/2}} \delta_{l0} \delta_{m0}. \tag{B.7.95}$$

Hence, we obtain

$$\Delta E_{nlm} = \frac{e^2 \hbar^2}{8\pi \epsilon_0 m_e^2 c^2 n^3 a_0^3} \delta_{l0} = -\frac{\alpha^2 E_n}{n} \delta_{l0}. \tag{B.7.96}$$

Thus,

$$\Delta E_{nlm} = -\frac{\alpha^2 E_n}{n} \tag{B.7.97}$$

for an $l = 0$ state, and

$$\Delta E_{nlm} = 0 \tag{B.7.98}$$

for an $l > 0$ state.

7.13 According to the analysis of Section 7.7, the modification to the energy of an energy eigenstate of a hydrogen-like atom, characterized by the standard quantum numbers n, l, and m, caused by spin-orbit coupling takes the form

$$\Delta E_{nlm} = \frac{1}{2\,m_e^2\,c^2} \left\langle \frac{1}{r}\frac{dV}{dr} \right\rangle \frac{l\,\hbar^2}{2} \tag{B.7.99}$$

for $j = l + 1/2$, and

$$\Delta E_{nlm} = -\frac{1}{2\,m_e^2\,c^2} \left\langle \frac{1}{r}\frac{dV}{dr} \right\rangle \frac{(l+1)\,\hbar^2}{2} \tag{B.7.100}$$

for $j = l - 1/2$, and

$$\Delta E_{nlm} = 0 \tag{B.7.101}$$

for the special case $l = 0$. Now, for a hydrogen atom, $V = -e^2/(4\pi\,\epsilon_0\,r)$, so we obtain

$$\Delta E_{nlm} = \frac{1}{2\,m_e^2\,c^2} \frac{e^2}{4\pi\,\epsilon_0} \left\langle \frac{1}{r^3} \right\rangle \frac{l\,\hbar^2}{2} \tag{B.7.102}$$

for $j = l + 1/2$, and

$$\Delta E_{nlm} = -\frac{1}{2\,m_e^2\,c^2} \frac{e^2}{4\pi\,\epsilon_0} \left\langle \frac{1}{r^3} \right\rangle \frac{(l+1)\,\hbar^2}{2} \tag{B.7.103}$$

for $j = l - 1/2$, and

$$\Delta E_{nlm} = 0 \tag{B.7.104}$$

for $l = 0$. According to Exercise 7.9,

$$\left\langle \frac{a_0^3}{r^3} \right\rangle = \frac{1}{l\,(l+1/2)\,(l+1)\,n^3}. \tag{B.7.105}$$

It follows that

$$\Delta E_{nlm} = -\frac{\alpha^2\,E_n}{n^2} \left[\frac{n}{2\,(l+1/2)\,(l+1)} \right] \tag{B.7.106}$$

for $j = l + 1/2$, and

$$\Delta E_{nlm} = \frac{\alpha^2\,E_n}{n^2} \left[\frac{n}{2\,(l+1/2)\,l} \right] \tag{B.7.107}$$

for $j = l - 1/2$, and

$$\Delta E_{nlm} = 0 \tag{B.7.108}$$

for $l = 0$. Here, use has been made of Equations (B.7.88) and (B.7.89).

7.14 Adding the energy-shifts calculated for the energy eigenstates of a hydrogen atom in the previous three exercises, we get

$$\Delta E_{nlm} = \frac{\alpha^2\,E_n}{n^2} \left(\frac{n}{l+1} - \frac{3}{4} \right) \tag{B.7.109}$$

for $j = l + 1/2$, and

$$\Delta E_{nlm} = \frac{\alpha^2 E_n}{n^2} \left(\frac{n}{l} - \frac{3}{4} \right) \tag{B.7.110}$$

for $j = l - 1/2$, and

$$\Delta E_{nlm} = \frac{\alpha^2 E_n}{n^2} \left(n - \frac{3}{4} \right) \tag{B.7.111}$$

for the special case $l = 0$ (in which $j = 1/2$). However, the previous three expressions are all equivalent to

$$\Delta E_{nlm} = \frac{\alpha^2 E_n}{n^2} \left(\frac{n}{j + 1/2} - \frac{3}{4} \right). \tag{B.7.112}$$

For the case of an $n = 2$ state of a hydrogen atom, $E_n = E_0/4$, where $E_0 = -13.6\,\text{eV}$. Hence, given that $\alpha \simeq 1/137$, the energy shift of the $j = 3/2$ state is

$$\Delta E = \frac{1}{4} \frac{E_0 \, \alpha^2}{16}, \tag{B.7.113}$$

whereas that of the $j = 1/2$ state is

$$\Delta E = \frac{5}{4} \frac{E_0 \, \alpha^2}{16}. \tag{B.7.114}$$

Thus, the energy of the $j = 3/2$ state exceeds that of the $j = 1/2$ state by

$$\delta E = \frac{|E_0| \, \alpha^2}{16} = 4.5 \times 10^{-5} \, \text{eV}. \tag{B.7.115}$$

7.15 From the previous exercise, the energy splitting of the $n = 2$ states of a hydrogen atom caused by fine structure is

$$\delta E_{FS} = \frac{|E_0| \, \alpha^2}{16}. \tag{B.7.116}$$

On the other hand, according to the analysis of Section 7.6, the energy splitting predicted by the linear Stark effect is

$$\delta E_S = 6 \, e \, a_0 \, |\mathbf{E}|. \tag{B.7.117}$$

We expect the expressions for the Stark energy shifts given in Section 7.6 to hold good when

$$\delta E_S \gg \delta E_{FS}, \tag{B.7.118}$$

or

$$|\mathbf{E}| \gg \frac{|E_0| \, \alpha^2}{96 \, e \, a_0} \sim 1 \times 10^5 \, \text{V m}^{-1}. \tag{B.7.119}$$

7.16 According to Equation (7.115),

$$\mathcal{Y}_{l\,m_j}^{l+1/2} = \left(\frac{l+m_j+1/2}{2l+1}\right)^{1/2} Y_{l\,m_j-1/2}\,\chi_+ + \left(\frac{l-m_j+1/2}{2l+1}\right)^{1/2} Y_{l\,m_j+1/2}\,\chi_-, \quad \text{(B.7.120)}$$

$$\mathcal{Y}_{l\,m_j}^{l-1/2} = -\left(\frac{l-m_j+1/2}{2l+1}\right)^{1/2} Y_{l\,m_j-1/2}\,\chi_+ + \left(\frac{l+m_j+1/2}{2l+1}\right)^{1/2} Y_{l\,m_j+1/2}\,\chi_-.$$
$$\text{(B.7.121)}$$

The previous two equations can be inverted to give

$$Y_{l\,m_j-1/2}\,\chi_+ = \left(\frac{l+m_j+1/2}{2l+1}\right)^{1/2} \mathcal{Y}_{l\,m_j}^{l+1/2} - \left(\frac{l-m_j+1/2}{2l+1}\right)^{1/2} \mathcal{Y}_{l\,m_j}^{l-1/2}, \quad \text{(B.7.122)}$$

$$Y_{l\,m_j+1/2}\,\chi_- = \left(\frac{l-m_j+1/2}{2l+1}\right)^{1/2} \mathcal{Y}_{l\,m_j}^{l+1/2} + \left(\frac{l+m_j+1/2}{2l+1}\right)^{1/2} \mathcal{Y}_{l\,m_j}^{l-1/2}. \quad \text{(B.7.123)}$$

Consolidating these expressions, we obtain

$$Y_{l m_l}\,\chi_\pm = \left(\frac{l\pm m_l+1}{2l+1}\right)^{1/2} \mathcal{Y}_{l\,m_l\pm1/2}^{l+1/2} \mp \left(\frac{l\mp m_l}{2l+1}\right)^{1/2} \mathcal{Y}_{l\,m_l\pm1/2}^{l-1/2}. \quad \text{(B.7.124)}$$

If $l=0$ and $m_l=0$ then Equation (B.7.124) yields

$$Y_{00}\,\chi_\pm = \mathcal{Y}_{0\,\pm1/2}^{1/2}. \quad \text{(B.7.125)}$$

If $l=1$ and $m_l=0$ then we obtain

$$Y_{10}\,\chi_\pm = \sqrt{\frac{2}{3}}\,\mathcal{Y}_{1\,\pm1/2}^{3/2} \mp \sqrt{\frac{1}{3}}\,\mathcal{Y}_{1\,\pm1/2}^{1/2}. \quad \text{(B.7.126)}$$

If $l=1$ and $m_l=\mp1$ then we get

$$Y_{1\mp1}\,\chi_\pm = \sqrt{\frac{1}{3}}\,\mathcal{Y}_{1\,\mp1/2}^{3/2} \mp \sqrt{\frac{2}{3}}\,\mathcal{Y}_{1\,\mp1/2}^{1/2}. \quad \text{(B.7.127)}$$

Finally, if $l=1$ and $m_l=\pm1$ then we obtain

$$Y_{1\pm1}\,\chi_\pm = \mathcal{Y}_{1\,\pm3/2}^{3/2}. \quad \text{(B.7.128)}$$

7.17 We can write the wavefunctions of the unperturbed (by the fine structure Hamiltonian) eigenstates in the form

$$\psi_{1\pm} = \sqrt{\frac{1}{2}}\, R_{20}\, Y_{00}\, \chi_{\pm} - \sqrt{\frac{1}{2}}\, R_{21}\, Y_{10}\, \chi_{\pm}, \tag{B.7.129}$$

$$\psi_{2\pm} = \sqrt{\frac{1}{2}}\, R_{20}\, Y_{00}\, \chi_{\pm} + \sqrt{\frac{1}{2}}\, R_{21}\, Y_{10}\, \chi_{\pm}, \tag{B.7.130}$$

$$\psi_{3\pm} = R_{21}\, Y_{11}\, \chi_{\pm}, \tag{B.7.131}$$

$$\psi_{4\pm} = R_{21}\, Y_{1-1}\, \chi_{\pm}. \tag{B.7.132}$$

Here, the $R_{nl}(r)$ are radial wavefunctions, the $Y_{lm}(\theta,\varphi)$ are spherical harmonics, and the $\chi_{\pm}$ are standard Pauli two-component spinors. However, making use of the results of the previous exercises, we can also write

$$\psi_{1\pm} = \sqrt{\frac{1}{2}}\, R_{20}\, \mathcal{Y}^{1/2}_{0\,\pm 1/2} - \sqrt{\frac{1}{3}}\, R_{21}\, \mathcal{Y}^{3/2}_{1\,\pm 1/2} \pm \sqrt{\frac{1}{6}}\, R_{21}\, \mathcal{Y}^{1/2}_{1\,\pm 1/2}, \tag{B.7.133}$$

$$\psi_{2\pm} = \sqrt{\frac{1}{2}}\, R_{20}\, \mathcal{Y}^{1/2}_{0\,\pm 1/2} + \sqrt{\frac{1}{3}}\, R_{21}\, \mathcal{Y}^{3/2}_{1\,\pm 1/2} \mp \sqrt{\frac{1}{6}}\, R_{21}\, \mathcal{Y}^{1/2}_{1\,\pm 1/2}, \tag{B.7.134}$$

$$\psi_{3+} = R_{21}\, \mathcal{Y}^{3/2}_{1\,3/2}, \tag{B.7.135}$$

$$\psi_{3-} = \sqrt{\frac{1}{3}}\, R_{21}\, \mathcal{Y}^{3/2}_{1\,1/2} + \sqrt{\frac{2}{3}}\, R_{21}\, \mathcal{Y}^{1/2}_{1\,1/2}, \tag{B.7.136}$$

$$\psi_{4+} = \sqrt{\frac{1}{3}}\, R_{21}\, \mathcal{Y}^{3/2}_{1\,-1/2} - \sqrt{\frac{2}{3}}\, R_{21}\, \mathcal{Y}^{1/2}_{1\,-1/2}, \tag{B.7.137}$$

$$\psi_{4-} = R_{21}\, \mathcal{Y}^{3/2}_{1\,-3/2}. \tag{B.7.138}$$

Here, the $\mathcal{Y}^{j}_{l m_j}$ are spin-angular functions.

According to the analysis of Exercise 7.13, the expectation value of H_{LS} is

$$\langle H_{LS}\rangle = -\alpha^2\, E_n \left[\frac{1}{2\,n\,(l+1/2)\,(l+1)}\right] \tag{B.7.139}$$

for an eigenstate whose wavefunction is $R_{nl}\, \mathcal{Y}^{l+1/2}_{l m_j}$. Likewise,

$$\langle H_{LS}\rangle = \alpha^2\, E_n \left[\frac{1}{2\,n\,(l+1/2)\,l}\right] \tag{B.7.140}$$

for an eigenstate whose wavefunction is $R_{nl}\, \mathcal{Y}^{l-1/2}_{l m_j}$. Finally,

$$\langle H_{LS}\rangle = 0 \tag{B.7.141}$$

for the special case of an eigenstate with $l = 0$. Thus, it follows from Equations (B.7.133)–(B.7.138) that

$$\langle H_{LS} \rangle_{1\pm} = 0, \tag{B.7.142}$$

$$\langle H_{LS} \rangle_{2\pm} = 0, \tag{B.7.143}$$

$$\langle H_{LS} \rangle_{3\pm} = \mp \frac{1}{12} \, \alpha^2 \, E_2, \tag{B.7.144}$$

$$\langle H_{LS} \rangle_{4\pm} = \pm \frac{1}{12} \, \alpha^2 \, E_2. \tag{B.7.145}$$

According to the analysis of Exercise 7.11, the expectation value of H_R is

$$\langle H_R \rangle = \frac{\alpha^2 \, E_n}{n^2} \left(\frac{n}{l + 1/2} - \frac{3}{4} \right) \tag{B.7.146}$$

for an eigenstate whose wavefunction is $R_{nl} \, Y_{lm}$. It follows from Equations (B.7.129)–(B.7.132) that

$$\langle H_R \rangle_{1\pm} = \frac{23}{48} \, \alpha^2 \, E_2, \tag{B.7.147}$$

$$\langle H_R \rangle_{2\pm} = \frac{23}{48} \, \alpha^2 \, E_2, \tag{B.7.148}$$

$$\langle H_R \rangle_{3\pm} = \frac{7}{48} \, \alpha^2 \, E_2, \tag{B.7.149}$$

$$\langle H_R \rangle_{4\pm} = \frac{7}{48} \, \alpha^2 \, E_2. \tag{B.7.150}$$

Finally, according to the analysis of Exercise 7.12, the expectation value of H_D is

$$\langle H_D \rangle = -\frac{\alpha^2 \, E_n}{n} \tag{B.7.151}$$

for an eigenstate wavefunction is $R_{n0} \, Y_{00}$, and

$$\langle H_D \rangle = 0 \tag{B.7.152}$$

for an $l > 0$ state. It follows from Equations (B.7.129)–(B.7.132) that

$$\langle H_D \rangle_{1\pm} = -\frac{1}{4} \, \alpha^2 \, E_2, \tag{B.7.153}$$

$$\langle H_D \rangle_{2\pm} = -\frac{1}{4} \, \alpha^2 \, E_2, \tag{B.7.154}$$

$$\langle H_D \rangle_{3\pm} = 0, \tag{B.7.155}$$

$$\langle H_D \rangle_{4\pm} = 0. \tag{B.7.156}$$

Thus, we deduce that

$$\langle H_{FS}\rangle_{1\pm} = \frac{11}{48}\,\alpha^2\,E_2, \tag{B.7.157}$$

$$\langle H_{FS}\rangle_{2\pm} = \frac{11}{48}\,\alpha^2\,E_2, \tag{B.7.158}$$

$$\langle H_{FS}\rangle_{3\pm} = \frac{7}{48}\,\alpha^2\,E_2 \mp \frac{4}{48}\,\alpha^2\,E_2, \tag{B.7.159}$$

$$\langle H_{FS}\rangle_{4\pm} = \frac{7}{48}\,\alpha^2\,E_2 \pm \frac{4}{48}\,\alpha^2\,E_2. \tag{B.7.160}$$

Hence, the perturbed energies of the various eigenstates are

$$E_{1\pm} = E_2\left(1 + \frac{11}{48}\,\alpha^2\right) + 3\,e\,a_0\,|\mathbf{E}|, \tag{B.7.161}$$

$$E_{2\pm} = E_2\left(1 + \frac{11}{48}\,\alpha^2\right) - 3\,e\,a_0\,|\mathbf{E}|, \tag{B.7.162}$$

$$E_{3-} = E_2\left(1 + \frac{11}{48}\,\alpha^2\right), \tag{B.7.163}$$

$$E_{4+} = E_2\left(1 + \frac{11}{48}\,\alpha^2\right), \tag{B.7.164}$$

$$E_{3+} = E_2\left(1 + \frac{11}{48}\,\alpha^2\right) - \frac{1}{6}\,\alpha^2\,E_2, \tag{B.7.165}$$

$$E_{4-} = E_2\left(1 + \frac{11}{48}\,\alpha^2\right) - \frac{1}{6}\,\alpha^2\,E_2. \tag{B.7.166}$$

7.18 We can write

$$H = H_0 + H_B + H_{LS} + H_R + H_D, \tag{B.7.167}$$

where H_0 is the unperturbed Hamiltonian, and H_B the magnetic Hamiltonian. The simultaneous eigenstates of H_0 and H_B are characterized by the standard quantum numbers n, l, m_l, and m_s. According to Section 7.8,

$$\langle H_B\rangle = \mu_B\,B\,(m_l + 2\,m_s). \tag{B.7.168}$$

Moreover, making use of the analysis of Exercise 7.13,

$$\langle H_{LS}\rangle = -\alpha^2\,E_n\left[\frac{m_l\,m_s}{n\,l\,(l+1/2)\,(l+1)}\right]. \tag{B.7.169}$$

However,

$$\langle H_{LS}\rangle = 0. \tag{B.7.170}$$

for the special case $l = 0$. According to the previous exercise,

$$\langle H_R \rangle = \frac{\alpha^2 E_n}{n^2} \left(\frac{n}{l + 1/2} - \frac{3}{4} \right). \tag{B.7.171}$$

Finally,

$$\langle H_D \rangle = -\frac{\alpha^2 E_n}{n} \tag{B.7.172}$$

for an $l = 0$ state, and

$$\langle H_D \rangle = 0 \tag{B.7.173}$$

otherwise. Of course, $\langle H_0 \rangle = E_n$. It follows that the energies of the various $n = 2$ states are

$$E_{2,1,1,1/2} = E_2 \left(1 + \frac{5}{16} \alpha^2 \right) + 2\mu_B B - \frac{1}{3} \alpha^2 E_2, \tag{B.7.174}$$

$$E_{2,1,0,1/2} = E_2 \left(1 + \frac{5}{16} \alpha^2 \right) + \mu_B B - \frac{1}{6} \alpha^2 E_2, \tag{B.7.175}$$

$$E_{2,0,0,1/2} = E_2 \left(1 + \frac{5}{16} \alpha^2 \right) + \mu_B B, \tag{B.7.176}$$

$$E_{2,1,1,-1/2} = E_2 \left(1 + \frac{5}{16} \alpha^2 \right), \tag{B.7.177}$$

$$E_{2,1,-1,1/2} = E_2 \left(1 + \frac{5}{16} \alpha^2 \right), \tag{B.7.178}$$

$$E_{2,0,0,-1/2} = E_2 \left(1 + \frac{5}{16} \alpha^2 \right) - \mu_B B, \tag{B.7.179}$$

$$E_{2,1,0,-1/2} = E_2 \left(1 + \frac{5}{16} \alpha^2 \right) - \mu_B B - \frac{1}{6} \alpha^2 E_2, \tag{B.7.180}$$

$$E_{2,1,-1,-1/2} = E_2 \left(1 + \frac{5}{16} \alpha^2 \right) - 2\mu_B B - \frac{1}{3} \alpha^2 E_2. \tag{B.7.181}$$

7.19 According to the previous exercise, the energy splitting associated with the strong field Zeeman effect is

$$\delta E_Z = \mu_B B. \tag{B.7.182}$$

On the other hand, the energy splitting associated with fine structure is of order

$$\delta E_{FS} \simeq \alpha^2 |E_0|, \tag{B.7.183}$$

where E_0 is the hydrogen ground-state energy. The Paschen-Back limit holds good when

$$\delta E_Z \gg \delta E_{FS}, \tag{B.7.184}$$

or

$$B \gg \frac{\alpha^2 |E_0|}{\mu_B} \sim \alpha^2 \frac{m_e e}{\epsilon_0 h a_0} \simeq 25 \text{ tesla}. \tag{B.7.185}$$

B.8 Chapter 8

8.1 Equations (8.11)–(8.13) yield

$$i \frac{dc_1}{dt} = \frac{e_{11}}{\hbar} c_1 + \frac{1}{2} \gamma \exp\left[+i (\omega - \omega_{21})\right] c_2, \tag{B.8.1}$$

$$i \frac{dc_2}{dt} = \frac{e_{22}}{\hbar} c_2 + \frac{1}{2} \gamma \exp\left[-i (\omega - \omega_{21})\right] c_1, \tag{B.8.2}$$

where $\omega_{21} = (E_2 - E_1)/\hbar$. Multiplying the first and second equations by $\exp(i\, e_{11}\, t/\hbar)$ and $\exp(i\, e_{22}\, t/\hbar)$, respectively, we obtain

$$i \frac{d\hat{c}_1}{dt} = \frac{\gamma}{2} \exp\left[+i (\omega - \hat{\omega}_{21}) t\right] \hat{c}_2, \tag{B.8.3}$$

$$i \frac{d\hat{c}_2}{dt} = \frac{\gamma}{2} \exp\left[-i (\omega - \hat{\omega}_{21}) t\right] \hat{c}_1, \tag{B.8.4}$$

where $\hat{c}_1 = c_1 \exp(i\, e_{11}\, t/\hbar)$, $\hat{c}_2 = c_2 \exp(i\, e_{22}\, t/\hbar)$, and

$$\hat{\omega}_{21} = \frac{E_2 + e_{22} - E_1 - e_{11}}{\hbar}. \tag{B.8.5}$$

However, the previous two equations are identical in form to Equations (8.18) and (8.19). Hence, by analogy with the analysis of Section 8.3, the solution which is such that $c_2 = 0$ at $t = 0$ is

$$\hat{c}_2(t) = \frac{-i\gamma}{[\gamma^2 + (\omega - \hat{\omega}_{21})^2]^{1/2}} \exp\left[-i (\omega - \hat{\omega}_{21}) \frac{t}{2}\right]$$
$$\sin\left(\left[\gamma^2 + (\omega - \hat{\omega}_{21})^2\right]^{1/2} \frac{t}{2}\right). \tag{B.8.6}$$

The probability of finding the system in state 2 is $P_2(t) = |c_2|^2 = |\hat{c}_2|^2$. Hence, we obtain

$$P_2(t) = \frac{\gamma^2}{\gamma^2 + (\omega - \hat{\omega}_{21})^2} \sin^2\left(\left[\gamma^2 + (\omega - \hat{\omega}_{21})^2\right]^{1/2} \frac{t}{2}\right). \tag{B.8.7}$$

8.2 (a) Reusing the analysis of Section 8.4, we find that

$$H_0 = -\omega_0 S_z, \tag{B.8.8}$$

$$H_1 = -\frac{\gamma}{2} \left[e^{-i\omega t} S^+ + e^{-i\omega t} S^-\right], \tag{B.8.9}$$

$$E_m = -m \hbar \omega_0, \tag{B.8.10}$$

where $\omega_0 = g\,\mu_N\,B_0/\hbar$ and $\omega_1 = g\,\mu_N\,B_1/\hbar$. Thus,

$$\langle s, m | H_1 | s, m' \rangle\, e^{i\omega_{mm'} t} = -\frac{\gamma\hbar}{2}\, e^{-i(\omega-\omega_0)t}\, [s(s+1) - m'(m'+1)]^{1/2}\, \delta_{m\,m'+1}$$

$$-\frac{\gamma\hbar}{2}\, e^{i(\omega-\omega_0)t}\, [s(s+1) - m'(m'-1)]^{1/2}\, \delta_{m\,m'-1},$$

$$\text{(B.8.11)}$$

where $\omega_{mm'} = (E_m - E_{m'})/\hbar$, and use has been made of the well-known properties of the raising and lowering operators, as well as the orthonormality of the spin eigenkets. Thus, it follows from Equation (8.11) that

$$\frac{dc_m}{dt} = \frac{i\gamma}{2}\Big([s(s+1) - m(m-1)]^{1/2}\, e^{i(\omega-\omega_0)t}\, c_{m-1}$$

$$+ [s(s+1) - m(m+1)]^{1/2}\, e^{-i(\omega-\omega_0)t}\, c_{m+1}\Big) \quad \text{(B.8.12)}$$

for $-s \le m \le s$.

(b) For $s = 1/2$ and $\omega = \omega_0$, we have

$$\frac{dc_{1/2}}{dt} = \frac{i\gamma}{2}\, c_{-1/2}, \tag{B.8.13}$$

$$\frac{dc_{-1/2}}{dt} = \frac{i\gamma}{2}\, c_{1/2}. \tag{B.8.14}$$

By inspection, the solution that satisfies $c_{1/2}(0) = 1$ is

$$c_{1/2}(t) = \cos(\gamma t/2), \tag{B.8.15}$$

$$c_{-1/2}(t) = i\,\sin(\gamma t/2). \tag{B.8.16}$$

(c) For $s = 1$ and $\omega = \omega_0$, we have

$$\frac{dc_1}{dt} = \frac{i\gamma}{\sqrt{2}}\, c_0, \tag{B.8.17}$$

$$\frac{dc_0}{dt} = \frac{i\gamma}{\sqrt{2}}\, (c_1 + c_{-1}), \tag{B.8.18}$$

$$\frac{dc_{-1}}{dt} = \frac{i\gamma}{\sqrt{2}}\, c_0. \tag{B.8.19}$$

However, it can be seen that if $c_1 = (c_{1/2})^2$, $c_0 = \sqrt{2}\, c_{1/2}\, c_{-1/2}$, and $c_{-1} = (c_{-1/2})^2$ then the previous set of equations reduce to Equations (B.8.13)–(B.8.14). Hence,

$$c_1(t) = \cos^2(\gamma t/2), \tag{B.8.20}$$

$$c_0(t) = i\,\sqrt{2}\,\cos(\gamma t/2)\,\sin(\gamma t/2), \tag{B.8.21}$$

$$c_{-1}(t) = -\sin^2(\gamma t/2). \tag{B.8.22}$$

(d) For $s = 3/2$ and $\omega = \omega_0$, we have

$$\frac{dc_{3/2}}{dt} = \frac{i\gamma}{2}\sqrt{3}\,c_{1/2}, \tag{B.8.23}$$

$$\frac{dc_{1/2}}{dt} = \frac{i\gamma}{2}\left(c_{-1/2} + \sqrt{3}\,c_{3/2}\right), \tag{B.8.24}$$

$$\frac{dc_{-1/2}}{dt} = \frac{i\gamma}{2}\left(\sqrt{3}\,c_{-3/2} + c_{1/2}\right), \tag{B.8.25}$$

$$\frac{dc_{-3/2}}{dt} = \frac{i\gamma}{2}\sqrt{3}\,c_{-1/2}. \tag{B.8.26}$$

However, it can be seen that if $c_{3/2} = (c_{1/2})^3$, $c_{1/2} = \sqrt{3}\,(c_{1/2})^2\,c_{-1/2}$, $c_{-1/2} = \sqrt{3}\,c_{1/2}\,(c_{-1/2})^2$, and $c_{-3/2} = (c_{-1/2})^3$ then the previous set of equations reduce to Equations (B.8.13)–(B.8.14). Hence,

$$c_{3/2}(t) = \cos^3(\gamma t/2), \tag{B.8.27}$$

$$c_{1/2}(t) = i\sqrt{3}\,\cos(\gamma t/2)\,\sin^2(\gamma t/2), \tag{B.8.28}$$

$$c_{-1/2}(t) = -\sqrt{3}\,\cos^2(\gamma t/2)\,\sin(\gamma t/2), \tag{B.8.29}$$

$$c_{-3/2}(t) = -i\,\sin^3(\gamma t/2). \tag{B.8.30}$$

At this stage, the generalization to higher spin systems is obvious.

8.3 We have

$$i\hbar\,\frac{\partial T}{\partial t} = (H_0 + H_1)\,T, \tag{B.8.31}$$

and

$$T = \exp\left[\frac{-i\,H_0\,(t-t_0)}{\hbar}\right]T_I. \tag{B.8.32}$$

Differentiation of the previous equation yields

$$i\hbar\,\frac{\partial T}{\partial t} = H_0\,\exp\left[\frac{-i\,H_0\,(t-t_0)}{\hbar}\right]T_I + i\hbar\,\exp\left[\frac{-i\,H_0\,(t-t_0)}{\hbar}\right]\frac{\partial T_I}{\partial t}. \tag{B.8.33}$$

However, Equations (B.8.31) and (B.8.32) give

$$i\hbar\,\frac{\partial T}{\partial t} = (H_0 + H_1)\,\exp\left[\frac{-i\,H_0\,(t-t_0)}{\hbar}\right]T_I. \tag{B.8.34}$$

Combining the previous two equations, we obtain

$$i\hbar\,\exp\left[\frac{-i\,H_0\,(t-t_0)}{\hbar}\right]\frac{\partial T_I}{\partial t} = H_1\,\exp\left[\frac{-i\,H_0\,(t-t_0)}{\hbar}\right]T_I, \tag{B.8.35}$$

which yields

$$i\hbar\,\frac{\partial T_I}{\partial t} = H_I\,T_I, \tag{B.8.36}$$

where

$$H_I = \exp\left[\frac{i\,H_0\,(t - t_0)}{\hbar}\right] H_1 \exp\left[\frac{-i\,H_0\,(t - t_0)}{\hbar}\right]. \tag{B.8.37}$$

8.4 Let us take the limit $t \to \infty$ and $\omega_{fi} = (E_f - E_i)/\hbar \to 0$ such that $x = \omega_{fi}\,t/2$ remains finite. Equation (8.82) becomes

$$c_f(t) = \frac{i\,t}{\hbar}\,e^{i\,x}\left[\left(H_{fi} + \sum_m \frac{H_{fm}\,H_{mi}}{E_m - E_i}\right) \text{sinc}(x)\right.$$
$$\left. - \sum_m \frac{H_{fm}\,H_{mi}}{E_m - E_i}\exp\left(\frac{i\,\omega_{im}\,t}{2}\right)\text{sinc}\left(\frac{\omega_{im}\,t}{2}\right)\right]_{E_f = E_i}. \tag{B.8.38}$$

However, if $E_m \neq E_i$ then $(\omega_{im}\,t/2) \to \infty$ and $\text{sinc}(\omega_{im}\,t/2) \to 0$. Hence,

$$c_f(t) = \frac{i\,t}{\hbar}\left(H_{fi} + \sum_m \frac{H_{fm}\,H_{mi}}{E_m - E_i}\right)_{E_f = E_i} e^{i\,x}\,\text{sinc}(x), \tag{B.8.39}$$

which implies that

$$P_{i \to f}(t) = |c_f(t)|^2 = \frac{t^2}{\hbar^2}\left|H_{fi} + \sum_m \frac{H_{fm}\,H_{mi}}{E_m - E_i}\right|^2_{E_f = E_i} \text{sinc}^2(x). \tag{B.8.40}$$

Thus,

$$P_{i \to [f]} = \int_{-\infty}^{\infty} dE_f\, P_{i \to f}\,\rho(E_f)$$
$$= \frac{2\,t}{\hbar}\left|H_{fi} + \sum_m \frac{H_{fm}\,H_{mi}}{E_m - E_i}\right|^2 \rho(E_f)\Bigg|_{E_f = E_i} \int_{-\infty}^{\infty} dx\,\text{sinc}^2(x), \tag{B.8.41}$$

which yields

$$P_{i \to [f]} = \frac{2\pi\,t}{\hbar}\overline{\left|H_{fi} + \sum_m \frac{H_{fm}\,H_{mi}}{E_m - E_i}\right|^2 \rho(E_f)}\Bigg|_{E_f = E_i}, \tag{B.8.42}$$

because $\int_{-\infty}^{\infty} dx\,\text{sinc}^2(x) = \pi$. Finally,

$$w_{i \to [f]} = \frac{dP_{i \to [f]}}{dt} = \frac{2\pi}{\hbar}\overline{\left|H_{fi} + \sum_m \frac{H_{fm}\,H_{mi}}{E_m - E_i}\right|^2 \rho(E_f)}\Bigg|_{E_f = E_i}. \tag{B.8.43}$$

8.5 The mean lifetime of the state is

$$\langle t \rangle = \frac{\int_0^\infty dt\, t\, P_{i\to i}(t)}{\int_0^\infty dt\, P_{i\to i}(t)} = \tau_i \frac{I_1}{I_0} = \tau_i, \tag{B.8.44}$$

where $I_n = n!$ is defined in Exercise 4.11.

8.6 We can write

$$\mathbf{p}\cdot\mathbf{A} - \mathbf{A}\cdot\mathbf{p} = p_i A_i - A_i p_i = [p_i, A_i] = -i\hbar\frac{\partial A_i}{\partial x_i} = -i\hbar\,\nabla\cdot\mathbf{A}, \tag{B.8.45}$$

where use has been made of Equation (3.34). It follows that if $\nabla\cdot\mathbf{A} = 0$ then

$$\mathbf{p}\cdot\mathbf{A} = \mathbf{A}\cdot\mathbf{p}. \tag{B.8.46}$$

The Hamiltonian in question takes the form

$$H = \frac{\mathbf{p}\cdot\mathbf{p}}{2m_e} + \frac{e\mathbf{A}\cdot\mathbf{p}}{m_e} + \frac{e^2\mathbf{A}\cdot\mathbf{A}}{2m_e} - e\phi + \Phi(r). \tag{B.8.47}$$

It follows that

$$H^\dagger = \frac{\mathbf{p}^\dagger\cdot\mathbf{p}^\dagger}{2m_e} + \frac{e\,\mathbf{p}^\dagger\cdot\mathbf{A}^\dagger}{m_e} + \frac{e^2\mathbf{A}^\dagger\cdot\mathbf{A}^\dagger}{2m_e} - e\phi^\dagger + \Phi^\dagger(r). \tag{B.8.48}$$

However, $\mathbf{p}^\dagger = \mathbf{p}$. Moreover, $\mathbf{A}^\dagger = \mathbf{A}$, $\phi^\dagger = \phi$, and $\Phi^\dagger = \Phi$, because $\mathbf{A}$, ϕ, and Φ are all real functions of the Hermitian operator $\mathbf{x}$. Hence,

$$H^\dagger = \frac{\mathbf{p}\cdot\mathbf{p}}{2m_e} + \frac{e\mathbf{p}\cdot\mathbf{A}}{m_e} + \frac{e^2\mathbf{A}\cdot\mathbf{A}}{2m_e} - e\phi + \Phi(r). \tag{B.8.49}$$

Equations (B.8.46), (B.8.47), and (B.8.49) imply that $H^\dagger = H$. In other words, the Hamiltonian (B.8.47) is Hermitian.

8.7 Without loss of generality, we can say that $\mathbf{d}_{if}$ is orientated along the z-axis, so that $\mathbf{d}_{if} = d_{if}\,\mathbf{e}_z$. The Cartesian components of ϵ are $(\sin\theta\cos\varphi, \sin\theta\sin\varphi, \cos\theta)$, where the polar angles θ and φ parameterize the direction of polarization of the incident radiation. Thus,

$$|\epsilon\cdot\mathbf{d}_{if}|^2 = |d_{if}|^2\cos^2\theta. \tag{B.8.50}$$

Averaging over all possible polarization directions (which is equivalent to averaging over all possible directions of the incident radiation), we obtain

$$\left\langle|\epsilon\cdot\mathbf{d}_{if}|^2\right\rangle = \frac{1}{4\pi}\oint d\Omega\,|\epsilon\cdot\mathbf{d}_{if}|^2 = \frac{|d_{if}|^2}{2}\int_0^\pi d\theta\sin\theta\cos^2\theta = \frac{|d_{if}|^2}{3}. \tag{B.8.51}$$

However, $|d_{if}|^2 = \mathbf{d}_{if}^*\cdot\mathbf{d}_{if}$.

8.8 According to Equation (8.165),

$$\oint d\Omega \, \frac{dw_{i\to f}^{\text{spn}}}{d\Omega} = \frac{\alpha \, \omega_{if}^2}{2\pi \, c^2} \oint d\Omega \sum_{j=1,2} |\boldsymbol{\epsilon}_j \cdot \mathbf{d}_{if}|^2. \tag{B.8.52}$$

However,

$$\mathbf{n} = (\sin\theta \cos\varphi, \, \sin\theta \sin\varphi, \, \cos\theta), \tag{B.8.53}$$

$$\boldsymbol{\epsilon}_1 = (\cos\theta \cos\varphi, \, \cos\theta \sin\varphi, \, -\sin\theta), \tag{B.8.54}$$

$$\boldsymbol{\epsilon}_2 = (-\sin\varphi, \, \cos\varphi, \, 0), \tag{B.8.55}$$

where $d\Omega = \sin\theta \, d\theta \, d\varphi$. Let $\mathbf{d}_{if} = d_x \, \mathbf{e}_x + d_y \, \mathbf{e}_y + d_z \, \mathbf{e}_z$. It follows that

$$\begin{aligned} |\boldsymbol{\epsilon}_1 \cdot \mathbf{d}_{if}|^2 = {}& \cos^2\theta \cos^2\varphi |d_x|^2 + \cos^2\theta \sin^2\varphi |d_y|^2 + \sin^2\theta |d_z|^2 \\ & + \cos^2\theta \cos\varphi \sin\varphi \, (d_x \, d_y^* + d_x^* \, d_y) \\ & - \cos\theta \sin\theta \cos\varphi \, (d_x \, d_z^* + d_x^* \, d_z) \\ & - \cos\theta \sin\theta \sin\varphi \, (d_y \, d_z^* + d_y^* \, d_z), \end{aligned} \tag{B.8.56}$$

and

$$|\boldsymbol{\epsilon}_2 \cdot \mathbf{d}_{if}|^2 = \sin^2\varphi |d_x|^2 + \cos^2\varphi |d_y|^2 - \cos\varphi \sin\varphi \, (d_x \, d_y^* + d_y^* \, d_x). \tag{B.8.57}$$

Hence,

$$\begin{aligned} \oint d\Omega \, |\boldsymbol{\epsilon}_1 \cdot \mathbf{d}_{if}|^2 &= \int_0^{2\pi} d\varphi \int_0^{\pi} d\theta \, \sin\theta \, |\boldsymbol{\epsilon}_1 \cdot \mathbf{d}_{if}|^2 \\ &= \frac{2\pi}{3} \left(|d_x|^2 + |d_y|^2 + 4 \, |d_z|^2 \right), \end{aligned} \tag{B.8.58}$$

$$\oint d\Omega \, |\boldsymbol{\epsilon}_2 \cdot \mathbf{d}_{if}|^2 = \int_0^{2\pi} d\varphi \int_0^{\pi} d\theta \, \sin\theta \, |\boldsymbol{\epsilon}_2 \cdot \mathbf{d}_{if}|^2 = 2\pi \left(|d_x|^2 + |d_y|^2 \right), \tag{B.8.59}$$

which implies that

$$\oint d\Omega \sum_{j=1,2} |\boldsymbol{\epsilon}_j \cdot \mathbf{d}_{if}|^2 = \frac{8\pi}{3} \left(|d_x|^2 + |d_y|^2 + |d_z|^2 \right) = \frac{8\pi}{3} \, |d_{if}|^2. \tag{B.8.60}$$

Thus,

$$\oint d\Omega \, \frac{dw_{i\to f}^{\text{spn}}}{d\Omega} = \frac{4 \, \alpha \, \omega_{if}^2 \, |d_{if}|^2}{3 \, c^2}. \tag{B.8.61}$$

8.9 The general matrix element that mediates a spontaneous transition between an initial state i and a final state f is written

$$\boldsymbol{\epsilon} \cdot \mathbf{d}_{if} = \frac{-\mathrm{i}}{m_e\, \omega_{if}} \left\langle i \left| \exp\left[\mathrm{i}\left(\frac{\omega}{c}\right) \mathbf{n} \cdot \mathbf{x} \right] \boldsymbol{\epsilon} \cdot \mathbf{p} \right| f \right\rangle. \tag{B.8.62}$$

Here, $\mathbf{n}$ is the normalized wavevector of the emitted photon, and $\boldsymbol{\epsilon}$ is its electric polarization vector. Moreover, $\boldsymbol{\epsilon} \cdot \mathbf{n} = 0$. If states i and f possess zero orbital angular momentum then they are both $l = 0$ states. Hence, their wavefunctions are functions of r only, and are independent of the polar and azimuthal angles, θ and φ, respectively. Using the Schrödinger representation, we can write $\mathbf{p} = -\mathrm{i}\hbar\nabla$. However, $\nabla f(r) = f'(r)\,\mathbf{x}/r$. Hence, the angular part of the integral appearing on the right-hand side of the previous equation is

$$\oint d\Omega\, \exp\left[\mathrm{i}\left(\frac{\omega}{c}\right) \mathbf{n} \cdot \mathbf{x} \right] \frac{\boldsymbol{\epsilon} \cdot \mathbf{x}}{r}, \tag{B.8.63}$$

where $d\Omega = \sin\theta\, d\theta\, d\varphi$. Without loss of generality, we can say that $\mathbf{n} = \mathbf{e}_z$. It follows that $\boldsymbol{\epsilon} = \cos\alpha\,\mathbf{e}_x + \sin\alpha\,\mathbf{e}_y$. Moreover, $\mathbf{x} = r\,(\sin\theta\cos\varphi, \sin\theta\sin\varphi, \cos\theta)$. Hence, the previous integral becomes

$$\int_0^\pi d\theta\, \sin^2\theta\, \exp\left[\mathrm{i}\left(\frac{\omega}{c}\right) r \cos\theta \right] \oint d\varphi\, \cos(\varphi - \alpha) = 0. \tag{B.8.64}$$

Thus, we deduce that the general matrix element $\boldsymbol{\epsilon} \cdot \mathbf{d}_{if}$ is zero for a spontaneous transition between two $l = 0$ states, which implies that such a transition is absolutely forbidden.

8.10 We can write

$$[L_z, x] = -y\,[p_x, x] = \mathrm{i}\hbar\, y, \tag{B.8.65}$$

$$[L_z, y] = x\,[p_y, y] = -\mathrm{i}\hbar\, x, \tag{B.8.66}$$

where use has been made of the results of Section 4.1. It follows that

$$\langle n, l, m | [L_z, x] | n', l', m' \rangle = \mathrm{i}\hbar\, \langle n, l, m | y | n', l', m' \rangle, \tag{B.8.67}$$

$$\langle n, l, m | [L_z, y] | n', l', m' \rangle = -\mathrm{i}\hbar\, \langle n, l, m | x | n', l', m' \rangle. \tag{B.8.68}$$

The previous expressions reduce to

$$(m - m')\, \langle n, l, m | x | n', l', m' \rangle = \mathrm{i}\, \langle n, l, m | y | n', l', m' \rangle, \tag{B.8.69}$$

$$(m - m')\, \langle n, l, m | y | n', l', m' \rangle = -\mathrm{i}\, \langle n, l, m | x | n', l', m' \rangle. \tag{B.8.70}$$

It follows that

$$(m - m' + 1)(m - m' - 1)\langle n, l, m| x |n', l', m'\rangle = 0, \qquad (B.8.71)$$

$$(m - m' + 1)(m - m' - 1)\langle n, l, m| y |n', l', m'\rangle = 0. \qquad (B.8.72)$$

Hence, we deduce that the matrix elements $\langle n, l, m| x |n', l', m'\rangle$ and $\langle n, l, m| y |n', l', m'\rangle$ are zero unless $m' = m \pm 1$.

According to Equation (7.52),

$$L^4 z - 2 L^2 z L^2 + z L^4 - 2\hbar^2 (L^2 z + z L^2) = 0, \qquad (B.8.73)$$

where $L^2 = L_x^2 + L_y^2 + L_z^2$. Cyclic permutations of the form $x, y, z \to y, z, x$ leave L^2 invariant, but transform the previous equation into

$$L^4 x - 2 L^2 x L^2 + x L^4 - 2\hbar^2 (L^2 x + x L^2) = 0, \qquad (B.8.74)$$

and

$$L^4 y - 2 L^2 y L^2 + y L^4 - 2\hbar^2 (L^2 y + t L^2) = 0. \qquad (B.8.75)$$

Equation (B.8.74) yields

$$\langle n, l, m| L^4 x - 2 L^2 x L^2 + x L^4 - 2\hbar^2 (L^2 x + x L^2) |n', l', m'\rangle = 0, \qquad (B.8.76)$$

which reduces to

$$(l + l' + 2)(l + l')(l - l' + 1)(l - l' - 1)\langle n, l, m| x |n', l', m'\rangle = 0. \qquad (B.8.77)$$

Hence, the matrix element $\langle n, l, m| x |n', l', m'\rangle$ is zero unless $l = l' = 0$ or $l' = l \pm 1$. However, if $l = l' = 0$ then $m' = m$. However, we have already seen that the matrix element is zero in this case. Hence, the matrix element $\langle n, l, m| x |n', l', m'\rangle$ is zero unless $l' = l \pm 1$. Similar arguments reveal that the matrix element $\langle n, l, m| y |n', l', m'\rangle$ is zero unless $l' = l \pm 1$.

8.11   The Hamiltonian is

$$H = \frac{p^2}{2 m_e} + V(r). \qquad (B.8.78)$$

Thus, it follows from Equation (3.33) that

$$[H, x_i] = -i\hbar \frac{\partial H}{\partial p_i} = -i \frac{\hbar p_i}{m_e}. \qquad (B.8.79)$$

Hence,

$$[[H, x_i], x_i] = -i \frac{\hbar}{m_e} [p_i, x_i] = -\frac{\hbar^2}{m_e}, \qquad (B.8.80)$$

where use has been made of the fundamental commutation relation $[x_i, p_i] = i\hbar$. It follows that

$$\langle n| [[H, x_i], x_i] |n\rangle = -\frac{\hbar^2}{m_e}, \qquad (B.8.81)$$

where use has been made of the orthonormality constraint $\langle n|m \rangle = \delta_{nm}$. However,

$$[[H, x_i], x_i] = H x_i^2 - 2 x_i H x_i + x_i^2 H, \tag{B.8.82}$$

which implies that

$$\langle n| [[H, x_i], x_i] |n \rangle = 2 E_n \langle n| x_i^2 |n \rangle - 2 \langle n| x_i H x_i |n \rangle. \tag{B.8.83}$$

Equations (B.8.81) and (B.8.83) yield

$$\langle n| x_i (H - E_n) x_i |n \rangle = \frac{\hbar^2}{2 m_e}. \tag{B.8.84}$$

Now, because the energy eigenkets span ket space, we have

$$\sum_m |m \rangle \langle m| = 1, \tag{B.8.85}$$

where the sum is over all eigenkets. It follows that

$$
\begin{aligned}
\langle n| x_i (H - E_n) x_i |n \rangle &= \sum_m \langle n| x_i |m \rangle \langle m| (H - E_n) x_i |n \rangle \\
&= \sum_m (E_m - E_n) \langle n| x_i |m \rangle \langle m| x_i |n \rangle \\
&= \sum_m (E_m - E_n) |\langle n| x_i |m \rangle|^2. \tag{B.8.86}
\end{aligned}
$$

Thus,

$$\sum_m (E_m - E_n) |\langle n| x_i |m \rangle|^2 = \frac{\hbar^2}{2 m_e}. \tag{B.8.87}$$

Let

$$F_{nm} = \frac{2 m_e \, \omega_{mn}}{\hbar} |\langle n| x_i |m \rangle|^2, \tag{B.8.88}$$

where $\omega_{mn} = (E_m - E_n)/\hbar$. The previous two equations can be combined to give

$$\sum_m F_{nm} = 1. \tag{B.8.89}$$

8.12 Let $|n, l, m \rangle$ be a properly normalized energy eigenket of the hydrogen atom characterized by the conventional quantum numbers n, l, and m. The corresponding wavefunction is written $\psi_{nlm}(r, \theta, \varphi) = R_{nl}(r) \, Y_{lm}(\theta, \varphi)$.

(a) According to the selection rules derived in Exercise 8.10, the matrix element $\langle n, l, m| x |n', l', m' \rangle$ is zero unless $l' = l \pm 1$ and $m' = m \pm 1$. Thus, the only possible non-zero $1s \to 2p$ matrix elements are $\langle 1, 0, 0 | x | 2, 1, \pm 1 \rangle$. Now, $x = r \sin\theta \cos\varphi$. Hence,

$$\langle 1, 0, 0 | x | 2, 1, \pm 1 \rangle = \int_0^\infty dr \, r^3 R_{10}(r) R_{21}(r) \oint d\Omega \sin\theta \cos\varphi \, Y_{00}^* Y_{1\pm 1}. \tag{B.8.90}$$

According to Section 4.4,

$$\oint d\Omega \sin\theta \cos\varphi \, Y_{00}^* Y_{1\pm 1} = \mp \sqrt{\frac{3}{2}} \frac{1}{4\pi} \int_0^{2\pi} d\varphi \, \cos\varphi \, (\cos\varphi \pm i \sin\varphi)$$
$$\int_0^\pi d\theta \, \sin^3\theta, \tag{B.8.91}$$

which reduces to

$$\oint d\Omega \sin\theta \cos\varphi \, Y_{00}^* Y_{1\pm 1} = \mp \sqrt{\frac{1}{6}}. \tag{B.8.92}$$

It follows from Exercise 4.12 that

$$\int_0^\infty dr \, r^3 R_{10} R_{21} = \frac{a_0}{\sqrt{6}} \left(\frac{2}{3}\right)^5 \int_0^\infty dy \, y^4 \, e^{-y}. \tag{B.8.93}$$

The integral evaluates to 24 (see Exercise 4.11), so

$$\int_0^\infty dr \, r^3 R_{10} R_{21} = \frac{24 \, a_0}{\sqrt{6}} \left(\frac{2}{3}\right)^5. \tag{B.8.94}$$

Hence,

$$\langle 1, 0, 0 | x | 2, 1, \pm 1 \rangle = \mp \frac{2^7}{3^5} a_0. \tag{B.8.95}$$

According to the derived in Exercise 8.10, the matrix element $\langle n, l, m | y | n', l', m' \rangle$ is zero unless $l' = l \pm 1$ and $m' = m \pm 1$. Thus, the only possible non-zero $1s \to 2p$ matrix elements are $\langle 1, 0, 0 | y | 2, 1, \pm 1 \rangle$. Now, $x = r \sin\theta \sin\varphi$. Hence,

$$\langle 1, 0, 0 | y | 2, 1, \pm 1 \rangle = \int_0^\infty dr \, r^3 R_{10}(r) R_{21}(r) \oint d\Omega \sin\theta \sin\varphi \, Y_{00}^* Y_{1\pm 1}. \tag{B.8.96}$$

According to Section 4.4,

$$\oint d\Omega \sin\theta \sin\varphi \, Y_{00}^* Y_{1\pm 1} = \mp \sqrt{\frac{3}{2}} \frac{1}{4\pi} \int_0^{2\pi} d\varphi \, \sin\varphi \, (\cos\varphi \pm i \sin\varphi)$$
$$\int_0^\pi d\theta \, \sin^3\theta, \tag{B.8.97}$$

which reduces to

$$\oint d\Omega \sin\theta \cos\varphi \, Y_{00}^* Y_{1\pm 1} = -i \sqrt{\frac{1}{6}}. \tag{B.8.98}$$

However, we have already seen that

$$\int_0^\infty dr \, r^3 \, R_{10} R_{21} = \frac{24 \, a_0}{\sqrt{6}} \left(\frac{2}{3}\right)^5. \tag{B.8.99}$$

It follows that

$$\langle 1, 0, 0 | \, y \, | 2, 1, \pm 1 \rangle = -\mathrm{i} \frac{2^7}{3^5} \, a_0. \tag{B.8.100}$$

According to the derived in Section 8.11, the matrix element $\langle n, l, m | \, z \, | n', l', m' \rangle$ is zero unless $l' = l \pm 1$ and $m' = m$. Thus, the only possible non-zero $1s \to 2p$ matrix element is $\langle 1, 0, 0 | z | 2, 1, 0 \rangle$. Now, $z = r \cos \theta$. Hence,

$$\langle 1, 0, 0 | z | 2, 1, 0 \rangle = \int_0^\infty dr \, r^3 \, R_{10}(r) R_{21}(r) \oint d\Omega \, \cos \theta \, Y_{00}^* \, Y_{10}. \tag{B.8.101}$$

According to Section 4.4,

$$\oint d\Omega \, \cos \theta \, Y_{00}^* \, Y_{10} = \mp \frac{\sqrt{3}}{4\pi} \int_0^{2\pi} d\varphi \int_0^\pi d\theta \, \sin \theta \, \cos^2 \theta, \tag{B.8.102}$$

which reduces to

$$\oint d\Omega \, \cos \theta \, Y_{00}^* \, Y_{10} = \frac{1}{\sqrt{3}}. \tag{B.8.103}$$

However, we have already seen that

$$\int_0^\infty dr \, r^3 \, R_{10} R_{21} = \frac{24 \, a_0}{\sqrt{6}} \left(\frac{2}{3}\right)^5. \tag{B.8.104}$$

It follows that

$$\langle 1, 0, 0 | z | 2, 1, 0 \rangle = \sqrt{2} \frac{2^7}{3^5} \, a_0. \tag{B.8.105}$$

(b) By analogy with the previous analysis, the only non-zero $2s \to 3p$ matrix elements are written

$$\langle 1, 0, 0 | \, x \, | 3, 1, \pm 1 \rangle = \mp \sqrt{\frac{1}{6}} \int_0^\infty dr \, r^3 \, R_{10}(r) R_{31}(r), \tag{B.8.106}$$

$$\langle 1, 0, 0 | \, y \, | 3, 1, \pm 1 \rangle = -\mathrm{i} \sqrt{\frac{1}{6}} \int_0^\infty dr \, r^3 \, R_{10}(r) R_{31}(r), \tag{B.8.107}$$

$$\langle 1, 0, 0 | z | 3, 1, 0 \rangle = \sqrt{\frac{1}{3}} \int_0^\infty dr \, r^3 \, R_{10}(r) R_{31}(r). \tag{B.8.108}$$

It follows from from Exercise 4.12 that

$$\int_0^\infty dr \, r^3 \, R_{10} R_{31} = \frac{a_0}{\sqrt{6}} \frac{3^2}{2^6} \int_0^\infty dy \, y^4 \left(1 - \frac{y}{8}\right) \mathrm{e}^{-y}. \tag{B.8.109}$$

The integral evaluates to 9 (see Exercise 4.11), so

$$\int_0^\infty dr\, r^3 R_{10} R_{31} = \frac{a_0}{\sqrt{6}} \frac{3^4}{2^6}. \tag{B.8.110}$$

Hence,

$$\langle 1,0,0| x |3,1,\pm1\rangle = \mp\frac{3^3}{2^7} a_0 \tag{B.8.111}$$

$$\langle 1,0,0| y |3,1,\pm1\rangle = -i\frac{3^3}{2^7} a_0 \tag{B.8.112}$$

$$\langle 1,0,0| z |3,1,0\rangle = \sqrt{2}\frac{3^3}{2^7} a_0. \tag{B.8.113}$$

(c) By analogy with the previous analysis, the only non-zero $2s \to 2p$ matrix elements are written

$$\langle 2,0,0| x |2,1,\pm1\rangle = \mp\sqrt{\frac{1}{6}} \int_0^\infty dr\, r^3 R_{20}(r) R_{21}(r), \tag{B.8.114}$$

$$\langle 2,0,0| y |2,1,\pm1\rangle = -i\sqrt{\frac{1}{6}} \int_0^\infty dr\, r^3 R_{20}(r) R_{21}(r), \tag{B.8.115}$$

$$\langle 2,0,0| z |2,1,0\rangle = \sqrt{\frac{1}{3}} \int_0^\infty dr\, r^3 R_{20}(r) R_{21}(r). \tag{B.8.116}$$

It follows from from Exercise 4.12 that

$$\int_0^\infty dr\, r^3 R_{20} R_{21} = \frac{a_0}{\sqrt{3}\,2^2} \int_0^\infty dy\, y^4 \left(1 - \frac{y}{5}\right) e^{-y}. \tag{B.8.117}$$

The integral evaluates to -36 (see Exercise 4.11), so

$$\int_0^\infty dr\, r^3 R_{20} R_{21} = -3\sqrt{3}\, a_0. \tag{B.8.118}$$

Hence,

$$\langle 2,0,0| x |2,1,\pm1\rangle = \pm\frac{3}{\sqrt{2}} a_0 \tag{B.8.119}$$

$$\langle 2,0,0| y |2,1,\pm1\rangle = i\frac{3}{\sqrt{2}} a_0 \tag{B.8.120}$$

$$\langle 2,0,0| z |2,1,0\rangle = -3 a_0. \tag{B.8.121}$$

8.13 (a) For transitions from the $2p(m = \pm 1)$ to the $1s$ state, we have

$$d_{21}^2 = |\langle 2, 1, \pm 1| x |0, 0, 1\rangle|^2 + |\langle 2, 1, \pm 1| y |0, 0, 1\rangle|^2 + |\langle 2, 1, \pm 1| z |0, 0, 1\rangle|^2$$

$$= \left|\mp \frac{2^7}{3^5} a_0\right|^2 + \left|-i \frac{2^7}{3^5} a_0\right|^2 + |0|^2 = \frac{2^{15}}{3^{10}} a_0^2, \tag{B.8.122}$$

where use has been made of the previous exercise. For transitions from the $2p(m = 0)$ to the $1s$ state, we have

$$d_{21}^2 = |0|^2 + |0|^2 + \left|\sqrt{2} \frac{2^7}{3^5} a_0\right|^2 = \frac{2^{15}}{3^{10}} a_0^2. \tag{B.8.123}$$

It follows that d_{21}^2 is the same for all three $2p$ states. Hence, the transition rates from these states to the $1s$ state are all the same. The generic spontaneous transition rate from a $2p$ to a $1s$ state can be written

$$\frac{w_{21} \hbar}{m_e c^2} = \frac{2^2}{3 \alpha} \left(\frac{\omega_{21} \hbar}{m_e c^2}\right)^3 \left(\frac{d_{21}^2}{a_0^2}\right). \tag{B.8.124}$$

Now, the hydrogen atom energy levels are such that

$$\frac{E_n}{m_e c^2} = -\frac{\alpha^2}{2 n^2}. \tag{B.8.125}$$

Hence,

$$\frac{\omega_{21} \hbar}{m_e c^2} = \frac{E_2 - E_1}{m_e c^2} = \frac{3 \alpha^2}{2^3}. \tag{B.8.126}$$

It follows that

$$\frac{w_{21} \hbar}{m_e c^2} = \frac{2^2}{3 \alpha} \left(\frac{3 \alpha^2}{2^3}\right)^3 \frac{2^{15}}{3^{10}}, \tag{B.8.127}$$

or

$$w_{21} = \left(\frac{2}{3}\right)^8 \alpha^5 \frac{m_e c^2}{\hbar} = 6.27 \times 10^8 \text{ s}^{-1}. \tag{B.8.128}$$

Now,

$$\frac{w_{21}}{\omega_{21}} = \frac{2^{11}}{3^9} \alpha^3. \tag{B.8.129}$$

However, we can interpret w_{21} as $\Delta\omega_{21}$ (i.e., as an uncertainty in the transition frequency ω_{21}). Moreover, $\Delta\lambda/\lambda = \Delta\omega/\omega$ for electromagnetic waves. Hence,

$$\frac{\Delta\lambda}{\lambda} = \frac{2^{11}}{3^9} \alpha^3 = 4.0 \times 10^{-8}. \tag{B.8.130}$$

(b) For transitions from the $3p(m = \pm 1)$ to the $1s$ state, we have

$$d_{31}^2 = |\langle 3, 1, \pm 1 | x | 0, 0, 1 \rangle|^2 + |\langle 3, 1, \pm 1 | y | 0, 0, 1 \rangle|^2 + |\langle 3, 1, \pm 1 | z | 0, 0, 1 \rangle|^2$$

$$= \left| \mp \frac{3^3}{2^7} a_0 \right|^2 + \left| -i \frac{3^3}{2^7} a_0 \right|^2 + |0|^2 = \frac{3^6}{2^{13}} a_0^2, \tag{B.8.131}$$

where use has been made of the previous exercise. For transitions from the $3p(m = 0)$ to the $1s$ state, we have

$$d_{31}^2 = |0|^2 + |0|^2 + \left| \sqrt{2} \frac{3^3}{2^7} a_0 \right|^2 = \frac{3^6}{2^{13}} a_0^2. \tag{B.8.132}$$

It follows that d_{31}^2 is the same for all three $3p$ states. The generic spontaneous transition rate from a $3p$ to a $1s$ state can be written

$$\frac{w_{31} \hbar}{m_e c^2} = \frac{2^2}{3 \alpha} \left(\frac{\omega_{31} \hbar}{m_e c^2} \right)^3 \left(\frac{d_{31}^2}{a_0^2} \right). \tag{B.8.133}$$

However,

$$\frac{\omega_{31} \hbar}{m_e c^2} = \frac{E_3 - E_1}{m_e c^2} = \frac{2^2 \alpha^2}{3^2}. \tag{B.8.134}$$

It follows that

$$\frac{w_{31} \hbar}{m_e c^2} = \frac{2^2}{3 \alpha} \left(\frac{2^2 \alpha^2}{3^2} \right)^3 \frac{3^6}{2^{13}}, \tag{B.8.135}$$

or

$$w_{31} = \frac{\alpha^5}{3 \, 2^5} \frac{m_e c^2}{\hbar} = 1.67 \times 10^8 \, \text{s}^{-1}. \tag{B.8.136}$$

Now,

$$\frac{w_{31}}{\omega_{31}} = \frac{3}{2^7} \alpha^3. \tag{B.8.137}$$

However, we can interpret w_{31} as $\Delta \omega_{31}$ (i.e., as an uncertainty in the transition frequency ω_{31}). Moreover, $\Delta \lambda / \lambda = \Delta \omega / \omega$ for electromagnetic waves. Hence,

$$\frac{\Delta \lambda}{\lambda} = \frac{3}{2^7} \alpha^3 = 9.1 \times 10^{-9}. \tag{B.8.138}$$

8.14 (a) Suppose that the radiation is polarized in the x-direction. The only non-zero contributions to the $1s \to 2p$ oscillator strength are

$$F_{1,0,0 \to 2,1,\pm 1} = \frac{2 m_e \omega_{21}}{\hbar} |\langle 1, 0, 0 | x | 2, 1, \pm 1 \rangle|^2, \tag{B.8.139}$$

where

$$\omega_{21} = \frac{E_2 - E_1}{\hbar} = -\frac{3}{4} \frac{E_0}{\hbar}. \tag{B.8.140}$$

Now, from Exercise 8.12,

$$|\langle 1, 0, 0| \, x \, |2, 1, \pm 1\rangle|^2 = \frac{2^{14}}{3^{10}} a_0^2. \tag{B.8.141}$$

Moreover,

$$E_0 = -\frac{\hbar^2}{2 \, m_e \, a_0^2}. \tag{B.8.142}$$

Hence, we obtain

$$F_{1,0,0 \to 2,1,\pm 1} = \frac{2^{12}}{3^9}. \tag{B.8.143}$$

Thus, the net oscillator strength is

$$F_{1s \to 2p} = F_{1,0,0 \to 2,1,1} + F_{1,0,0 \to 2,1,-1} = \frac{2^{13}}{3^9} = 0.4162. \tag{B.8.144}$$

Suppose that the radiation is polarized in the y-direction. The only non-zero contributions to the $1s \to 2p$ oscillator strength are

$$F_{1,0,0 \to 2,1,\pm 1} = \frac{2 \, m_e \, \omega_{21}}{\hbar} |\langle 1, 0, 0| \, y \, |2, 1, \pm 1\rangle|^2. \tag{B.8.145}$$

However, from Exercise 8.12,

$$|\langle 1, 0, 0| \, y \, |2, 1, \pm 1\rangle|^2 = |\langle 1, 0, 0| \, x \, |2, 1, \pm 1\rangle|^2. \tag{B.8.146}$$

Hence, we deduce that

$$F_{1s \to 2p} = F_{1,0,0 \to 2,1,1} + F_{1,0,0 \to 2,1,-1} = \frac{2^{13}}{3^9} = 0.4162. \tag{B.8.147}$$

Suppose, finally, that is polarized in the z-direction. The only non-zero contribution to the $1s \to 2p$ oscillator strength is

$$F_{1,0,0 \to 2,1,0} = \frac{2 \, m_e \, \omega_{21}}{\hbar} |\langle 1, 0, 0| \, z \, |2, 1, 0\rangle|^2. \tag{B.8.148}$$

However, from Exercise 8.12,

$$|\langle 1, 0, 0| \, z \, |2, 1, 0\rangle|^2 = 2 \, |\langle 1, 0, 0| \, x \, |2, 1, \pm 1\rangle|^2. \tag{B.8.149}$$

Hence, we deduce that

$$F_{1s \to 2p} = F_{1,0,0 \to 2,1,0} = \frac{2^{13}}{3^9} = 0.4162. \tag{B.8.150}$$

It is clear that the $1s \to 2p$ oscillator strength is independent of the direction of polarization of the radiation.

(b) Suppose that the radiation is polarized in the x-direction. The only non-zero contributions to the $1s \rightarrow 3p$ oscillator strength are

$$F_{1,0,0\rightarrow3,1,\pm1} = \frac{2\,m_e\,\omega_{31}}{\hbar}\,|\langle 1,0,0|\,x\,|3,1,\pm1\rangle|^2, \tag{B.8.151}$$

where

$$\omega_{31} = \frac{E_3 - E_1}{\hbar} = -\frac{8}{9}\frac{E_0}{\hbar}. \tag{B.8.152}$$

Now, from Exercise 8.12,

$$|\langle 1,0,0|\,x\,|3,1,\pm1\rangle|^2 = \frac{3^6}{2^{14}}\,a_0^2. \tag{B.8.153}$$

Hence, we obtain

$$F_{1,0,0\rightarrow3,1,\pm1} = \frac{3^4}{2^{11}}. \tag{B.8.154}$$

Thus, the net oscillator strength is

$$F_{1s\rightarrow3p} = F_{1,0,0\rightarrow3,1,1} + F_{1,0,0\rightarrow3,1,-1} = \frac{4^4}{2^{10}} = 0.07910. \tag{B.8.155}$$

Exercise 8.12 also yields

$$|\langle 1,0,0|\,y\,|3,1,\pm1\rangle|^2 = |\langle 1,0,0|\,z\,|3,1,\pm1\rangle|^2, \tag{B.8.156}$$

$$|\langle 1,0,0|\,z\,|3,1,0\rangle|^2 = 2\,|\langle 1,0,0|\,z\,|3,1,\pm1\rangle|^2. \tag{B.8.157}$$

Hence, analogous arguments to those given in part (a) reveal that the $1s \rightarrow 3p$ oscillator strength is independent of the direction of polarization of the radiation.

8.15 The analysis of Section 7.7 reveals that

$$R_{10}\,\mathcal{Y}_{0\ +1/2}^{1/2} = R_{10}\,Y_{00}\,\chi_+, \tag{B.8.158}$$

$$R_{10}\,\mathcal{Y}_{0\ -1/2}^{1/2} = R_{10}\,Y_{00}\,\chi_-, \tag{B.8.159}$$

$$R_{21}\,\mathcal{Y}_{1\ +1/2}^{1/2} = R_{21}\left(\sqrt{\frac{1}{3}}\,Y_{10}\,\chi_- - \sqrt{\frac{2}{3}}\,Y_{11}\,\chi_-\right), \tag{B.8.160}$$

$$R_{21}\,\mathcal{Y}_{1\ -1/2}^{1/2} = R_{21}\left(\sqrt{\frac{2}{3}}\,Y_{1-1}\,\chi_+ - \sqrt{\frac{1}{3}}\,Y_{10}\,\chi_-\right), \tag{B.8.161}$$

$$R_{21}\,\mathcal{Y}_{1\ +3/2}^{3/2} = R_{21}\,Y_{11}\chi_+, \tag{B.8.162}$$

$$R_{21}\,\mathcal{Y}_{1\ +1/2}^{3/2} = R_{21}\left(\sqrt{\frac{2}{3}}\,Y_{10}\,\chi_+ + \sqrt{\frac{1}{3}}\,Y_{11}\,\chi_-\right), \tag{B.8.163}$$

$$R_{21}\,\mathcal{Y}_{1\ -1/2}^{3/2} = R_{21}\left(\sqrt{\frac{1}{3}}\,Y_{1-1}\,\chi_+ + \sqrt{\frac{2}{3}}\,Y_{10}\,\chi_-\right), \tag{B.8.164}$$

$$R_{21}\,\mathcal{Y}_{1\,-3/2}^{3/2} = R_{21}\,Y_{1-1}\chi_-,\tag{B.8.165}$$

where the Y_{lm} are spherical harmonics, and the $\chi_\pm$ conventional spinors. Now, according to the analysis of Exercise 8.13, the spontaneous decay rate between the various $2p$ and $1s$ states can be written

$$w\left(\mathcal{Y}_{1\,m_j}^j \to \mathcal{Y}_{0\,m}^{1/2}\right) = \frac{3^2}{2^7}\,\alpha^5\left(\frac{d_x^2 + d_y^2 + d_z^2}{a_0^2}\right)\frac{m_e\,c^2}{\hbar},\tag{B.8.166}$$

where

$$d_i^2 = \left|\left\langle R_{21}\,\mathcal{Y}_{1\,m_j}^j\,|\,x_i\,|\,R_{10}\,\mathcal{Y}_{0\,m}^{1/2}\right\rangle\right|^2.\tag{B.8.167}$$

However, the same exercise also reveals that the only non-zero $2p \to 1s$ electric dipole matrix elements are

$$|\langle R_{21}\,Y_{1\pm1}\,\chi_\pm\,|\,x\,|\,R_{10}\,Y_{00}\,\chi_\pm\rangle|^2 = \frac{C}{2},\tag{B.8.168}$$

$$|\langle R_{21}\,Y_{1\pm1}\,\chi_\pm\,|\,y\,|\,R_{10}\,Y_{00}\,\chi_\pm\rangle|^2 = \frac{C}{2},\tag{B.8.169}$$

$$|\langle R_{21}\,Y_{10}\,\chi_\pm\,|\,z\,|\,R_{10}\,Y_{00}\,\chi_\pm\rangle|^2 = C,\tag{B.8.170}$$

where

$$C = \frac{2^{15}}{3^{10}}\,a_0^2.\tag{B.8.171}$$

The previous equations can be combined to give

$$w\left(\mathcal{Y}_{1\,+3/2}^{3/2} \to \mathcal{Y}_{0\,+1/2}^{1/2}\right) = w_0,\tag{B.8.172}$$

$$w\left(\mathcal{Y}_{1\,+3/2}^{3/2} \to \mathcal{Y}_{0\,-1/2}^{1/2}\right) = 0,\tag{B.8.173}$$

$$w\left(\mathcal{Y}_{1\,+1/2}^{3/2} \to \mathcal{Y}_{0\,+1/2}^{1/2}\right) = \frac{2}{3}\,w_0,\tag{B.8.174}$$

$$w\left(\mathcal{Y}_{1\,+1/2}^{3/2} \to \mathcal{Y}_{0\,-1/2}^{1/2}\right) = \frac{1}{3}\,w_0,\tag{B.8.175}$$

$$w\left(\mathcal{Y}_{1\,-1/2}^{3/2} \to \mathcal{Y}_{0\,+1/2}^{1/2}\right) = \frac{1}{3}\,w_0,\tag{B.8.176}$$

$$w\left(\mathcal{Y}_{1\,-1/2}^{3/2} \to \mathcal{Y}_{0\,-1/2}^{1/2}\right) = \frac{2}{3}\,w_0,\tag{B.8.177}$$

$$w\left(\mathcal{Y}_{1\,-3/2}^{3/2} \to \mathcal{Y}_{0\,+1/2}^{1/2}\right) = 0,\tag{B.8.178}$$

$$w\left(\mathcal{Y}_{1\,-3/2}^{3/2} \to \mathcal{Y}_{0\,-1/2}^{1/2}\right) = w_0,\tag{B.8.179}$$

$$w\left(\mathcal{Y}_{1\,+1/2}^{1/2} \to \mathcal{Y}_{0\,+1/2}^{1/2}\right) = \frac{1}{3}\,w_0,\tag{B.8.180}$$

$$w\left(\mathcal{Y}_{1\,+1/2}^{1/2} \to \mathcal{Y}_{0\,-1/2}^{1/2}\right) = \frac{2}{3}\,w_0,\tag{B.8.181}$$

$$w\left(\mathcal{Y}^{1/2}_{1\,-1/2} \to \mathcal{Y}^{1/2}_{0\,+1/2}\right) = \frac{2}{3}\, w_0, \tag{B.8.182}$$

$$w\left(\mathcal{Y}^{1/2}_{1\,-1/2} \to \mathcal{Y}^{1/2}_{0\,-1/2}\right) = \frac{1}{3}\, w_0, \tag{B.8.183}$$

where

$$w_0 = \left(\frac{2}{3}\right)^8 \alpha^5 \, \frac{m_e\, c^2}{\hbar} = 6.27 \times 10^8 \, \mathrm{s}^{-1}. \tag{B.8.184}$$

In thermal equilibrium, we expect all of the $2p$ states to be equally populated, because the energy difference between the $(2p)_{3/2}$ and the $(2p)_{1/2}$ states is relatively small. It follows from Equations (B.8.172)–(B.8.183) that if there are N atoms in each state then the number of $(2p)_{3/2} \to (1s)_{1/2}$ transitions per second is $N w_0 + (2/3) N w_0 + (1/3) N w_0 + (1/3) N w_0 + (2/3) N w_0 + N w_0 = 4 N w_0$. On the other hand, the number of $(2p)_{1/2} \to (1s)_{1/2}$ transitions per second is $(1/3) N w_0 + (2/3) N w_0 + (2/3) N w_0 + (1/3) N w_0 = 2 N w_0$. The fact that there are twice as many $(2p)_{3/2} \to (1s)_{1/2}$ as $(2p)_{1/2} \to (1s)_{1/2}$ transitions per second implies that the spectral line associated with the former transitions is twice as bright as that associated with the latter transitions.

8.16 The analysis of Section 7.7 reveals that

$$R_{20}\, \mathcal{Y}^{1/2}_{0\,+1/2} = R_{20}\, Y_{00}\, \chi_+, \tag{B.8.185}$$

$$R_{20}\, \mathcal{Y}^{1/2}_{0\,-1/2} = R_{20}\, Y_{00}\, \chi_-, \tag{B.8.186}$$

$$R_{21}\, \mathcal{Y}^{3/2}_{1\,+3/2} = R_{21}\, Y_{11}\chi_+, \tag{B.8.187}$$

$$R_{21}\, \mathcal{Y}^{3/2}_{1\,+1/2} = R_{21}\left(\sqrt{\frac{2}{3}}\, Y_{10}\, \chi_+ + \sqrt{\frac{1}{3}}\, Y_{11}\, \chi_-\right), \tag{B.8.188}$$

$$R_{21}\, \mathcal{Y}^{3/2}_{1\,-1/2} = R_{21}\left(\sqrt{\frac{1}{3}}\, Y_{1-1}\, \chi_+ + \sqrt{\frac{2}{3}}\, Y_{10}\, \chi_-\right), \tag{B.8.189}$$

$$R_{21}\, \mathcal{Y}^{3/2}_{1\,-3/2} = R_{21}\, Y_{1-1}\chi_-, \tag{B.8.190}$$

where the Y_{lm} are spherical harmonics, and the $\chi_\pm$ conventional spinors. According to Exercise 7.14, the energy of the $(2p)_{3/2}$ states exceeds that of the $(2p)_{1/2}$ states by

$$E_i - E_f = \frac{\alpha^5}{2^5}\, m_e\, c^2. \tag{B.8.191}$$

Hence, the spontaneous decay rate between the various $(2p)_{3/2}$ and $(2s)_{1/2}$ states can be written

$$w(\mathcal{Y}^{3/2}_{1\,m_j} \to \mathcal{Y}^{1/2}_{0\,m}) = \frac{4\,\alpha\,\omega_{if}^3}{3\,c^2}\,(d_x^2 + d_y^2 + d_z^2) = \frac{\alpha^{11}}{2^{13}\,3}\left(\frac{d_x^2 + d_y^2 + d_z^2}{a_0^2}\right)\frac{m_e\,c^2}{\hbar}, \tag{B.8.192}$$

where $\omega_{if} = (E_i - E_f)/\hbar$, and

$$d_i^2 = \left| \left\langle R_{21} \, \mathcal{Y}_{1m_j}^{3/2} \, | \, x_i \, | \, R_{20} \, \mathcal{Y}_{0m}^{1/2} \right\rangle \right|^2 . \tag{B.8.193}$$

However, it follows from Exercise 8.12 that the only non-zero $2p \to 2s$ electric dipole matrix elements are

$$|\langle R_{21} \, Y_{1\pm1} \, \chi_\pm \, | \, x \, | \, R_{20} \, Y_{00} \, \chi_\pm \rangle|^2 = \frac{C}{2}, \tag{B.8.194}$$

$$|\langle R_{21} \, Y_{1\pm1} \, \chi_\pm \, | \, y \, | \, R_{20} \, Y_{00} \, \chi_\pm \rangle|^2 = \frac{C}{2}, \tag{B.8.195}$$

$$|\langle R_{21} \, Y_{10} \, \chi_\pm \, | \, z \, | \, R_{20} \, Y_{00} \, \chi_\pm \rangle|^2 = C, \tag{B.8.196}$$

where

$$C = 3^2 \, a_0^2. \tag{B.8.197}$$

Thus, by analogy with the analysis of the previous exercise, the various $(2p)_{3/2} \to (2s)_{1/2}$ transition rates are

$$w\left(\mathcal{Y}_{1\,+3/2}^{3/2} \to \mathcal{Y}_{0\,+1/2}^{1/2} \right) = w_0, \tag{B.8.198}$$

$$w\left(\mathcal{Y}_{1\,+3/2}^{3/2} \to \mathcal{Y}_{0\,-1/2}^{1/2} \right) = 0, \tag{B.8.199}$$

$$w\left(\mathcal{Y}_{1\,+1/2}^{3/2} \to \mathcal{Y}_{0\,+1/2}^{1/2} \right) = \frac{2}{3} \, w_0, \tag{B.8.200}$$

$$w\left(\mathcal{Y}_{1\,+1/2}^{3/2} \to \mathcal{Y}_{0\,-1/2}^{1/2} \right) = \frac{1}{3} \, w_0, \tag{B.8.201}$$

$$w\left(\mathcal{Y}_{1\,-1/2}^{3/2} \to \mathcal{Y}_{0\,+1/2}^{1/2} \right) = \frac{1}{3} \, w_0, \tag{B.8.202}$$

$$w\left(\mathcal{Y}_{1\,-1/2}^{3/2} \to \mathcal{Y}_{0\,-1/2}^{1/2} \right) = \frac{2}{3} \, w_0, \tag{B.8.203}$$

$$w\left(\mathcal{Y}_{1\,-3/2}^{3/2} \to \mathcal{Y}_{0\,+1/2}^{1/2} \right) = 0, \tag{B.8.204}$$

$$w\left(\mathcal{Y}_{1\,-3/2}^{3/2} \to \mathcal{Y}_{0\,-1/2}^{1/2} \right) = w_0, \tag{B.8.205}$$

where

$$w_0 = \frac{3 \, \alpha^{11}}{2^{13}} \, \frac{m_e \, c^2}{\hbar} = 3.55 \times 10^{-6} \, \text{s}^{-1}. \tag{B.8.206}$$

8.17 (a) As in Exercise 8.8, we have

$$\mathbf{n} = (\sin\theta \, \cos\varphi, \, \sin\theta \, \sin\varphi, \, \cos\theta), \tag{B.8.207}$$

$$\boldsymbol{\epsilon}_1 = (\cos\theta \, \cos\varphi, \, \cos\theta \, \sin\varphi, \, -\sin\theta), \tag{B.8.208}$$

$$\boldsymbol{\epsilon}_2 = (-\sin\varphi, \, \cos\varphi, \, 0), \tag{B.8.209}$$

where $\mathbf{n}$ is the normalized wavevector of the emitted photon, and $\epsilon_{1,2}$ are its two independent polarization directions. According to Equation (8.165), the angular distribution of the spontaneously emitted photons is

$$\frac{dw_{i \to f}^{\text{spn}}}{d\Omega} = \frac{\alpha\,\omega_{if}^3}{2\pi\,c^2} \sum_{j=1,2} |\epsilon_j \cdot \mathbf{d}_{if}|^2. \qquad (B.8.210)$$

However, for a $2p \to 1s$ transition in a hydrogen atom,

$$\frac{\omega_{if}\,\hbar}{m_e\,c^2} = \frac{3\,\alpha^2}{2^3}. \qquad (B.8.211)$$

(See Exercise 8.13.) Moreover, according to Exercise 8.12, for a $2, 1, \pm 1 \to 1, 0, 0$ transition,

$$\mathbf{d}_{if} = d\,(\mp 1,\, -\mathrm{i},\, 0), \qquad (B.8.212)$$

where

$$d = \frac{2^7}{3^5}\,a_0. \qquad (B.8.213)$$

Hence,

$$\epsilon_1 \cdot \mathbf{d}_{if} = \mp d\,\cos\theta\,\mathrm{e}^{\pm\mathrm{i}\varphi}, \qquad (B.8.214)$$

$$\epsilon_2 \cdot \mathbf{d}_{if} = -\mathrm{i}\,d\,\mathrm{e}^{\pm\mathrm{i}\varphi}, \qquad (B.8.215)$$

which implies that

$$|\epsilon_1 \cdot \mathbf{d}_{if}|^2 = d^2\,\cos^2\theta, \qquad (B.8.216)$$

$$|\epsilon_2 \cdot \mathbf{d}_{if}|^2 = d^2. \qquad (B.8.217)$$

Thus, given that $(m_e\,c^2\,a_0/\hbar\,c)^2 = \alpha^{-2}$, we obtain

$$\frac{dw_{2,1,\pm 1 \to 1,0,0}^{\text{spn}}}{d\Omega} = \frac{2^4\,\alpha^5}{3^7\,\pi}\,\frac{m_e\,c^2}{\hbar}\,(1 + \cos^2\theta). \qquad (B.8.218)$$

(b) According to Exercise 8.12, for a $2, 1, 0 \to 1, 0, 0$ transition,

$$\mathbf{d}_{if} = d\,(0,\, 0,\, \sqrt{2}). \qquad (B.8.219)$$

Hence,

$$\epsilon_1 \cdot \mathbf{d}_{if} = -\sqrt{2}\,d\,\sin\theta, \qquad (B.8.220)$$

$$\epsilon_2 \cdot \mathbf{d}_{if} = 0, \qquad (B.8.221)$$

which implies that

$$|\boldsymbol{\epsilon}_1 \cdot \mathbf{d}_{if}|^2 = 2\,d^2\,\sin^2\theta, \tag{B.8.222}$$

$$|\boldsymbol{\epsilon}_2 \cdot \mathbf{d}_{if}|^2 = 0. \tag{B.8.223}$$

Thus, we obtain

$$\frac{dw_{2,1,0\to1,0,0}^{\text{spn}}}{d\Omega} = \frac{2^5\,\alpha^5}{3^7\,\pi}\,\frac{m_e\,c^2}{\hbar}\,\sin^2\theta. \tag{B.8.224}$$

(c) In thermal equilibrium, we expect the degenerate $2,1,+1$, $2,1,0$, and $2,1,-1$ states to be equally populated. Thus, the net angular distribution of spontaneously emitted photon is the average of that associated with the three $2p$ states. In other words,

$$\frac{dw_{2p\to1s}^{\text{spn}}}{d\Omega} = \frac{1}{3}\frac{dw_{2,1,+1\to1,0,0}^{\text{spn}}}{d\Omega} + \frac{1}{3}\frac{dw_{2,1,0\to1,0,0}^{\text{spn}}}{d\Omega} + \frac{1}{3}\frac{dw_{2,1,-1\to1,0,0}^{\text{spn}}}{d\Omega}, \tag{B.8.225}$$

which implies that

$$\frac{dw_{2p\to1s}^{\text{spn}}}{d\Omega} = \frac{2^6\,\alpha^5}{3^8\,\pi}\,\frac{m_e\,c^2}{\hbar}. \tag{B.8.226}$$

8.18 (a) According to Section 7.9,

$$E_i - E_f = \frac{8}{3}\,g_p\,\alpha^2\,\frac{m_e}{m_p}\,|E_0|, \tag{B.8.227}$$

where E_0 is the hydrogen ground-state energy. However, $E_0 = (1/2)\,\alpha^2\,m_e\,c^2$, so

$$\omega_{if} = \frac{8}{3}\,g_p\,\alpha^4\,\frac{m_e}{m_p}\,\frac{m_e\,c^2}{\hbar}. \tag{B.8.228}$$

(b) If the $|\pm\rangle_{e,p}$ states are properly normalized then $\langle\pm|\pm\rangle_{e,p} = 1$ and $\langle\pm|\mp\rangle_{e,p} = 0$. Furthermore, $2S_{ez}|\pm\rangle_e = \pm\hbar|\pm\rangle_e$, $2S_{ex}|\pm\rangle_e = \hbar|\mp\rangle_e$, and $2S_{ey}|\pm\rangle_e = \pm i\,\hbar|\mp\rangle_e$. Thus, if the initial state is $|1,1\rangle$ and the final state is $|0,0\rangle$ then

$$\mathbf{M}_{if} = \frac{\hbar}{\sqrt{2}}\left(\mathbf{e}_x - i\,\mathbf{e}_y\right). \tag{B.8.229}$$

On the other hand, if the initial state is $|1,-1\rangle$ and the final state is $|0,0\rangle$ then

$$\mathbf{M}_{if} = -\frac{\hbar}{\sqrt{2}}\left(\mathbf{e}_x + i\,\mathbf{e}_y\right). \tag{B.8.230}$$

Finally, if the initial state is $|1,0\rangle$ and the final state is $|0,0\rangle$ then

$$\mathbf{M}_{if} = -\hbar\,\mathbf{e}_z. \tag{B.8.231}$$

(c) The two independent photon polarization vectors are

$$\epsilon_1 = (\cos\theta \cos\varphi, \cos\theta \sin\varphi, -\sin\theta), \tag{B.8.232}$$

$$\epsilon_2 = (-\sin\varphi, \cos\varphi, 0), \tag{B.8.233}$$

where $\mathbf{n} = (\sin\theta \cos\varphi, \sin\theta \sin\varphi, \cos\theta)$ is the normalized wavevector of the emitted photon. Thus, for the $|1, \pm 1\rangle \to |1, 0\rangle$ transition,

$$\epsilon_1 \cdot \mathbf{M}_{if} = \mp \frac{\hbar}{\sqrt{2}} \cos\theta\, e^{\pm i\varphi}, \tag{B.8.234}$$

$$\epsilon_2 \cdot \mathbf{M}_{if} = -i \frac{\hbar}{\sqrt{2}} e^{\pm i\varphi}. \tag{B.8.235}$$

Hence,

$$\sum_{j=1,2} |\epsilon_1 \cdot \mathbf{M}_{if}|^2 = \frac{\hbar^2}{2} (1 + \cos^2\theta). \tag{B.8.236}$$

On the other hand, for the $|1, 0\rangle \to |0, 0\rangle$ transition,

$$\epsilon_1 \cdot \mathbf{M}_{if} = \hbar \sin\theta, \tag{B.8.237}$$

$$\epsilon_2 \cdot \mathbf{M}_{if} = 0. \tag{B.8.238}$$

Hence,

$$\sum_{j=1,2} |\epsilon_1 \cdot \mathbf{M}_{if}|^2 = \hbar^2 \sin^2\theta. \tag{B.8.239}$$

Now, according to Equation (8.206), the angular distribution of the emitted radiation is

$$\frac{dw_{i\to f}^{\text{spn}}}{d\Omega} = \frac{\alpha\,\omega_{if}^3}{8\pi\, m_e^2\, c^4} \sum_{j=1,2} |\epsilon_j \cdot \mathbf{M}_{if}|^2. \tag{B.8.240}$$

It follows that

$$\frac{dw_{i\to f}^{\text{spn}}}{d\Omega} = \frac{2^2}{3^3\,\pi}\, g_p^3\, \alpha^{13} \left(\frac{m_e}{m_p}\right)^3 \frac{m_e c^2}{\hbar} (1 + \cos^2\theta) \tag{B.8.241}$$

for the $|1, \pm 1\rangle \to |1, 0\rangle$ transition, and

$$\frac{dw_{i\to f}^{\text{spn}}}{d\Omega} = \frac{2^3}{3^3\,\pi}\, g_p^3\, \alpha^{13} \left(\frac{m_e}{m_p}\right)^3 \frac{m_e c^2}{\hbar} \sin^2\theta \tag{B.8.242}$$

for the $|1, 0 \to |1, 0\rangle$ transition.

(d) According to Equation (8.207), the net transition rate is

$$w_{i\to f}^{\text{spn}} = \frac{\alpha\,\omega_{if}^3\, |M_{if}|^2}{3\, m_e^2\, c^4}. \tag{B.8.243}$$

However, $|M_{if}|^2 = \hbar^2$ for all three possible transitions. Hence, we obtain

$$w_{i \to f}^{\text{spn}} = \frac{2^6}{3^4} g_p^3 \alpha^{13} \left(\frac{m_e}{m_p}\right)^3 \frac{m_e c^2}{\hbar} = 2.88 \times 10^{-15} \text{ s}^{-1}. \qquad \text{(B.8.244)}$$

8.19 (a) According to Equation (8.217),

$$\frac{dw_{3d \to 1s}^{\text{spn}}}{d\Omega} = \frac{\alpha \, \omega_{3d \to 1s}^5}{8\pi \, c^4} \sum_{j=1,2} \left|\boldsymbol{\epsilon}_j \cdot \mathbf{Q}_{3d \to 1s} \cdot \mathbf{n}\right|^2. \qquad \text{(B.8.245)}$$

However,

$$\omega_{3d \to 1s} = \frac{E_{3d} - E_{1s}}{\hbar} = \frac{8}{9} \frac{|E_0|}{\hbar} = \frac{4}{9} \alpha^2 \frac{m_e c^2}{\hbar}, \qquad \text{(B.8.246)}$$

where E_0 is the hydrogen ground-state energy. Furthermore, $a_0 = \hbar/(\alpha \, m_e \, c)$. Hence, we obtain

$$\frac{dw_{3d \to 1s}^{\text{spn}}}{d\Omega} = \frac{\alpha^7}{8\pi} \left(\frac{2}{3}\right)^{10} \left(\sum_{j=1,2} \frac{\left|\boldsymbol{\epsilon}_j \cdot \mathbf{Q}_{3d \to 1s} \cdot \mathbf{n}\right|^2}{a_0^4}\right) \frac{m_e c^2}{\hbar}. \qquad \text{(B.8.247)}$$

(b) The electric quadrupole matrix elements for the $3d(m = 0) \to 1s$ transition are written

$$Q_{ij} = \int_0^\infty dr \, r^4 \, R_{10} \, R_{32} \oint d\Omega \, Y_{20} \left(x_i x_j/r^2 - \delta_{ij}\right) Y_{00}, \qquad \text{(B.8.248)}$$

where $x_1/r = \sin\theta \cos\varphi$, $x_2/r = \sin\theta \sin\varphi$, and $x_3/r = \cos\theta$. Here,

$$Y_{00}(\theta, \varphi) = \frac{1}{\sqrt{4\pi}}, \qquad \text{(B.8.249)}$$

$$Y_{20}(\theta, \varphi) = \sqrt{\frac{5}{16\pi}} \, (3 \cos^3\theta - 1). \qquad \text{(B.8.250)}$$

Thus, given that $\oint d\Omega \, Y_{20} \, Y_{00} = 0$, we can write

$$A_{ij} = \oint d\Omega \, Y_{20} \left(x_i x_j/r^2 - \delta_{ij}\right) Y_{00}$$

$$= \frac{\sqrt{5}}{4} \left\langle \int_0^\pi d\theta \, \sin\theta \, (3 \cos^2\theta - 1) \left(\frac{x_i x_j}{r^2}\right) \right\rangle, \qquad \text{(B.8.251)}$$

where $\langle \cdots \rangle$ denotes an average over φ. It is evident that the only non-zero A_{ij} values are A_{11}, A_{22}, and A_{33}. (Because the other values average to zero in φ.) In fact,

$$A_{11} = A_{22} = \frac{\sqrt{5}}{8} \int_0^\pi d\theta \, \sin\theta \, (3\cos^2\theta - 1) \sin^2\theta, \tag{B.8.252}$$

$$A_{33} = \frac{\sqrt{5}}{4} \int_0^\pi d\theta \, \sin\theta \, (3\cos^2\theta - 1) \cos^2\theta. \tag{B.8.253}$$

However, $\int_0^\pi d\theta \, \sin\theta \, (3\cos^2\theta - 1) = 0$, which implies that $A_{11} = A_{22} = -A_{33}/2$. Finally,

$$A_{33} = \frac{2}{3\sqrt{5}}. \tag{B.8.254}$$

Now, from Exercise 4.12,

$$\int_0^\infty dr \, r^4 \, R_{10} \, R_{32} = \sqrt{\frac{2}{15}} \, \frac{3^3}{2^{12}} \, a_0^2 \int_0^\infty y^6 \, \exp(-y). \tag{B.8.255}$$

The integral evaluates to 720 (see Exercise 4.11). Hence, we obtain

$$\int_0^\infty dr \, r^4 \, R_{10} \, R_{32} = \sqrt{\frac{10}{3}} \, \frac{3^5}{2^8} \, a_0^2. \tag{B.8.256}$$

It follows, from the previous analysis, the only non-zero $3d(m = 0) \to 1s$ electric quadrupole matrix elements takes the values $Q_{xx} = -Q_{zz}/2$, $Q_{yy} = -Q_{zz}/2$, and

$$Q_{zz} = \sqrt{\frac{2}{3}} \, \frac{3^4}{2^7} \, a_0^2. \tag{B.8.257}$$

(c) Now,

$$\mathbf{n} = (\sin\theta \cos\varphi, \, \sin\theta \sin\varphi, \, \cos\theta), \tag{B.8.258}$$

$$\boldsymbol{\epsilon}_1 = (\cos\theta \cos\varphi, \, \cos\theta \sin\varphi, \, -\sin\theta), \tag{B.8.259}$$

$$\boldsymbol{\epsilon}_2 = (-\sin\varphi, \, \cos\varphi, \, 0). \tag{B.8.260}$$

Hence,

$$\boldsymbol{\epsilon}_1 \cdot \mathbf{Q}_{3d \to 1s} \cdot \mathbf{n} = -\frac{3}{2} \, Q_{zz} \cos\theta \sin\theta, \tag{B.8.261}$$

$$\boldsymbol{\epsilon}_2 \cdot \mathbf{Q}_{3d \to 1s} \cdot \mathbf{n} = 0. \tag{B.8.262}$$

Thus, Equation (B.8.247) yields

$$\frac{dw_{3d \to 1s}^{\text{spn}}}{d\Omega} = \frac{\alpha^7}{2^8 \, 3\pi} \, \frac{m_e \, c^2}{\hbar} \, \cos^2\theta \, \sin^2\theta. \tag{B.8.263}$$

Finally,

$$w_{3d \to 1s}^{\mathrm{spn}} = \oint d\Omega \, \frac{dw_{3d \to 1s}^{\mathrm{spn}}}{d\Omega} = \frac{\alpha^7}{2^5 \, 3^2 \, 5} \, \frac{m_e \, c^2}{\hbar} = 594.1 \, \mathrm{s}^{-1}. \tag{B.8.264}$$

8.20 We have

$$J_n = \int_0^\infty dy \, y^n \, e^{-\beta y} \, \sin(b \, y). \tag{B.8.265}$$

However,

$$(-1)^n \left(\frac{\partial}{\partial \beta} \right)^n e^{-\beta y} = y^n \, e^{-\beta y}. \tag{B.8.266}$$

Hence, we deduce that

$$J_n = (-1)^n \left(\frac{\partial}{\partial \beta} \right)^n J_0. \tag{B.8.267}$$

Now,

$$J_0 = \frac{1}{2i} \int_0^\infty dy \left[e^{-(\beta - ib)y} - e^{-(\beta + ib)y} \right] = \frac{1}{2i} \left[\frac{e^{-(\beta + ib)y}}{\beta + ib} - \frac{e^{-(\beta + ib)y}}{\beta - ib} \right]_0^\infty$$

$$\doteq \frac{b}{\beta^2 + b^2}. \tag{B.8.268}$$

Hence,

$$J_1 = -\frac{\partial J_0}{\partial \beta} = \frac{2 \, b \, \beta}{(\beta^2 + b^2)^2}, \tag{B.8.269}$$

and

$$J_2 = -\frac{\partial J_1}{\partial \beta} = \frac{8 \, b \, \beta^2}{(\beta^2 + b^2)^3} - \frac{2 \, b}{(\beta^2 + b^2)^2}. \tag{B.8.270}$$

8.21 Treating β as a small parameter, and neglecting terms of order β^2, we can write

$$I = \oint d\Omega \, \frac{\sin^2 \theta \, \cos^2 \varphi}{(1 - \beta \cos \theta)^4} \simeq \oint d\Omega \, \sin^2 \theta \, \cos^2 \varphi \, (1 + 4\beta \cos \theta). \tag{B.8.271}$$

Hence,

$$I \simeq \oint \cos^2 \varphi \, d\varphi \int_{-1}^1 d\mu \, (1 - \mu^2) \, (1 + 4\beta \mu), \tag{B.8.272}$$

where $\mu = \cos \theta$. It follows that

$$I \simeq \pi \left[\mu - \frac{\mu^3}{3} + 2\beta \mu^2 - \beta \mu^4 \right]_{-1}^1 = \frac{4\pi}{3}. \tag{B.8.273}$$

8.22 If $\epsilon = \mathbf{e}_y$ then $\epsilon \cdot \mathbf{k}_f = k_f \sin \theta \sin \varphi$. Hence, Equation (8.251) becomes

$$\frac{d\sigma_{i \to f}^{\mathrm{abs}}}{d\Omega} = 2^6 \, \alpha \, \hat{I} \, (k_f \, a_0)^3 \, \frac{\sin^2 \theta \, \sin^2 \varphi}{[1 + (q \, a_0)^2]^4} \, a_0^2, \tag{B.8.274}$$

which leads to

$$\frac{d\sigma_{i\to f}^{\text{abs}}}{d\Omega} \simeq 2^6 \, \alpha \, \hat{I}^{7/2} \, (1-\hat{I})^{3/2} \, a_0^2 \, \frac{\sin^2\theta \, \sin^2\varphi}{(1-\beta\cos\theta)^4}. \tag{B.8.275}$$

Now, the photoionization cross-section for unpolarized radiation is

$$\left. \frac{d\sigma_{i\to f}^{\text{abs}}}{d\Omega} \right|_{\text{unpolz}} = \frac{1}{2} \left. \frac{d\sigma_{i\to f}^{\text{abs}}}{d\Omega} \right|_{\boldsymbol{\epsilon}=\mathbf{e}_x} + \frac{1}{2} \left. \frac{d\sigma_{i\to f}^{\text{abs}}}{d\Omega} \right|_{\boldsymbol{\epsilon}=\mathbf{e}_y}. \tag{B.8.276}$$

In other words, the cross-section for unpolarized radiation is the average of Equations (8.261) and (B.8.275). Thus, given that $\cos^2\varphi + \sin^2\varphi = 1$, we obtain

$$\left. \frac{d\sigma_{i\to f}^{\text{abs}}}{d\Omega} \right|_{\text{unpolz}} \simeq 2^5 \, \alpha \, \hat{I}^{7/2} \, (1-\hat{I})^{3/2} \, a_0^2 \, \frac{\sin^2\theta}{(1-\beta\cos\theta)^4}. \tag{B.8.277}$$

8.23 For the photoionization of the $2s$ state of the hydrogen atom, the initial wavefunction is

$$\psi_i(\mathbf{x}) = \frac{1}{\sqrt{\pi}\,2^{3/2}\,a_0^{3/2}} \left(1 - \frac{r}{2a_0}\right) \exp\left(-\frac{r}{2a_0}\right). \tag{B.8.278}$$

Hence, Equation (8.244) is replaced by

$$\left\langle i \left| \boldsymbol{\epsilon}\cdot\mathbf{p}\,\mathrm{e}^{-\mathrm{i}\mathbf{k}\cdot\mathbf{x}} \right| f \right\rangle = \frac{\hbar\,\boldsymbol{\epsilon}\cdot\mathbf{k}_f}{\sqrt{\pi}\,\sqrt{V}\,2^{3/2}\,a_0^{3/2}} \int_V d^3\mathbf{x} \left(1 - \frac{r}{2a_0}\right) \mathrm{e}^{-r/2a_0}\,\mathrm{e}^{-\mathrm{i}\mathbf{q}\cdot\mathbf{x}}. \tag{B.8.279}$$

However,

$$\int_V d^3\mathbf{x} \left(1 - \frac{r}{2a_0}\right) \mathrm{e}^{-r/2a_0}\,\mathrm{e}^{-\mathrm{i}\mathbf{q}\cdot\mathbf{x}} = \frac{2^4\,\pi\,a_0^2}{q} \int_0^\infty dy\, y\,(1-y)\,\sin(2\,q\,a_0\,y)$$

$$= -2^7\,\pi\,a_0^3\,\frac{\left[1 - (2\,q\,a_0)^2\right]}{\left[1 + (2\,q\,a_0)^2\right]^3}, \tag{B.8.280}$$

where use has been made of Exercise 8.20. Thus,

$$\left\langle i \left| \boldsymbol{\epsilon}\cdot\mathbf{p}\,\mathrm{e}^{-\mathrm{i}\mathbf{k}\cdot\mathbf{x}} \right| f \right\rangle = -\frac{2^7\,\sqrt{\pi}\,\hbar\,(\boldsymbol{\epsilon}\cdot\mathbf{k}_f)\,a_0^{3/2}}{\sqrt{V}\,2^{3/2}} \frac{\left[1 - (2\,q\,a_0)^2\right]}{\left[1 + (2\,q\,a_0)^2\right]^3}. \tag{B.8.281}$$

It follows from Equation (8.241) that

$$\frac{d\sigma_{i\to f}^{\text{abs}}}{d\Omega} = \frac{2^{10}\,\alpha\,\hbar}{m_e\,k\,c}\,k_f\,(\boldsymbol{\epsilon}\cdot\mathbf{k}_f)^2\,a_0^3\,\frac{\left[1 - (2\,q\,a_0)^2\right]^2}{\left[1 + (2\,q\,a_0)^2\right]^6}. \tag{B.8.282}$$

8.24 (a) Following the analysis of Section 8.15, the number of photon states contained in a volume $d^3\mathbf{k}$ of wavevector space is $\rho(\mathbf{k})\,d^3\mathbf{k}$, where

$$\rho(\mathbf{k}) = \frac{V}{(2\pi)^3}. \tag{B.8.283}$$

Hence, the number of photon states for which the wavevector has a magnitude lying between k and $k + dk$, and is directed into the range of solid angles $d\Omega$, is $\rho(k)\,dk = \rho(\mathbf{k})\,d^3\mathbf{k}$, where $d^3\mathbf{k} = k^2\,dk\,d\Omega$. Thus,

$$\rho(k) = \frac{V k^2}{(2\pi)^3}\,d\Omega. \tag{B.8.284}$$

Finally, the number of photon states whose angular frequency lies between ω and $\omega + d\omega$. and whose direction of motion is into the range of solid angles $d\Omega$, is $\rho(\omega)\,d\omega = \rho(k)\,(dk/d\omega)\,d\omega$. However, $d\omega/dk = c$ for a photon state. Hence,

$$\rho(\omega) = \frac{V k^2}{(2\pi)^3\,c}\,d\Omega. \tag{B.8.285}$$

(b) The initial electron state is

$$\psi_i(\mathbf{x}) = \frac{1}{\sqrt{V}}\,e^{i\mathbf{k}_i\cdot\mathbf{x}}, \tag{B.8.286}$$

where $\mathbf{p}_j = \hbar\mathbf{k}_i$ is the initial momentum. Obviously, the initial energy is

$$E_i = \frac{p_i^2}{2\,m_e}. \tag{B.8.287}$$

The final electron state is

$$\psi_f(\mathbf{x}) = \frac{1}{\sqrt{\pi}\,a_0^{3/2}}\,e^{-r/a_0}, \tag{B.8.288}$$

where a_0 is the Bohr radius. Thus, the final energy is $E_f = -I$, where $I = (1/2)\,\alpha^2\,m_e\,c^2$.

It follows, from the given expression for $d\sigma_{i\to f}^{\text{stm}}$, as well as that just derived for $\rho(\omega)$, that

$$\frac{d\sigma_{i\to f}^{\text{stm}}}{d\Omega} = \frac{\alpha V k}{2\pi\,m_e^2\,c^2}\sum_{j=1,2}\left|\langle f\,|\,\boldsymbol{\epsilon}\cdot\mathbf{p}\,e^{-i\mathbf{k}\cdot\mathbf{x}}\,|\,i\rangle\right|^2. \tag{B.8.289}$$

Here, use has been made of the electromagnetic dispersion relation $\omega = kc$. Now,

$$\langle f\,|\,\boldsymbol{\epsilon}_j\cdot\mathbf{p}\,e^{-i\mathbf{k}\cdot\mathbf{x}}\,|\,i\rangle = \frac{1}{\sqrt{\pi}\,\sqrt{V}\,a_0^{3/2}}\int_V d^3\mathbf{x}\,e^{-r/a_0}\,\boldsymbol{\epsilon}_j\cdot\mathbf{p}\,e^{-i\mathbf{q}\cdot\mathbf{x}}, \tag{B.8.290}$$

where $\mathbf{q} = \mathbf{k} - \mathbf{k}_i$. Hence, reusing the analysis of Section 8.15, we obtain

$$\langle f \,|\, \boldsymbol{\epsilon}_j \cdot \mathbf{p}\, e^{-i\mathbf{k}\cdot\mathbf{x}} \,|\, i \rangle = \frac{2^3\,\sqrt{\pi}\,\hbar\,\boldsymbol{\epsilon}_j \cdot \mathbf{k}_i\, a_0^{3/2}}{[1 + (q\,a_0)^2]^2}, \tag{B.8.291}$$

which implies that

$$\frac{d\sigma_{i\to f}^{\text{stm}}}{d\Omega} = \frac{2^5\,\alpha\,\hbar^2}{m_e^2\,c^2} \sum_{j=1,2} \frac{k\,(\boldsymbol{\epsilon}_j \cdot \mathbf{k}_i)^2\, a_0^3}{[1 + (q\,a_0)^2]^4}. \tag{B.8.292}$$

Now, we can write

$$\mathbf{k}_i = k_i\, \mathbf{e}_z, \tag{B.8.293}$$

$$\mathbf{k} = k\,(\sin\theta\,\cos\varphi,\, \sin\theta\,\sin\varphi,\, \cos\theta), \tag{B.8.294}$$

$$\boldsymbol{\epsilon}_1 = (\cos\theta\,\cos\varphi,\, \cos\theta\,\sin\varphi,\, -\sin\theta), \tag{B.8.295}$$

$$\boldsymbol{\epsilon}_2 = (-\sin\varphi,\, \cos\varphi,\, 0), \tag{B.8.296}$$

where $\mathbf{k} \cdot \mathbf{k}_i = k\,k_i\,\cos\theta$. It follows that $\boldsymbol{\epsilon}_1 \cdot \mathbf{k}_i = -k_i\,\sin\theta$, and $\boldsymbol{\epsilon}_2 \cdot \mathbf{k}_i = 0$. Thus,

$$\frac{d\sigma_{i\to f}^{\text{stm}}}{d\Omega} = \frac{2^5\,\alpha\,\hbar^2}{m_e^2\,c^2} \frac{k\,k_i^2\,a_0^3\,\sin^2\theta}{[1 + (q\,a_0)^2]^4}. \tag{B.8.297}$$

(c) We have

$$\hbar\,\omega = E_i + I = \frac{\hbar^2\,k_i^2}{2\,m_e} + I. \tag{B.8.298}$$

Hence,

$$(k_i\,a_0)^2 = \frac{1 - \hat{I}}{\hat{I}}, \tag{B.8.299}$$

$$\frac{k_i}{k} = \frac{2}{\beta}\,(1 - \hat{I}), \tag{B.8.300}$$

$$1 + (q\,a_0)^2 = \hat{I}^{-1}\,(1 - \beta\,\cos\theta), \tag{B.8.301}$$

where

$$\hat{I} = \frac{I}{\hbar\,\omega} = \frac{I}{E_i + I}, \tag{B.8.302}$$

$$\beta = \frac{\hbar\,k_i}{m_e\,c} = \left(\frac{2\,E_i}{m_e\,c^2}\right)^{1/2}. \tag{B.8.303}$$

Thus,

$$\frac{d\sigma_{i\to f}^{\text{stm}}}{d\Omega} = 2^4\,\alpha^3\,\beta\,\frac{E_i^{1/2}\,I^{5/2}}{(E_i + I)^3}\,a_0^2\,\frac{\sin^2\theta}{(1 - \beta\,\cos\theta)^4}. \tag{B.8.304}$$

B.9 Chapter 9

9.1 Let $|\lambda\rangle$ denote an eigenket of the permutation operator P_{12} corresponding to the eigenvalue λ. Thus,

$$P_{12}|\lambda\rangle = \lambda|\lambda\rangle. \qquad (B.9.1)$$

However, according to Equation (9.6),

$$P_{12}^2 = 1. \qquad (B.9.2)$$

Hence,

$$P_{12}^2|\lambda\rangle = \lambda P_{12}|\lambda\rangle = \lambda^2|\lambda\rangle = |\lambda\rangle. \qquad (B.9.3)$$

Assuming that the $|\lambda\rangle$ form a complete set (see the following exercise), the previous equation implies that

$$\lambda^2 = 1, \qquad (B.9.4)$$

which only has the solutions $\lambda = \pm 1$.

9.2 A general two-particle state ket in which one particle has the set of eigenvalues k', and the other has the set of eigenvalues k'', is written

$$|k'\,k''\rangle = c_1\,|k'\rangle\,|k''\rangle + c_2\,|k''\rangle\,|k'\rangle, \qquad (B.9.5)$$

where c_1 and c_2 are arbitrary complex numbers. Now, according to the previous exercise, the two eigenkets of the two-particle permutation operator are such that $P_{12}\,|k'\,k''\rangle_\pm = \pm|k'\,k''\rangle_\pm$. Here, $\pm$ correspond to the eigenvalues ± 1, respectively. This implies that

$$P_{12}\,|k'\,k''\rangle_\pm = c_1\,|k''\rangle\,|k'\rangle + c_2\,|k'\rangle\,|k''\rangle = \pm|k'\,k''\rangle_\pm = \pm c_1\,|k'\rangle\,|k''\rangle \pm c_2\,|k''\rangle\,|k'\rangle, \qquad (B.9.6)$$

because $P_{12}\,|k'\rangle\,|k''\rangle = |k''\rangle\,|k'\rangle$, et cetera. Thus,

$$c_1 = \pm c_2. \qquad (B.9.7)$$

Hence, assuming that $\langle k'|k''\rangle = \delta_{k'\,k''}$, the properly normalized eigenkets of P_{12} are

$$|k'\,k''\rangle_+ = \frac{1}{\sqrt{2}}\,(|k'\rangle\,|k''\rangle + |k''\rangle\,|k'\rangle), \qquad (B.9.8)$$

$$|k'\,k''\rangle_- = \frac{1}{\sqrt{2}}\,(|k'\rangle\,|k''\rangle - |k''\rangle\,|k'\rangle). \qquad (B.9.9)$$

Note that the $|k'\,k''\rangle_\pm$ span ket space (because the $|k'\rangle\,|k''\rangle$ span ket space).

9.3 Consider, for example, a system of three identical particles. We can write

$$\langle k'|\,\langle k''|\,\langle k'''|l'\rangle\,|l''\rangle\,|l'''\rangle = \delta_{k'\,l'}\,\delta_{k''\,l''}\,\delta_{k'''\,l'''}. \qquad (B.9.10)$$

Moreover,

$$P_{12}\,|l'\rangle\,|l''\rangle\,|l'''\rangle = |l''\rangle\,|l'\rangle\,|l'''\rangle \qquad (B.9.11)$$

and

$$\langle l'|\langle l''|\langle l'''|\,P_{12}^\dagger = \langle l''|\langle l'|\langle l'''|. \tag{B.9.12}$$

The previous equation also implies that

$$\langle k'|\langle k''|\langle k'''|\,P_{12}^\dagger = \langle k''|\langle k'|\langle k'''|. \tag{B.9.13}$$

Thus, Equations (B.9.11) and (B.9.13) yield

$$\langle k'|\langle k''|\langle k'''|\,P_{12}\,|l'\rangle\,|l''\rangle\,|l'''\rangle = \delta_{k'\,l''}\,\delta_{k''\,l'}\,\delta_{k'''\,l'''}, \tag{B.9.14}$$

and

$$\langle k'|\langle k''|\langle k'''|\,P_{12}^\dagger\,|l'\rangle\,|l''\rangle\,|l'''\rangle = \delta_{k''\,l'}\,\delta_{k'\,l''}\,\delta_{k'''\,l'''} = \delta_{k'\,l''}\,\delta_{k''\,l'}\,\delta_{k'''\,l'''}, \tag{B.9.15}$$

respectively. It follows that

$$\langle k'|\langle k''|\langle k'''|\,P_{12}^\dagger\,|l'\rangle\,|l''\rangle\,|l'''\rangle = \langle k'|\langle k''|\langle k'''|\,P_{12}\,|l'\rangle\,|l''\rangle\,|l'''\rangle. \tag{B.9.16}$$

The fact that P_{12} and $P_{12}^\dagger$ have identical matrix elements implies that $P_{12}^\dagger = P_{12}$. This proof can be extended to deal with a system containing an arbitrary number of identical particles.

9.4   Making use of Equations (9.26) and (9.33), we have

$$P_{123}\,H\,|k'\rangle\,|k''\rangle\,|k'''\rangle = E\,P_{123}\,|k'\rangle\,|k''\rangle\,|k'''\rangle = E\,|k'''\rangle\,|k'\rangle\,|k''\rangle. \tag{B.9.17}$$

However, Equations (9.28) and (9.33) yield

$$H\,P_{123}\,|k'\rangle\,|k''\rangle\,|k'''\rangle = H\,|k'''\rangle\,|k'\rangle\,|k''\rangle = E\,|k'''\rangle\,|k'\rangle\,|k''\rangle. \tag{B.9.18}$$

The previous two equations imply that

$$P_{123}\,H\,|k'\rangle\,|k''\rangle\,k'''\rangle = H\,P_{123}\,|k'\rangle\,|k''\rangle\,|k'''\rangle. \tag{B.9.19}$$

Because the $|k'\rangle\,|k''\rangle\,|k'''\rangle$ form a complete set, we deduce that

$$P_{123}\,H = H\,P_{123}. \tag{B.9.20}$$

9.5   We have, from Equation (9.34),

$$P_{123} = P_{12}\,P_{31} = P_{23}\,P_{12} = P_{31}\,P_{23}. \tag{B.9.21}$$

Hence,

$$P_{123}^2 = P_{23}\,P_{12}\,P_{12}\,P_{31} = P_{23}\,P_{31}, \tag{B.9.22}$$

because $P_{12}^2 = 1$. [See Equation (9.25).] But,

$$P_{123}^\dagger = P_{23}^\dagger\,P_{31}^\dagger = P_{23}\,P_{31}, \tag{B.9.23}$$

because the two-particle permutation operators are Hermitian. (See Exercise 9.3.) So,

$$P_{123}^2 = P_{123}^\dagger. \tag{B.9.24}$$

Finally,

$$P_{123}^3 = P_{123}^2 P_{123} = P_{23} P_{31} P_{31} P_{23} = 1, \tag{B.9.25}$$

because $P_{31}^2 = P_{23}^2 = 1$. [See Equation (9.25).] Hence,

$$P_{123} P_{123}^\dagger = 1. \tag{B.9.26}$$

9.6 Consider a one-dimensional box extending from $x = 0$ to $x = L$. The possible single-particle energies and properly normalized spatial wavefunctions are [Fitzpatrick (2013)]

$$E_n = \frac{\hbar^2 \pi^2}{2 m L^2} n^2, \tag{B.9.27}$$

and

$$\psi_n(x) = \left(\frac{2}{L}\right)^{1/2} \sin\left(n \pi \frac{x}{L}\right), \tag{B.9.28}$$

respectively, where m is the particle mass, and n a positive integer.

Generalizing to three dimensions, the single-particle energies and wavefunctions are

$$E_{n_1 n_2 n_3} = \frac{\hbar^2 \pi^2}{2 m L^2} (n_1^2 + n_2^2 + n_3^2), \tag{B.9.29}$$

and

$$\psi_{n_1 n_2 n_3}(\mathbf{x}) = \left(\frac{2}{L}\right)^{3/2} \sin\left(n_1 \pi \frac{x}{L}\right) \sin\left(n_2 \pi \frac{y}{L}\right) \sin\left(n_3 \pi \frac{z}{L}\right), \tag{B.9.30}$$

respectively, where n_1, n_2, and n_3 are positive integers.

Consider the case of two identical non-interacting spin-1/2 particles. Let n_1, n_2, n_3 be the quantum numbers of one particle, and n_1', n_2', n_3' the quantum numbers of the other. Likewise, let $\mathbf{x}$ be the position vector of one particle, and $\mathbf{x}'$ the position vector of the other. For the singlet (i.e., overall spin zero) case, the spatial wavefunction must be symmetric with respect to interchange of particles. For the triplet (i.e., overall spin one) case, the spatial wavefunction must be antisymmetric. (See Section 9.4.)

For the singlet state, the spatial wavefunction is

$$\psi(\mathbf{x}, \mathbf{x}') = \frac{1}{\sqrt{2}} \left[\psi_{n_1 n_2 n_3}(\mathbf{x}) \psi_{n_1' n_2' n_3'}(\mathbf{x}') + \psi_{n_1' n_2' n_3'}(\mathbf{x}) \psi_{n_1 n_2 n_3}(\mathbf{x}') \right], \tag{B.9.31}$$

and the associated energy is written

$$E = \frac{\hbar^2 \pi^2}{2 m L^2} \left(n_1^2 + n_2^2 + n_3^2 + n_1'^2 + n_2'^2 + n_3'^2 \right). \tag{B.9.32}$$

There is no restriction on the values that the quantum numbers can take (other that that they must all be positive integers).

For the triplet state, the spatial wavefunction is

$$\psi(\mathbf{x}, \mathbf{x}') = \frac{1}{\sqrt{2}} \left[\psi_{n_1 n_2 n_3}(\mathbf{x}) \, \psi_{n_1' n_2' n_3'}(\mathbf{x}') - \psi_{n_1' n_2' n_3'}(\mathbf{x}) \, \psi_{n_1 n_2 n_3}(\mathbf{x}') \right], \qquad \text{(B.9.33)}$$

and the associated energy is written

$$E = \frac{\hbar^2 \pi^2}{2 m L^2} \left(n_1^2 + n_2^2 + n_3^2 + {n_1'}^2 + {n_2'}^2 + {n_3'}^2 \right). \qquad \text{(B.9.34)}$$

In this case, there is an additional restriction that the set of quantum numbers n_1, n_2, and n_3 cannot be identical to the set n_1', n_2', n_3', because the corresponding spatial wavefunction would be null.

9.7 Two non-identical spin-1 particles possessing no orbital angular momentum can form a system of total angular momentum 2, 1, or 0. (See Chapter 6.) In the case of a total angular-momentum-2 state, the eigenvalue of the total angular momentum is $6\hbar^2$, and the possible eigenvalues of the projection of the overall angular momentum along the z-axis are $2\hbar$, $\hbar$, 0, $-\hbar$, and $-2\hbar$. (See Chapter 4.) In the case of a total angular-momentum-1 state, the eigenvalue of the total angular momentum is $2\hbar^2$, and the possible eigenvalues of the projection of the overall angular momentum along the z-axis are $\hbar$, 0, and $-\hbar$. Finally, in the case of an total angular-momentum-0 state, the eigenvalue of the total angular momentum is 0, and the only possible eigenvalue of the projection of the overall angular momentum along the z-axis is 0.

If the two particles are identical then the overall wavefunction must be symmetric with respect to interchange of particles. However, the single-particle spatial wavefunctions of both particles are identical (because they both have the same s-state wavefunction). Thus, it is not possible to form an antisymmetric overall spatial wavefunction: that is, the spatial wavefunction is necessarily symmetric. Thus, the overall spin wavefunction must also be symmetric. This is the case for the overall spin-2 and spin-0 spin wavefunctions, but not for the overall spin-1 spin wavefunction (which is antisymmetric). Thus, if the two particles are identical then they cannot be in a state of overall spin-1.

9.8 Essentially, we need to redo the analysis of Section 4.6, making the substitution $e^2 \rightarrow Z e^2$. However, it is evident from Appendix A that this substitution is equivalent to $\alpha \rightarrow Z\alpha$, $a_0 \rightarrow a_0/Z$, and $E_0 \rightarrow Z^2 E_0$.

(a) With the substitution $E_0 \rightarrow Z^2 E_0$, Equation (4.132) yields

$$E_n = \frac{Z^2 E_0}{n^2}. \qquad \text{(B.9.35)}$$

(b) With the substitution $a_0 \to a_0/Z$, Exercise 4.12, parts (a), (b), and (c), yield

$$R_{10}(r) = \frac{2Z^{3/2}}{a_0^{3/2}} \exp\left(-\frac{Zr}{a_0}\right), \tag{B.9.36}$$

$$R_{20}(r) = \frac{2Z^{3/2}}{(2a_0)^{3/2}} \left(1 - \frac{Zr}{2a_0}\right) \exp\left(-\frac{Zr}{2a_0}\right), \tag{B.9.37}$$

$$R_{21}(r) = \frac{Z^{3/2}}{\sqrt{3}(2a_0)^{3/2}} \frac{Zr}{a_0} \exp\left(-\frac{Zr}{2a_0}\right), \tag{B.9.38}$$

respectively.

9.9 It is evident that

$$\int dx_2 \, x_2^n \, e^{-\beta x_2} \equiv (-1)^n \frac{\partial^n}{\partial \beta^n} \int dx_2 \, e^{-\beta x_2}. \tag{B.9.39}$$

Hence,

$$
\begin{aligned}
\int_0^{x_1} dx_2 \, x_2^n \, e^{-\beta x_2} &= (-1)^n \frac{\partial^n}{\partial \beta^n} \left[\beta^{-1}\left(1 - e^{-\beta x_1}\right)\right] \\
&= (-1)^n \frac{\partial^n}{\partial \beta^n} \left(\beta^{-1}\right) - (-1)^n \frac{\partial^n}{\partial \beta^n} \left(\beta^{-1} e^{-\beta x_1}\right) \\
&= (-1)^n \frac{\partial^n}{\partial \beta^n} \left(\beta^{-1}\right) \\
&\quad - (-1)^n \sum_{k=0,n} \frac{n!}{k!\,(n-k)!} \frac{\partial^{n-k}}{\partial \beta^{n-k}} \left(\beta^{-1}\right) \frac{\partial^k}{\partial \beta^k} \left(e^{-\beta x_1}\right).
\end{aligned}
\tag{B.9.40}
$$

However,

$$\frac{\partial^n}{\partial \beta^n}\left(\beta^{-1}\right) = \frac{(-1)^n \, n!}{\beta^{n+1}}, \tag{B.9.41}$$

$$\frac{\partial^n}{\partial \beta^n}\left(e^{-\beta x_1}\right) = (-1)^n \, x_1^n \, e^{-\beta x_1}. \tag{B.9.42}$$

Thus, we obtain

$$\int_0^{x_1} dx_2 \, x_2^n \, e^{-\beta x_2} = \frac{n!}{\beta^{n+1}} - \frac{e^{-\beta x_1}}{\beta^{n+1}} \sum_{k=0,n} \frac{n!}{k!} (\beta x_1)^k. \tag{B.9.43}$$

Taking the limit $x_1 \to \infty$, it is clear that

$$\int_0^\infty dx_2 \, x_2^n \, e^{-\beta x_2} = \frac{n!}{\beta^{n+1}}. \tag{B.9.44}$$

Thus,

$$\int_{x_1}^{\infty} dx_2\, x_2^n\, e^{-\beta x_2} = \int_0^{\infty} dx_2\, x_2^n\, e^{-\beta x_2} - \int_0^{x_1} dx_2\, x_2^n\, e^{-\beta x_2}$$

$$= \frac{e^{-\beta x_1}}{\beta^{n+1}} \sum_{k=0,n} \frac{n!}{k!} (\beta x_1)^k. \tag{B.9.45}$$

9.10 It follows from the previous exercise that

$$F(\beta, x_1) \equiv \int_0^{x_1} dx_2\, x_2^2\, e^{-\beta x_2} + x_1 \int_{x_1}^{\infty} dx_2\, x_2\, e^{-\beta x_2}$$

$$= \frac{2}{\beta^3} - \frac{1}{\beta^3} \left[2 + 2\beta x_1 + (\beta x_1)^2 \right] e^{-\beta x_1} + \frac{x_1}{\beta^2} (1 + \beta x_1) e^{-\beta x_1}$$

$$= \frac{2}{\beta^3} - \frac{1}{\beta^3} (2 + \beta x_1) e^{-\beta x_1}. \tag{B.9.46}$$

Thus,

$$\frac{\Delta E}{|E_0|} = 2^5 Z \int_0^{\infty} dx_1\, x_1\, F(2, x_1)\, e^{-2 x_1}$$

$$= 2^3 Z \int_0^{\infty} dx_1 \left[x_1\, e^{-2 x_1} - x_1 (1 + x_1) e^{-4 x_1} \right]$$

$$= 2^3 Z \left[\left(\frac{1}{2^2} - \frac{1}{2^4} \right) I_1 - \frac{I_2}{2^6} \right] = \frac{5 Z}{4}. \tag{B.9.47}$$

Here, $I_n = n!$ is defined in Exercise 4.11.

9.11 The ground-state energy of a singly-ionized helium ion is the same as that of the ground-state energy of a hydrogen atom calculated with nuclear charge $2e$. Thus, it follows from Exercise 9.8 that

$$E_{\text{He+}} = 4 E_0 = -54.42 \,\text{eV}, \tag{B.9.48}$$

where E_0 is the conventional hydrogen ground-state energy. The ground-state ionization energy is simply the difference between the energy of the ground-state helium ion and that of the ground-state helium atom. The latter energy is $E_{\text{He}} = -78.98 \,\text{eV}$. Thus,

$$E_{\text{ion}} = E_{\text{He+}} - E_{\text{He}} = -54.42 + 78.98 = 24.56 \,\text{eV}. \tag{B.9.49}$$

9.12 We can write

$$E(2p\, 2p) = 2 E_{210} + \left\langle \frac{e^2}{4\pi\, \epsilon_0\, r_{12}} \right\rangle, \tag{B.9.50}$$

where $E_{210} = E_0/4$, and E_0 is the hydrogen ground-state energy. Here, E_{nlm} denotes the energy of a hydrogen atom characterized by the quantum numbers n, l, and m. The wavefunction takes the form

$$\phi(\mathbf{x}_1, \mathbf{x}_2) = \psi_{210}(\mathbf{x}_1)\,\psi_{210}(\mathbf{x}_2), \tag{B.9.51}$$

where

$$\psi_{210}(\mathbf{x}) = \frac{2}{\sqrt{3}\,a_0^{3/2}}\,\frac{r}{a_0}\,\exp\left(-\frac{r}{a}\right). \tag{B.9.52}$$

Here, use has been made of Exercise 9.8. Thus,

$$\frac{E(2p\,2p)}{|E_0|} = -2 + 2\left\langle\frac{a_0}{r_>}\right\rangle. \tag{B.9.53}$$

Now,

$$2\left\langle\frac{a_0}{r_>}\right\rangle = \frac{2^5}{3^2\,a_0^5}\int_0^\infty dr_1\,r_1\left(\frac{r_1}{a_0}\right)^2 e^{-2r_1/a_0}\left[\int_0^{r_2} dr_2\,r_2^2\left(\frac{r_2}{a_0}\right)^2 e^{-2r_2/a_0}\right.$$
$$\left. +r_1\int_{r_1}^\infty dr_2\,r_2\left(\frac{r_2}{a_0}\right)^2 e^{-2r_2/a_0}\right], \tag{B.9.54}$$

which yields

$$2\left\langle\frac{a_0}{r_>}\right\rangle = \frac{1}{2^4\,3^2}\int_0^\infty dx_1\,x_1^3\,e^{-x_1}\left[\int_0^{x_2} dx_2\,x_2^4\,e^{-x_2} + x_1\int_{x_1}^\infty dx_2\,x_2^3\,e^{-x_2}\right], \tag{B.9.55}$$

where $x_1 = 2r_1/a_0$ and $x_2 = 2r_2/a_0$. Making use of Exercise 9.9, the previous expression reduces to

$$2\left\langle\frac{a_0}{r_>}\right\rangle = \frac{1}{2^4\,3^2}\int_0^\infty dx_1\,x_1^3\,e^{-x_1}\left[24 - (24 + 18\,x_1 + 6\,x_1^2 + x_1^3)e^{-x_1}\right]. \tag{B.9.56}$$

Hence,

$$2\left\langle\frac{a_0}{r_>}\right\rangle = \frac{1}{2^4\,3^2}\left[24\,I_3 - \left(\frac{24\,I_3}{2^4} + \frac{18\,I_4}{2^5} + \frac{6\,I_5}{2^6} + \frac{I_6}{2^7}\right)\right], \tag{B.9.57}$$

where $I_n = n!$ is defined in Exercise 4.11. Thus,

$$2\left\langle\frac{a_0}{r_>}\right\rangle = \frac{3\cdot 31}{2^7}, \tag{B.9.58}$$

which gives

$$E(2p\,2p) = \left(-2 + \frac{3\cdot 31}{2^7}\right)|E_0| = -17.33\,\text{eV}. \tag{B.9.59}$$

9.13 (a) We have, from Chapter 4 and Section 9.5,

$$\psi_{000}(\mathbf{x}_{1,2}) = \frac{1}{\sqrt{4\pi}}\, R_{10}(r_{1,2}), \tag{B.9.60}$$

$$\psi_{nlm}(\mathbf{x}_{1,2}) = Y_{lm}(\theta_{1,2}, \varphi_{1,2})\, R_{nl}(r_{1,2}), \tag{B.9.61}$$

$$\frac{1}{r_{12}} = \sum_{l'=0,\infty}^{m'=-l',+l'} \frac{4\pi}{2\,l'+1} \frac{r_<^{l'}}{r_>^{l'+1}}\, Y_{l'\,m'}^*(\theta_1,\varphi_1)\, Y_{l'\,m'}(\theta_2,\varphi_2), \tag{B.9.62}$$

$$\delta_{ll'}\, \delta_{mm'} = \oint d\Omega\, Y_{lm}^*(\theta,\varphi)\, Y_{l'\,m'}(\theta,\varphi) = \oint d\Omega\, Y_{lm}(\theta,\varphi)\, Y_{l'\,m'}^*(\theta,\varphi), \tag{B.9.63}$$

$$\sqrt{4\pi}\, \delta_{l0}\, \delta_{m0} = \oint d\Omega\, Y_{lm}(\theta,\varphi) = \oint d\Omega\, Y_{lm}^*(\theta,\varphi). \tag{B.9.64}$$

Now, from Equation (9.67),

$$\frac{I}{|E_0|} = \sum_{l'=0,\infty}^{m'=-l',+l'} \frac{2\,a_0}{2\,l'+1} \int_0^\infty dr_1\, r_1^2 \oint d\Omega_1 \int_0^\infty dr_2\, r_2^2 \oint d\Omega_2$$

$$[R_{10}(r_1)]^2\, |Y_{lm}(\theta_2,\varphi_2)|^2\, [R_{nl}(r_2)]^2 \frac{r_<^{l'}}{r_>^{l'+1}}\, Y_{l'\,m'}^*(\theta_1,\varphi_1)\, Y_{l'\,m'}(\theta_2,\varphi_2), \tag{B.9.65}$$

where use has been made of Equations (B.9.60)–(B.9.62). It follows from Equation (B.9.64) that

$$\frac{I}{|E_0|} = 2\,a_0 \int_0^\infty dr_1\, r_1^2 \int_0^\infty dr_2\, r_2^2 \oint d\Omega_2$$

$$[R_{10}(r_1)]^2\, |Y_{lm}(\theta_2,\varphi_2)|^2\, [R_{nl}(r_2)]^2 \frac{1}{r_>}\, \sqrt{4\pi}\, Y_{00}(\theta_2,\varphi_2). \tag{B.9.66}$$

But, $\sqrt{4\pi}\, Y_{00}(\theta_2,\varphi_2) = 1$. Hence, Equation (B.9.63) yields

$$\frac{I}{|E_0|} = 2\,a_0 \int_0^\infty dr_1\, r_1^2 \int_0^\infty dr_2\, r_2^2\, [R_{10}(r_1)]^2\, [R_{nl}(r_2)]^2 \frac{1}{r_>}, \tag{B.9.67}$$

which can also be written

$$\frac{I}{|E_0|} = 2\,a_0 \int_0^\infty dr_1\, r_1\, R_{10}(r_1)\, R_{10}(r_1) \left[\int_0^{r_1} dr_2\, r_2^2\, R_{nl}(r_2)\, R_{nl}(r_2) \right.$$

$$\left. + \int_{r_1}^\infty dr_2\, r_1\, r_2\, R_{nl}(r_2)\, R_{nl}(r_2) \right]. \tag{B.9.68}$$

Now, from Equation (9.68),

$$\frac{J}{|E_0|} = \sum_{l'=0,\infty}^{m'=-l',+l'} \frac{2\,a_0}{2\,l'+1} \int_0^\infty dr_1\, r_1^2 \oint d\Omega_1 \int_0^\infty dr_2\, r_2^2 \oint d\Omega_2$$

$$R_{10}(r_1)\, Y_{lm}(\theta_2,\varphi_2)\, R_{nl}(r_2)\, R_{10}(r_2)\, Y_{lm}^*(\theta_1,\varphi_1)\, R_{nl}(r_1)$$

$$\frac{r_<^{l'}}{r_>^{l'+1}}\, Y_{l'm'}(\theta_1,\varphi_1)\, Y_{l'm'}^*(\theta_2,\varphi_2), \tag{B.9.69}$$

where use has been made of Equations (B.9.60)–(B.9.62). It follows from Equation (B.9.63) that

$$\frac{J}{|E_0|} = \frac{2\,a_0}{2\,l+1} \int_0^\infty dr_1\, r_1^2 \int_0^\infty dr_2\, r_2^2\, R_{10}(r_1)\, R_{nl}(r_2)\, R_{10}(r_2)\, R_{nl}(r_1)\, \frac{r_<^l}{r_>^{l+1}}, \tag{B.9.70}$$

which can also be written

$$\frac{J}{|E_0|} = \frac{2\,a_0}{2\,l+1} \int_0^\infty dr_1\, r_1\, R_{10}(r_1)\, R_{nl}(r_1)\left[\int_0^{r_1} dr_2\, r_2^2 \left(\frac{r_2}{r_1}\right)^l R_{10}(r_2)\, R_{nl}(r_2)\right.$$

$$\left.+ \int_{r_1}^\infty dr_2\, r_1\, r_2 \left(\frac{r_1}{r_2}\right)^l R_{10}(r_2)\, R_{nl}(r_2)\right]. \tag{B.9.71}$$

(b) According to Exercise 9.8, with $Z = 2$, we have

$$R_{10}(r) = \frac{2^2\sqrt{2}}{a_0^{3/2}}\, e^{-2r/a_0}, \tag{B.9.72}$$

$$R_{21}(r) = \frac{2}{\sqrt{3}\, a_0^{3/2}}\, \frac{r}{a_0}\, e^{-r/a_0}. \tag{B.9.73}$$

Thus, it follows from Part (a) that

$$\frac{I}{|E_0|} = \frac{2^8}{3\,a_0^5} \int_0^\infty dr_1\, r_1\, e^{-4r_1/a_0}\left[\int_0^{r_1} dr_2\, r_2^2 \left(\frac{r_2}{a_0}\right)^2 e^{-2r_2/a_0}\right.$$

$$\left.+ r_1 \int_{r_1}^\infty dr_2\, r_2 \left(\frac{r_2}{a_0}\right)^2 e^{-2r_2/a_0}\right]. \tag{B.9.74}$$

Let $x_1 = 2\,r_1/a_0$ and $x_2 = 2\,r_2/a_0$. The previous expression reduces to

$$\frac{I}{|E_0|} = \frac{2}{3} \int_0^\infty dx_1\, x_1\, e^{-2x_1}\left[\int_0^{x_1} dx_2\, x_2^4\, e^{-x_2} + x_1 \int_{x_1}^\infty dx_2\, x_2^3\, e^{-x_2}\right]. \tag{B.9.75}$$

Making use of Exercise 9.9, we obtain

$$\frac{I}{|E_0|} = \frac{2}{3} \int_0^\infty dx_1 \, x_1 \, e^{-2x_1} \left[24 - (24 + 18\,x_1 + 6\,x_1^2 + x_1^3)\,e^{-x_1} \right], \qquad (B.9.76)$$

which leads to

$$\frac{I}{|E_0|} = 2^2 \, I_1 - \frac{2}{3^3} \left[2^3 \cdot 3\,I_1 + 2 \cdot 3\,I_2 + \frac{2}{3}\,I_3 + \frac{I_4}{3^3} \right], \qquad (B.9.77)$$

where $I_n = n!$ is defined in Exercise 4.11. Thus,

$$\frac{I}{|E_0|} = \frac{2^2 \cdot 59}{3^5} = 0.9712. \qquad (B.9.78)$$

It follows from Part (a) that

$$\frac{J}{|E_0|} = \frac{2^8}{3^2 \, a_0^5} \int_0^\infty dr_1 \, \frac{r_1}{a_0} \, e^{-3r_1/a_0} \left[\int_0^{r_1} dr_2 \, r_2^3 \, \frac{r_2}{a_0} \, e^{-3r_2/a_0} \right.$$
$$\left. + r_1^3 \int_{r_1}^\infty dr_2 \, \frac{r_2}{a_0} \, e^{-3r_2/a_0} \right]. \qquad (B.9.79)$$

Let $x_1 = 3\,r_1/a_0$ and $x_2 = 3\,r_2/a_0$. The previous expression reduces to

$$\frac{J}{|E_0|} = \frac{2^8}{3^9} \int_0^\infty dx_1 \, x_1 \, e^{-x_1} \left[\int_0^{x_1} dx_2 \, x_2^4 \, e^{-x_2} + x_1^3 \int_{x_1}^\infty dx_2 \, x_2 \, e^{-x_2} \right]. \qquad (B.9.80)$$

According to Exercise 9.9, we can write

$$\frac{J}{|E_0|} = \frac{2^8}{3^9} \int_0^\infty dx_1 \, x_1 \, e^{-x_1} \left[24 - \left(24 + 24\,x_1 + 12\,x_1^2 + 3\,x_1^3 \right) e^{-x_1} \right], \qquad (B.9.81)$$

which leads to

$$\frac{J}{|E_0|} = \frac{2^{11}}{3^8} \, I_1 - \frac{2^6}{3^9} \left(24\,I_1 + 12\,I_2 + 3\,I_3 + \frac{3}{8}\,I_4 \right), \qquad (B.9.82)$$

where $I_n = n!$ is defined in Exercise 4.11. Thus,

$$\frac{J}{|E_0|} = \frac{2^6 \cdot 7}{3^8} = 0.0683. \qquad (B.9.83)$$

Finally, it follows from Equations (9.65) and (9.66) that

$$E(1s\,2p)_\pm = (-4 - 1 + 0.9712 \pm 0.0683)\,|E_0|$$
$$= (-4.0288 \pm 0.0683)\,|E_0| = (-54.81 \pm 0.93)\,\text{eV}, \qquad (B.9.84)$$

where the upper sign corresponds to parahelium.

9.14 According to Exercise 9.8, with $Z = 2$ and $Z = 1$, respectively, we have

$$R_{10}(r) = \frac{2^2 \sqrt{2}}{a_0^{3/2}} e^{-2r/a_0}, \tag{B.9.85}$$

$$R_{21}(r) = \frac{1}{\sqrt{3}\, 2^{3/2}\, a_0^{3/2}} \frac{r}{a_0} e^{-r/2 a_0}. \tag{B.9.86}$$

Thus, it follows from the previous exercise that

$$\frac{J}{|E_0|} = \frac{2^3}{3^2\, a_0^5} \int_0^\infty dr_1 \frac{r_1}{a_0} e^{-5 r_1/2 a_0} \left[\int_0^{r_1} dr_2\, r_2^3 \frac{r_2}{a_0} e^{-5 r_2/a_0} \right.$$
$$\left. + r_1^3 \int_{r_1}^\infty dr_2 \frac{r_2}{a_0} e^{-5 r_2/a_0} \right]. \tag{B.9.87}$$

Let $x_1 = 5 r_1/2 a_0$ and $x_2 = 5 r_2/2 a_0$. The previous expression reduces to

$$\frac{J}{|E_0|} = \frac{2^{10}}{3^2\, 5^7} \int_0^\infty dx_1\, x_1 e^{-x_1} \left[\int_0^{x_1} dx_2\, x_2^4\, e^{-x_2} + x_1^3 \int_{x_1}^\infty dx_2 x_2 e^{-x_2} \right]. \tag{B.9.88}$$

Comparing with Equations (B.9.80) and (B.9.83), we obtain

$$\frac{J}{|E_0|} = \frac{2^8 \cdot 7}{5^7 \cdot 3} = 0.00765. \tag{B.9.89}$$

9.15 We have

$$J_0(\beta) = \int_{-\infty}^\infty dx\, e^{-\beta x^2} = \frac{1}{\sqrt{\beta}} \int_{-\infty}^\infty dy\, e^{-y^2}, \tag{B.9.90}$$

where $y = \sqrt{\beta}\, x$. Thus,

$$J_0^2 = \frac{1}{\beta} \int_{-\infty}^\infty dx\, e^{-x^2} \int_{-\infty}^\infty dy\, e^{-y^2} = \frac{1}{\beta} \int_{-\infty}^\infty \int_{-\infty}^\infty dx\, dy\, e^{-(x^2+y^2)}. \tag{B.9.91}$$

Let $r = \sqrt{x^2 + y^2}$. It follows that $dx\, dy = 2\pi\, r\, dr$. Hence,

$$J_0^2 = \frac{2\pi}{\beta} \int_0^\infty dr\, r\, e^{-r} = \frac{\pi}{\beta} I_1 = \frac{\pi}{\beta}, \tag{B.9.92}$$

where $I_n = n!$ is defined in Exercise 4.11. Thus,

$$J_0(\beta) = \left(\frac{\pi}{\beta} \right)^{1/2}. \tag{B.9.93}$$

Now,

$$J_n(\beta) = (-1)^n \left(\frac{\partial}{\partial \beta} \right)^n \int_{-\infty}^\infty dx\, e^{-\beta x^2} = (-1)^n \frac{\partial^n J_0}{\partial \beta^n} = \pi^{1/2} (-1)^n \frac{\partial^n \beta^{-1/2}}{\partial \beta^n}. \tag{B.9.94}$$

It follows that

$$J_{n>0}(\beta) = \frac{1 \cdot 3 \cdot 5 \cdots (2n-1)}{2^n} \left(\frac{\pi}{\beta^{2n+1}} \right)^{1/2}. \qquad (B.9.95)$$

9.16 Let the E_n and the $|n\rangle$ be the true eigenvalues and eigenkets of H, respectively. That is,

$$H |n\rangle = E_n |n\rangle, \qquad (B.9.96)$$

where

$$E_0 < E_1 < E_2 < \cdots. \qquad (B.9.97)$$

The $|n\rangle$ are assumed to be orthonormal, so that $\langle n|m\rangle = \delta_{nm}$. We can write

$$|\psi\rangle = \sum_n c_n |n\rangle, \qquad (B.9.98)$$

where the c_n are complex numbers. It follows that

$$\langle \psi|\psi\rangle = \sum_n |c_n|^2, \qquad (B.9.99)$$

$$\langle \psi| H |\psi\rangle = \sum_n E_n |c_n|^2. \qquad (B.9.100)$$

Hence,

$$\langle H \rangle = \frac{\langle \psi| H |\psi\rangle}{\langle \psi|\psi\rangle} = \frac{\sum_n E_n |c_n|^2}{\sum_n |c_n|^2}, \qquad (B.9.101)$$

which implies that

$$\langle H \rangle - E_0 = \frac{\sum_{n>0}(E_n - E_0)|c_n|^2}{\sum_n |c_n|^2} \geq 0. \qquad (B.9.102)$$

9.17 Now,

$$\int_{-\infty}^{\infty} dx \, [\psi_0(x)]^2 = \frac{\alpha}{\pi^{1/2}} \int_{-\infty}^{\infty} dx \, e^{-\alpha^2 x^2} = \frac{\alpha}{\pi^{1/2}} J_0(\alpha^2) = \frac{\alpha}{\pi^{1/2}} \frac{\pi^{1/2}}{\alpha} = 1, \quad (B.9.103)$$

where $J_n(\beta)$ is defined in Exercise 9.15. Furthermore,

$$\int_{-\infty}^{\infty} dx \, [\psi_1(x)]^2 = \frac{2\beta^{3/2}}{\pi^{1/2}} \int_{-\infty}^{\infty} dx \, x^2 \, e^{-\beta^2 x^2} = \frac{2\beta^3}{\pi^{1/2}} J_1(\beta^2)$$

$$= \frac{2\beta^3}{\pi^{1/2}} \frac{\pi^{1/2}}{2\beta^3} = 1. \qquad (B.9.104)$$

Note, incidentally, that we can be sure that $\psi_1(x)$ is orthogonal to the true ground-state wavefunction of the system, because we expect the ground-state wavefunction to be even in x, whereas $\psi_1(x)$ is odd. (See Exercise 3.3.)

The Hamiltonian of the system is written

$$H = -\frac{\hbar^2}{2m} \frac{d^2}{dx^2} + \lambda x^4. \qquad (B.9.105)$$

Thus, the expectation value of the Hamiltonian, calculated with the trial wavefunction $\psi_0(x)$, is

$$E(\alpha) = \frac{\alpha}{\pi^{1/2}} \int_{-\infty}^{\infty} dx \, e^{-\alpha^2 x^2/2} \left(-\frac{\hbar^2}{2m} \frac{d^2}{dx^2} + \lambda x^4 \right) e^{-\alpha^2 x^2/2}, \qquad (B.9.106)$$

which yields

$$E(\alpha) = \frac{\alpha}{\pi^{1/2}} \int_{-\infty}^{\infty} dx \, e^{-\alpha^2 x^2/2} \left[-\frac{\hbar^2}{2m} (\alpha^4 x^2 - \alpha^2) + \lambda x^4 \right] e^{-\alpha^2 x^2/2}$$

$$= \frac{\alpha}{\pi^{1/2}} \left\{ -\frac{\hbar^2}{2m} \left[\alpha^4 J_1(\alpha^2) - \alpha^2 J_0(\alpha^2) \right] + \lambda J_2(\alpha^2) \right\}. \qquad (B.9.107)$$

Now, according to Exercise 9.15,

$$J_0(\alpha^2) = \frac{\pi^{1/2}}{\alpha}, \qquad (B.9.108)$$

$$J_1(\alpha^2) = \frac{\pi^{1/2}}{2\alpha^3}, \qquad (B.9.109)$$

$$J_2(\alpha^2) = \frac{3\pi^{1/2}}{4\alpha^5}. \qquad (B.9.110)$$

Hence, we obtain

$$E(\alpha) = \frac{\hbar^2 \alpha^2}{4m} + \frac{3\lambda}{4\alpha^4}. \qquad (B.9.111)$$

The value of α that minimizes $E(\alpha)$ is

$$\alpha_0 = \left(\frac{6m\lambda}{\hbar^2} \right)^{1/6}. \qquad (B.9.112)$$

Thus, our estimate for the ground-state energy of the system is

$$E_0 = E(\alpha_0) = \frac{3^{1/3}}{4} \left(\frac{\hbar^2}{2m} \right)^{2/3} \lambda^{1/3}. \qquad (B.9.113)$$

The expectation value of the Hamiltonian, calculated with the trial wavefunction $\psi_1(x)$, is

$$E(\beta) = \frac{2\beta^3}{\pi^{1/2}} \int_{-\infty}^{\infty} dx \, x \, e^{-\beta^2 x^2/2} \left(-\frac{\hbar^2}{2m} \frac{d^2}{dx^2} + \lambda x^4 \right) x \, e^{-\beta^2 x^2/2}, \qquad (B.9.114)$$

which yields

$$E(\beta) = \frac{2\beta^3}{\pi^{1/2}} \int_{-\infty}^{\infty} dx\, x\, e^{-\beta^2 x^2/2} \left[-\frac{\hbar^2}{2m} (\beta^4 x^2 - 3\beta^2) + \lambda x^4 \right] x\, e^{-\beta^2 x^2/2}$$

$$= \frac{2\beta^3}{\pi^{1/2}} \left\{ -\frac{\hbar^2}{2m} \left[\beta^4 J_2(\beta^2) - 3\beta^2 J_1(\beta^2) \right] + \lambda J_3(\beta^2) \right\}. \qquad \text{(B.9.115)}$$

However, according to Exercise 9.15,

$$J_1(\beta^2) = \frac{\pi^{1/2}}{2\beta^3}, \qquad \text{(B.9.116)}$$

$$J_2(\beta^2) = \frac{3\pi^{1/2}}{4\beta^5}, \qquad \text{(B.9.117)}$$

$$J_3(\beta^2) = \frac{15\pi^{1/2}}{8\beta^7}. \qquad \text{(B.9.118)}$$

Hence, we obtain

$$E(\beta) = \frac{3\hbar^2 \beta^2}{4m} + \frac{15\lambda}{4\beta^4}. \qquad \text{(B.9.119)}$$

The value of β that minimizes $E(\beta)$ is

$$\beta_1 = \left(\frac{10\, m\, \lambda}{\hbar^2} \right)^{1/6}. \qquad \text{(B.9.120)}$$

Thus, our estimate for the energy of the first excited state is

$$E_1 = E(\beta_1) = \frac{9 \cdot 5^{1/3}}{4} \left(\frac{\hbar^2}{2m} \right)^{2/3} \lambda^{1/3}. \qquad \text{(B.9.121)}$$

Incidentally, we can be sure that this is the energy of an excited state, rather than the ground state, because $E_1 > E_0$.

9.18 For the case of a two-electron atom of nuclear charge $Z_0\, e$, we can reuse the analysis of Section 9.7, except that we must replace Equation (9.87) by

$$H_1 = \frac{(Z - Z_0)\, e^2}{4\pi\, \epsilon_0\, r_1} + \frac{(Z - Z_0)\, e^2}{4\pi\, \epsilon_0\, r_2} + \frac{e^2}{4\pi\, \epsilon_0\, r_{12}}. \qquad \text{(B.9.122)}$$

The energy of the atom is

$$E(Z) = 2 Z^2\, E_0 + \langle H_1 \rangle, \qquad \text{(B.9.123)}$$

where E_0 is the ground-state energy of a hydrogen atom, and the expectation value is calculated using the trial wavefunction (9.83). The analysis of Section 9.7 implies that

$$\frac{\langle H_1 \rangle}{|E_0|} = 4 Z (Z - Z_0) + \frac{5Z}{4}. \qquad \text{(B.9.124)}$$

Hence,

$$E(Z) = \left[\frac{(16Z_0 - 5)Z}{4} - 2Z^2\right]E_0. \tag{B.9.125}$$

The value of Z that minimizes this expression is

$$Z_c = Z_0 - \frac{5}{16}. \tag{B.9.126}$$

Hence, our estimate for the ground-state energy of a two-electron atom of nuclear charge $Z_0 e$ is

$$E = E(Z_c) = \frac{(16Z_0 - 5)^2}{2^7}E_0. \tag{B.9.127}$$

9.19 We have

$$\phi(r_1, r_2) = \frac{(Z_1 Z_2)^{3/2}}{\sqrt{2}\,\pi a_0^3}\left[e^{-(Z_1 r_1 + Z_2 r_2)/a_0} + \epsilon\, e^{-(Z_2 r_1 + Z_1 r_2)/a_0}\right]. \tag{B.9.128}$$

Furthermore, we can write

$$H = K + V_1 + V_2, \tag{B.9.129}$$

where

$$\frac{K}{|E_0|} = -a_0^2\frac{\partial^2}{\partial r_1^2} - \frac{2a_0^2}{r_1}\frac{\partial}{\partial r_1} - a_0^2\frac{\partial^2}{\partial r_2^2} - \frac{2a_0^2}{r_2}\frac{\partial}{\partial r_2}, \tag{B.9.130}$$

and

$$\frac{V_1}{|E_0|} = -2Z\frac{a_0}{r_1} - 2Z\frac{a_0}{r_2}, \tag{B.9.131}$$

with

$$\frac{V_2}{|E_0|} = 2\frac{a_0}{r_>}. \tag{B.9.132}$$

Finally,

$$\langle H \rangle = \frac{\langle K \rangle + \langle V_1 \rangle + \langle V_2 \rangle}{\langle 1 \rangle}. \tag{B.9.133}$$

We can write

$$\langle 1 \rangle = 2^3\,(Z_1 Z_2)^3\left[\int_0^\infty dx_1\, x_1^2\, e^{-2Z_1 x_1}\int_0^\infty dx_2\, x_2^2\, e^{-2Z_2 x_2}\right.$$
$$+ \int_0^\infty dx_1\, x_1^2\, e^{-2Z_2 x_1}\int_0^\infty dx_2\, x_2^2\, e^{-2Z_1 x_2}$$
$$\left.+ 2\,\epsilon\int_0^\infty dx_1\, x_1^2\, e^{-(Z_1 + Z_2)x_1}\int_0^\infty dx_2\, x_2^2\, e^{-(Z_1 + Z_2)x_2}\right], \tag{B.9.134}$$

where $x_1 = r_1/a_0$ and $x_2 = r_2/a_0$. Thus,

$$\langle 1 \rangle = 2^3\,(Z_1 Z_2)^3\left[\frac{2\,I_2^2}{(2Z_1)^3\,(2Z_2)^3} + \frac{2\,\epsilon\,I_2^2}{(Z_1 + Z_2)^6}\right], \tag{B.9.135}$$

where $I_n = n!$ is defined in Exercise 4.11. Hence,

$$\langle 1 \rangle = 1 + \frac{\epsilon \, 2^6 \, (Z_1 Z_2)^3}{(Z_1 + Z_2)^6}. \tag{B.9.136}$$

We can write

$$\begin{aligned}
\frac{\langle K \rangle}{|E_0|} = {}&-(Z_1^2 + Z_2^2) \langle 1 \rangle \\
&+ 2^3 \, (Z_1 Z_3)^3 \Bigg[2 Z_1 \int_0^\infty dx_1 \, x_1 \, e^{-2Z_1 x_1} \int_0^\infty dx_2 \, x_2^2 \, e^{-2Z_2 x_2} \\
&+ 2 Z_1 \int_0^\infty dx_1 \, x_1^2 \, e^{-2Z_2 x_1} \int_0^\infty dx_2 \, x_2 \, e^{-2Z_1 x_2} \\
&+ 2 Z_2 \int_0^\infty dx_1 \, x_1^2 \, e^{-2Z_1 x_1} \int_0^\infty dx_2 \, x_2 \, e^{-2Z_2 x_2} \\
&+ 2 Z_2 \int_0^\infty dx_1 \, x_1 \, e^{-2Z_2 x_1} \int_0^\infty dx_2 \, x_2^2 \, e^{-2Z_1 x_2} \\
&+ \epsilon \, 2 \, (Z_1 + Z_2) \int_0^\infty dx_1 \, x_1 \, e^{-(Z_1 + Z_2) x_1} \int_0^\infty dx_2 \, x_2^2 \, e^{-(Z_1 + Z_2) x_2} \\
&+ \epsilon \, 2 \, (Z_1 + Z_2) \int_0^\infty dx_1 \, x_1^2 \, e^{-(Z_1 + Z_2) x_1} \int_0^\infty dx_2 \, x_2 \, e^{-(Z_1 + Z_2) x_2} \Bigg], \tag{B.9.137}
\end{aligned}$$

which reduces to

$$\begin{aligned}
\frac{\langle K \rangle}{|E_0|} = {}&-(Z_1^2 + Z_2^2) \langle 1 \rangle \\
&+ 2^3 \, (Z_1 Z_3)^3 \left[\frac{4 Z_1 I_1 I_2}{(2 Z)_1^2 \, (2 Z_2)^3} + \frac{4 Z_2 I_1 I_2}{(2 Z)_1^3 \, (2 Z_2)^2} + \frac{4 \epsilon I_1 I_2}{(Z_1 + Z_2)^4} \right] \\
= {}&Z_1^2 + Z_2^2 + \frac{\epsilon \, 2^7 \, (Z_1 Z_2)^4}{(Z_1 + Z_2)^6}. \tag{B.9.138}
\end{aligned}$$

We can write

$$\begin{aligned}
\frac{\langle V_1 \rangle}{|E_0|} = {}&-2^4 \, Z \, (Z_1 Z_3)^3 \Bigg[\int_0^\infty dx_1 \, x_1 \, e^{-2Z_1 x_1} \int_0^\infty dx_2 \, x_2^2 \, e^{-2Z_2 x_2} \\
&+ \int_0^\infty dx_1 \, x_1^2 \, e^{-2Z_2 x_1} \int_0^\infty dx_2 \, x_2 \, e^{-2Z_1 x_2} \\
&+ \int_0^\infty dx_1 \, x_1^2 \, e^{-2Z_1 x_1} \int_0^\infty dx_2 \, x_2 \, e^{-2Z_2 x_2} \\
&+ \int_0^\infty dx_1 \, x_1 \, e^{-2Z_2 x_1} \int_0^\infty dx_2 \, x_2^2 \, e^{-2Z_1 x_2}
\end{aligned}$$

$$+ 2\,\epsilon \int_0^\infty dx_1\, x_1\, e^{-(Z_1+Z_2)\,x_1} \int_0^\infty dx_2\, x_2^2\, e^{-(Z_1+Z_2)\,x_2}$$

$$+ 2\,\epsilon \int_0^\infty dx_1\, x_1^2\, e^{-(Z_1+Z_2)\,x_1} \int_0^\infty dx_2\, x_2\, e^{-(Z_1+Z_2)\,x_2} \Bigg], \tag{B.9.139}$$

which reduces to

$$\frac{\langle V_1 \rangle}{|E_0|} = -2^4\, Z\,(Z_1\,Z_3)^3 \left[\frac{2\,I_1\,I_2}{(2\,Z)_1^2\,(2\,Z_2)^3} + \frac{2\,I_1\,I_2}{(2\,Z)_1^3\,(2\,Z_2)^2} + \frac{4\,\epsilon\,I_1\,I_2}{(Z_1+Z_2)^5} \right]$$

$$= -2\,Z \left[Z_1 + Z_2 + \frac{\epsilon\,2^6\,(Z_1\,Z_2)^3}{(Z_1+Z_2)^5} \right]. \tag{B.9.140}$$

We can write

$$\frac{\langle V_2 \rangle}{|E_0|} = 2^4\,(Z_1\,Z_3)^3 \Bigg\{ \int_0^\infty dx_1\, x_1\, e^{-2Z_1\,x_1} \left[\int_0^{x_1} dx_2\, x_2^2\, e^{-2Z_2\,x_2} + x_1 \int_{x_1}^\infty dx_2\, x_2\, e^{-2Z_2\,x_2} \right]$$

$$+ \int_0^\infty dx_1\, x_1\, e^{-2Z_2\,x_1} \left[\int_0^{x_1} dx_2\, x_2^2\, e^{-2Z_1\,x_2} + x_1 \int_{x_1}^\infty dx_2\, x_2\, e^{-2Z_1\,x_2} \right]$$

$$+ 2\,\epsilon \int_0^\infty dx_1\, x_1\, e^{-(Z_1+Z_2)\,x_1} \left[\int_0^{x_1} dx_2\, x_2^2\, e^{-(Z_1+Z_2)\,x_2} + x_1 \int_{x_1}^\infty dx_2\, x_2\, e^{-(Z_1+Z_2)\,x_2} \right] \Bigg\}, \tag{B.9.141}$$

which reduces to

$$\frac{\langle V_2 \rangle}{|E_0|} = 2^4\,(Z_1\,Z_3)^3 \Bigg\{ \frac{2}{(2\,Z_2)^3} \int_0^\infty dx_1\, x_1\, e^{-2Z_1\,x_1} - \frac{1}{(2\,Z_2)^3} \int_0^\infty dx_1\, x_1\,(2 + 2\,Z_2\,x_1)\, e^{-2\,(Z_1+Z_2)\,x_1}$$

$$+ \frac{2}{(2\,Z_1)^3} \int_0^\infty dx_1\, x_1\, e^{-2Z_2\,x_1} - \frac{1}{(2\,Z_1)^3} \int_0^\infty dx_1\, x_1\,(2 + 2\,Z_1\,x_1)\, e^{-2\,(Z_1+Z_2)\,x_1}$$

$$+ \frac{2^2}{(Z_1+Z_2)^3} \int_0^\infty dx_1\, x_1\, e^{-(Z_1+Z_2)\,x_1}$$

$$- \frac{2\,\epsilon}{(Z_1+Z_2)^3} \int_0^{x_1} dx_1\, x_1\,[2 + (Z_1+Z_2)\,x_1]\, e^{-2\,(Z_1+Z_2)\,x_1} \Bigg\} \tag{B.9.142}$$

with the aid of Exercise 9.10. Hence, we obtain

$$\frac{\langle V_2 \rangle}{|E_0|} = 2^4\,(Z_1\,Z_3)^3 \Bigg\{ \frac{2\,I_1}{(2\,Z_2)^3\,(2\,Z_1)^2} - \frac{1}{(2\,Z_2)^3} \left[\frac{2\,I_1}{(2\,[Z_1+Z_2])^2} + \frac{2\,Z_2\,I_2}{(2\,[Z_1+Z_2])^3} \right]$$

$$+ \frac{2\,I_1}{(2\,Z_1)^3\,(2\,Z_2)^2} - \frac{1}{(2\,Z_1)^3} \left[\frac{2\,I_1}{(2\,[Z_1+Z_2])^2} + \frac{2\,Z_1\,I_2}{(2\,[Z_1+Z_2])^3} \right]$$

$$+ \frac{2^2\,\epsilon\,I_1}{(Z_1+Z_2)^5} - \frac{2\,\epsilon}{(Z_1+Z_2)^3} \left[\frac{2\,I_1}{(2\,[Z_1+Z_2])^2} + \frac{(Z_1+Z_2)\,I_2}{(2\,[Z_1+Z_2])^3} \right] \Bigg\}, \tag{B.9.143}$$

which reduces to

$$\frac{\langle V_2 \rangle}{|E_0|} = 2^4 (Z_1 Z_2)^3 \left[\frac{5\,\epsilon}{2\,(Z_1 + Z_2)^5} + \frac{Z_1^2 + 3\,Z_1 Z_2 + Z_2^2}{2^3 \, Z_1^2 \, Z_2^2 \, (Z_1 + Z_2)^3} \right]. \tag{B.9.144}$$

Thus, with $x = Z_1 + Z_2$ and $y = 2\sqrt{Z_1 Z_2}$, we get

$$\frac{\langle H \rangle}{|E_0|} = \left[x^8 - 2\,Z\,x^7 - \frac{1}{2}\,x^6 y^2 + \frac{1}{2}\,x^5 y^2 + \frac{1}{8}\,x^3 y^4 \right.$$
$$\left. -\epsilon \left(2Z - \frac{5}{8} \right) x\,y^6 + \frac{1}{2}\,\epsilon\,y^8 \right] \bigg/ \left(x^6 + \epsilon\,y^6 \right). \tag{B.9.145}$$

9.20 Equation (9.107) can be written

$$J = \frac{2}{y}\,e^{-y} \left[-(1 + y) \int_0^\infty dx\,x\,e^{-2x} - \int_0^\infty dx\,x^2\,e^{-2x} + (1 + y) \int_0^y dx\,x \right.$$
$$\left. - \int_0^y dx\,x^2 \right] + \frac{2}{y}\,e^y \left[(1 - y) \int_y^\infty dx\,x\,e^{-2x} + \int_y^\infty dx\,x^2\,e^{-2x} \right]. \tag{B.9.146}$$

Making use of Exercise 9.9, we obtain

$$J = \frac{2}{y}\,e^{-y} \left[-\frac{I_1}{4}\,(1 + y) - \frac{I_2}{8} + \frac{y^2}{2}\,(1 + y) - \frac{y^3}{3} \right.$$
$$\left. + \frac{1}{4}\,(1 - y)(1 + 2y) + \frac{1}{8}\,(2 + 4y + 4y^2) \right], \tag{B.9.147}$$

which reduces to

$$J = e^{-y} \left(1 + y + \frac{y^2}{3} \right). \tag{B.9.148}$$

Here, $I_n = n!$ is defined in Exercise 4.11.

9.21 Equation (9.115) can be written

$$D = \sum_{l = 0, \infty} 2 \int_0^\infty dx\,x^2\,\frac{r_<^l}{r_>^{l+1}}\,e^{-2x} \int_0^\pi d\theta \, \sin\theta\,P_l(\cos\theta), \tag{B.9.149}$$

where use has been made of Equation (9.60), and $r_>$ $(r_<)$ is the larger (smaller) of x and y. However [Abramowitz and Stegun (1965)],

$$\int_0^\pi d\theta \, \sin\theta\,P_l(\cos\theta) = \int_{-1}^1 d\mu\,P_l(\mu) = 2\,\delta_{l0}. \tag{B.9.150}$$

Hence, we obtain

$$D = \frac{4}{y} \int_0^y dx\,x^2\,e^{-2x} + 4 \int_y^\infty dx\,x\,e^{-2x}, \tag{B.9.151}$$

which reduces to

$$D = \frac{1}{2y}\left[2 - e^{-2y}(2 + 2y)\right] = \frac{1}{y}\left[1 - (1 + y)e^{-2y}\right] \tag{B.9.152}$$

with the aid of Exercise 9.9.

9.22 A comparison of Equations (9.105), (9.107), and (9.118) reveals that

$$E = \frac{2}{y}e^{-y}\left[-(1 + y)\int_0^\infty dx\, e^{-2x} - \int_0^\infty dx\, x e^{-2x} + (1 + y)\int_0^y dx\right.$$
$$\left. - \int_0^y dx\, x\right] + \frac{2}{y}e^y\left[(1 - y)\int_y^\infty dx\, e^{-2x} + \int_y^\infty dx\, x e^{-2x}\right]. \tag{B.9.153}$$

Making use of Exercise 9.9, we obtain

$$E = \frac{2}{y}e^{-y}\left[-\frac{I_0}{2}(1 + y) - \frac{I_1}{4} + y(1 + y) - \frac{y^2}{2}\right.$$
$$\left. + \frac{1}{2}(1 - y) + \frac{1}{4}(1 + 2y)\right], \tag{B.9.154}$$

which reduces to

$$E = (1 + y)e^{-y}. \tag{B.9.155}$$

Here, $I_n = n!$ is defined in Exercise 4.11.

9.23 Let $a = a_0/Z$ and $y = R/a$, where a_0 is the Bohr radius, and R the proton separation. The trial wavefunction is

$$\psi_+(\mathbf{x}) = A\left[\psi_0(\mathbf{x}_1) + \psi_1(\mathbf{x}_2)\right], \tag{B.9.156}$$

where

$$\psi_0(\mathbf{x}) = \frac{1}{\sqrt{\pi}\, a^{3/2}}\, e^{-r/a}, \tag{B.9.157}$$

and $r = |\mathbf{x}|$. Using the results of the previous exercise, we can normalized the wavefunction by setting $A = I^{-1/2}$, where

$$I = 2(1 + J), \tag{B.9.158}$$

and

$$J = e^{-y}\left(1 + y + \frac{y^2}{3}\right). \tag{B.9.159}$$

The electron Hamiltonian can be written

$$H_e = \frac{\mathbf{p}^2}{2m_e} - \frac{Ze^2}{4\pi\,\epsilon_0}\left(\frac{1}{r_1} + \frac{1}{r_2}\right) + (Z - 1)\frac{e^2}{4\pi\,\epsilon_0}\left(\frac{1}{r_1} + \frac{1}{r_2}\right). \tag{B.9.160}$$

Moreover,

$$\left(\frac{\mathbf{p}^2}{2\,m_e} - \frac{Z\,e^2}{4\pi\,\epsilon_0\,r}\right)\psi_0(\mathbf{x}) = Z^2\,E_0\,\psi_0(\mathbf{x}),\tag{B.9.161}$$

where E_0 is the hydrogen ground-state energy. Thus, defining

$$D = \left\langle\psi_0(\mathbf{x}_1)\left|\frac{a}{r_2}\right|\psi_0(\mathbf{x}_1)\right\rangle,\tag{B.9.162}$$

$$E = \left\langle\psi_0(\mathbf{x}_1)\left|\frac{a}{r_1}\right|\psi_0(\mathbf{x}_2)\right\rangle,\tag{B.9.163}$$

$$F = \left\langle\psi_0(\mathbf{x}_1)\left|\frac{a}{r_1}\right|\psi_0(\mathbf{x}_1)\right\rangle,\tag{B.9.164}$$

we find that

$$\frac{\langle\psi_+|\,H_e\,|\psi_+\rangle}{|E_0|} = -Z^2 - 4\,Z\,A^2\,[D + (2 - Z)\,E + (1 - Z)\,F].\tag{B.9.165}$$

Thus, the total expectation value of the molecular energy is

$$E_{\text{total}} = \langle H_e\rangle + \frac{e^2}{4\pi\,\epsilon_0\,R} = \langle H_e\rangle + \frac{2\,Z}{y}.\tag{B.9.166}$$

Reusing the analysis of the previous exercise, we deduce that

$$D = \frac{1}{y}\left[1 - (1 + y)\,\mathrm{e}^{-2y}\right],\tag{B.9.167}$$

$$E = (1 + y)\,\mathrm{e}^{-y}.\tag{B.9.168}$$

Moreover,

$$F = \frac{4}{a^3}\int_0^\infty dr\,r^2\,\mathrm{e}^{-2r/a}\frac{a}{r} = 4\int_0^\infty dx\,x\,\mathrm{e}^{-2x} = I_1 = 1,\tag{B.9.169}$$

where $I_n = n!$ is defined in Exercise 4.11. Thus, we obtain

$$E_{\text{total}} = -F_+(R/a)\,E_0,\tag{B.9.170}$$

where

$$F_+(y) = -Z^2\tag{B.9.171}$$

$$+ \frac{2\,Z}{y}\left[\frac{(1 + y)\,\mathrm{e}^{-2y} + (1 - 2\,y^2/3)\,\mathrm{e}^{-y} + (Z - 1)\,y\,(1 + [1 + y]\,\mathrm{e}^{-y})}{1 + (1 + y + y^2/3)\,\mathrm{e}^{-y}}\right].$$

B.10   Chapter 10

10.1 Equations (10.2) and (10.4) yield

$$\left(-\frac{p^2}{2m} + E\right)|\psi\rangle = H_1 |\psi\rangle. \tag{B.10.1}$$

Making use of Equation (2.78), we can write

$$\langle \mathbf{x}|\left(-\frac{p^2}{2m} + E\right)|\psi\rangle = \left(\frac{\hbar^2}{2m}\nabla^2 + E\right)\psi(\mathbf{x}) = \frac{\hbar^2}{2m}(\nabla^2 + k^2)\,\psi(\mathbf{x}), \tag{B.10.2}$$

where $E = \hbar^2 k^2/2m$. Hence, Equation (B.10.1) becomes

$$(\nabla^2 + k^2)\,\psi(\mathbf{x}) = \frac{2m}{\hbar^2}\,\langle \mathbf{x}| H_1 |\psi\rangle. \tag{B.10.3}$$

10.2 Now, $\phi(\mathbf{x})$ is a solution of Equation (10.3), which implies that

$$(\nabla^2 + k^2)\,\phi(\mathbf{x}) = \mathbf{0}. \tag{B.10.4}$$

Hence, from Equation (10.7),

$$\begin{aligned}
(\nabla^2 + k^2)\,\psi(\mathbf{x}) &= \frac{2m}{\hbar^2}\int d^3\mathbf{x}'\,(\nabla^2 + k^2)\,G(\mathbf{x},\mathbf{x}')\,\langle \mathbf{x}'| H_1 |\psi\rangle \\
&= \frac{2m}{\hbar^2}\int d^3\mathbf{x}'\,\delta^3(\mathbf{x} - \mathbf{x}')\,\langle \mathbf{x}'| H_1 |\psi\rangle \\
&= \frac{2m}{\hbar^2}\,\langle \mathbf{x}'| H_1 |\psi\rangle,
\end{aligned} \tag{B.10.5}$$

where use has been made of Equation (10.8). Thus, Equation (10.7) is indeed a solution of Equation (10.5).

10.3 (a) We need to find the solution of

$$(\nabla^2 + k^2)\,G(R) \equiv \frac{1}{R^2}\frac{d}{dR}\left(R^2 \frac{dG}{dR}\right) + k^2 G = 0 \tag{B.10.6}$$

in the region $R > 0$. Let us try

$$G(R) = -\frac{e^{\pm ikR}}{4\pi R}. \tag{B.10.7}$$

It follows that

$$R^2 \frac{dG}{dR} = \frac{\mp i k R\, e^{\pm ikR} + e^{\pm ikR}}{4\pi}, \tag{B.10.8}$$

and, hence, that

$$\frac{1}{R^2}\frac{d}{dR}\left(R^2 \frac{dG}{dR}\right) = \frac{k^2\, e^{\pm ikR}}{4\pi R} = -k^2 G. \tag{B.10.9}$$

Thus, we deduce that (B.10.7) is a solution of Equation (B.10.6).

(b) We have

$$\int_V dV \, \nabla^2 \, G(R) = \int_V dV \, \nabla \cdot \nabla G = \oint_S d\mathbf{S} \cdot \nabla G = \left[4\pi R^2 \, \frac{\partial G}{\partial R} \right]_{\lim R \to 0}, \quad \text{(B.10.10)}$$

where V is a sphere of vanishing radius, centered on $R = 0$, whereas S is its bounding surface. Thus, with $G(R)$ taking the form (B.10.7), we get

$$\int_V dV \, (\nabla^2 + k^2) \, G(R) = \left[\mp k R \, e^{\pm ikR} + e^{\pm ikR} - k^2 R \, e^{\pm ikR} \right]_{\lim R \to 0} = 1. \quad \text{(B.10.11)}$$

It follows that

$$\int_V dV \, (\nabla^2 + k^2) \, G(R) = 1. \quad \text{(B.10.12)}$$

The only possible simultaneous solution of Equation (B.10.6) and the previous equation is

$$(\nabla^2 + k^2) \, G(R) = \delta^3(\mathbf{R}). \quad \text{(B.10.13)}$$

10.4 (a) Consider

$$F(k) \equiv \frac{1}{2\pi} \int_{-\infty}^{\infty} dx \, \exp\left(-\frac{x^2}{2\sigma_x^2}\right) \exp(i k x)$$

$$= \frac{1}{2\pi} \int_{-\infty}^{\infty} dx \, \exp\left(-\frac{x^2}{2\sigma_x^2} + i k x\right). \quad \text{(B.10.14)}$$

Completing the square,

$$-\frac{x^2}{2\sigma_x^2} + i k x = \frac{-(x - i\sigma_x^2 k)^2 - \sigma_x^4 k^2}{2\sigma_x^2}, \quad \text{(B.10.15)}$$

so

$$F(k) = \exp\left(-\frac{k^2}{2\sigma_k^2}\right) \frac{1}{2\pi} \int_{-\infty}^{\infty} dx \, \exp\left[-\frac{(x - i\sigma_x^2 k)^2}{2\sigma_x^2}\right], \quad \text{(B.10.16)}$$

where $\sigma_k = 1/\sigma_x$. Let $z = (x - i\sigma_x^2 k)/\sqrt{2}\,\sigma_x$. It follows that

$$F(k) = \exp\left(-\frac{k^2}{2\sigma_k^2}\right) \frac{\sqrt{2}\,\sigma_x}{2\pi} \int_{-\infty}^{\infty} dz \, e^{-z^2}. \quad \text{(B.10.17)}$$

However, $\int_{-\infty}^{\infty} dz \, e^{-z^2} = \sqrt{\pi}$ (see Exercise 9.15), so

$$\frac{1}{2\pi} \int_{-\infty}^{\infty} dx \, \exp\left(-\frac{x^2}{2\sigma_x^2}\right) \exp(i k x) = \frac{1}{(2\pi \sigma_k^2)^{1/2}} \, \exp\left(-\frac{k^2}{2\sigma_k^2}\right). \quad \text{(B.10.18)}$$

The real part of the previous equation yields

$$\frac{1}{2\pi} \int_{-\infty}^{\infty} dx \, \exp\left(-\frac{x^2}{2\,\sigma_x^2}\right) \cos(k\,x) = \frac{1}{(2\pi\,\sigma_k^2)^{1/2}} \, \exp\left(-\frac{k^2}{2\,\sigma_k^2}\right). \qquad \text{(B.10.19)}$$

(b) Consider

$$I \equiv \frac{1}{(2\pi\,\sigma_k^2)^{1/2}} \int_{-\infty}^{\infty} dk \, \exp\left(-\frac{k^2}{2\,\sigma_k^2}\right). \qquad \text{(B.10.20)}$$

Let $z = k/\sqrt{2}\,\sigma_k$. It follows that

$$I = \frac{\sqrt{2}\,\sigma_k}{(2\pi\,\sigma_k^2)^{1/2}} \int_{-\infty}^{\infty} dz \, e^{-z^2} = \frac{\sqrt{2}\,\sigma_k}{(2\pi\,\sigma_k^2)^{1/2}} \, \sqrt{\pi} = 1. \qquad \text{(B.10.21)}$$

(c) Consider

$$G(k) \equiv \frac{1}{(2\pi\,\sigma_k^2)^{1/2}} \, \exp\left(-\frac{k^2}{2\,\sigma_k^2}\right). \qquad \text{(B.10.22)}$$

It is clear from part (b) that

$$\int_{-\infty}^{\infty} dk \, G(k) = 1. \qquad \text{(B.10.23)}$$

Thus, in the limit $\sigma_k \to 0$, $G(k)$ becomes a curve that is zero everywhere, apart from the immediate vicinity of $k = 0$, but has unit area under it. In other words,

$$\lim_{\sigma_k \to 0} G(k) = \delta(k), \qquad \text{(B.10.24)}$$

where $\delta(k)$ is a Dirac delta function. Thus, taking the limit $\sigma_x = 1/\sigma_k \to 0$ in Equation (B.10.19), we deduce that

$$\frac{1}{2\pi} \int_{-\infty}^{\infty} dx \, \cos(k\,x) = \delta(x). \qquad \text{(B.10.25)}$$

10.5 Let

$$\psi^{\pm}(\mathbf{x}) = \frac{\exp(\pm i\,k\,r)}{r}, \qquad \text{(B.10.26)}$$

where $r = |\mathbf{x}|$. Now, the associated probability current is

$$\mathbf{j} = \frac{\hbar}{m} \, \text{Im}(\psi^* \, \nabla \psi). \qquad \text{(B.10.27)}$$

Thus,

$$\mathbf{j} = \frac{\hbar}{m} \, \text{Im}\left(-\frac{1}{r^3} \pm \frac{i\,k}{r^2}\right) \nabla r = \pm \frac{\hbar\,k}{m} \frac{\nabla r}{r^2}. \qquad \text{(B.10.28)}$$

It can be seen that $\psi^+(\mathbf{x})$ has a probability current that is everywhere directed radially outward from the origin. In other words, $\psi^+(\mathbf{x})$ corresponds to an outgoing radial wave. Likewise, $\psi^-(\mathbf{x})$ corresponds to an ingoing radial wave.

10.6 Let

$$I = \int_0^\infty dr\, e^{-\mu r}\, \sin(q\, r). \tag{B.10.29}$$

It follows that

$$
\begin{aligned}
I &= \frac{1}{2\mathrm{i}} \int_0^\infty dr \left[e^{-(\mu - \mathrm{i}q)r} - e^{-(\mu + \mathrm{i}q)r} \right] \\
&= \frac{1}{2\mathrm{i}} \left[-\frac{e^{-(\mu - \mathrm{i}q)r}}{\mu - \mathrm{i}q} + \frac{e^{-(\mu + \mathrm{i}q)r}}{\mu + \mathrm{i}q} \right]_0^\infty = \frac{q}{\mu^2 + q^2}.
\end{aligned}
\tag{B.10.30}
$$

10.7 We have

$$\sigma_{\text{total}} = \oint d\Omega\, \frac{d\sigma}{d\Omega}. \tag{B.10.31}$$

Moreover, from Equation (10.38),

$$\frac{d\sigma}{d\Omega} \simeq \left(\frac{2\, m\, V_0}{\hbar^2\, \mu} \right)^2 \frac{1}{\left[4\, k^2\, \sin^2(\theta/2) + \mu^2 \right]^2}. \tag{B.10.32}$$

Hence,

$$\sigma_{\text{total}} = \left(\frac{2\, m\, V_0}{\hbar^2\, \mu} \right)^2 2\pi \int_0^\pi \frac{d\theta\, \sin\theta}{\left[4\, k^2\, \sin^2(\theta/2) + \mu^2 \right]^2}. \tag{B.10.33}$$

Let $x = 1 - \cos\theta$. It follows that $\sin^2(\theta/2) = x/2$. Thus,

$$\sigma_{\text{total}} = \left(\frac{2\, m\, V_0}{\hbar^2\, \mu} \right)^2 2\pi \int_0^2 \frac{dx}{(2\, k^2\, x + \mu^2)^2}. \tag{B.10.34}$$

Let $y = 2\, k^2\, x/\mu^2$. The previous integral reduces to

$$\sigma_{\text{total}} = \left(\frac{2\, m\, V_0}{\hbar^2\, \mu} \right)^2 \frac{\pi}{\mu^2\, k^2} \int_0^{4k^2/\mu^2} \frac{dy}{(1 + y)^2} = \left(\frac{2\, m\, V_0}{\hbar^2} \right)^2 \frac{4\pi}{\mu^4\, (4\, k^2 + \mu^2)}. \tag{B.10.35}$$

10.8 According to the Born approximation,

$$f(\mathbf{k}, \mathbf{k}') = -\frac{2\, m}{\hbar^2\, q} \int_0^\infty dr\, r\, V(r)\, \sin(q\, r), \tag{B.10.36}$$

where $\mathbf{q} = |\mathbf{k} - \mathbf{k}'| = 2\, k\, \sin(\theta/2)$, and θ is the scattering angle. Given that $V(r) =$

$V_0 \exp(-r^2/a^2)$, we get

$$
\begin{aligned}
f(\mathbf{k}, \mathbf{k}') &= -\frac{2\,m\,V_0}{\hbar^2\,q} \int_0^\infty dr\, r \, \exp(-r^2/a^2)\, \sin(q\,r) \\
&= -\frac{2\,m\,V_0\,a^2}{\hbar^2\,q} \int_0^\infty dy\, y\, \exp(-y^2)\, \sin[(q\,a)\,y] \\
&= -\frac{2\,m\,V_0\,a^2}{\hbar^2\,q}\, \frac{\sqrt{\pi}}{4}\, q\,a\, \exp\left[-\frac{(q\,a)^2}{4}\right]
\end{aligned}
\tag{B.10.37}
$$

[Gradshteyn *et al.* (1980)]. Hence,

$$
\frac{d\sigma}{d\Omega} = |f(\mathbf{k}, \mathbf{k}')|^2 = \left(\frac{\sqrt{\pi}\,m\,V_0\,a^3}{2\,\hbar^2}\right)^2 \exp\left[-2\,(k\,a)^2\,\sin^2(\theta/2)\right].
\tag{B.10.38}
$$

Now,

$$
\sigma_{\text{total}} = \left(\frac{\sqrt{\pi}\,m\,V_0\,a^3}{2\,\hbar^2}\right)^2 2\pi \int_0^\pi d\theta\, \sin\theta\, \exp\left[-2\,(k\,a)^2\,\sin^2(\theta/2)\right].
\tag{B.10.39}
$$

Let $\mu = \cos\theta$. It follows that $\sin^2(\theta/2) = (1 - \mu)/2$. Thus, we obtain

$$
\sigma_{\text{total}} = \left(\frac{\sqrt{\pi}\,m\,V_0\,a^3}{2\,\hbar^2}\right)^2 2\pi\, e^{-(ka)^2} \int_{-1}^1 d\mu\, e^{-(ka)^2\,\mu},
\tag{B.10.40}
$$

which reduces to

$$
\sigma_{\text{total}} = \left(\frac{\sqrt{\pi}\,m\,V_0\,a^3}{2\,\hbar^2}\right)^2 2\pi \left[\frac{1 - e^{-2\,(ka)^2}}{(k\,a)^2}\right].
\tag{B.10.41}
$$

10.9 The incident state is written

$$
|\mathbf{k}\rangle = \frac{e^{i\mathbf{k}\cdot\mathbf{x}}}{(2\pi)^{3/2}}\, \psi_0(\mathbf{x}')
\tag{B.10.42}
$$

[see Equation (10.14)], where $\mathbf{k}$ is the wavenumber of the incident electron, $\mathbf{x}$ the position of the incident/scattered electron, $\mathbf{x}'$ the position of the atomic electron, and $\psi_0(\mathbf{x}')$ the properly normalized ground-state atomic wavefunction. The scattered state is written

$$
|\mathbf{k}'\rangle = \frac{e^{i\mathbf{k}'\cdot\mathbf{x}}}{(2\pi)^{3/2}}\, \psi_0(\mathbf{x}'),
\tag{B.10.43}
$$

where $\mathbf{k}'$ is the wavenumber of the scattered electron. Because the scattering is elastic, $|\mathbf{k}'| = |\mathbf{k}|$, and the initial and final atomic wavefunctions are identical. In the Born approximation, which is appropriate to high energy scattering,

$$
f(\mathbf{k}', \mathbf{k}) = -\frac{(2\pi)^2\, m_e}{\hbar^2}\, \langle\mathbf{k}'|\, V\, |\mathbf{k}\rangle
\tag{B.10.44}
$$

[see Equation (10.21)], where

$$V(\mathbf{x}, \mathbf{x}') = -\frac{e^2}{4\pi\,\epsilon_0\,|\mathbf{x}|} + \frac{e^2}{4\pi\,\epsilon_0\,|\mathbf{x} - \mathbf{x}'|} \qquad (\text{B}.10.45)$$

is the scattering potential. Note that, for fast incident electrons, we can neglect the fact that the incident and atomic electrons are identical particles, because their wavefunctions have very little overlap.

It follows, from the previous expressions, that

$$f(\mathbf{q}) = \frac{m_e\,e^2}{8\pi^2\,\epsilon_0\,\hbar^2} \int d^3\mathbf{x}' \int d^3\mathbf{x}\, e^{\,i\,\mathbf{q}\cdot\mathbf{x}}\, |\psi_0(\mathbf{x}')|^2 \left(\frac{1}{|\mathbf{x}|} - \frac{1}{|\mathbf{x} - \mathbf{x}'|}\right), \qquad (\text{B}.10.46)$$

where $\mathbf{q} = \mathbf{k} - \mathbf{k}'$. Thus,

$$f(\mathbf{q}) = \frac{m_e\,e^2}{8\pi^2\,\epsilon_0\,\hbar^2}\,(A - B), \qquad (\text{B}.10.47)$$

where

$$A = \int d^3\mathbf{x}\,\frac{e^{\,i\,\mathbf{q}\cdot\mathbf{x}}}{|\mathbf{x}|} \int d^3\mathbf{x}'\,|\psi_0(\mathbf{x}')|^2 = \int d^3\mathbf{x}\,\frac{e^{\,i\,\mathbf{q}\cdot\mathbf{x}}}{|\mathbf{x}|}, \qquad (\text{B}.10.48)$$

$$B = \int d^3\mathbf{x} \int d^3\mathbf{x}'\,\frac{e^{\,i\,\mathbf{q}\cdot\mathbf{x}}}{|\mathbf{x} - \mathbf{x}'|}\,|\psi_0(\mathbf{x}')|^2. \qquad (\text{B}.10.49)$$

Let $\mathbf{x}'' = \mathbf{x} - \mathbf{x}'$. It follows that

$$B = \int d^3\mathbf{x}'' \int d^3\mathbf{x}'\,\frac{e^{\,i\,\mathbf{q}\cdot(\mathbf{x}'+\mathbf{x}'')}}{|\mathbf{x}''|}\,|\psi_0(\mathbf{x}')|^2, \qquad (\text{B}.10.50)$$

which implies that

$$B = A\,C, \qquad (\text{B}.10.51)$$

where

$$C = \int d^3\mathbf{x}\,|\psi_0(\mathbf{x})|^2\,e^{\,i\,\mathbf{q}\cdot\mathbf{x}}. \qquad (\text{B}.10.52)$$

Hence,

$$f(\mathbf{q}) = \frac{m_e\,e^2}{8\pi^2\,\epsilon_0\,\hbar^2}\,A\,(1 - C). \qquad (\text{B}.10.53)$$

In evaluating the integrals, we can assume, without loss of generality, that $\mathbf{q} = q\,\mathbf{e}_z$. Thus,

$$A = 2\pi \int_0^\infty dr\,r^2 \int_0^\pi d\theta\,\sin\theta\,\frac{e^{\,i\,q\,r\,\cos\theta}}{r} = 2\pi \int_0^\infty dr\,r \int_{-1}^1 d\mu\,e^{\,i\,q\,r\,\mu}$$

$$= \frac{4\pi}{q} \int_0^\infty dr\,\sin(q\,r) = \frac{4\pi}{q^2}. \qquad (\text{B}.10.54)$$

Now, the ground-state wavefunction is spherically symmetric, so

$$C = 2\pi \int_0^r dr\, r^2\, |\psi_0(r)|^2 \int_0^\pi d\theta\, \sin\theta\, e^{iqr\sin\theta}$$

$$= \frac{4\pi}{q} \int_0^\infty dr\, r\, |\psi_0(r)|^2\, \sin(qr). \tag{B.10.55}$$

In fact,

$$\psi_0(r) = \frac{\sqrt{\pi}}{a_0^{3/2}}\, e^{-r/a_0}, \tag{B.10.56}$$

so

$$C = \frac{4}{q\,a_0} \int_0^\infty dy\, y\, e^{-2y}\, \sin(q\,a_0\,y). \tag{B.10.57}$$

Thus,

$$C = \frac{16}{[4 + (q\,a_0)^2]^2} \tag{B.10.58}$$

[Gradshteyn *et al.* (1980)]. Hence,

$$f(q) = \frac{2\,m_e\,e^2}{4\pi\,\epsilon_0\,\hbar^2\,q^2} \left(1 - \frac{16}{[4 + (q\,a_0)^2]^2}\right), \tag{B.10.59}$$

and

$$\frac{d\sigma}{d\Omega} = |f(q)|^2 = \left(\frac{2\,m_e\,e^2}{4\pi\,\epsilon_0\,\hbar^2\,q^2}\right)^2 \left(1 - \frac{16}{[4 + (q\,a_0)^2]^2}\right)^2. \tag{B.10.60}$$

10.10 Adopting the standard spherical coordinates, r, θ, φ, and assuming that $\psi = \psi(r,\theta)$, the free-space Schrödinger equation,

$$(\nabla^2 + k^2)\,\psi = 0, \tag{B.10.61}$$

becomes [Riley *et al.* (2013)]

$$\left[r^2\,\frac{\partial^2}{\partial r^2} + 2\,r\,\frac{\partial}{\partial r} + \frac{1}{\sin\theta}\,\frac{\partial}{\partial\theta}\left(\sin\theta\,\frac{\partial}{\partial\theta}\right) + k^2\,r^2\right]\psi(r,\theta) = 0. \tag{B.10.62}$$

Let us search for a separable solution of the form

$$\psi(r,\theta) = R(r)\,P(\cos\theta). \tag{B.10.63}$$

We obtain

$$\frac{1}{R}\left[r^2\,\frac{d^2}{dr^2} + 2\,r\,\frac{d}{dr} + k^2\,r^2\right]R = -\frac{1}{P}\,\frac{d}{d\mu}\left[(1-\mu^2)\,\frac{dP}{d\mu}\right], \tag{B.10.64}$$

where $\mu = \cos\theta$. Thus, we can write

$$r^2\,\frac{d^2R_l}{dr^2} + 2\,r\,\frac{dR_l}{dr} + \left[k^2\,r^2 - l(l+1)\right]R_l = 0, \tag{B.10.65}$$

and

$$\frac{d}{d\mu}\left[(1-\mu^2)\frac{dP_l}{d\mu}\right] + l(l+1)P_l = 0, \tag{B.10.66}$$

where l is independent of r and μ. As is well known, the previous equation only possesses solutions that are well behaved in the interval $-1 \leq \mu \leq 1$ provided that l is a non-negative integer [Jackson (1975)]. These solutions are known as the Legendre polynomials. Thus, the most general solution of Equation (B.10.61) that is independent of the azimuthal angle, φ, takes the form

$$\psi(r,\theta) = \sum_{l=0,\infty} R_l(r)\,P_l(\cos\theta), \tag{B.10.67}$$

where the radial function, $R_l(r)$, satisfies the differential equation (B.10.65).

10.11 (a) The Born approximation yields

$$f(\mathbf{k}',\mathbf{k}) = -\frac{m}{2\pi\hbar^2}\int d^3x'\,e^{i(\mathbf{k}-\mathbf{k}')\cdot\mathbf{x}'}\,V(\mathbf{x}'). \tag{B.10.68}$$

Hence,

$$\frac{d\sigma}{d\Omega} = |f|^2 = \left(\frac{m}{2\pi\hbar^2}\right)^2\int d^3x\int d^3x'\,e^{i(\mathbf{k}-\mathbf{k}')\cdot(\mathbf{x}-\mathbf{x}')}\,V(\mathbf{x})\,V(\mathbf{x}'). \tag{B.10.69}$$

Now,

$$\sigma_{\text{total}} = \oint d\Omega\,\frac{d\sigma}{d\Omega}, \tag{B.10.70}$$

where the integral is over all possible directions of $\mathbf{k}'$. In fact,

$$\oint d\Omega\,e^{-i\mathbf{k}'\cdot(\mathbf{x}-\mathbf{x}')} = 2\pi\int_0^\pi d\theta\,\sin\theta\,e^{-ik|\mathbf{x}-\mathbf{x}'|\cos\theta}, \tag{B.10.71}$$

where θ is the angle subtended between $\mathbf{k}'$ and $\mathbf{x}-\mathbf{x}'$. We have also made use of the fact that $k' = k$. Thus, we obtain

$$\oint d\Omega\,e^{-i\mathbf{k}'\cdot(\mathbf{x}-\mathbf{x}')} = 4\pi\,\frac{\sin[k\,|\mathbf{x}-\mathbf{x}'|]}{k\,|\mathbf{x}-\mathbf{x}'|}, \tag{B.10.72}$$

which implies that

$$\sigma_{\text{total}} = \frac{m^2}{\pi\hbar^4}\int d^3x\int d^3x'\,e^{i\mathbf{k}\cdot(\mathbf{x}-\mathbf{x}')}\,\frac{\sin[k\,|\mathbf{x}-\mathbf{x}'|]}{k\,|\mathbf{x}-\mathbf{x}'|}\,V(\mathbf{x})\,V(\mathbf{x}'). \tag{B.10.73}$$

Suppose that $V(\mathbf{x}) = V(r)$. In this case, we can average the previous expression over all possible directions of $\mathbf{x}-\mathbf{x}'$ without changing its value. Now,

$$\left\langle e^{i\mathbf{k}\cdot(\mathbf{x}-\mathbf{x}')}\right\rangle = \frac{1}{2}\int_0^\pi d\theta\,\sin\theta\,e^{ik|\mathbf{x}-\mathbf{x}'|\cos\theta}, \tag{B.10.74}$$

where θ is the angle subtended between $\mathbf{k}$ and $\mathbf{x} - \mathbf{x}'$. Thus, we obtain

$$\left\langle e^{i\mathbf{k}\cdot(\mathbf{x}-\mathbf{x}')}\right\rangle = \frac{\sin[k\,|\mathbf{x}-\mathbf{x}'|]}{k\,|\mathbf{x}-\mathbf{x}'|}. \tag{B.10.75}$$

It follows that

$$\sigma_{\text{total}} = \langle\sigma_{\text{total}}\rangle = \frac{m^2}{\pi\hbar^4}\int d^3x\int d^3x'\,\frac{\sin^2[k\,|\mathbf{x}-\mathbf{x}'|]}{k^2\,|\mathbf{x}-\mathbf{x}'|^2}\,V(r)\,V(r'). \tag{B.10.76}$$

(b) According to the optical theorem,

$$\sigma_{\text{total}} = \frac{4\pi}{k}\,\text{Im}\,[f(\mathbf{k},\mathbf{k})]. \tag{B.10.77}$$

Retaining the first two terms in the Born expansion, we have

$$f(\mathbf{k},\mathbf{k}) = f^{(1)}(\mathbf{k},\mathbf{k}) + f^{(2)}(\mathbf{k},\mathbf{k}), \tag{B.10.78}$$

where

$$f^{(1)}(\mathbf{k},\mathbf{k}) = -\frac{m}{2\pi\hbar^2}\int d^3x\,V(r), \tag{B.10.79}$$

$$f^{(2)}(\mathbf{k},\mathbf{k}) = \left(\frac{m}{2\pi\hbar^2}\right)^2\int d^3x\int d^3x'\,e^{-i\mathbf{k}\cdot(\mathbf{x}-\mathbf{x}')}\,\frac{e^{ik|\mathbf{x}-\mathbf{x}'|}}{|\mathbf{x}-\mathbf{x}'|}\,V(r)\,V(r'). \tag{B.10.80}$$

As previously, we can average the preceding expression over all possible directions of $\mathbf{x}-\mathbf{x}'$ without changing its value. Thus, making use of Equation (B.10.75), we obtain

$$f^{(2)}(\mathbf{k},\mathbf{k}) = \langle f^{(2)}(\mathbf{k},\mathbf{k})\rangle$$
$$= \left(\frac{m}{2\pi\hbar^2}\right)^2\int d^3x\int d^3x'\,\frac{\sin[k\,|\mathbf{x}-\mathbf{x}'|]}{k\,|\mathbf{x}-\mathbf{x}'|}\,\frac{e^{ik|\mathbf{x}-\mathbf{x}'|}}{|\mathbf{x}-\mathbf{x}'|}\,V(r)\,V(r'). \tag{B.10.81}$$

It follows that

$$\text{Im}\left[f^{(1)}(\mathbf{k},\mathbf{k})\right] = 0, \tag{B.10.82}$$

$$\text{Im}\left[f^{(2)}(\mathbf{k},\mathbf{k})\right] = \left(\frac{m}{2\pi\hbar^2}\right)^2\int d^3x\int d^3x'\,\frac{\sin^2[k\,|\mathbf{x}-\mathbf{x}'|]}{k\,|\mathbf{x}-\mathbf{x}'|^2}\,V(r)\,V(r'). \tag{B.10.83}$$

Hence, the optical theorem yields

$$\sigma_{\text{total}} = \frac{m^2}{\pi\hbar^4}\int d^3x\int d^3x'\,\frac{\sin^2[k\,|\mathbf{x}-\mathbf{x}'|]}{k^2\,|\mathbf{x}-\mathbf{x}'|^2}\,V(r)\,V(r'). \tag{B.10.84}$$

444

Quantum Mechanics

10.12 Equations (10.5) and (10.12) imply that

$$(\nabla^2 + k^2)\,\psi = \frac{2\,m}{\hbar^2}\,V(r). \tag{B.10.85}$$

Adopting the standard spherical coordinates, r, θ, φ, and assuming that $\psi = \psi(r,\theta)$, the previous equation becomes [Riley *et al.* (2013)]

$$\left[r^2 \frac{\partial^2}{\partial r^2} + 2\,r\,\frac{\partial}{\partial r} + \frac{1}{\sin\theta}\,\frac{\partial}{\partial\theta}\left(\sin\theta\,\frac{\partial}{\partial\theta}\right) + k^2\,r^2 - \frac{2\,m}{\hbar^2}\,r^2\,V\right]\psi(r,\theta) = 0. \tag{B.10.86}$$

Reusing the analysis of Exercise 10.10, the most general well-behaved solution of the preceding equation takes the form

$$\psi(r,\theta) = \sum_{l=0,\infty} R_l(r)\,P_l(\cos\theta), \tag{B.10.87}$$

where

$$r^2 \frac{d^2 R_l}{dr^2} + 2\,r\,\frac{dR_l}{dr} + \left[k^2\,r^2 - l(l+1) - \frac{2\,m}{\hbar^2}\,r^2\,V\right]R_l = 0. \tag{B.10.88}$$

Finally, writing $R_l(r) = u_l(r)/r$, we obtain

$$\frac{d^2 u_l}{dr^2} + \left[k^2 - \frac{2\,m}{\hbar^2}\,V - \frac{l(l+1)}{r^2}\right]u_l = 0. \tag{B.10.89}$$

10.13 According to partial wave theory,

$$\frac{d\sigma}{d\Omega} = |f(\theta)|^2, \tag{B.10.90}$$

where θ is the scattering angle, and

$$f(\theta) = \sum_{l=0,\infty}(2\,l+1)\,\frac{\exp(i\,\delta_l)}{k}\,\sin\delta_l\,P_l(\cos\theta). \tag{B.10.91}$$

The radial wavefunction takes the form

$$A_l(r) = e^{i\delta_l}\left[\cos\delta_l\,j_l(k\,r) - \sin\delta_l\,\eta_l(k\,r)\right] \tag{B.10.92}$$

for $r > R$, and satisfies

$$r^2 \frac{d^2 A_l}{dr^2} + 2\,r\,\frac{dA_l}{dr} + [k'^2\,r^2 - l(l+1)]\,A_l = 0 \tag{B.10.93}$$

for $r < R$, where

$$k'^2 = k^2 - \frac{2\,m}{\hbar^2}\,V_0. \tag{B.10.94}$$

The solution of Equation (B.10.93) that is well behaved at $r = 0$ is

$$A_l(r) = a_l \, j_l(k' \, r), \tag{B.10.95}$$

where a_l is a constant. Matching the logarithmic derivatives of Equations (B.10.92) and (B.10.95) at $r = R$, we obtain

$$\frac{(k' \, R) \, j_l'(k' \, R)}{j_l(k' \, R)} = \frac{(k \, R) \, j_l'(k \, R) - \tan \delta_l \, (k \, R) \, \eta_l'(k \, R)}{j_l(k \, R) - \tan \delta_l \, \eta_l(k \, R)}. \tag{B.10.96}$$

However,

$$z \frac{df_l}{dz} = l \, f_l(z) - z \, f_{l+1}(z), \tag{B.10.97}$$

where $f_l(z)$ stands for either $j_l(z)$ or $\eta_l(z)$ [Abramowitz and Stegun (1965)]. Thus,

$$\frac{(k' \, R) \, j_{l+1}(k' \, R)}{j_l(k' \, R)} = \frac{(k \, R) \, j_{l+1}(k \, R)}{j_l(k \, R)} \left[1 - \tan \delta_l \, \frac{\eta_{l+1}(k \, R)}{j_{l+1}(k \, R)} \right]$$

$$\left[1 - \tan \delta_l \, \frac{\eta_l(k \, R)}{j_l(k \, R)} \right]^{-1}. \tag{B.10.98}$$

Suppose that $k \, R \ll 1$ and $k' \, R \ll 1$. This implies that $|V_0|/E \ll 1$, where $E = \hbar^2 \, k^2 / 2 \, m$. It follows that, to lowest order,

$$\frac{(k \, R) \, j_{l+1}(k \, R)}{j_l(k \, R)} \simeq (k \, R)^2 \, \frac{(2 \, l + 1)!!}{(2 \, l + 3)!!}, \tag{B.10.99}$$

$$\frac{\eta_l(k \, R)}{j_l(k \, R)} \simeq -\frac{(2 \, l + 1)!! \, (2 \, l - 1)!!}{(k \, R)^{2l+1}} \tag{B.10.100}$$

[Abramowitz and Stegun (1965)]. Hence, assuming that $|\delta_l| \ll 1$, we get

$$\delta_l \simeq \frac{(k \, R)^{2l+3}}{(2 \, l + 3)!! \, (2 \, l + 1)!!} \left(\frac{k'^2}{k^2} - 1 \right) = \frac{(k \, R)^{2l+3}}{(2 \, l + 3)!! \, (2 \, l + 1)!!} \, \frac{V_0}{E}, \tag{B.10.101}$$

which is indeed small. Thus,

$$\delta_0 \simeq \frac{(k \, R)^3}{3} \left(\frac{V_0}{E} \right), \tag{B.10.102}$$

$$\delta_1 \simeq \frac{(k \, R)^5}{45} \left(\frac{V_0}{E} \right), \tag{B.10.103}$$

and so on. It follows that

$$\frac{d\sigma}{d\Omega} = \left(\frac{4 \, m^2 \, V_0^2 \, R^6}{9 \, \hbar^4} \right) \left[1 + \frac{2}{5} \, (k \, R)^2 \, \cos \theta + O(k \, R)^4 \right]. \tag{B.10.104}$$

Hence,

$$\sigma_{\text{total}} = \oint d\Omega \frac{d\sigma}{d\Omega} = \left(\frac{16\pi}{9}\right)\left(\frac{m^2 V_0^2 R^6}{\hbar^4}\right)\left[1 + O(kR)^4\right]. \tag{B.10.105}$$

10.14 The $l = 0$ wavefunction can be written $\psi_0(r) = u_0(r)/r$, where

$$\frac{d^2 u_0}{dr^2} + \left[k^2 - \gamma\,\delta(r-a)\right]u_0 = 0. \tag{B.10.106}$$

The solution in the region $r < a$ that is well behaved at $r = 0$ is

$$u_0(r) = \alpha\,\sin(kr). \tag{B.10.107}$$

In the region, $r > a$, we can write

$$u_0(r) = \beta\,\sin(kr + \delta_0), \tag{B.10.108}$$

where δ_0 is the S-wave phase-shift. Here, α and β are constants. We require the radial function to be continuous at $r = a$: that is,

$$\alpha\,\sin(ka) = \beta\,\sin(ka + \delta_0). \tag{B.10.109}$$

Also, integration of Equation (B.10.106) across $r = a$ yields

$$\left[\frac{du_0}{dr}\right]_{r=a-}^{r=a+} = \gamma\,u_0(a), \tag{B.10.110}$$

or

$$\beta\,\cos(ka + \delta_0) - \alpha\,\cos(ka) = \frac{\gamma}{k}\,\beta\,\sin(ka + \delta_0). \tag{B.10.111}$$

Equations (B.10.109) and (B.10.111) can be combined to give

$$\cot(ka + \delta_0) - \cot(ka) = \frac{\gamma}{k}, \tag{B.10.112}$$

or

$$\delta_0 = -ka + \tan^{-1}\left[\frac{1}{\cot(ka) + \gamma/k}\right]. \tag{B.10.113}$$

Suppose that $\gamma/k \gg 1$ and $\cot(ka) \sim O(1)$. In this case, we can write the solution of the previous equation as

$$\delta_0 \simeq -ka + \frac{1}{\cot(ka) + \gamma/k} + O\left(\frac{k}{\gamma}\right)^3, \tag{B.10.114}$$

which reduces to

$$\delta_0 \simeq -ka + \frac{k}{\gamma} - \left(\frac{k}{\gamma}\right)^2\cot(ka) + O\left(\frac{k}{\gamma}\right)^3 \tag{B.10.115}$$

[Abramowitz and Stegun (1965)].

Let

$$ka = n\pi + \theta, \tag{B.10.116}$$

where $|\theta| \ll 1$. It follows that

$$\cot(ka) \simeq \frac{1}{\theta}. \tag{B.10.117}$$

Hence,

$$\delta_0 \simeq -n\pi + \tan^{-1}\zeta, \tag{B.10.118}$$

where

$$\zeta = \frac{1}{1/\theta + \gamma/k}. \tag{B.10.119}$$

Thus,

$$\sin^2\delta_0 \simeq \cos^2(n\pi)\,\sin^2(\tan^{-1}\zeta) = \frac{\zeta^2}{1+\zeta^2} \tag{B.10.120}$$

[Abramowitz and Stegun (1965)]. If

$$\theta = -\frac{k}{\gamma} + \frac{k^2}{\gamma^2}\,y \tag{B.10.121}$$

then

$$\frac{1}{\theta} \simeq -\frac{\gamma}{k} - y, \tag{B.10.122}$$

which implies that

$$y = -\frac{1}{\zeta}. \tag{B.10.123}$$

Hence, we deduce that if $k = k_n$, where

$$k_n\,a = n\pi - \frac{k}{\gamma} + \frac{k^2}{\gamma^2}\,y \tag{B.10.124}$$

then

$$\sigma_0 = \frac{4\pi}{k_n^2}\,\sin^2\delta_0. \tag{B.10.125}$$

The nth resonance is at $k = k_n$, where

$$k_n \simeq \frac{n\pi}{a}\left(1 - \frac{1}{\gamma a}\right). \tag{B.10.126}$$

Thus, the associated resonant energy is

$$E_n = \frac{\hbar^2 k_n^2}{2m} \simeq \frac{n^2\pi^2\hbar^2}{2ma^2}. \tag{B.10.127}$$

Now, we can write

$$y \simeq \frac{(k-k_n)\,a}{k_n^2/\gamma^2} \tag{B.10.128}$$

However,

$$\frac{E - E_n}{E_n} = 2 \frac{k - k_n}{k_n}, \tag{B.10.129}$$

so

$$y = \frac{(E - E_n)\gamma^2 a}{2 k_n E_n} \simeq \frac{(E - E_n)\gamma^2 a^2}{2 n \pi E_n} = \frac{(E - E_n)}{\Gamma_n/2}, \tag{B.10.130}$$

where

$$\Gamma_n = \frac{4 n \pi E_n}{(\gamma a)^2}. \tag{B.10.131}$$

Hence, in the vicinity of the nth resonance, Equation (B.10.125) yields

$$\sigma_0 \simeq \frac{4\pi}{k_n^2} \frac{\Gamma_n^2/4}{(E - E_n)^2 + \Gamma_n^2/4}. \tag{B.10.132}$$

The net S-wave contribution to the total scattering cross-section is the sum of the non-resonant contribution, which is

$$\sigma_0 \simeq \frac{4\pi}{k^2} \sin^2(k a), \tag{B.10.133}$$

because the non-resonant phase-shift is approximately $k a$, and all of the resonant contributions. Thus, we obtain

$$\sigma_0 \simeq \frac{4\pi}{k^2} \left[\sin^2(k a) + \sum_{n=1,\infty} \frac{\Gamma_n^2/4}{(E - E_n)^2 + \Gamma_n^2/4} \right]. \tag{B.10.134}$$

10.15 According to Section 10.3, the scattering amplitude for a particle of mass m, electric charge e, and wavenumber k, scattered from a fixed Coulomb potential generated by a point charge e located at the origin, is

$$f(\theta) \simeq -\frac{m e^2}{8\pi \epsilon_0 \hbar^2 k^2} \frac{1}{\sin^2(\theta/2)}, \tag{B.10.135}$$

where use has been made of the Born approximation. For the case of two counter-propagating proton beams, of momentum $\hbar k$, in the center of mass frame, the effective mass, m, is the reduced mass, $m_p/2$, where m_p is the proton mass. (See Section 10.1.) Hence,

$$f(\theta) \simeq -\frac{m_p e^2}{16\pi \epsilon_0 \hbar^2 k^2} \frac{1}{\sin^2(\theta/2)}. \tag{B.10.136}$$

According to Equation (10.171), if the two proton beams are unpolarized than the differential scattering cross-section is given by

$$\left(\frac{d\sigma}{d\Omega} \right) = |f(\theta)|^2 + |f(\pi - \theta)|^2 - \frac{1}{2} \left[f^*(\theta) f(\pi - \theta) + f(\theta) f^*(\pi - \theta) \right]. \tag{B.10.137}$$

Hence, we obtain

$$\frac{d\sigma}{d\Omega} \simeq \left(\frac{m_p\,e^2}{16\pi\,\epsilon_0\,\hbar^2\,k^2}\right)^2\left[\frac{1}{\sin^4(\theta/2)} + \frac{1}{\cos^4(\theta/2)} - \frac{1}{\sin^2(\theta/2)\,\cos^2(\theta/2)}\right]. \quad (B.10.138)$$

B.11 Chapter 11

11.1 Let $\partial_\mu \equiv \partial/\partial x^\mu$ and $\partial^\mu \equiv \partial/\partial x_\mu$. Now,

$$\frac{\partial}{\partial x^{\mu'}} = \frac{\partial x^\nu}{\partial x^{\mu'}}\frac{\partial}{\partial x^\nu}, \quad (B.11.1)$$

or

$$\partial_{\mu'} = \frac{\partial x^\nu}{\partial x^{\mu'}}\partial_\nu. \quad (B.11.2)$$

However, according to Equation (11.56),

$$x^\nu = a_\mu{}^\nu\,x^{\mu'}, \quad (B.11.3)$$

where the $a_\mu{}^\nu$ are independent of the x^ν. The previous equation implies that

$$\frac{\partial x^\nu}{\partial x^{\mu'}} = a_\mu{}^\nu. \quad (B.11.4)$$

Hence, from Equation (B.11.2),

$$\partial_{\mu'} = a_\mu{}^\nu\,\partial_\nu, \quad (B.11.5)$$

which is the correct transformation rule for the components of a covariant 4-vector. [See Equation (11.53).] Now,

$$\frac{\partial}{\partial x_\mu} = \frac{\partial x^\nu}{\partial x_\mu}\frac{\partial}{\partial x^\nu}, \quad (B.11.6)$$

or

$$\partial^\mu = \frac{\partial x^\nu}{\partial x_\mu}\partial_\nu. \quad (B.11.7)$$

However,

$$x^\nu = g^{\nu\mu}\,x_\mu, \quad (B.11.8)$$

which implies that

$$\frac{\partial x^\nu}{\partial x_\mu} = g^{\nu\mu} = g^{\mu\nu}. \quad (B.11.9)$$

Hence, Equation (B.11.7) gives

$$\partial^\mu = g^{\mu\nu}\,\partial_\nu, \quad (B.11.10)$$

which demonstrates that ∂^μ transforms as the contravariant component of a 4-vector.

11.2 According to Equations (11.26) and (11.27), $\beta^2 = 1$ and $\gamma^0 = \beta$. It follows that

$$\{\gamma^0, \gamma^0\} = \beta^2 + \beta^2 = 2 = 2g^{00}. \tag{B.11.11}$$

According to Equations (11.25) and (11.28), $\alpha_i\beta + \beta\alpha_i = 0$ and $\alpha_i = \beta\gamma^i$. It follows that

$$\beta\gamma^i\beta + \beta^2\gamma^i = \beta\gamma^i\beta + \gamma^i = 0. \tag{B.11.12}$$

Left-multiplying by β, making use of the fact that $\beta^2 = 1$, we obtain

$$\gamma^i\beta + \beta\gamma^i = \{\gamma^i, \gamma^0\} = 0 = g^{i0}. \tag{B.11.13}$$

This also implies that

$$\{\gamma^0, \gamma^i\} = g^{0i}. \tag{B.11.14}$$

Equation (11.24) yields

$$\alpha_i\alpha_j + \alpha_j\alpha_i = \beta\gamma^i\beta\gamma^j + \beta\gamma^j\beta\gamma^i$$
$$= -\beta^2\gamma^i\gamma^j - \beta^2\gamma^j\gamma^i = -\{\gamma^i, \gamma^j\} = 2\delta_{ij}, \tag{B.11.15}$$

where use has been made of the facts that $\gamma^i\beta = -\beta\gamma^i$ and $\beta^2 = 1$. The previous equation implies that

$$\{\gamma^i, \gamma^j\} = -2\delta_{ij} = 2g^{ij}. \tag{B.11.16}$$

Hence, it follows from Equations (B.11.11), (B.11.13), (B.11.14), and (B.11.16) that

$$\{\gamma^\mu, \gamma^\nu\} = 2g^{\mu\nu}. \tag{B.11.17}$$

11.3 If A, B, C are matrices then

$$\text{Tr}(ABC) = \text{Tr}(BCA) = \text{Tr}(CAB) \tag{B.11.18}$$

[Riley *et al.* (2013)]. Here, $\text{Tr}(A)$ denotes the trace of matrix A (i.e., the product of its diagonal elements). Given that $\{\alpha_i, \beta\} \equiv \alpha_i\beta + \beta\alpha_i = 0$, and $\beta^2 = 1$, we deduce that

$$\alpha_i = -\beta\alpha_i\beta. \tag{B.11.19}$$

Hence,

$$\text{Tr}(\alpha_i) = -\text{Tr}(\beta\alpha_i\beta) = -\text{Tr}(\alpha_i\beta^2) = -\text{Tr}(\alpha_i), \tag{B.11.20}$$

which implies that

$$\text{Tr}(\alpha_i) = 0. \tag{B.11.21}$$

Given that $\{\alpha_i, \alpha_i\} = 2\alpha_i^2 = 2$, which implies that $\alpha_i^2 = 1$, and $\alpha_i\beta + \beta\alpha_i = 0$, we deduce that

$$\beta = -\alpha_i\beta\alpha_i. \tag{B.11.22}$$

Hence,

$$\text{Tr}(\beta) = -\text{Tr}(\alpha_i\beta\alpha_i) = -\text{Tr}(\beta\alpha_i^2) = -\text{Tr}(\beta), \tag{B.11.23}$$

which implies that

$$\text{Tr}(\beta) = 0. \tag{B.11.24}$$

Now, the eigenvalue, λ, and eigenvector, ψ, of some matrix A satisfy

$$A\psi = \lambda\psi. \tag{B.11.25}$$

It is easily demonstrated that if A is Hermitian then its eigenvalues are real numbers [Riley *et al.* (2013)]. Suppose that

$$A^2 = 1. \tag{B.11.26}$$

It follows that

$$A^2\psi = \lambda A\psi = \lambda^2\psi = \psi. \tag{B.11.27}$$

Hence, we deduce that $\lambda^2 = 1$. Because λ is real, it follows that $\lambda = \pm 1$. In other words, the eigenvalues of an Hermitian matrix, A, that satisfies $A^2 = 1$ can only take the values $+1$ or -1. Now, the α_i and β are all Hermitian matrices whose squares are equal to the identity matrix. Thus, it follows that their eigenvalues can only take the values $+1$ and -1. As is well known, if the α_i and β are n-dimensional matrices then they have n eigenvalues [Riley *et al.* (2013)]. Furthermore, the trace of a matrix is equal to the sum of its eigenvalues [Riley *et al.* (2013)]. Because the α_i and β are all traceless matrices, it follows that the sums of their eigenvalues are all zero. However, given that the eigenvalues can only take the values ± 1, we deduce that the dimension of the α_i and β matrices must be even. In fact, if the α_i and β are $2n$-dimensional then they each must have n eigenvalues that take the value $+1$, and n eigenvalues that take the value -1.

11.4 We have

$$\{\gamma^0, \gamma^0\} = 2\gamma^0\gamma^0 = 2\begin{pmatrix} 1, & 0 \\ 0, & -1 \end{pmatrix}\begin{pmatrix} 1, & 0 \\ 0, & -1 \end{pmatrix} = \begin{pmatrix} 1, 0 \\ 0, 1 \end{pmatrix} = 2 = 2g^{00}. \tag{B.11.28}$$

Furthermore,

$$\{\gamma^0, \gamma^i\} = \gamma^0\gamma^i + \gamma^i\gamma^0 = \begin{pmatrix} 1, & 0 \\ 0, & -1 \end{pmatrix}\begin{pmatrix} 0, & \sigma_i \\ -\sigma_i, & 0 \end{pmatrix} + \begin{pmatrix} 0, & \sigma_i \\ -\sigma_i, & 0 \end{pmatrix}\begin{pmatrix} 1, & 0 \\ 0, & -1 \end{pmatrix}$$

$$= \begin{pmatrix} 0, & \sigma_i \\ \sigma_i, & 0 \end{pmatrix} + \begin{pmatrix} 0, & -\sigma_i \\ -\sigma_i, & 0 \end{pmatrix} = \begin{pmatrix} 0, 0 \\ 0, 0 \end{pmatrix} = 0 = 2g^{0i}. \tag{B.11.29}$$

Similarly,

$$\{\gamma^i, \gamma^0\} = 2g^{i0}. \tag{B.11.30}$$

Finally,

$$\{\gamma^i, \gamma^j\} = \gamma^i \gamma^i + \gamma^j \gamma^i = \begin{pmatrix} 0, & \sigma_i \\ -\sigma_i, & 0 \end{pmatrix} \begin{pmatrix} 0, & \sigma_j \\ -\sigma_j, & 0 \end{pmatrix} + \begin{pmatrix} 0, & \sigma_j \\ -\sigma_j, & 0 \end{pmatrix} \begin{pmatrix} 0, & \sigma_i \\ -\sigma_i, & 0 \end{pmatrix}$$

$$= \begin{pmatrix} -\{\sigma_i, \sigma_j\}, & 0 \\ 0, & -\{\sigma_i, \sigma_j\} \end{pmatrix} = \begin{pmatrix} -2\delta_{ij}, & 0 \\ 0, & -2\delta_{ij} \end{pmatrix} = -2\delta_{ij} = 2g^{ij}. \quad \text{(B.11.31)}$$

Here, use has been made of the standard result,

$$\{\sigma_i, \sigma_j\} = 2\delta_{ij}. \quad \text{(B.11.32)}$$

Equations (B.11.28), (B.11.29), (B.11.30), and (B.11.31) can be combined to give

$$\{\gamma^\mu, \gamma^\nu\} = 2g^{\mu\nu}. \quad \text{(B.11.33)}$$

11.5 We have

$$\{\alpha^i, \alpha^j\} = \alpha^i \alpha^j + \alpha^j \alpha^i = \begin{pmatrix} 0, & \sigma_i \\ \sigma_i, & 0 \end{pmatrix} \begin{pmatrix} 0, & \sigma_j \\ \sigma_j, & 0 \end{pmatrix} + \begin{pmatrix} 0, & \sigma_j \\ \sigma_j, & 0 \end{pmatrix} \begin{pmatrix} 0, & \sigma_i \\ \sigma_i, & 0 \end{pmatrix}$$

$$= \begin{pmatrix} \sigma_i \sigma_j, & 0 \\ 0, & \sigma_j \sigma_i \end{pmatrix} + \begin{pmatrix} \sigma_j \sigma_i, & 0 \\ 0, & \sigma_j \sigma_i \end{pmatrix} = \begin{pmatrix} \{\sigma_i, \sigma_j\}, & 0 \\ 0 & \{\sigma_i, \sigma_j\} \end{pmatrix}$$

$$= \begin{pmatrix} 2\delta_{ij}, & 0 \\ 0, & 2\delta_{ij} \end{pmatrix} = 2\delta_{ij}, \quad \text{(B.11.34)}$$

where use has been made of the standard result

$$\{\sigma_i, \sigma_j\} = 2\delta_{ij}. \quad \text{(B.11.35)}$$

Furthermore,

$$\{\alpha^i, \beta\} = \alpha^i \beta + \beta \alpha^i = \begin{pmatrix} 0, & \sigma_i \\ \sigma_i, & 0 \end{pmatrix} \begin{pmatrix} 1, & 0 \\ 0, & -1 \end{pmatrix} + \begin{pmatrix} 1, & 0 \\ 0, & -1 \end{pmatrix} \begin{pmatrix} 0, & -\sigma_i \\ \sigma_i, & 0 \end{pmatrix}$$

$$= \begin{pmatrix} 0, & -\sigma_i \\ \sigma_i, & 0 \end{pmatrix} + \begin{pmatrix} 0, & \sigma_i \\ -\sigma_i, & 0 \end{pmatrix} = \begin{pmatrix} 0, & 0 \\ 0, & 0 \end{pmatrix} = 0. \quad \text{(B.11.36)}$$

Finally,

$$\beta^2 = \begin{pmatrix} 1, & 0 \\ 0, & -1 \end{pmatrix} \begin{pmatrix} 1, & 0 \\ 0, & -1 \end{pmatrix} = \begin{pmatrix} 1, & 0 \\ 0, & 1 \end{pmatrix} = 1. \quad \text{(B.11.37)}$$

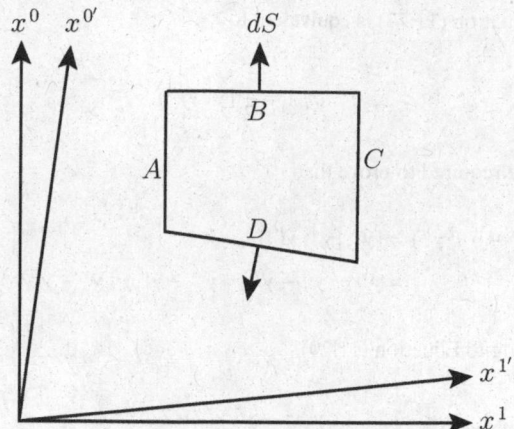

11.6 We are told that

$$\partial_\mu j^\mu = 0, \tag{B.11.38}$$

and that the components of the 4-vector j^μ tend to zero as $|\mathbf{x}| \to \infty$. Now, the 4-dimensional generalization of the divergence theorem [Riley *et al.* (2013)], states that

$$\int_V d^4\mathbf{x} \, \partial_\mu j^\mu = \oint_S dS_\mu \, j^\mu \tag{B.11.39}$$

where dS_μ is an element of the 3-surface, S, bounding the 4-volume, V. Hence, the previous two equations yield

$$\oint_S dS_\mu \, j^\mu = 0. \tag{B.11.40}$$

The particular 4-volume over which the integration is performed is indicated in the previous figure. The 3-surfaces A and C are located at $|\mathbf{x}| \to \infty$, so the spatial components of j^μ vanish on A and C. The 3-surface B is chosen normal to the x^0-axis, whereas the 3-surface D is chosen normal to the $x^{0'}$-axis. Here, the x^μ and the $x^{\mu'}$ are coordinates in two arbitrarily chosen inertial frames. It follows from Equation (B.11.40) that

$$\int dS_0 \, j^0 + \int dS_{0'} \, j^{0'} = 0. \tag{B.11.41}$$

Here, we have made use of the fact that $dS_\mu j^\mu$ is a scalar and, therefore, has the same value in all inertial frames. Because $dS_0 = d^3\mathbf{x}$ and $dS_{0'} = -d^3\mathbf{x}'$ it follows that

$$\int d^3\mathbf{x}' \, j^{0'} = \int d^3\mathbf{x} j^0. \tag{B.11.42}$$

In other words, $\int d^3\mathbf{x} \, j^0$ is invariant under Lorentz transformation.

11.7 Equation (11.76) can also be written

$$2\,\mathrm{i}\,(g^{\nu\alpha}\,\gamma^\beta - g^{\nu\beta}\,\gamma^\alpha) = [\gamma^\nu, \sigma^{\alpha\beta}]. \tag{B.11.43}$$

Moreover, Equation (11.77) is equivalent to

$$\sigma^{\mu\nu} = \frac{i}{2}[\gamma^\mu, \gamma^\nu].$$

(B.11.44)

Hence, we are required to prove that

$$4(g^{\nu\alpha}\gamma^\beta - g^{\nu\beta}\gamma^\alpha) = [\gamma^\nu, [\gamma^\alpha, \gamma^\beta]]$$
$$= \gamma^\nu\gamma^\alpha\gamma^\beta - \gamma^\nu\gamma^\beta\gamma^\alpha - \gamma^\alpha\gamma^\beta\gamma^\nu + \gamma^\beta\gamma^\alpha\gamma^\nu.$$

(B.11.45)

Now, according to Equation (11.29),

$$\gamma^\alpha\gamma^\beta = -\gamma^\beta\gamma^\alpha + 2g^{\alpha\beta}.$$

(B.11.46)

Thus,

$$\gamma^\nu\gamma^\alpha\gamma^\beta = -\gamma^\alpha\gamma^\nu\gamma^\beta - 2g^{\nu\alpha}\gamma^\beta,$$

(B.11.47)

$$-\gamma^\nu\gamma^\beta\gamma^\alpha = \gamma^\beta\gamma^\nu\gamma^\alpha - 2g^{\nu\beta}\gamma^\alpha,$$

(B.11.48)

$$-\gamma^\alpha\gamma^\beta\gamma^\nu = \gamma^\alpha\gamma^\nu\gamma^\beta - 2g^{\nu\beta}\gamma^\alpha,$$

(B.11.49)

$$\gamma^\beta\gamma^\alpha\gamma^\nu = -\gamma^\beta\gamma^\nu\gamma^\alpha + 2g^{\nu\alpha}\gamma^\beta,$$

(B.11.50)

where we have made use of the fact that $g^{\alpha\beta} = g^{\beta\alpha}$ and also that $g^{\alpha\beta}$ commutes with the γ^ν. Summing the previous four equations, we obtain

$$[\gamma^\nu, [\gamma^\alpha, \gamma^\beta]] = 4(g^{\nu\alpha}\gamma^\beta - g^{\nu\beta}\gamma^\alpha).$$

(B.11.51)

11.8 We have

$$I^\mu_\nu = \begin{pmatrix} 0, -1, 0, 0 \\ -1, 0, 0, 0 \\ 0, 0, 0, 0 \\ 0, 0, 0, 0 \end{pmatrix},$$

(B.11.52)

and

$$(I^2)^\mu_\nu = \begin{pmatrix} 1, 0, 0, 0 \\ 0, 1, 0, 0 \\ 0, 0, 0, 0 \\ 0, 0, 0, 0 \end{pmatrix},$$

(B.11.53)

with

$$I^3 = I.$$

(B.11.54)

Now,

$$a^\mu_\nu = \lim_{N\to\infty} \left(g^\mu_\nu + \frac{\omega}{N} I^\mu_\nu \right)^N = \exp\left(\omega\, I^\mu_\nu \right)$$

$$= g^\mu_\nu + \omega\, I^\mu_\nu + \frac{\omega^2}{2!} (I^2)^\mu_\nu + \frac{\omega^3}{3!} I^\mu_\nu + \frac{\omega^4}{4!} (I^2)^\mu_\nu + \cdots$$

$$= g^\mu_\nu - (I^2)^\mu_\nu + \sinh\omega\, I^\mu_\nu + \cosh\omega\, (I^2)^\mu_\nu. \tag{B.11.55}$$

Hence,

$$a^\mu_\nu = \begin{pmatrix} \cosh\omega, & -\sinh\omega, & 0, & 0 \\ -\sinh\omega, & \cosh\omega, & 0, & 0 \\ 0, & 0, & 1, & 0 \\ 0, & 0, & 0, & 1 \end{pmatrix}. \tag{B.11.56}$$

Thus, given that $x^{\mu'} = a^\mu_\nu x^\nu$, we obtain

$$x^{0'} = \cosh\omega \left(x^0 - \tanh\omega\, x^1 \right), \tag{B.11.57}$$

$$x^{1'} = \cosh\omega \left(x^1 - \tanh\omega\, x^0 \right), \tag{B.11.58}$$

$$x^{2'} = x^2, \tag{B.11.59}$$

$$x^{3'} = x^3. \tag{B.11.60}$$

A canonical Lorentz transformation in which frame S' moves relative to frame S at speed v parallel to the x-axis yields $x' = \gamma(x - vt)$, where $\gamma = (1 - v^2/c^2)^{-1/2}$ [Rindler (1966)]. Comparison with Equation (B.11.58) reveals that

$$v = c\,\tanh\omega. \tag{B.11.61}$$

Furthermore, it is easily verified that $\cosh\omega = (1 - v^2/c^2)^{-1/2}$.

Writing $\psi' = A\psi$, and making use of Equation (11.78), the spinor transformation matrix for a finite Lorentz transformation becomes

$$A = \lim_{N\to 0} \left(1 + \frac{1}{8} [\gamma_\mu, \gamma_\nu] I^{\mu\nu} \frac{\omega}{N} \right)^N = \exp\left(\frac{1}{8} [\gamma_\mu, \gamma_\nu] I^{\mu\nu}\, \omega \right). \tag{B.11.62}$$

However, for a Lorentz transformation parallel to the x^1-axis, $I^{01} = -I^{10} = 1$, with all other $I^{\mu\nu}$ taking the value zero. Hence,

$$A = \exp\left(\frac{1}{4} [\gamma_0, \gamma_1]\, \omega \right). \tag{B.11.63}$$

Now,

$$[\gamma_0, \gamma_1] = -[\gamma^0, \gamma^1] = -2\,\alpha_1, \tag{B.11.64}$$

which implies that

$$A = \exp\left(-\frac{1}{2}\alpha_1\omega\right). \tag{B.11.65}$$

Given that $\alpha_1^2 = 1$, we find that

$$\exp\left(-\frac{1}{2}\alpha_1\omega\right) = 1 - \alpha_1\frac{\omega}{2} + \frac{1}{2!}\left(\frac{\omega}{2}\right)^2 - \frac{\alpha_1}{3!}\left(\frac{\omega}{2}\right)^3 + \cdots$$

$$= \cosh\left(\frac{\omega}{2}\right) - \sinh\left(\frac{\omega}{2}\right)\alpha_1, \tag{B.11.66}$$

so

$$A = \cosh\left(\frac{\omega}{2}\right) - \sinh\left(\frac{\omega}{2}\right)\alpha_1, \tag{B.11.67}$$

which is equivalent to

$$A = \begin{pmatrix} \cosh(\omega/2), & 0, & 0, & -\sinh(\omega/2) \\ 0, & \cosh(\omega/2), & -\sinh(\omega/2), & 0 \\ 0, & -\sinh(\omega/2), & \cosh(\omega/2), & 0 \\ -\sinh(\omega/2), & 0, & 0, & \cosh(\omega/2) \end{pmatrix}. \tag{B.11.68}$$

11.9 Let $\mathbf{n} = (\cos\theta_1, \cos\theta_2, \cos\theta_3)$. The infinitesimal Lorentz transform between frames S and S' can then be written $x^{\mu'} = a^\mu_\nu x^\nu$, where

$$a^\mu_\nu = g^\mu_\nu + \delta\omega\, I^\mu_\nu, \tag{B.11.69}$$

and

$$I^\mu_\nu = \begin{pmatrix} 0, & -\cos\theta_1, & -\cos\theta_2, & -\cos\theta_3 \\ -\cos\theta_1, & 0, & 0, & 0 \\ -\cos\theta_2, & 0, & 0, & 0 \\ -\cos\theta_3, & 0, & 0, & 0 \end{pmatrix}. \tag{B.11.70}$$

According to Equation (11.78), the corresponding transformation rule for spinors is $\psi' = A\psi$, where

$$A = 1 + \frac{1}{8}[\gamma_\mu, \gamma_\nu]\, I^{\mu\nu}\, \delta\omega, \tag{B.11.71}$$

and

$$I^{\mu\nu} = \begin{pmatrix} 0, & \cos\theta_1, & \cos\theta_2, & \cos\theta_3 \\ -\cos\theta_1, & 0, & 0, & 0 \\ -\cos\theta_2, & 0, & 0, & 0 \\ -\cos\theta_3, & 0, & 0, & 0 \end{pmatrix}. \tag{B.11.72}$$

It follows that

$$[\gamma_\mu, \gamma_\nu]\, I^{\mu\nu} = 2\,[\gamma_0, \gamma_1]\cos\theta_1 + 2\,[\gamma_0, \gamma_2]\cos\theta_2 + 2\,[\gamma_0, \gamma_3]\cos\theta_3. \tag{B.11.73}$$

However,

$$[\gamma_0, \gamma_i] = -[\gamma^0, \gamma^i] = -2\,\alpha_i, \tag{B.11.74}$$

so

$$[\gamma_\mu, \gamma_\nu]\, I^{\mu\nu} = -4\,(\cos\theta_1\,\alpha_1 + \cos\theta_2\,\alpha_2 + \cos\theta_3\,\alpha_3) = -4\,\mathbf{n}\cdot\boldsymbol{\alpha}. \tag{B.11.75}$$

Thus,

$$A = 1 - \frac{1}{2}\,(\mathbf{n}\cdot\boldsymbol{\alpha})\,\delta\omega. \tag{B.11.76}$$

The transformation matrix for a finite Lorentz transformation is

$$A = \lim_{N\to\infty}\left[1 - \frac{1}{2}\,(\mathbf{n}\cdot\boldsymbol{\alpha})\,\frac{\omega}{N}\right]^N = \exp\left[-\frac{1}{2}\,(\mathbf{n}\cdot\boldsymbol{\alpha})\,\omega\right]. \tag{B.11.77}$$

It is easily demonstrated that $(\mathbf{n}\cdot\boldsymbol{\alpha})^2 = 1$. Hence,

$$\exp\left[-\frac{1}{2}\,(\mathbf{n}\cdot\boldsymbol{\alpha})\,\omega\right]. = 1 - \frac{\omega}{2}\,(\mathbf{n}\cdot\boldsymbol{\alpha}) + \frac{1}{2!}\left(\frac{\omega}{2}\right)^2 - \frac{1}{3!}\left(\frac{\omega}{2}\right)^3(\mathbf{n}\cdot\boldsymbol{\alpha}) + \cdots$$

$$= \cosh\left(\frac{\omega}{2}\right) - \sinh\left(\frac{\omega}{2}\right)(\mathbf{n}\cdot\boldsymbol{\alpha}), \tag{B.11.78}$$

which implies that

$$A = \cosh\left(\frac{\omega}{2}\right) - \sinh\left(\frac{\omega}{2}\right)(\mathbf{n}\cdot\boldsymbol{\alpha}). \tag{B.11.79}$$

By analogy with the solution to the previous exercise, $\omega = \tanh^{-1}(v/c)$.

11.10 We can write

$$\psi^r = e^{-i\,\epsilon_r\,(m_e\,c^2/\hbar)\,t}\,w^r(0). \tag{B.11.80}$$

Now, energy is represented by the operator $E = i\hbar\,\partial/\partial t$, whereas momentum is represented by $\mathbf{p} = -i\hbar\,\partial/\partial\mathbf{x}$. Thus, it is clear from Equation (B.11.80) that $E\psi^r = \epsilon_r\,m_e\,c^2\,\psi^r$ and $\mathbf{p}\,\psi^r = \mathbf{0}\,\psi^r$. In other words, the ψ^r are eigenstates of the energy operator corresponding to the energy $\epsilon_r\,m_e\,c^2$, and are eigenstates of the momentum operator corresponding to the momentum $\mathbf{0}$.

The ψ^r satisfy the free-space Dirac equation provided that

$$i\hbar\gamma^\mu\,\partial_\mu\psi^r \equiv i\,\frac{\hbar}{c}\,\gamma^0\,\frac{\partial\psi^r}{\partial t} \equiv \epsilon_r\,m_e\,c\,\gamma^0\,\psi^r = m_e\,c\,\psi^r. \tag{B.11.81}$$

Here, we have made use of the fact that the ψ^r are independent of $\mathbf{x}$. The previous equation is satisfied as long as

$$\gamma^0\,w^r(0) = \epsilon_r\,w^r(0), \tag{B.11.82}$$

which is easily seen to be the case.

The operator that represents the component of spin angular momentum parallel to the x_3-axis is $S_3 = (\hbar/2)\,\Sigma_3$. However, it is easily demonstrated that $\Sigma_3\,w^r(0) = \zeta_r\,w^r(0)$. Hence, the ψ^r are eigenstates of S_3 corresponding to the eigenvalues $\zeta_r\,(\hbar/2)$.

11.11 Given that the Dirac equation, as well as its solutions, are Lorentz invariant, we can obtain the solutions for an electron of momentum $\mathbf{p}$ moving in free space by taking the corresponding solutions calculated for $\mathbf{p} = \mathbf{0}$, from the previous exercise, and transforming to a frame moving with respect to our original frame with velocity $\mathbf{v} = v\mathbf{n}$, where $\mathbf{n} = -\mathbf{p}/p$. Given that $p = \gamma m_e v$, where $\gamma = (1-v^2/c^2)^{-1/2}$, the appropriate value of v is specified by $v/c = \tanh\omega$, where $\sinh\omega = p/m_e c$ and $\cosh\omega = E/m_e c^2$.

According to the previous exercise, when expressed in Lorentz-invariant form, the zero momentum solutions to the free-space Dirac equation become

$$\psi^r = e^{-i\,\epsilon_r\,(p_\mu x^\mu/\hbar)}\, w^r(0). \tag{B.11.83}$$

The finite momentum solutions are thus

$$\psi^r = e^{-i\,\epsilon_r\,(p_\mu x^\mu/\hbar)}\, w^r(\mathbf{p}), \tag{B.11.84}$$

where, according to Exercise 11.9,

$$w^r(\mathbf{p}) = A\, w^r(\mathbf{0}), \tag{B.11.85}$$

and

$$A = \cosh\left(\frac{\omega}{2}\right)\left[1 - \tanh\left(\frac{\omega}{2}\right)(\mathbf{n}\cdot\boldsymbol{\alpha})\right]. \tag{B.11.86}$$

Now,

$$\tanh\left(\frac{\omega}{2}\right) = \frac{\tanh\omega}{1 + \sqrt{1+\tanh^2\omega}} = \frac{v/c}{1+\sqrt{1+(v/c)^2}} = \frac{pc}{E+m_e c^2}, \tag{B.11.87}$$

$$\cosh\left(\frac{\omega}{2}\right) = \sqrt{\frac{1+\cosh\omega}{2}} = \sqrt{\frac{E+m_e c^2}{2m_e c^2}}. \tag{B.11.88}$$

Hence,

$$A = \sqrt{\frac{E+m_e c^2}{2m_e c^2}}\left[1 + \frac{(\mathbf{p}\cdot\boldsymbol{\alpha})c}{E+m_e c^2}\right]. \tag{B.11.89}$$

It is easily demonstrated that

$$\mathbf{p}\cdot\boldsymbol{\alpha} = \begin{pmatrix} 0, & 0, & p_z, & p_- \\ 0, & 0, & p_+, & -p_z \\ p_z, & p_-, & 0, & 0 \\ p_+, & -p_z, & 0, & 0 \end{pmatrix}, \tag{B.11.90}$$

which implies that

$$A = \sqrt{\frac{E + m_e c^2}{2\,m_e c^2}}
\begin{pmatrix}
1, & 0, & \frac{p_z c}{E+m_e c^2}, & \frac{p_- c}{E+m_e c^2} \\
0, & 1, & \frac{p_+ c}{E+m_e c^2}, & \frac{-p_z c}{E+m_e c^2} \\
\frac{p_z c}{E+m_e c^2}, & \frac{p_- c}{E+m_e c^2}, & 1, & 0 \\
\frac{p_+ c}{E+m_e c^2}, & \frac{-p_z c}{E+m_e c^2}, & 0, & 1
\end{pmatrix}. \tag{B.11.91}$$

Thus,

$$w^1(\mathbf{p}) = \sqrt{\frac{E + m_e c^2}{2\,m_e c^2}}
\begin{pmatrix}
1 \\ 0 \\ \frac{p_z c}{E+m_e c^2} \\ \frac{p_+ c}{E+m_e c^2}
\end{pmatrix}, \quad
w^2(\mathbf{p}) = \sqrt{\frac{E + m_e c^2}{2\,m_e c^2}}
\begin{pmatrix}
0 \\ 1 \\ \frac{p_- c}{E+m_e c^2} \\ \frac{-p_z c}{E+m_e c^2}
\end{pmatrix}, \tag{B.11.92}$$

$$w^3(\mathbf{p}) = \sqrt{\frac{E + m_e c^2}{2\,m_e c^2}}
\begin{pmatrix}
\frac{p_z c}{E+m_e c^2} \\ \frac{p_+ c}{E+m_e c^2} \\ 1 \\ 0
\end{pmatrix}, \quad
w^4(\mathbf{p}) = \sqrt{\frac{E + m_e c^2}{2\,m_e c^2}}
\begin{pmatrix}
\frac{p_- c}{E+m_e c^2} \\ \frac{-p_z c}{E+m_e c^2} \\ 0 \\ 1
\end{pmatrix}. \tag{B.11.93}$$

Finally, in the limit $p \to 0$, we have $E \to m_e c^2$, in which case the $w^r(\mathbf{p})$ clearly asymptote to the $w^r(0)$. Hence, the ψ^r asymptote to the zero momentum ψ^r specified in the previous exercise.

11.12 We have

$$\Sigma_i \Sigma_j = \begin{pmatrix} \sigma_i, & 0 \\ 0, & \sigma_i \end{pmatrix}\begin{pmatrix} \sigma_j, & 0 \\ 0, & \sigma_j \end{pmatrix} = \begin{pmatrix} \sigma_i \sigma_j, & 0 \\ 0, & \sigma_i \sigma_j \end{pmatrix}. \tag{B.11.94}$$

Hence,

$$\{\Sigma_i, \Sigma_j\} = \begin{pmatrix} \{\sigma_i, \sigma_j\}, & 0 \\ 0, & \{\sigma_i, \sigma_j\} \end{pmatrix} = \begin{pmatrix} 2\,\delta_{ij}, & 0 \\ 0, & 2\,\delta_{ij} \end{pmatrix} = 2\,\delta_{ij}, \tag{B.11.95}$$

where use has been made of

$$\{\sigma_i, \sigma_j\} = 2\,\delta_{ij}. \tag{B.11.96}$$

Bibliography

Abramowitz, M. and Stegun, I. A. (1965). *Handbook of Mathematical Functions: With Formulas, Graphs, and Mathematical Tables* (Dover, New York NY).

Aharonov, Y. and Bohm, D. (1959). Significance of electromagnetic potentials in quantum theory, *Physical Review* **115**, p. 485.

Anderson, C. D. (1933). The positive electron, *Physical Review* **43**, p. 491.

Bailey, V. A. and Townsend, J. S. (1921). The motion of electrons in gases, *Philosophical Magazine* **42**, p. 873.

Baker, H. (1901). Further applications of matrix notation to integration problems, *Proceedings of the London Mathematical Society* **34**, p. 347.

Bardeen, J., Cooper, L. N., and Schreiffer, J. (1957). Theory of superconductivity, *Physical Review* **108**, p. 1175.

Beringer, J. *et al.* (2012). Review of particle physics, *Physical Review D* **86**, p. 010001.

Bethe, H. A. (1935). Theory of disintegration of nuclei by neutrons, *Physical Review* **47**, p. 747.

Bjorken, B. D. and Drell, S. D. (1964). *Relativistic Quantum Mechanics* (McGraw-Hill, New York NY).

Bohr, N. (1913). On the constitution of atoms and molecules, Part I, *Philosophical Magazine* **26**, p. 1.

Born, M. (1926). Zur Quantenmechanik der Strossvorgänge, *Zeitschrift für Physik* **37**, p. 863.

Born, M. and Jordan, P. (1925). Zur Quantenmechanik, *Zeitschrift für Physik* **34**, p. 858.

Born, M. and Oppenheimer, J. R. (1927). Zur Qantuntheorie der Molekeln, *Annalen der Physik* **389**, p. 457.

Bose, S. N. (1924). Plancks Gesetz und Lichtquantenhypothese, *Zeitschrift für Physik* **26**, p. 178.

Breit, G. and Teller, E. (1940). Metastability of hydrogen and helium levels, *Astrophysical Journal* **91**, p. 215.

Breit, G. and Wigner, E. (1936). Capture of slow neutrons, *Physical Review* **49**, p. 519.

Bubb, F. (1924). Direction of ejection of photo-electrons by polarized x-rays, *Physical Review* **23**, p. 137.

Burrau, Ø. (1927). Berechnung des Energiewertes des Wasserstoffmolekel-Ions (H2+) im Normalzustand, *Kongelige Danske Videnskabernes Selskab Matematisk-Fysiske Meddeleser* **7**, p. 1.

Quantum Mechanics

Bushell, P. J. (1972). On the convergence of the Born series, *Journal of Mathematical Physics* **13**, p. 1540.

Campbell, J. (1896). On a law of combination of operators bearing on the theory of continuous transformation groups, *Proceedings of the London Mathematical Society* **28**, p. 381.

Chandrasekhar, S. (1944). Some remarks on the negative hydrogen ion and its absorption coefficient, *Astrophysical Journal* **100**, p. 176.

Compton, A. H. (1923). The spectrum of scattered X-rays, *Physical Review* **21**, p. 483.

Corney, A. (1977). *Atomic and Laser Spectroscopy* (Oxford University Press, Oxford UK).

Darwin, C. G. (1928). The wave equations of the electron, *Proceedings of the Royal Society of London: Series A* **118**, p. 654.

Davisson, C. (1928). The diffraction of electrons by a crystal of nickel, *Bell System Technical Journal* **7**, p. 90.

de Broglie, L. (1925). Recherches sur la théorie des quanta, *Annals of Physics (Paris)* **3**, p. 22.

Deaver, B. S. and Fairbank, W. M. (1961). Experimental evidence for quantized flux in superconducting cylinders, *Physical Review Letters* **7**, p. 43.

Dirac, P. A. M. (1926). On the theory of quantum mechanics, *Proceedings of the Royal Society: Series A* **112**, p. 661.

Dirac, P. A. M. (1927). The quantum theory of emission and absorption of radiation, *Proceedings of the Royal Society of London: Series A* **114**, p. 767.

Dirac, P. A. M. (1928). The quantum theory of the electron, *Proceedings of the Royal Society: Series A* **117**, p. 610.

Dirac, P. A. M. (1930). A theory of electrons and protons, *Proceedings of the Royal Society: Series A* **126**, p. 360.

Dirac, P. A. M. (1958). *The Principles of Quantum Mechanics*, 4th edn. (Oxford University Press, Oxford UK).

Döll, R. and Naubauer, M. (1961). Experimental proof of magnetic flux quantization in a superconducting ring, *Physical Review Letters* **7**, p. 51.

Dyson, F. (1949). The radiation theories of Tomonaga, Schwinger, and Feynman, *Physical Review* **75**, p. 486.

Ehrenfest, P. (1911). Welsche Züge der Lichtquantenhypothesese spielen in der Theorie der Wärmestrahlung eine wesentlische Role, *Annalen der Physik* **36**, p. 91.

Ehrenfest, P. (1927). Bemerkung über die angenäherte Gültigkeit der klassischen Mechanik innerhalb der Quantenmechanik, *Zeitschrift für Physik* **45**, p. 455.

Einstein, A. (1905). Über einen die Erzeugung und Verwandlung des Lichtes betreffenden heuristischen Gesichtspunkt, *Annalen der Physik* **17**, p. 132.

Einstein, A. (1916). Zur Quantentheorie der Strahlung, *Mitteilungen der Physikalischen Gesellschaft Zürich* **18**, p. 47.

Einstein, A. (1924). Quantentheorie des einatomigen idealen Gases, *Sitzungsberichte der Preussischen Akademie der Wissenschaften* **22**, p. 261.

Einstein, A. and Stern, O. (1913). Einige Argumente für die Annahme einer molekularen Agitation beim absoluten Nullpunk, *Annalen der Physik* **40**, p. 551.

Epstein, P. S. (1926). The Stark effect from the point of view of Schroedinger's quantum theory, *Physical Review* **28**, p. 695.

Faxen, H. and Holtzmark, J. (1927). Beitrag zur Theorie des Durchganges langsamer Elektronen durch Gase, *Zeitschrift für Physik* **45**, p. 307.

Fermi, E. (1926). Sulla quantizzazione del gas perfetto monoatomico, *Rendiconti Lincei* **3**, p. 145.

Fermi, E. (1950). *Nuclear Physics* (University of Chicago Press, Chicago IL).

Feynman, R. P. (1939). Forces in molecules, *Physical Review* **56**, p. 340.

Feynman, R. P., Leighton, R. B., and Sands, M. (1963). *The Feynman Lectures on Physics* (Addison-Wesley, Reading MA).

Fierz, M. (1939). Über die relativistische Theorie kräftefreier Teilchen mit beliebigem Spin, *Helvetica Physica Acta* **12**, p. 3.

Finkelstein, B. N. and Horowitz, G. E. (1928). Über die Energie des He-atoms und des positiven H2-ions im Normalzustande, *Zeitschrift für Physik* **48**, p. 118.

Fitzpatrick, R. (2008). *Maxwell's Equations And The Principles Of Electromagnetism* (Jones & Bartlett Learning, Burlington MA).

Fitzpatrick, R. (2012). *An Introduction to Celestial Mechanics* (Cambridge University Press, Cambridge UK).

Fitzpatrick, R. (2013). *Oscillations and Waves: An Introduction* (CRC Press, Baca Raton FL).

Flanigan, F. J. (2010). *Complex Variables: Harmonic and Analytic Functions* (Dover, New York, NY).

Gasiorowicz, S. (1996). *Quantum Physics*, 2nd edn. (John Wiley & Sons, New York NY).

Gerlach, W. and Stern, O. (1922). Das magnetische Moment des Silberatoms, *Zeitschrift für Physik* **9**, p. 353.

Goldstein, H., Poole, C., and Safko, J. (2002). *Classical Mechanics*, 3rd edn. (Addison-Wesley, San Francisco CA).

Göppert-Mayer, M. (1931). Über Elementarakte mit zwei Quantensprüngen, *Annalen der Physik* **9**, p. 939.

Gordon, W. (1928). Die Energieniveaus des Wasserstoffatoms nach der Diracschen Quantentheorie des Elektrons, *Zeitschrift für Physik* **48**, p. 11.

Goudsmit, S. A. and Uhlenbeck, G. E. (1925). Ersetzung der Hypothese vom unmechanischen Zwang durch eine Forderung bezüglich des inneren Verhaltens jedes einzelnen Elektrons, *Naturwissenschaften* **13**, p. 953.

Gradshteyn, I. S., Ryzhik, I. M., and Jeffrey, A. (1980). *Table of Integrals, Series, and Products* (Academic Press, San Diego CA).

Greene, G. L., Ramsey, N. F., Mampe, W., Pendlebury, J. M., Smith, K., Dress, W. B., Miller, P. D., and Perrin, P. (1979). Measurement of the neutron magnetic moment, *Physical Review D* **20**, p. 2139.

Griffiths, D. J. (2005). *Introduction to Quantum Mechanics*, 2nd edn. (Pearson Prentice Hall, Upper Saddle River NJ).

Hausdorff, F. (1906). Die symbolische Exponentialformel in der Gruppentheorie, *Berichte über die Verhandlungen der Sächsischen Gesellschaft der Wissenschaften zu Leipzig* **58**, p. 19.

Haynes, W. M. and Lide, D. R. (2011). *CRC Handbook of Chemistry and Physics*, 92nd edn. (CRC Press, Boca Raton FL).

Hecht, E. and Zajac, A. (1974). *Optics* (Addison-Wesley, Menlo Park CA).

Heisenberg, W. (1927). Über den anschaulichen Inhalt der quantentheoretischen Kinematik und Mechanik, *Zeitschrift für Physik* **43**, p. 172.

Høgaasen, H., Richard, J.-M., and Sorba, P. (2010). Two-electron atoms, ions and molecules, *American Journal of Physics* **78**, p. 86.

Jackson, J. D. (1975). *Classical Electrodynamics*, 2nd edn. (John Wiley & Sons, New York NY).

Kellner, G. W. (1927). Die Ionisierungsspannung des Heliums nach der Schrödingerschen Theorie, *Zeitschrift für Physik* **44**, p. 91.

Kohn, W. (1954). On the convergence of the Born expansion, *Reviews of Modern Physics* **26**, p. 472.

Kuhn, W. (1925). Über die Gesamtstärke der von einem Zustande ausgehenden Absorptionslinien, *Zeitschrift für Physik* **33**, p. 408.

Landé, A. (1921). Über den anomalen Zeemaneffekt (Teil I), *Zeitschrift für Physik* **5**, p. 231.

Meissner, W. and Ochsenfeld, R. (1933). Ein neuer Effect bei Eintritt der Supraleitfähigkeit, *Naturwissenschaften* **21**, p. 787.

Mie, G. (1908). Contributions to the optics of turbid media, particularly of colloidal metal solutions, *Annalen der Physik (Leipzig)* **25**, p. 377.

Millikan, R. A. (1916). Einstein's photoelectric equation and contact electromotive force, *Physical Review* **7**, p. 18.

Moore, C. E. (1949). *Atomic Energy Levels, Vol. I (Hydrogen through Vanadium)*, Circular of the National Bureau of Standards 467 (U.S. Government Printing Office, Washington DC).

Morse, P. M. and Feschbach, H. (1953). *Methods of Theoretical Physics* (McGraw-Hill, New York NY).

Park, D. (1974). *Introduction to the Quantum Theory*, 2nd edn. (McGraw-Hill, New York NY).

Paschen, F. and Back, E. (1921). Liniengruppen magnetisch vervollständigt, *Physica* **1**, p. 261.

Pauli, W. (1925). Über den Zusammenhang des Abschlusses der Elektronengruppen im Atom mit der Komplexstruktur der Spektren, *Zeitschrift für Physik* **33**, p. 879.

Pauli, W. (1927). Zur Quantenmechanik des magnetischen Elektrons, *Zeitschrift für Physik* **43**, p. 601.

Pauli, W. (1940). The connection between spin and statistics, *Physical Review* **58**, p. 716.

Planck, M. (1900a). Über eine Verbesserung der Wienschen Spektralgleichung, *Verhandlungen der Deutschen Physikalischen Gesellschaft* **2**, p. 202.

Planck, M. (1900b). Zur Theorie des Gesetzes der Energieverteilung im Normalspektrum, *Verhandlungen der Deutschen Physikalischen Gesellschaft* **2**, p. 237.

Plemelj, J. (1908). Riemannsche Funktionenscharen mit gegebener Monodromiegruppe, *Montsche für Mathematik und Physik* **19**, p. 211.

Procopiu, S. (1911). Sur les éléments d'énergie, *Annales scientifiques de l'Université de Jassy* **7**, p. 280.

Rabi, I. I. (1937). Space quantization in a gyrating magnetic field, *Physical Review* **51**, p. 652.

Rabi, I. I., Millman, S., Kusch, P., and Zacharias, J. R. (1939). The molecular beam resonance method for measuring nuclear magnetic moments. The magnetic moments of $_3\mathrm{Li}^6$, $_3\mathrm{Li}^7$ and $_9\mathrm{F}^{19}$, *Physical Review* **55**, p. 526.

Ramsauer, C. (1921). Über den Wirkungsquerschnitt der Gasmoleküle gegenüber langsamen Elektronen, *Annalen der Physik* **64**, p. 513.

Rauch, H., Zeilinger, A., Badurek, G., and Wilfing, A. (1975). Verification of coherent spinor rotation of fermions, *Physics Letters A* **54**, p. 456.

Reiche, F. and Thomas, W. (1925). Über die Zahl der Dispersionselektronen, die einem stationären Zustande zugeordnet sind, *Zeitschrift für Physik* **34**, p. 510.

Reif, F. (1965). *Fundamentals of Statistical and Thermal Physics* (McGraw-Hill International, Auckland, New Zealand).

Riley, K. F., Hobson, M. P., and Bence, S. J. (2013). *Mathematical Methods for Physics and Engineering: A Comprehensive Guide*, 3rd edn. (Cambridge University Press, Cambridge UK).

Rindler, W. (1966). *Special Relativity*, 2nd edn. (Interscience, New York NY).

Ritz, W. (1909). Über eine neue Methode zur Lösung gewisser Variationsprobleme der mathematischen Physik, *Journal für die Reine und Angewandte Mathematik* **135**, p. 1.

Sakurai, J. J. and Napolitano, J. (2011). *Modern Quantum Mechanics*, 2nd edn. (Addison-Wesley, Reading MA).

Sauter, F. (1931a). Über den atomaren Photoeffekt bei grosser Härte der anregenden Strahlung, *Annalen der Physik* **9**, p. 217.

Sauter, F. (1931b). Über den atomaren Photoeffekt in der K-Schale nach der relativistischen Wellenmechanik Diracs, *Annalen der Physik* **11**, p. 454.

Schrödinger, E. (1926a). Quantisierung als Eigenwertproblem, *Annalen der Physik* **80**, p. 437.

Schrödinger, E. (1926b). An undulatory theory of the mechanics of atoms and molecules, *Physical Review* **26**, p. 1049.

Schrödinger, E. (1930). Über die kräftefreie Bewegung in der relativistischen Quantenmechanik, *Sitzungsberichte der Preussischen Akademie der Wissenschaften. Physikalisch-mathematische Klasse* **24**, p. 418.

Schwinger, J. (1948). On quantum-electrodynamics and the magnetic moment of the electron, *Physical Review* **73**, p. 416.

Shapiro, J. and Breit, G. (1959). Metastability of $2s$ states of hydrogenic atoms, *Physical Review* **113**, p. 179.

Slater, J. C. (1929). The theory of complex spectra, *Physical Review* **34**, p. 1298.

Sobolev, S. L. (1936). Méthode nouvelle à résoudre le problème de Cauchy pour les équations linéaires hyperboliques normales, *Matematicheskii Sbornik* **1**, p. 39.

Stark, J. (1914). Beobachtungen über den Effekt des elektrischen Feldes auf Spektrallinien i. Quereffekt, *Annalen der Physik* **50**, p. 489.

Strobbe, M. (1930). Zur Quantenmechanik photoelektrischer Prozesse, *Annalen der Physik* **7**, p. 661.

Strutt, J. W. (1871). On the light from the sky, its polarization and colour, *Philosophical Magazine* **41**, p. 107.

Taylor, G. I. (1909). Interference fringes with feeble light, *Proceedings of the Cambridge Philosophical Society* **15**, p. 114.

Thomas, L. H. (1926). Motion of the spinning electron, *Nature* **117**, p. 514.

Thomas, W. (1925). Über die Zahl der Dispersionselektronen, die einem stationären Zustande zugeordnet sind. (Vorläufige Mitteilung), *Naturwissenschaften* **13**, p. 627.

Thomson, G. P. (1928). Experiments on the diffraction of cathode rays, *Proceedings of the Royal Society of London: Series A* **117**, p. 600.

Tonomura, A., Osakabe, N., Matsuda, T., Kawasaki, T., Endo, J., Yano, S., and Yamada, H. (1982). Evidence for Aharonov-Bohm effect with magnetic field completely shielded from electron wave, *Physical Review Letters* **48**, p. 1443.

Unsöld, A. (1927). Beiträge zur Quantenmechanik der Atome, *Annalen der Physik* **387**, p. 355.

van de Hulst, H. C., Muller, C., and Oort, J. (1954). The spiral structure of the outer part of the galactic system derived from the hydrogen emission at 21-cm wavelength, *Bulletin of the Astronomical Institutes of the Netherlands* **12**, p. 117.

Waller, I. (1926). Der Starkeffekt zweiter Ordnung bei Wasserstoff und die Rydbergkorrektion der Spektra von He und Li+, *Zeitschrift für Physik* **38**, p. 635.

Werner, S. A., Colella, R., Overhauser, A. W., and Eagen, C. F. (1975). Observation of the phase shift of a neutron due to precession in a magnetic field, *Physical Review Letters* **35**, p. 1053.

Winkler, P. F., Kleppner, D., Myint, T., and Walther, F. G. (1972). Magnetic moment of the proton in Bohr magnetons, *Physical Review A* **5**, p. 83.

Yukawa, H. (1935). On the interaction of elementary particles, *Proceedings of the Physico-Mathematical Society of Japan* **17**, p. 48.

Zeeman, P. (1897). On the influence of magnetism on the nature of the light emitted by a substance, *Philosophical Magazine* **43**, p. 226.

Index

Printed in the United States
By Bookmasters